REMOTE SENSING
ENERGY-RELATED STUDIES

ADVANCES IN THERMAL ENGINEERING

Editors:
JAMES P. HARTNETT
THOMAS F. IRVINE, JR.

I	Blackshear	• Heat Transfer in Fires: Thermophysics, Social Aspects, Economic Impact*
II	Afgan and Beer	• Heat Transfer in Flames*
III	de Vries and Afgan	• Heat and Mass Transfer in the Biosphere: Part I Transfer Processes in Plant Environment*
IV	Vargaftik	• Tables on the Thermophysical Properties of Liquids and Gases
V	Veziroglu	• Remote Sensing: Energy-Related Studies
	Begell	• Glossary of Terms in Heat Transfer: English, Russian, German, French, and Japanese (in preparation)
	Chang	• Control of Flow Separation (in preparation)
	Chi	• Heat Pipe Theory and Practice (in preparation)
	Denton	• Future Energy Production* (in preparation)
	Eckert and Goldstein	• Heat Transfer Measurement 2nd revised and augmented edition (in preparation)
	Ginoux	• Two-Phase Flows with Application to Nuclear Reactor Design Problems† (in preparation)
	Gutfinger	• Topics in Transport Phenomena: Bioprocesses, Mathematical Treatment, Mechanisms (in preparation)
	Hartnett	• Heat and Mass Transfer Problems in Future Energy Production* (in preparation)
	Hsu and Graham	• Transport Processes in Boiling and Two-Phase Systems, Including Near-Critical Fluids (in preparation)
	Pfender	• High-Temperature Phenomena in Electric Arcs (in preparation)
	Richards	• Measurement of Unsteady Fluid Dynamic Phenomena† (in preparation)
	Sieverding	• Two-Phase Fluid Dynamics in Turbines: Engineering, Instrumentation, Theory† (in preparation)
	Yovanovich	• Advanced Heat Conduction (in preparation)

*A publication of the International Centre for Heat and Mass Transfer, Belgrade.
†A von Karman Institute Book, Brussels.

REMOTE SENSING
ENERGY-RELATED STUDIES

EDITED BY

T. Nejat Veziroglu

Director
Clean Energy Research Institute
School of Engineering and Environmental Design
University of Miami, Coral Gables, Florida

Hemisphere Publishing Corporation
Washington   London

A HALSTED PRESS BOOK

John Wiley & Sons

New York   London   Sydney   Toronto

Copyright © 1975 by Hemisphere Publishing Corporation. All rights reserved. No part of this book may be reproduced in any form, by photostat, microform, retrieval system, or any other means, without the prior written permission of the publisher.

Hemisphere Publishing Corporation
1025 Vermont Ave., N.W., Washington, D.C. 20005

Distributed solely by Halsted Press, a Division of John Wiley & Sons, Inc., New York.

Library of Congress Cataloging in Publication Data

Remote Sensing Applied to Energy Related Problems
   Symposium, Miami, Fla., 1974.
   Remote sensing, energy related studies.

   "Presented by the Clean Energy Research Institute, School of Engineering and Environmental Design, University of Miami, Coral Gables, Florida; sponsored by the School of Continuing Studies, University of Miami."
   1. Remote sensing systems—Congresses. 2. Power resources—Congresses. I. Veziroglu, T. Nejat. II. Miami, University of, Coral Gables, Fla. Clean Energy Research Institute. III. Miami, University of, Coral Gables, Fla. School of Continuing Studies. IV. Title.
G70.4.R45   1974   620'.46   75-23018
ISBN 0-470-90665-0

Printed in the United States of America

CONTENTS

Keynote Address
   Charles W. Mathews 1

Our National Energy Future—The Role of Remote Sensing
   Harrison H. Schmitt 5

PART I. ATMOSPHERIC AND HYDROSPHERIC MEASUREMENTS

The Evolution of Atmospheric Measurements from Satellites
   M. Tepper 15

Seasat: A Spacecraft Views the Marine Environment with Microwave Sensors
   J. R. Apel 47

Remote Sensing of Oceans Using Microwave Sensors
   K. Krishen 61

PART II. ACTIVE SENSOR APPLICATIONS

Some Applications of Weather Radar to Problems of Energy Production
   E. Kessler 103

The Use of Lidar for Atmospheric Measurements
   M. P. McCormick 113

Practical Considerations to Use of Microwave Sensing from Space Platforms
   R. P. Eisenberg 129

Laser Measure of Sea Temperature and Turbidity in Depth
   J. G. Hirschberg, A. W. Wouters and J. D. Byrne 157

PART III. LAND USE MONITORING

Remote Sensing for Western Coal and Oil Shale Development Planning and Environmental Analysis
   H. D. Parker 171

Exploration for Nuclear and Fossil Fuels by Remote Sensing Techniques
   N. M. Short 189

Surveillance of the Missouri River Basin Using Remote Sensing
   J. H. Senne 233

Determining Potential Solar Power Sites in Western Hemisphere Ocean and Land Areas Based Upon Satellite Observations of Cloud Cover
   H. W. Hiser, H. V. Senn 247

PART IV. ENVIRONMENTAL QUALITY MONITORING

    Satellite Detection of Air Pollutants
        W. A. Lyons 263

    Remote Sensing by ERTS Satellite of Vegetational Resources Believed to be
    Under Possible Threat of Environmental Stress
        P. Poonai, W. J. Floyd and R. Hall 291

    Remote Sensing Applied to Thermal Pollution
        S. S. Lee, T. N. Veziroglu, S. Sengupta and N. L. Weinberg 303

    Remote Sensing Applied to Numerical Modelling
        S. Sengupta, S. S. Lee, T. N. Veziroglu and R. Bland 335

PART V. SPECIAL TOPICS

    The Satellite Solar Power Station Option
        P. E. Glaser 367

    Space-Acquired Imagery: A Versatile Tool in the Development of
    Energy Sources
        V. R. Frierson, D. L. Amsbury 395

    Multispectral Data Systems for Energy Related Problems
        C. L. Wilson, R. H. Rogers 403

    Locating Remotely Sensed Data on the Ground
        R. C. Malhotra, M. L. Radar 431

    The Remote Sensing of Small Temperature Difference
        N. J. Clinton, C. E. Campbell 437

    German Program in Remote Sensing
        A. Sieber 445

PART VI. WORKSHOP REPORTS

    Active Sensor Applications
        C. L. Wilson, H. V. Senn 471

    Environmental Quality Monitoring
        J. D. Lawrence, S. Sengupta 473

    Land Use and Resource Monitoring
        A. T. Joyce, A. L. Higer 475

    Oceanographic and Hydrologic Measurements from Satellites
        S. Fritz, N. L. Weinberg 479

AUTHOR INDEX 485
SUBJECT INDEX 489

LIST OF CONTRIBUTORS

DAVID L. AMSBURY, NASA, Johnson Space Center, Houston, Texas 77586.

DR. JOHN R. APEL, U.S. Department of Commerce, NOAA, AOML, Rickenbacker Causeway, Miami, Florida.

ROY BLAND, NASA, Kennedy Space Center, Kennedy Space Center, Florida 32899.

JAMES BYRNE, Department of Mechanical Engineering, University of Miami, P.O. Box 248294, Coral Gables, Florida 33124.

C. E. CAMPBELL, Lockheed Electronics Co., Inc., Aerospace Systems Division, Houston, Texas 77058.

N. J. CLINTON, Lockheed Electronics, Inc., Aerospace Systems Division, Houston, Texas 77058.

R. P. EISENBERG, Microwave Sensor Development, General Electric Valley Forge Space Center, King of Prussia, Pennsylvania.

W. J. FLOYD, Bethune Cookman College, Daytona Beach, Florida 32015.

V. R. FRIERSON, NASA, Johnson Space Center, Houston, Texas 77586.

DR. S. FRITZ, National Environmental Satellite Center, NOAA, Suitland, Maryland.

P. E. GLASER, Arthur D. Little, Inc., Cambridge, Massachusetts.

ROYCE HALL, Bethune Cookman College, Daytona Beach, Florida 32015.

A. L. HIGER, U.S. Geological Survey, Miami, Florida.

DR. JOSEPH G. HIRSCHBERG, Department of Physics, University of Miami, Coral Gables, Florida 33124.

DR. HOMER W. HISER, Director, Remote Sensing Laboratory, University of Miami, Coral Gables, Florida 33124.

DR. ARMOND T. JOYCE, NASA, Earth Resources Laboratory, Bay St. Louis, Missouri 39520.

DR. EDWIN KESSLER, National Severe Storms Laboratory, NOAA, Norman, Oklahoma.

KUMAR KRISHEN, Lockheed Electronics Co., 16811 El Camino Real, Houston, Texas 77058.

JAMES D. LAWRENCE, JR., NASA Langley Research Center, Hampton, Virginia 23665.

DR. SAMUEL S. LEE, Mechanical Engineering, P.O. Box 248294, University of Miami, Coral Gables, Florida 33124.

Dr. W. A. LYONS, Air Pollution Analysis Laboratory, University of Wisconsin, Milwaukee, Wisconsin.

R. C. MALHOTRA, Lockheed Electronics Co., Inc., Aerospace Systems Division, Houston, Texas 77058.

CHARLES W. MATHEWS, Associate Administrator for Applications, NASA, Washington, D.C.

M. P. McCORMICK, NASA Langley Research Ctr., Hampton, Virginia 23665.

H. D. PARKER, Ecosystems Analysis & Remote Sensing Applications, Western Scientific Services, Inc., Ft. Collins, Colorado.

LIST OF CONTRIBUTORS

DR. PREMSUKH POONAI, Bethune-Cookman College, Daytona Beach, Florida 32015.

M. L. RADAR, Lockheed Electronics Co., Inc., Aerospace Division, Houston, Texas 77058.

DR. HARRISON H. SCHMITT, Assistant Administrator, Office of Energy Programs, NASA, Washington, D.C.

DR. SUBRATA SENGUPTA, Mechanical Engineering, P.O. Box 248294, Coral Gables, Florida 33124.

DR. HARRY V. SENN, Remote Sensing Laboratory, University of Miami, Coral Gables, Florida 33124.

DR. J. H. SENNE, Department of Civil Engineering, University of Missouri, Rolla, Missouri.

DR. N. M. SHORT, Earth Resources Branch, NASA, Goddard Space Center, Greenbelt, Maryland.

DR. A. SIEBER, DVFLR/GfW, Porz-Wahn, Linder Hohe, West Germany.

DR. MORRIS TEPPER, Earth Observations Program, NASA, Washington, D.C.

DR. T. N. VEZIROGLU, Director of Clean Energy Research Institute, P.O. Box 248294, University of Miami, Coral Gables, Florida 33124.

DR. N. L. WEINBERG, Department of Electrical Engineering, University of Miami, Coral Gables, Florida 33124.

C. L. WILSON, Bendix Aerospace Systems Division, 3300 Plymount Road, Ann Arbor, Michigan 48107.

DR. A. W. WOUTERS, Department of Physics, University of Miami, Coral Gables, Florida 33124.

COMMITTEE

Co-Chairmen	**T. Nejat Veziroglu** Director, Clean Energy Research Institute Chairman, Department of Mechanical Engineering
	Homer W. Hiser Director, Remote Sensing Laboratory University of Miami
Special Consultants	J. Phillip Claybourne Chief, Science Technology and Applications Office NASA, Kennedy Space Center
	Samuel S. Lee Director, Thermal Pollution Laboratory Department of Mechanical Engineering
	Aaron L. Higer Research Hydrologist U.S. Geological Survey, Miami
Coordinator	Raynell Chungo Executive Secretary Clean Energy Research Institute

LIST OF CONTRIBUTORS

STAFF

Arrangements	Tony Pajares
	Arlene Jacobsohn
	The School of Continuing Studies
Program for Accompaniers	Mary Gene Knopf
	Conference Services
Special Assistants	Carol Pascalis
	Eileen Kavlock
	Dorothea Rondeau
	The School of Engineering and Environmental Design

SESSION OFFICIALS

Session 1. General Session

    Session Chairman: W. L. Webb
Chief, Meteorological Satellite Technical Area
Atmospheric Sciences Laboratory
White Sands Missile Range, N.M.

    Session Co-Chairman: J. P. Claybourne
Chief, Science, Technology and Applications Office
NASA, Kennedy Space Center, Florida

Session 2. Atmospheric and Hydrospheric Measurements from Satellites

    Session Chairman: S. Fritz
National Environmental Satellite Center
National Oceanic and Atmospheric Administration
Suitland, Maryland

    Session Co-Chairman: N. L. Weinberg
Professor of Electrical Engineering
University of Miami

Session 3. Active Sensor Applications

    Session Chairman: C. L. Wilson
Manager, Earth Resource Applications
Bendix Aerospace Systems Division
Ann Arbor, Michigan

    Session Co-Chairman: H. V. Senn
Associate Professor Mechanical Engineering
University of Miami

Session 4. Land Use Monitoring

    Session Chairman: A. T. Joyce
Chief, Land Remote Sensing Applications Group
Earth Resources Lab., NASA
Bay St. Louis, Mississippi

    Session Co-Chairman: A. L. Higer Research Hydrologist
U.S. Geological Survey, Miami

LIST OF CONTRIBUTORS

Session 5. Environmental Quality Monitoring

    Session Chairman: J. DeNoyer
                                     Director, EROS Program
                                     U.S. Geological Survey
                                     Reston, Virginia

    Session Co-Chairman: S. Sengupta
                                       Research Associate Professor
                                     University of Miami

Session 6. Working Groups and Plenary Session

    Session Chairman: W. B. Foster
                                     Deputy Assistant Administrator for Monitoring Systems
                                     U. S. Environmental Protection Agency
                                     Washington, D.C.

    Session Co-Chairman: H. W. Hiser
                                       Director, Remote Sensing Laboratory
                                     University of Miami

FOREWORD

Remote sensing involves the use of surface, airborne, or spacecraft instrumentation capable of scanning large areas of the earth, its atmosphere, and its oceans, to provide accurate high-resolution data for a wide range of useful applications. The quantity of data that can be gathered by advanced passive and active remote sensing systems is vastly larger than is possible with traditional data-gathering approaches. Remote sensing can help solve many old and new problems associated with the development of clean new energy sources. Remote sensing of thermal air and water pollution are typical examples. Thermal pollution and local potential for cooling can be important to nuclear power stations and other new energy-generating systems. The harnessing of the vast resources of the ocean via thermal energy conversion processes may depend on the success we have with planned oceanographic satellites such as "Seasat," to be launched by NASA in 1978.

There are important energy-related applications for remote sensing of land use and water resources. It can, for example, assist in providing optimum locations and distributions for new energy systems. Remote sensing of meteorological parameters, particularly rainfall and clouds, is very important. Remote sensing of clouds can be used to find optimum locations for solar power stations. Rainfall data are needed for existing hydroelectric facilities. They are also needed for determining cooling efficiencies, scrubbing, and dilution. These are but a few of the many applications of remote sensing to the development and use of new energy sources.

At the Remote Sensing Applied to Energy-Related Problems Symposium, Miami, December 2-4, 1974, the lecturers and authors, selected from among the leaders in this new and growing field, have very ably covered both the fundamental information and the latest developments. In view of the large number of requests we have received from all over the world, we find it especially rewarding to follow up this very successful symposium by presenting this formal edition of the proceedings. In this way, the information on remote sensing for energy problems that was presented at the symposium will become a permanent contribution to the technical literature on this subject of growing importance.

T. Nejat Veziroglu
Symposium Co-chairman and Editor

Homer W. Hiser
Symposium Co-chairman

PREFACE

It was indeed a most rewarding experience for the Remote Sensing Applied to Energy-Related Problems Symposium Committee and Staff to welcome approximately 100 representatives from industry, government, the university community and others to Miami in December, 1974. A preliminary soft cover edition of the Symposium Proceedings was issued to the attendees as they assembled for their perusal and "real time" reference purposes. Despite excellent cooperation from our authors, a number of papers could not be included in that edition and appeared as abstracts or titles only. There were, of course, the usual editorial shortcomings in the texts which did appear. Also the Symposium Keynote Address by Charles W. Mathews, the Banquet Address by Harrison H. Schmitt and the Workshop Reports could not be included in the earlier edition. These transcripts along with a full complement of papers, updated, retyped and checked by the authors, are presented in the current edition.

In the present edition, we have rearranged the order in which the papers appear. We are presenting them and the Workshop Reports under six basic headings rather than by the numerical ordering of sessions following the program of the Symposium itself. The original listing of sessions, along with the session officials, appears in the earlier page entitled "Session Officials."

We would like to again express our gratitude to the authors and presenters and their sponsoring organizations. Cooperation has been splendid all around throughout the extended period of preparation. I am especially grateful to our Clean Energy Research Institute staff members in general and to Ms. Raynell Chungo in particular for their additional measure of hard work in getting the text ready for our publishers.

T. Nejat Veziroglu, Editor
Director, Clean Energy Research
Institute

ACKNOWLEDGMENTS

We gratefully acknowledge the financial support provided by the sponsoring organization, the School of Continuing Studies, University of Miami.

We also wish to extend our appreciation to Mr. Charles W. Mathews, associate administrator for applications, NASA, for the keynote address, and to Dr. Harrison H. Schmitt, assistant administrator, Office of Energy Programs, NASA, for the banquet address.

Thanks also are due to our authors and presenters, who provided the substance of the symposium, related as published papers in the present proceedings, and also to the workshop group leaders for conducting the workshops and preparing the reports included here.

We owe a special debt of appreciation to the session chairmen and co-chairmen for organizing and executing the technical session. In acknowledgment we list these session officials on pages ix and x.

Finally, we wish to thank our publisher, the Hemisphere Publishing Corporation, for its basic role in printing and distributing the present formal edition of the proceedings.

The Symposium Committee

REMOTE SENSING
ENERGY-RELATED STUDIES

KEYNOTE ADDRESS

CHARLES W. MATHEWS
Associate Administrator for Applications
National Aeronautics and Space Administration
Washington, D.C.

It is indeed a pleasure to share with you some thoughts and ideas about our earth and the contributions that our country's space program has made and will continue to make toward the well being of this most beautiful of all planets.

Mankind has always been curious and has studiously observed his surroundings, whether to the rim of the valley or out to the planets, the sun, and the stars. As the centuries have passed, the tools with which man could observe and measure his surroundings have permitted ever-increasing precision, scope and reach of observation. Even the ancient Greeks were conscious of the limitations of one's vantage point. To paraphrase Socrates, speaking about 400 B.C., "We who inhabit the earth dwell like frogs at the bottom of a pool. Only if man could rise above the summit of the air could we behold the true earth, the world in which we live." This we can now do.

The possibilities of using earth satellites for such purposes have been foreseen by fiction writers for many years. "The Brick Moon" by Edward Everett Hale, first published in the Atlantic Monthly Magazine over one hundred years ago describes one such project which was to establish a brightly reflective satellite in polar orbit to permit more accurate navigation on the high seas. Unfortunately, a number of accidents happened at the launching site and premature launch occurred while the "Moon" was still occupied by its builders, thus establishing the first (if fictional) manned space station. Its uses proved many-fold: communication, as well as navigation and geodesy. In contrast, there have been many forces at work against exploration, experimentation, and research. A satire on these negative forces appears in a book written by Edwin A. Abbot entitled "Flatland" also published nearly one hundred years ago. This book describes a two-dimensional world and the society that developed in it. The particular quotation that seems most pertinent reads as follows: "Windows there are none in our houses; for the light comes to us alike in our homes and out of them, by day and by night, equally at all times and in all places, Whence we know not. It was in old days, with our learned men, an interesting and oft-investigated question: 'What is the origin of light?' and the solution of it has been repeatedly attempted, with no other result than to crowd our lunatic asylums with the would-be solvers." In spite of such "nay-sayers" over the centuries man's successful quest for knowledge of himself, his earth, and his universe continued.

At the present point in time our quest for knowledge in space endeavors is logically shifting emphasis from exploration and the search for purely scientific knowledge to the achievement of easier access to near space and to the application of our relatively newly-found capabilities to direct and near term uses. Indeed, applications in space communications and meteorology have already achieved a very important first level of maturity.

Other space applications are just now beginning to be pursued with substantial effort. They involve either operations in space or application of space technology right here on the ground.

There are numerous possibilities for space applications -- solar power generation in space and materials processing in space have recently received considerable attention in the Congress and elsewhere. Still, I personally feel that the capabilities for earth observations from space, in combination with global communications, are likely to continue to produce the most important applications in the near term. Earth orbit is a tremendous vantage point to view and gain knowledge of world conditions and activities rapidly and on a global basis. Because of this, the associated applications will be of critical importance to the nation, and the world, as well.

Why is this so important? The environment and resources of the earth's atmosphere, its continents, its coastal shelves, and its oceans, seas, and lakes have rightly become of great concern to people everywhere. This concern is in recognition of the closed ecology and finite resources of the earth, just as in the spaceships we build.

In one sense, the world has become small with complex interactions between activities of nations, continents and hemispheres. In another sense, the world is still large, involving tens of billions of acres of land area and with oceans many times larger. We know humans have impacted much of this immense area, but how do we establish a baseline and how do we measure changes and differentiate between man-caused and natural changes. To accomplish this end, large amounts of data must be sensed and gathered from the various regions of the world -- in many cases, on a global basis. These data must flow to centralized points where it can be processed into really interpretable information. The information must then flow to appropriate decision makers. But equally important, a parallel flow of information must go to the public to aid in their preparedness and understanding and to achieve their support. This human interface is probably more challenging than the technical problems involved, for considerable effort must be spent in establishing an understanding of the benefits of complex actions, their economic viability, and in avoiding the concerns of vested interests.

A significant step toward the establishment of such an information base about the earth's resources and surface environment came about with the launch of the first Earth Resources Technology Satellite in July 1972. This mission is an experimental one, but through its operation we already have developed understandable and highly useful applications associated with the very unique capabilities of this satellite. ERTS can obtain images from any area on the globe every eighteen days and, in fact, it provides repetitive coverage of every segment of the United States and of many selected areas in other parts of the world. Equally important, these images are very special -- not just color photographs, but electronically derived, multispectral images amenable to many kinds of interpretive techniques developed recently by various agencies, universities, and industrial organizations.

The ERTS program is typical of space observation activities carried out through a technique called remote sensing. Remote sensing has always been with us, the eye being nature's outstanding example of a sensor employing this technique. With the eye, one not only can discriminate geometry, but also texture, color, tone, hue, contrast, brightness, intensity, and motion, among other things. With this marvelous instrument one can easily differentiate fabrics from metals, or animate from inanimate objects, and such perception can go to lower levels of detail -- for example, that a metal is aluminum, or that a tree is an oak.

As is frequently the case, man has been able to draw on nature's example and provide expanded capabilities encompassing entirely new regimes of operation. Such extensions provide the basis for the practical application of regional and global surveillance of the earth's environment. Of course, other things are required too, such as aircraft and spacecraft as observing platforms and satellite communications for high speed data flow and computer hardware and software.

Through use of the ERTS remote sensing techniques, signature information produced in the various spectral regions is made applicable to many areas of endeavor. As we move from the very short wavelengths (ultraviolet) to the visible, topographic and geographic information becomes available, allowing for photo-interpretation of geometry. But by sensing of multi-spectral reflectance in these same regions, other interpretations can be made, such as vegetation type. Including the near infrared region of the spectrum greatly augments this interpretability, even providing information on plant stress or disease. Moving further into the infrared band affords temperature information about the surface -- discrimination of surface water, ice, or snow cover, soil moisture thermal pollution and temperature behavior of earth materials. To emphasize with a more specific example, the ability to inventory the world's food supplies on a rapid basis is practical using this approach.

Turning now to the area of oceanography, NASA is in the process of development of a SEASAT satellite with the expectation that it will fly in 1978. This satellite would be provided with a complement of microwave instruments that have been selected by working group representatives from all the various user disciplines in the field of oceanography.

From a scientific viewpoint, in addition to major contributions to geodetic science, significant advances in the fundamental understanding will be obtained in areas of wind-wave dynamics and propagation phenomena, ocean-storm interactions, ice-air-sea interactions, dynamic behavior of major ocean currents, materials transport, tide behavior in the open ocean, and Tsunami propagation.

The specific role of SEASAT in this effort would be to provide precision measurements of wave height and wave directional spectrum, data for frequent update of maps of current patterns and ocean temperature, data to establish the fine structure of the ocean geoid, and charts of ice fields and ice leads.

When these data flow into the hands of the appropriate investigators, we would expect to see outputs affording not only an understanding of the physical conditions of the sea surface and its air and land interface, but the ability to predict or forecast future effects of these observations. If this in fact be the case, we would foresee at some future date a great many very practical applications involved with faster ship routing, improved design of ships and off-shore structures, storm damage avoidance, coastal disaster warning, improved coastal protection and development; and better deep water port development, pollution dispersion, fishing predictions, iceberg avoidance, navigation in ice fields and weather prediction, to name a few. The SEASAT program, in effect, takes this rapidly evolving microwave capability and integrates it aboard a spacecraft to obtain answers in direct support of the stated needs of oceanographers.

Now, turning briefly to a third area of remote sensing involved with that fragile thin layer of air that surrounds and protects this planet, another satellite under development, NIMBUS G, will monitor and survey the constituents in the atmosphere, not just over this town or over this nation, but over the entire globe. Trace contaminant quantities will be measured everywhere, every day. Monitoring of the distribution and changes of the stratosphere ozone content which serves as a radiation shield is typical of what can be accomplished easily and effectively.

What then are the major challenges that we face? We still have much to do in learning how to extract the greatest benefits from our technology. We have to learn how to extract, display, and use the information our tools can provide, rapidly and easily and cheaply. We must develop and expand natural and predictive models of the environment so as to provide the kind of real assessments so necessary in real-world decision-making.

These are NASA-type jobs -- research, development, pioneering, showing what can be done and how to do it best. We are working hard on them, and many of you here have lent much support to the progress of United States' efforts.

An even larger job, however, is the development of competent local, regional, national, and global institutions that will build and operate such systems, that can assimilate and act on the new class of information, that can manage and control with wisdom the resources and environment of mankind. In this task, the NASA program and the NASA accomplishments can serve as catalysts -- but the responsibility for action really lies with everyone. The individual, his local and state governments, his federal department, the private sector -- all have major roles to play in assuring the proper employment of technology by man for the sake of mankind.

That, I believe, is the real challenge that the space age has posed for the world. If we all accept this challenge, there is no question but that the space capabilities of the nation will provide a payoff and a payback in enhancing the quality of life for everyone.

OUR NATIONAL ENERGY FUTURE-
THE ROLE OF REMOTE SENSING

HARRISON H. SCHMITT

Assistant Administrator Office of Energy Programs
National Aeronautics and Space Administration
Washington, D.C.
Division of Geological and Planetary Sciences
California Institute of Technology
Pasadena, CA

There is obviously great need for a long-range approach upon which to base our present actions related to energy. Any one of a number of such long-range approaches probably would be adequate; however, it is imperative that we choose one general path now in order to focus present action and to provide the foundation for improvement in the future.

Our present knowledge suggests that the Nation's energy future through the early part of the 21st century will consist of three specific, although interlocking phases. First, there is the present phase of crisis in energy supply which certainly will last as a real or internationally controlled political factor until at least 1980 and possibly until 1985 depending on our ability to find early relief. Second, there will be a transition phase which we must initiate in the present and which will last until ultimate alternatives to fossil and fission fuels are available. Finally, there is a phase which will see the gradual increase in the utilization of sources of renewable energy that can eventually replace fossil and fission fuels.

Several factors will exert major, if not dominant, controls on the duration of these three phases and the detailed character they assume. These factors are as follows:

1. The political need to become independent, or to demonstrate the <u>real</u> ability to be independent, of foreign supplies and foreign prices of fossil fuels.

2. The rate at which new domestic petroleum reserves are quickly identified and from which the Nation can demonstrate its capacity to be as self-sufficient as required.

3. The human need to preserve those of our resources now used as fossil fuels for far more valuable uses as petro-chemicals, fertilizers, and other products.

4. The economic and operational rates at which new plants related to the production of electricity and synthetic fuels can be brought into being and the lifetime of those plants required for amortization of the initial investment.

5. The degree to which conservation and preservation of the environment becomes possible either through voluntary, regulatory, or research and development efforts.

6. The rate at which the technology of futuristic energy sources advances toward practicality.

PHASE I: CRISIS

The nature of the crisis phase of our energy future will be either one of shortage or of dependence on uncertain foreign supplies. The duration of the crisis phase is dependent on the level of energy conservation we reach, the rate at which we produce from old or identify new sources of petroleum, and the rate of increase in energy produced from non-petroleum sources.

Reduced use of present forms of energy is the obvious first step in conservation; however, reduced consumption is possible only to the point at which levels of intolerable unemployment and/or inconvenience are reached. Conservation through the benefits of present research and development is clearly important if only to free future financial resources from the creation of energy.

Present research and development in conservation can take many possible routes. Given our present limitations and knowledge, the most promising and most quantitatively significant paths are in more efficient ground propulsion, improved residential and industrial insulation, solar heating and cooling of buildings, improved heating and cooling systems, and integrated utility systems for large population units. Also, if we are successful in the development of good and inexpensive technology in these areas, their value in the international marketplace will be high for many decades to come. It should be remembered, however, the practicality of major investments of capital in conservation measures is dependent on a significant rate of return on those investments. A good idea may not be a useful idea, particularly in times of inflation and high interest rates.

The most critical aspect of the crisis phase is the need for early relief from potential shortages of energy supply without unacceptable dependence on uncertain foreign supplies. Domestic reserves of petroleum, that is, oil and natural gas, are the only energy sources that can potentially provide this early relief. Other important future sources of energy, such as coal, oil shale, geothermal, solar, and nuclear, even in aggregate, cannot give enough total energy over the next ten years to provide more than a part of that required for independence from foreign manipulations.

It is clear to many geologists that great and deliverable quantities of petroleum can be found in on- and off-shore areas directly accessible to the Nation. If rapidly and publicly assessed and then put under exploration by industry, the demonstration of the availability of this petroleum can be made in time to provide relief by the early 1980's. Production as sources of fuel may never be necessary except as foreign pressure dictates. It is important that we take steps to find this petroleum and to be ready to refine it if that becomes necessary. The present lack of a national inventory of our potential resources of petroleum is the major factor preventing a rational approach to foreign economic policy. We just do not know what our domestic petroleum production might be if we were required to develop it fully.

PHASE II: TRANSITION

The Transition Phase of our energy future is dominated by two major factors: one is the need to gradually terminate the use of petroleum as an energy source and the second is the need to create the alternatives to petroleum which will carry us into the 21st century.

The basis for policy decisions relative to transition initially should be the assumption of success in the creation of inexhaustible alternatives to all fossil fuels. With this working assumption, our research and development needs aimed at transition should concentrate on energy sources that can (1) provide for the transition away from petroleum as a fuel, and (2) provide for regional and potentially National sources of competitive clean energy. There is no fundamental reason why these transition energy sources cannot be less expensive than petroleum if we do our job right.

Coal appears to be one of the most likely energy sources to ease the transition away from petroleum. Coal gasification, liquefaction, and central power plants not only will help make the transition but they can also provide the raw materials for future agricultural and chemical needs.

Oil shale is another possible transition energy source; however, it is not yet completely clear that extensive research, development, and production related to this resource is necessary. Although worth much additional study, before committing to a major development of oil shale, we should assess fully our financial resources, the many undesirable environmental implications of present oil shale technology, and the already existing technological base for the use of coal. Leaving such a great future chemical resource as oil shale in the ground for a few generations will probably not hurt us.

In transition, there is also, of course, the potential use of advanced methods of energy production through nuclear <u>fission</u>. Because of its controversial nature, however, and some real disadvantages, it will always be desirable to minimize the use of nuclear fission provided other non-fossil fuel alternatives are available. On the other hand, there is no question that we must accept some nuclear fission plants as being necessary facts of life until alternatives to them are available at specific locations. We must also solve the major problems of nuclear fission, problems of fuel production and handling, long-term waste disposal, security, plant siting, and reliability. There is no good reason why solutions cannot be found, but we must be sure they will be found.

PHASE III: RENEWABLE ENERGY

The Renewable Energy Phase, as mentioned above, is the period when there is one or more inexhaustible clean sources of energy that permit the preservation of natural hydrocarbons for the agricultural and chemical requirements of future generations. With preservation also comes the enhancement of the environment for all generations. There are several energy sources pertinent to renewable energy for which there are reasonably clear technological paths known and which can provide relatively clean electricity indefinitely.

The most desirable means of producing clean energy appear to be from hydroelectric, geothermal, solar, ocean thermal, and nuclear <u>fusion</u> sources, although large investments in research and development remain to be made before the full economic feasibilities of each have been demonstrated fully. It will probably be necessary to conduct research and development programs in each of the geothermal, solar, and ocean thermal areas because their applicability is limited geographically; however, if successfully developed and if properly exploited, in aggregate they represent essentially inexhaustible future sources of energy for many if not most regions of the Nation.

In addition to the regional use of natural energy sources, the energy from nuclear fusion presently appears to be the most promising "ultimate" source of energy. On the other hand, a significant level of fundamental scientific and engineering research should continue that may identify other, potentially even more desirable alternatives. Engineering studies of solar power stations in space and theoretical investigations into the nature of gravity are obvious candidates for such sustaining research. The value of such research as usual will have unanticipated benefits and ramifications even outside the questions of energy sources.

The concept of renewable energy also implies long-term research in other areas if the generation of electricity (or hydrogen or methane) will ultimately dominate the energy scene. For example, a hydrogen-based energy economy is a very attractive future possibility and should be thoroughly investigated. Also, futuristic means of ground and air propulsion that use new types of portable fuels eventually must be developed. Such means of transportation probably should begin to be in widespread use about the year 2000.

REMOTE SENSING IN OUR ENERGY FUTURE

The term "remote sensing" is gradually becoming a part of our everyday vocabulary, although it is not yet exactly a household phrase. Remote sensing as an art in science has been around for a long time; first and foremost in the form of man's eyes. Even though we are very far from duplicating with instruments and systems the capabilities of those eyes and their unbelievably sophisticated data processing systems, we have learned over the past decade in near-earth space to vastly extend their reach. It is this extension of our inherent ability to use our eyes and brain to observe, integrate, synthesize, interpret and apply that is what we have come to know as remote sensing. In a broad way we include in this technology the separate and combined techniques that use aircraft and satellites.

Remote sensing in our energy future will have potential applications in two major fields; fuel production and energy conversion. In fuel production, remote sensing will be broadly applicable so long as we are dependent on the use of large amounts of fossil fuels. This applicability will continue to some degree even into our renewable energy future until there are alternatives to petroleum as the feedstock for the creation of fertilizer and as a fuel for farm implements. In energy conversion, remote sensing will be applied so long as we must monitor and regulate the waste energy and undesirable or dangerous by-products from the conversion of one form of energy to another.

The most clearly defined primary role for remote sensing in fuel production will be in the increased efficiency by which we explore for and identify potential petroleum reserves in frontier areas.

Most workers in this field would agree, I believe, that several inherent features of various types of remotely acquired data, particularly those from satellites, potentially permit the rapid assessment of possible targets for ground inspection. The inherent features to which we can refer are 1) broad area, repetitive coverage under uniform and multi-spectral conditions that can be used for rapid reconnaissance mapping, monitoring and planning, 2) broad and local area coverage that reduces or eliminates the effects of variations because of illumination, clouds and vegetation, and 3) comparisons of the spectral signatures of productive versus unexplored areas.

Once targets for ground inspection are identified using combinations of remote sensing techniques, then the sequential use of rock and soil alteration studies, isotope geochemistry, amplitude enhancement of geophysical data, and finally test drilling can be used to progressively confirm the nature and extent of possible reserves.

It is clear that at least portions of this idea exploration sequence, which begins with remotely sensed data for reconnaissance, are in use today within the competitive energy industry. It is unfortunate that comparable activity is not underway to accelerate a public assessment of our petroleum reserves. Such an assessment or inventory, would allow the examination of national options of plenty which <u>cannot</u> now be considered and would eliminate other options of paucity which <u>must</u> now be considered.

Remotely acquired data can be applied to other specialized problems of fuel production. Of particular interest are the identification of rock structures that can effect the mechanics of coal production and the delineation of natural oil seeps offshore. In addition, the detailed assessment of the true potential of specific sites for large solar energy plants also appears feasible.

There has not yet been identified a clear utility for remotely acquired data in the area of uranium exploration, although several interesting possibilities have been suggested. The primary use will probably be in the long term, more efficient expansion of detailed reconnaissance mapping in potentially mineralized regions. In this process, we can expect to be able to concentrate our attentions on continental and regional structures that may correlate with geochemical provinces rich in uranium. Such integrated studies that employ ERTS and Skylab photography are just beginning in the southwestern United States.

The efficient and broad scale monitoring and predictive capabilities of remote sensing techniques have yet to be extensively applied in other potentially fertile fields related to energy production. I would mention here the monitoring of watersheds and the flood control, hydroelectric and irrigation systems that tap such watersheds. With remote and repetitive monitoring of watersheds and reservoirs, and predictions based upon that monitoring, it will be possible to optimize energy production, irrigation and flood control. In addition, sea, air and pipeline transport of energy and goods can be optimized through continuous control and monitoring of natural impediments and hazards, pollution effects and energy-efficient routes.

There is also, I believe, a very clear role for remote sensing in the monitoring and regulation of waste heat and undesirable by-products produced by energy conversion. However, we have yet to systematically exploit this potential. The list of possible new services and areas for cost savings is long, ranging from the monitoring of thermal and effluent pollution to the planning of land-use. It is to be hoped that the results of many current experiments by many investigators will make this environmental role for remote sensing obvious to all who must approve and implement such efforts.

This is one more great potential benefit of remote sensing technology that is often forgotten or is considered with only short term insight. This is the bridge of long-term friendships and mutual benefit that can be built between the United States and the emerging nations of the world upon the foundation of space technology. The decade of Apollo and Skylab have convinced many peoples and their leaders that through space technology they can bring themselves into the twentieth century. My travels around the world have shown me that these peoples now believe that they can participate in the future along side the present industrialized community of nations.

The potential availability of remote sensing and communication satellite systems during the next several decades is one of the basic ingredients of rapid international advancement and the mutual interdependence the world so desperately needs. With access to these systems, the chronic problems of resource management and education can be solved by the emerging nations. Agricultural monitoring and predictions, education related to birth control, and explorations for energy and minerals can all be considered possible with the establishment of the appropriate national and international infra-structures.

The construction of the bridge of international friendship and mutual benefit through space technology will not be easy. It will take far more foresight than that to which most of us are accustomed. On the other hand, the bonds of hope and good-will are already there to be used. Let us hope that we do not stand here ten years from now and wish we had done something.

CONCLUSION

As indicated at the onset, the energy policy guidelines I have suggested are probably just one of many possible schemes under which the states and the Federal Government could define specific programs. These guidelines do, however, represent some of the more logical extrapolations of our present knowledge of technology and trends. Later modifications to this scheme will certainly be necessary as knowledge expands, but to wait on that knowledge is to perpetuate crisis. To proceed on the basis of present knowledge, however, is to lay foundations for the future of the Nation and, indeed, the future of the world.

PART I

ATMOSPHERIC AND HYDROSPHERIC MEASUREMENTS

THE EVOLUTION OF ATMOSPHERIC MEASUREMENTS FROM SATELLITES

MORRIS TEPPER

National Aeronautics and Space Administration

ABSTRACT

Meteorological satellites are described from the first successful launch in 1960 to the current and future planned satellites. Two streams of activity have been established in space meteorology - a basic research and development program within NASA and an operational satellite program within NOAA.

Various sensors on the meteorological satellites have been developed. Camera imagers for viewing global cloud cover have operated both in stored-data mode and in a direct read-out mode. Scanners in the form of radiometers have provided observations of radiation, cloud cover, storm tracking, cloud top measurements, surface temperature, tropospheric water vapor, and mean stratospheric temperature and circulation with increased improvements from medium to very high resolution. Sounders have been developed to obtain radiometric observations from which vertical temperature profiles of the atmosphere could be deduced. Sounding systems at the end of this decade will have improved resolution, reach higher into the stratosphere and will be able to penetrate practically all conditions of cloudiness, exclusive of precipitating clouds.

Observations from space have become a major factor in the improvement of our understanding weather and its prediction and have played a major role in the establishment and activities of the Global Atmospheric Research Program (GARP). Principal future efforts will be in applying space meteorology to local severe storms and to the climate variability program.

I. INTRODUCTION

The remote sensing from space of the atmosphere and the weather phenomena that are embedded in it is one of the most mature applications of space research. The history of usage of meteorological satellites goes back to the beginning of the 1960's when a series of spinning TIROS satellites were launched and returned pictures of cloud cover to earth receiving stations. Since those early days, there has been a steady progression of modifications and improvements to the satellite systems and the sensors which they carry.

Most of these meteorological satellite systems were launched into polar orbit--specifically, into a sun-synchronous polar orbit, so that each satellite passed over any geographical point at the same local time. As the earth rotated beneath the satellite, total global coverage is achieved.

Additionally, geostationary satellites have been utilized which, since they are located in an orbit at a distance of 22,000 miles from the earth, appear stationary with respect to the sub-satellite point. It is possible to use sensors on such geostationary platforms to look repetitively and frequently at the unfolding weather patterns below, either over the entire disk of the earth or at selected portions of the disk.

II. EVOLUTION OF SATELLITE SYSTEM

The utilization of the data derived from meteorological satellites has proven to be so very useful to our National Weather Services that as early as the mid-1960's the U. S. Weather Bureau (now NOAA) instituted a National Operational Meteorological Satellite System, according to which the beneficial products of satellite meteorology could be made available in routine daily use.

In this manner, there have evolved two parallel streams of development. One, a basic research and development program conducted primarily by NASA; and the other, a program of operational satellite systems funded and managed by NOAA. Figure #1 shows the evolution of both of these programs. The top line shows the early experimental TIROS satellites extending over the period of the early 1960's. The next line shows a continuation of the research and development with the Nimbus systems. The third line lists the NOAA satellites which are used operationally. These operational satellites derive their technological developments in the NASA R&D satellite effort. Each new family in the operational satellite sequence is introduced by a prototype operational satellite funded and developed by NASA. The next group, the geostationary satellites, contains the Applications Technology Satellite (ATS) experimental satellites and the Synchronous Meteorological Satellite/Geostationary Operational Environmental Satellite (SMS/GOES)--the operational satellites.

By means of a shading code, we emphasize the basic missions of various groups of satellite systems. For example, the early TIROS and the early Nimbus satellites, right-slanted shading, were primarily concerned with cloud cover observations, global and local, as well as both day and night. (As we shall see later, the nighttime observations were made possible by the inclusion of an IR scanner.)

The next series of satellites, left-slanted shading and including the Nimbus 3 and 4 and the second generation operational satellites (ITOS), was primarily for the development of measurements of vertical profiles of atmospheric temperature and moisture. These first atmospheric sounder developments were severely limited by the presence of cloudiness in the field of view.

The next series, depicted in screen shading and including the most recent Nimbus satellites as well as the next generation of operational satellites beginning with the operational prototype TIROS-N, will provide the same kind of profile sounding capability; however, they are improved not only in the

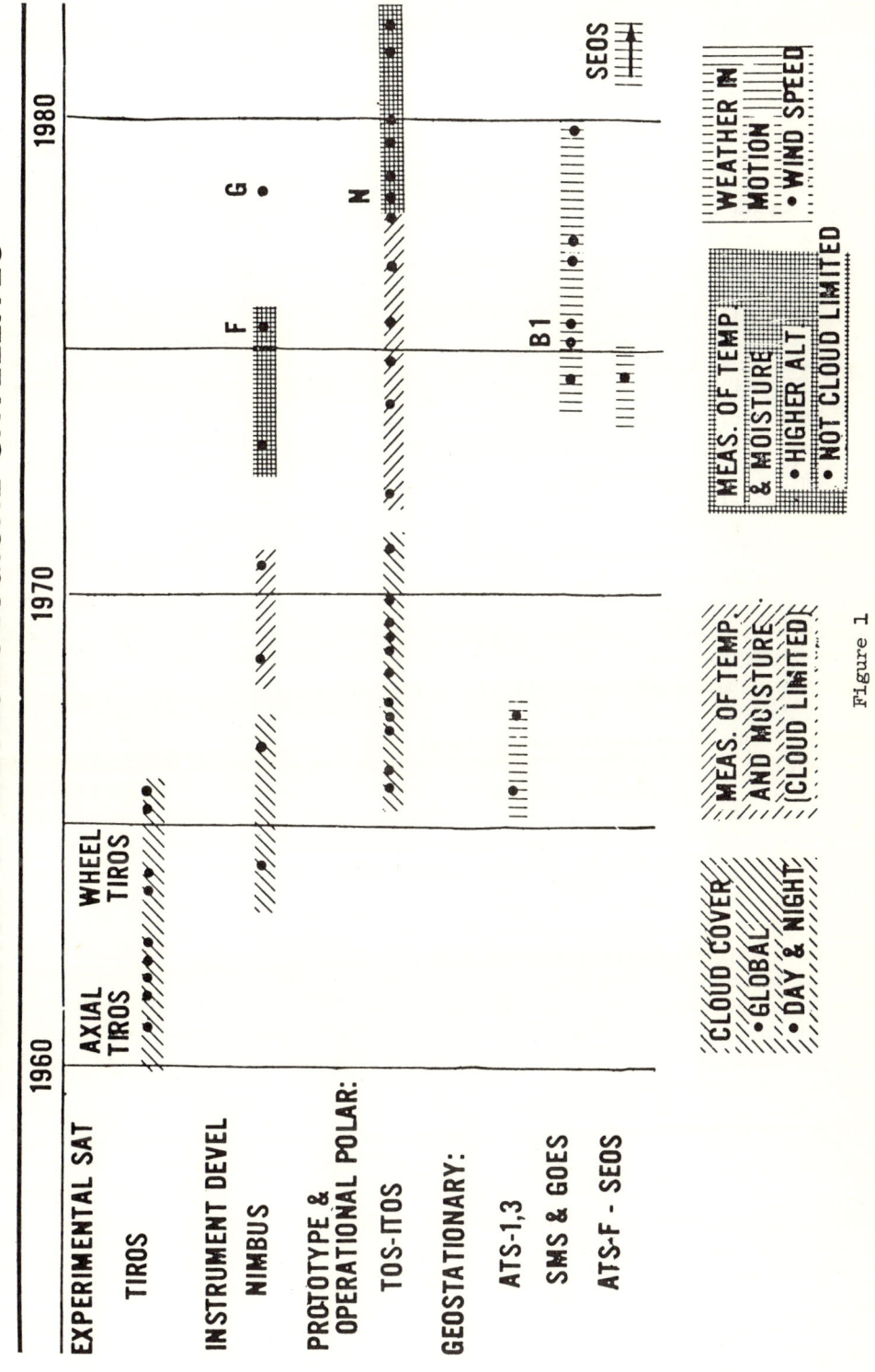

THE EVOLUTION OF ATMOSPHERIC MEASUREMENTS FROM SATELLITES 17

Figure 1

accuracy of the derived meteorological parameters, but also in the fact that the measurements are influenced very little by the presence of most cloudiness conditions.

Finally, the geostationary satellites are shown to contribute to the observation of "weather in motion" (vertical-line shading) and, more specifically, the computation of wind velocity from sequential cloud motions. These satellite systems, as we shall see later, have an important application in the area of severe storm observations and monitoring.

III. EVOLUTION OF METEOROLOGICAL SATELLITE SENSORS

The early satellites carried camera imagers (vidicon tubes) to photograph cloud cover. These instruments were later replaced by scanners which did not take pictures, but rather "painted" out their observations in a cross orbit motion as the spacecraft moved forward in its orbit. The other major group of instrumentation developed in the meteorological satellite program involved radiometers from which it was possible to derive soundings of temperature and moisture profiles using radiative transfer theory applied to the atmosphere and its constituents.

A. IMAGERS

Figure #2 shows the evolution of the meteorological satellite camera systems from TIROS 1 (1960) through Nimbus 4 (1970). The development of the early TIROS satellites produced imaging systems having the capability of viewing global cloud cover in either the stored-data mode for global analysis or in a direct-readout mode for local usage. The television pictures taken by the spinning TIROS-1 (see Figure #3), which was launched into a near-earth polar orbit on April 1, 1960, were oblique and often not optimally illuminated by the sun. Nevertheless, they showed details of the organization of quasi-global weather systems that were previously unknown. The value of such data was immediately recognized and within a few days of the TIROS-1 launch, the new data was available at the National Meteorological Center for operational use.

In the ensuing years, the meteoeological satellite program produced new developments in rapid order. In 1965, the spinning TIROS system was converted to a rolling cartwheel mode wherein the television cameras were pointed outward radially so that they were able to view the earth vertically. It was thus possible to acquire full earth coverage. ESSA-1, the operational version of this system, was launched on February 3, 1966.

The Nimbus satellites were earth-oriented polar-orbiting spacecraft and were the test bed for future operational ESSA sensors such as the Advanced Vidicon Camera System (AVCS) and the Automatic Picture Transmission (APT) System. Figure #4 is an example of data from the 3-camera Nimbus 2 AVCS taken on 17 June 1966 over Africa and the Middle East. In its operational version of ESSA-9, only a single-camera AVCS was used.

The APT cameras for Nimbus-1 and 2 used special long-storage vidicon tubes whose dielectric surfaces were capable of holding the image for the

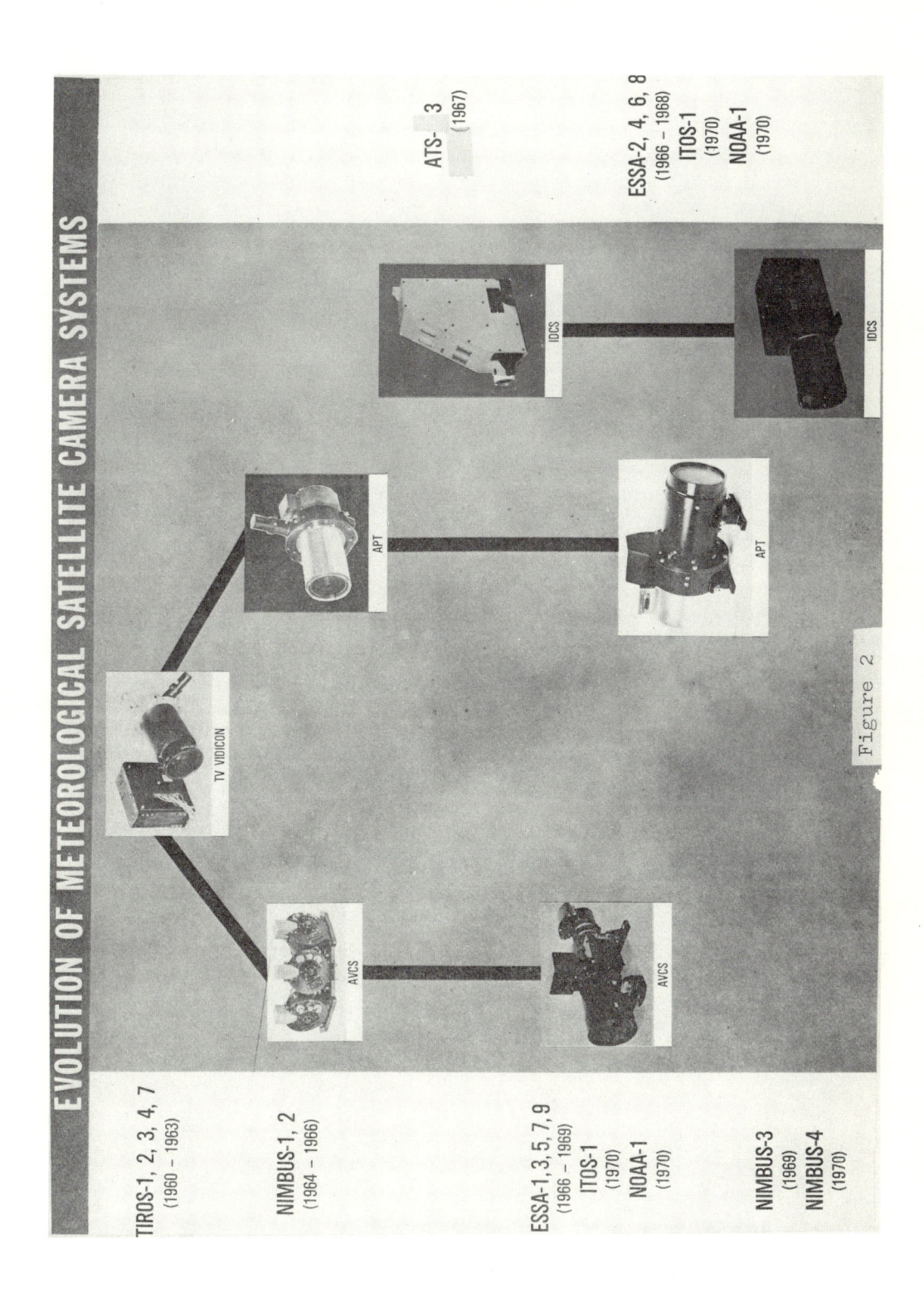

THE EVOLUTION OF ATMOSPHERIC MEASUREMENTS FROM SATELLITES 19

EVOLUTION OF METEOROLOGICAL SATELLITE CAMERA SYSTEMS

TIROS-1, 2, 3, 4, 7
(1960 – 1963)

NIMBUS-1, 2
(1964 – 1966)

ESSA-1, 3, 5, 7, 9
(1966 – 1969)
ITOS-1
(1970)
NOAA-1
(1970)

NIMBUS-3
(1969)
NIMBUS-4
(1970)

ATS-3
(1967)

ESSA-2, 4, 6, 8
(1966 – 1968)
ITOS-1
(1970)
NOAA-1
(1970)

Figure 2

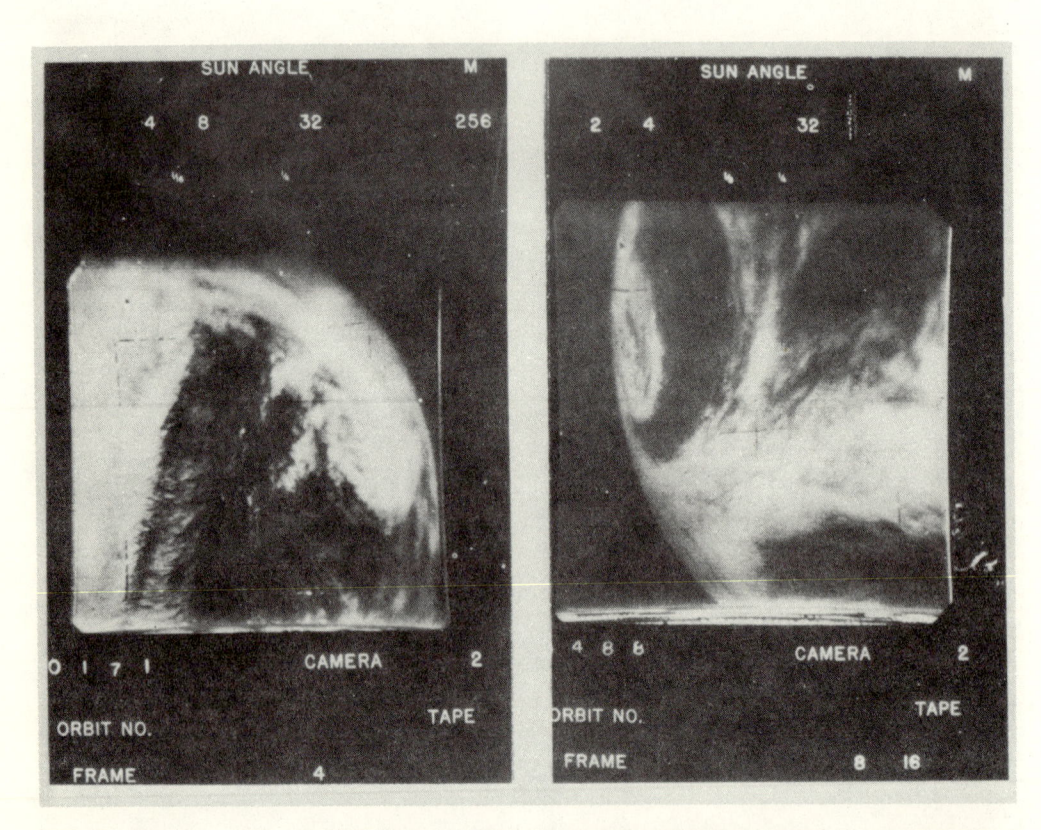

TYPICAL TV PICTURES RECORDED BY TIROS-1 (1960)

Figure 3

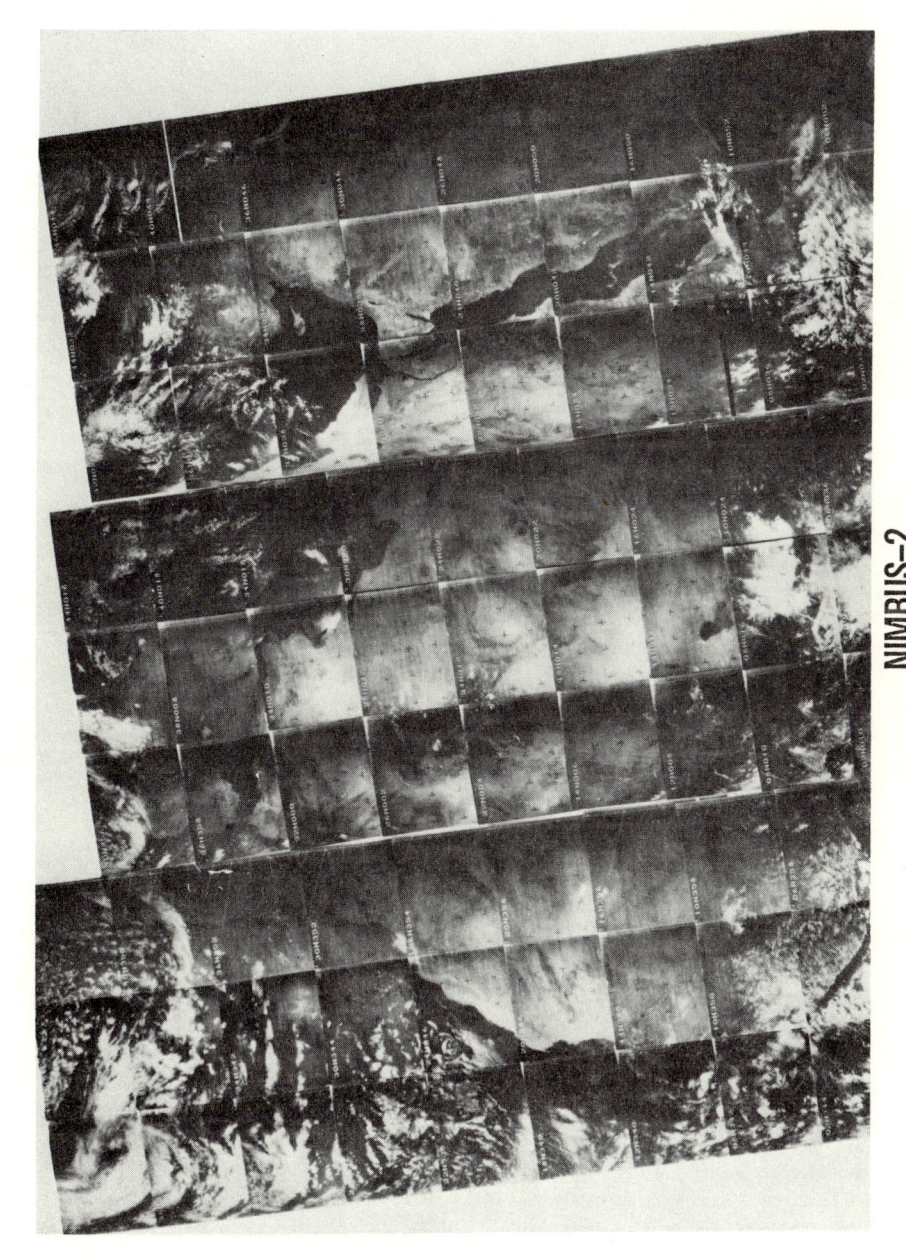

NIMBUS-2
AVCS (3 CAMERA) PICTURE OF AFRICA AND MIDDLE EAST
17 JUNE, 1966
Figure 4

required 200 seconds readout period. Over 300 ground stations, including 78 stations in 43 foreign countries, received APT coverage from Nimbus-1 and 2. An APT photograph covered an area 1200 x 1200 n. miles. The entire area within a radius of 1860 n. miles from each APT ground station was thus covered by three successive passes of the satellite and provided an immediate analysis of meteorological conditions by local stations around the world, using relatively inexpensive ground equipment.

Figure #5 was taken over Hurricane Alma by Nimbus-2 APT on June 9, 1966, over the Florida peninsula. Note the light gray area along the west coast of Florida which indicates the action of strong hurricane winds on shallow coastal waters.

The Image Dissector Camera Systems (IDCS) was flown on Nimbus-3 and 4 and was designed to replace the APT and AVCS to provide cloud cover pictures in real time and by stored playback mode. This instrument was also flown on the ATS-3 geostationary satellite. It was not used operationally on the NOAA satellites due to development in the scanning systems, as seen below. Figure #6 shows a Nimbus-4 IDCS picture over Alaska on March 29, 1971, in which snow fields and mountain ranges are clearly delineated.

B. SCANNERS

Figure #7 shows the evolution of meteorological satellites scanning systems from the early TIROS days to date. TIROS-2 (1960), 3 (1961), 4 (1962), and 7 (1963) carried aloft the first five-channel Medium Resolution Infrared Radiometer (MRIR). The data from this instrument have been used in research studies of radiation balance, cloud cover mapping, storm tracking, cloud top measurements, surface temperatures, tropospheric water vapor, mean stratospheric temperatures and circulation. An example of a quasi-global analysis of TIROS-3, MRIR (8-12 μm) data on July 16, 1961, is shown in Figure #8. The photo, whose original is in color, is shown here in black and white only and illustrates the global coverage of the MRIR data. Its original appears in NASA Special Publication No. 53, 1964, entitled "Quasi-Global Presentation of TIROS-III Radiation Data".

Nearly an order of magnitude increase in linear resolution over previous TIROS radiometers was achieved with the launch of the first High Resolution Infrared Radiometer (HRIR) aboard Nimbus-1 on August 28, 1964. Figure #9 shows a Nimbus-1 HRIR (3.4-4.2 μm) picture of Hurricane Gladys on September 18, 1964. The analog trace of the HRIR signal through the storm to the eye is also shown. High cirrus clouds are cold (white) in the 210° to 230°K range. The hurricane eye and ocean areas appear gray and black in the 290° to 310°K range.

The MRIR on TIROS and Nimbus was the forerunner of the HRIR and Temperature-Humidity Infrared Radiometer (THIR) on the Nimbus series (1964-1973), the operational Scanning Radiometer (SR) and Very High Resolution Radiometer (VHRR) on the NOAA-2, 3 (1972-73), and the planned Advanced Very High Resolution Radiometer (AVHRR) for the TIROS-N (1978).

The THIR on Nimbus-4 was a 2-channel instrument sensing infrared radiation in the "atmospheric Window" (10.5-12.5 μm) and water vapor channel

NIMBUS–2
APT PICTURE OF HURRICANE ALMA OVER FLORIDA
9 JUNE. 1966

Figure 5

NIMBUS-4
IDCS PICTURE SHOWING MOUNTAIN RANGES AND SNOWFIELDS OF ALASKA
29 MARCH, 1971
Figure 6

Figure 7

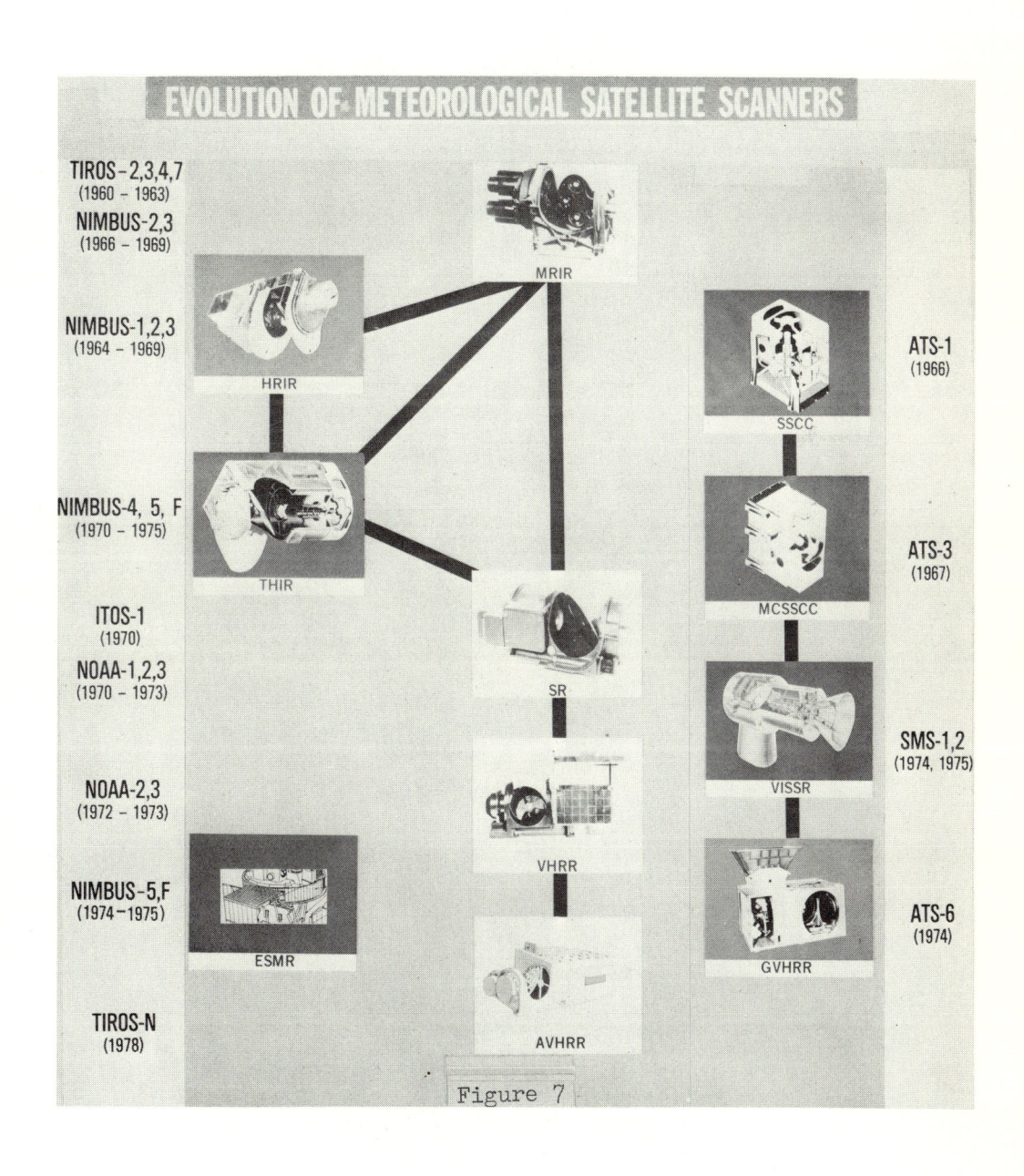

Figure 8

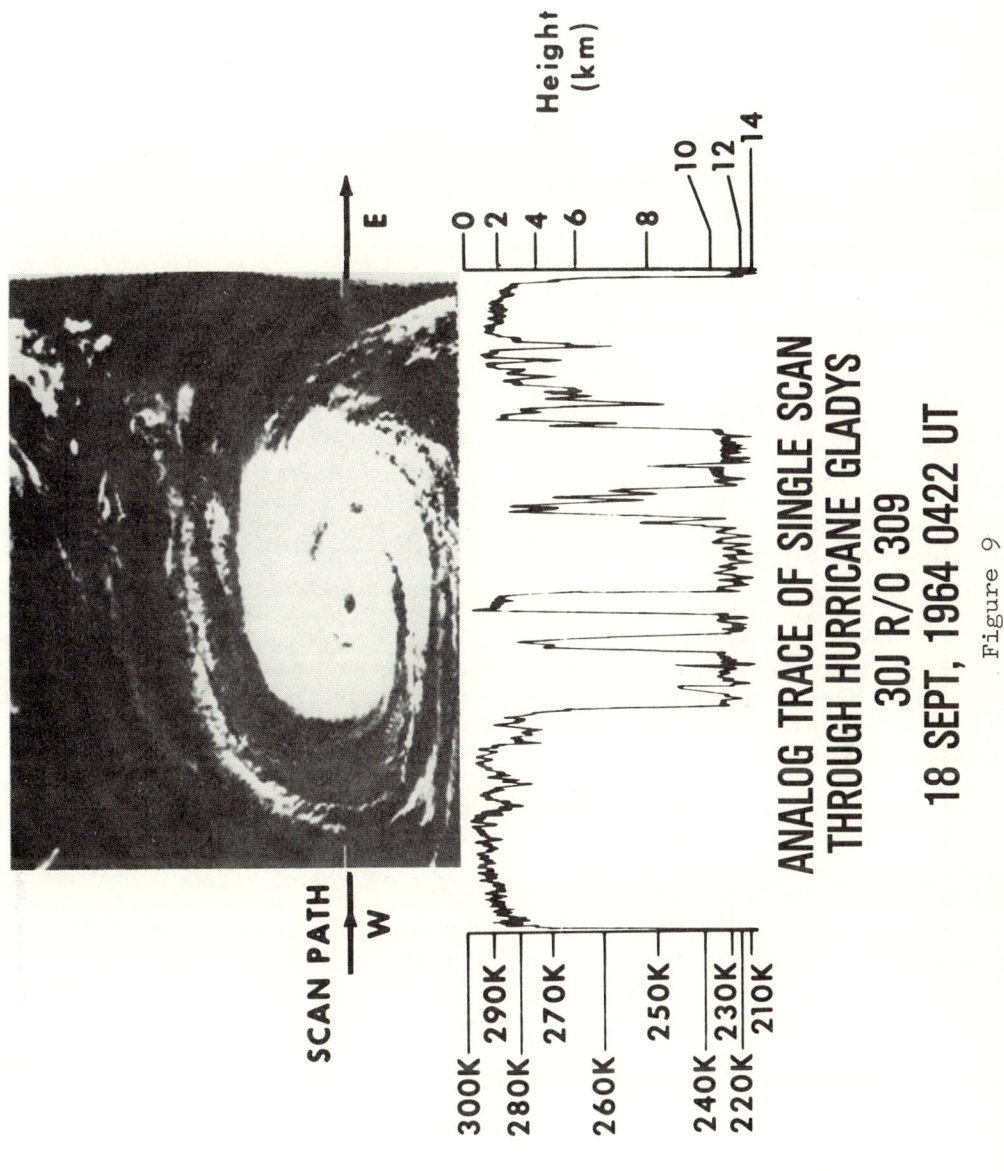

ANALOG TRACE OF SINGLE SCAN
THROUGH HURRICANE GLADYS
30J R/O 309
18 SEPT, 1964 0422 UT

Figure 9

(6.3-7.0 μm). The "atmospheric window" channel provided night and day cloud top or ground temperatures with a ground resolution of 5 n. miles. The water vapor absorption channel sensed the integrated moisture content of the upper atmosphere from 250 mb (10.5 km) down to 500 mb (5.5 km) with a ground resolution of 13 n. miles at the sub-satellite point. Figure #10 shows the product of both THIR channels recorded over the eastern Pacific Ocean on October 16, 1970. The right hand "window" strip shows the relatively clear skies over the ocean and the ground details in western U. S. While the lefthand 6.7 μm strip from a spectral band sensitive to water vapor shows the anticyclonic flow of moisture aloft outlined by the light and dark comma-shaped figure at the top of the strip. Clouds appear white in both pictures. The water vapor channel is particularly useful in cirrus cloud detection, location of jet streams, inferring vertical motion fields and indicating developing and dissipating upper level cyclones and anticyclones.

The Scanning Radiometer (SR) replaced the AVCS television camera system used earlier on ESSA satellites and became the operational system for the ITOS-1 (1970) and NOAA (1972-74) satellite series. This instrument contains a visual (0.52-0.73 μm) channel and an infrared (10.5-12.5 μm) channel, both of which can be transmitted in real time to APT ground stations. Figure #11 shows the ITOS-1 SR direct readout IR <u>nighttime</u> image of the eastern U. S. on February 20, 1970, with ground features indicated. Since the data were digitally available, it was not too difficult to combine them into hemispheric mosaics. One such example for NOAA-2 is given in Figure #12 (July 26, 1973), which shows in one picture all of the weather patterns in the area, most dramatic of which is Hurricane Ava located at $105°W$, $12°N$ at a time when it has reached super-hurricane strength (137 kt. winds).

NOAA-2 (1972) and 3 (1973) both flew the Very High Resolution Radiometer (VHRR) which contained visible and IR detectors sensitive at 0.6 to 0.7 μm and 10.5 to 12.5 μm, respectively, with 1 km ground resolution at nadir. In addition to a limited storage mode, the VHRR can also be transmitted via an automatic transmission mode. However, the ground equipment for this radiometer is more complex than for the standard APT transmission and is consequently more expensive than those stations receiving SR data. In Figure #13 we see that the visible channel VHRR imagery contains sun-glint over a major portion of the eastern seaboard waters of the U. S. on April 29, 1973. The abrupt change in surface roughness on the west (dark) side of the Gulf Stream results from the opposition of smoother waves inshore (light gray, cooler sea temperatures) propagating southward against the northward flow of the warmer (dark) Gulf Stream. The data from the IR channel delineates the Gulf Stream itself and thus have been used in the construction of near-operational charts of the Gulf Stream.

The Advanced Very High Resolution Radiometer (AVHRR) planned for TIROS-N (1978) is a 4-channel scanning radiometer with 2 visual and 2 infrared "window" channels, all having 0.6 n. mile ground resolution at nadir. The data will be digitized on-board the satellite, which insures less degradation in accuracy in the telemetry link. The system also had direct-readout capability. The simultaneous viewing by the two "window" channels is expected to provide global sea surface temperatures with an accuracy of about $\pm 1°C$, under clear sky conditions. The TIROS-N is the prototype polar-orbiting spacecraft for the third generation NOAA-A to G series planned for the 1978-1984 time period.

Figure 10

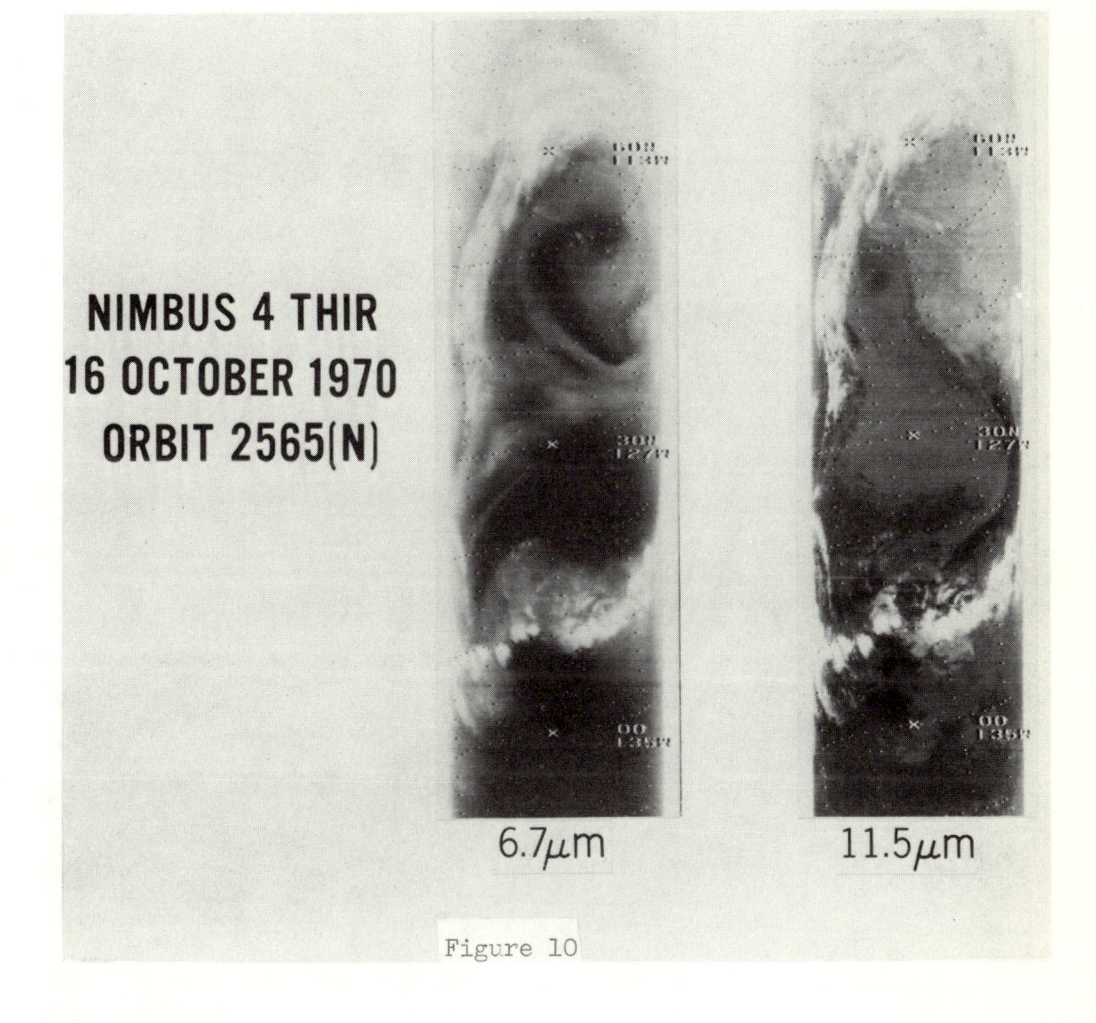

ITOS 1

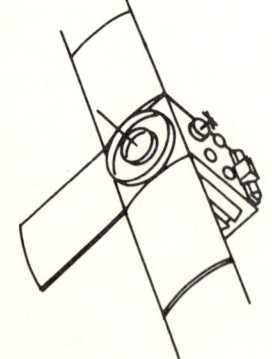

SCANNING RADIOMETER
REAL-TIME INFRARED NIGHTTIME
IMAGE 4 A.M. FEB. 14, 1970.

PROMINENT TERRAIN FEATURES

① GREAT LAKES
② JAMES BAY (Lower part of Hudson Bay)
③ ST. LAWRENCE RIVER
④ CAPE COD
⑤ LONG ISLAND
⑥ FLORIDA
⑦ CUBA
⑧ YUCATAN PENINSULA

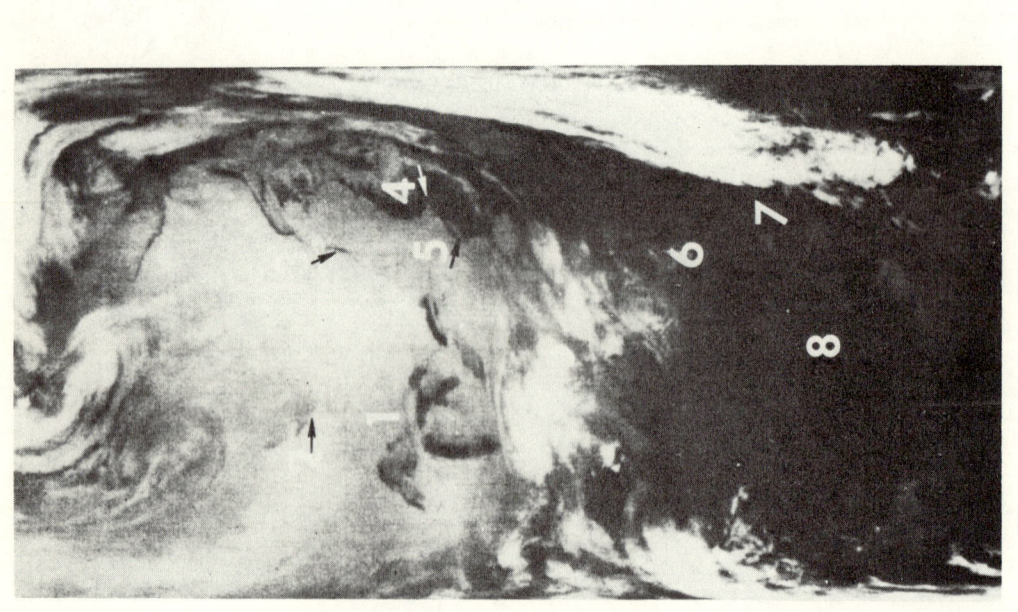

Figure 11

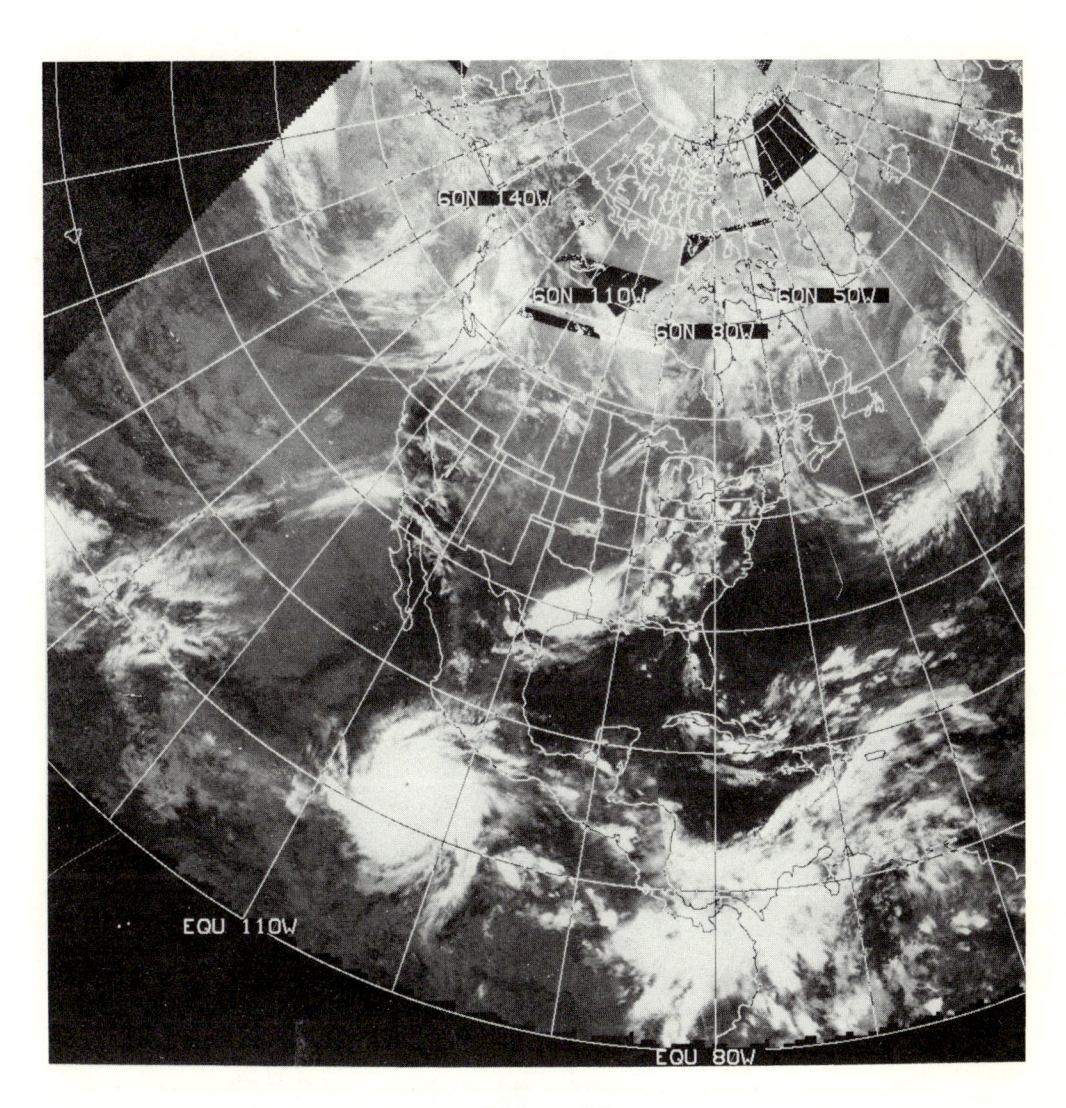

Figure 12

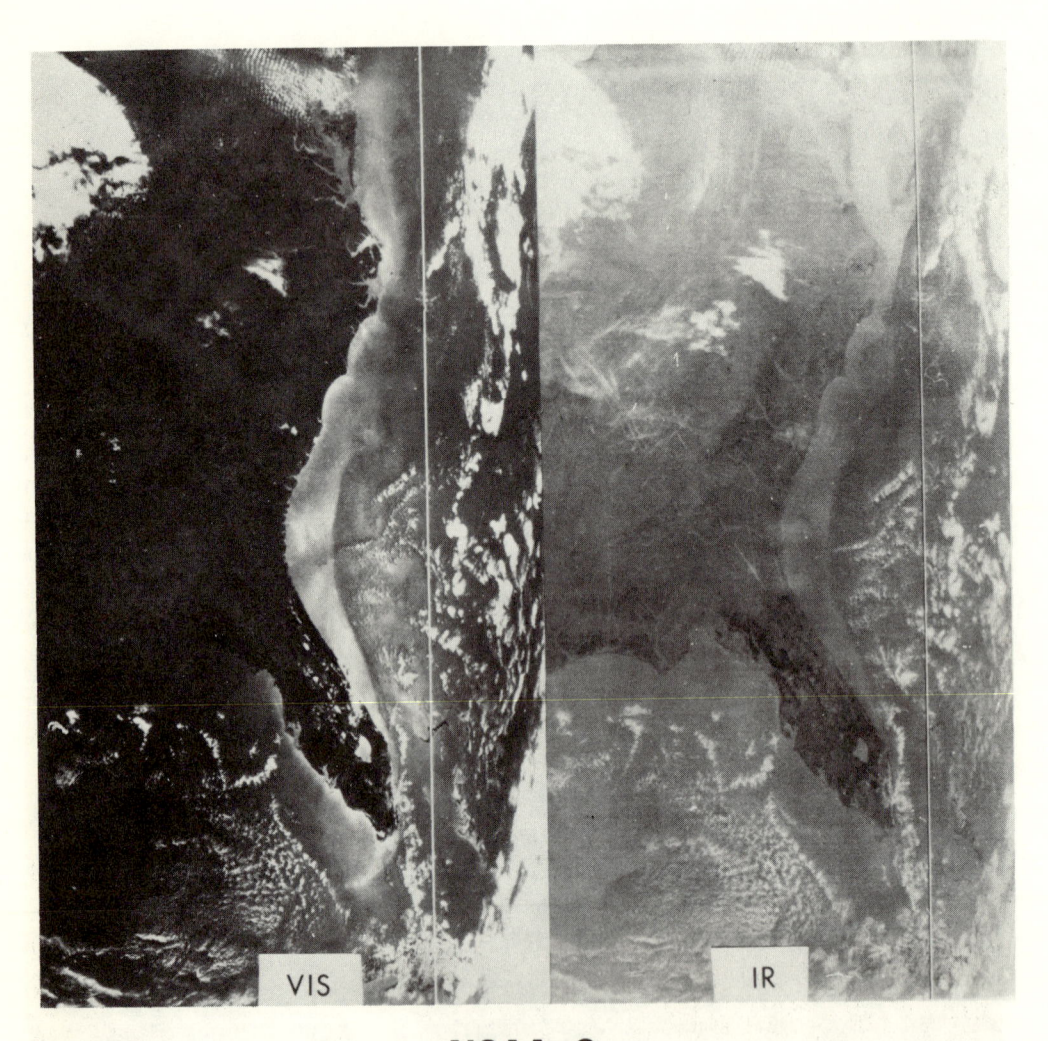

NOAA-2
VHRR ORBIT 2453 29 APRIL, 1973 15Z

Figure 13

The Electrically Scanning Microwave Radiometer (ESMR) flown on Nimbus-5 (1974) was designed to measure earth and atmospheric radiation in the 19.35 GHz (1.55 cm) region. A direct application of the ESMR data was to map the extent of sea ice even through clouds in the polar regions. Figure #14 shows the change in Antarctic sea ice from January 7, 1963, to August 24, 1973 (summer to winter season). The ocean surface appears cold (white) while the sea ice emits the warmest (darkest) microwave brightness temperature. These images were used operationally by the U. S. Navy Fleet Weather Facility to prepare maps of sea ice for resupply ship routing and for later studies of sea ice dynamics. In addition, semi-quantitative measurements of light, moderate and heavy rainfall over the oceans have been made using ESMR data. Figure #15 shows Nimbus-5 ESMR data over several frontal systems with light, moderate heavy rain (increasing dark grey tones) in the Pacific and Atlantic Oceans during December 1972.

The ATS-1 and 3, which were geostationary satellites (22,000 miles, orbital altitude), were equipped with Spin Scan Cloud Cover (SSCC) cameras that provided full disk pictures of the earth's cloud cover on a 24-minute basis. Figure #16 shows a gridded ATS-1, SSCC picture over the eastern Pacific Ocean on 15 December 1966. Note the perturbations in the Inter-tropical Zone of Convergence (ITC) at the center of the photo, frontal activity in the Gulf of Alaska and east of New Zealand. The ATS-3 carried a multi-color version of the SSCC in which fiber optics were used to carry information to three separate photo-multipliers in the red, blue, and green color. Figure #17 shows a color picture recorded by ATS-3 on 19 November 1967 over South America and the North and South Atlantic Ocean. Note the convective storm clouds over South America, the Inter-tropical Zone of Convergence clouds extending between Africa and South America, and frontal systems in the North and South Atlantic Ocean.

Film loops of this type of data have been made which dramatically depict "weather in motion." Several man-computer interactive techniques have been developed to infer wind velocities and divergence patterns from cloud motions on the film loops and from digitized ATS data tapes. Figure #18 shows a derived wind field from ATS-1 sequential cloud motion data on 12 August 1972.

The Visible and Infrared Spin Scan Radiometer (VISSR), which was aboard SMS-1 provided earth-disk images of the western hemisphere every 30 minutes, day and night, with a 0.9 km daytime and 9 km nighttime resolution. The Geosynchronous Very High Resolution Radiomer (GVHRR) on ATS-6 contained lower resolution visual and IR sensors with 5.5 km and 11 km ground resolution, respectively, at nadir. However, this was the first utilization of a scanning radiometer on a 3-axis stabilized platform.

C. SOUNDERS

Figure #19 shows the evolution of satellite instrumentation for the purpose of sounding the vertical characteristics of the atmosphere.

We have already seen that the early beginnings of such techniques could be found in the Nimbus-2 satellite which carried a multispectral Medium Resolution Infrared Radiometer capable of providing temperature information about several levels in the atmosphere, the earth's surface, cloud tops, moisture levels, and even in the stratosphere.

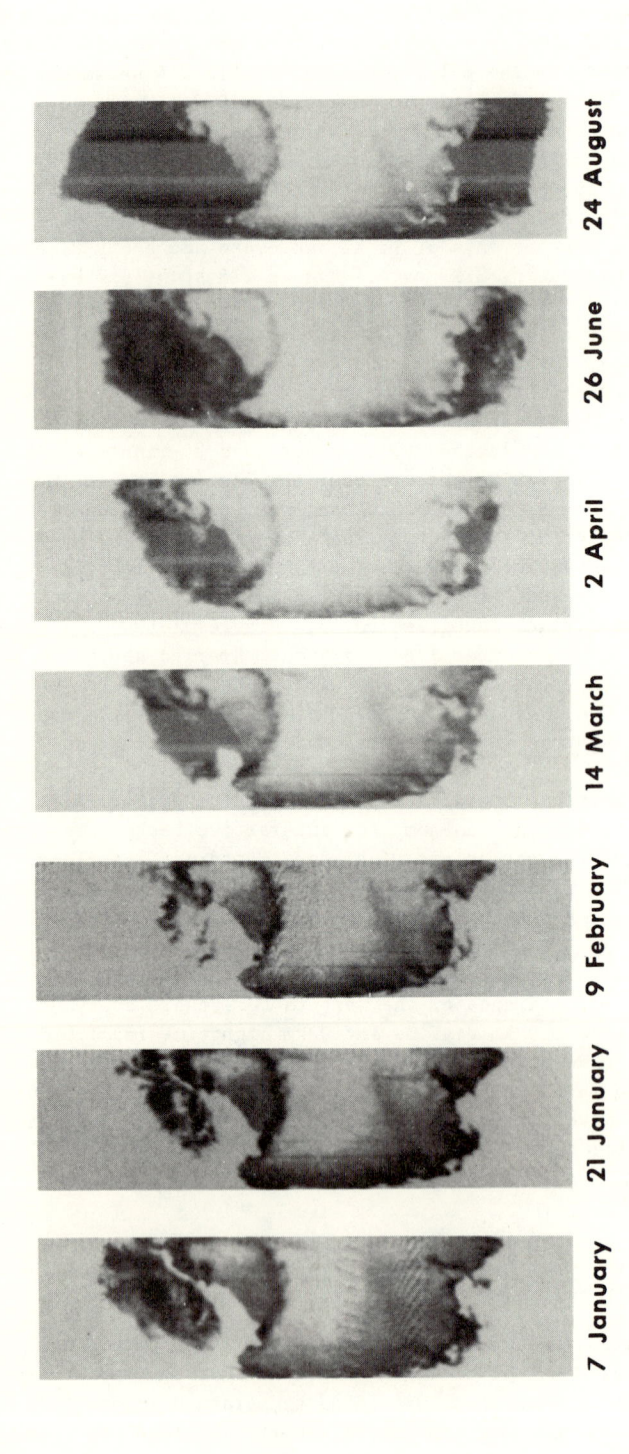

NIMBUS-5 ESMR
MICROWAVE OBSERVATIONS OF ANTARCTIC
SEA ICE CHANGES IN 1973

Figure 14

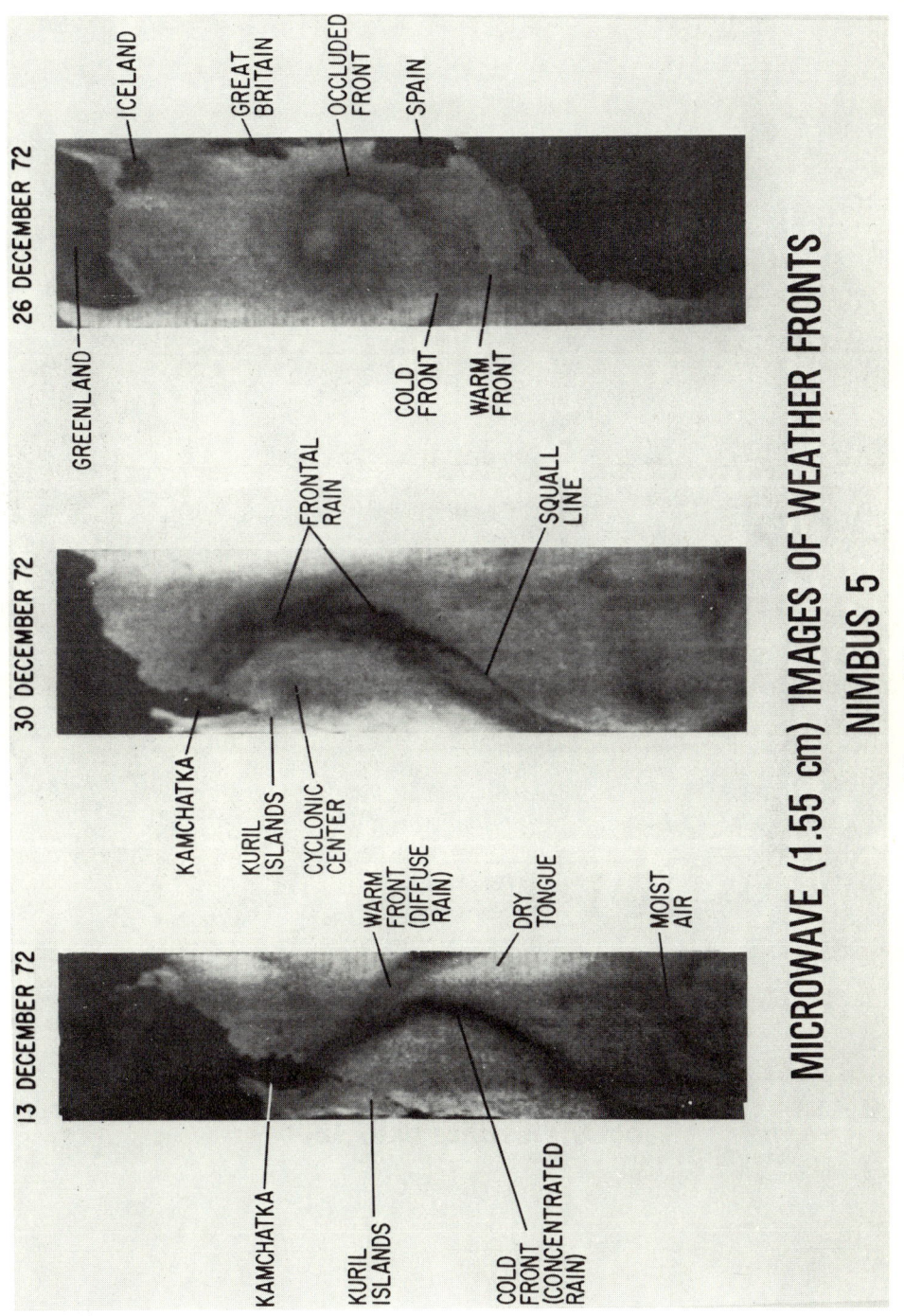

MICROWAVE (1.55 cm) IMAGES OF WEATHER FRONTS
NIMBUS 5

Figure 15

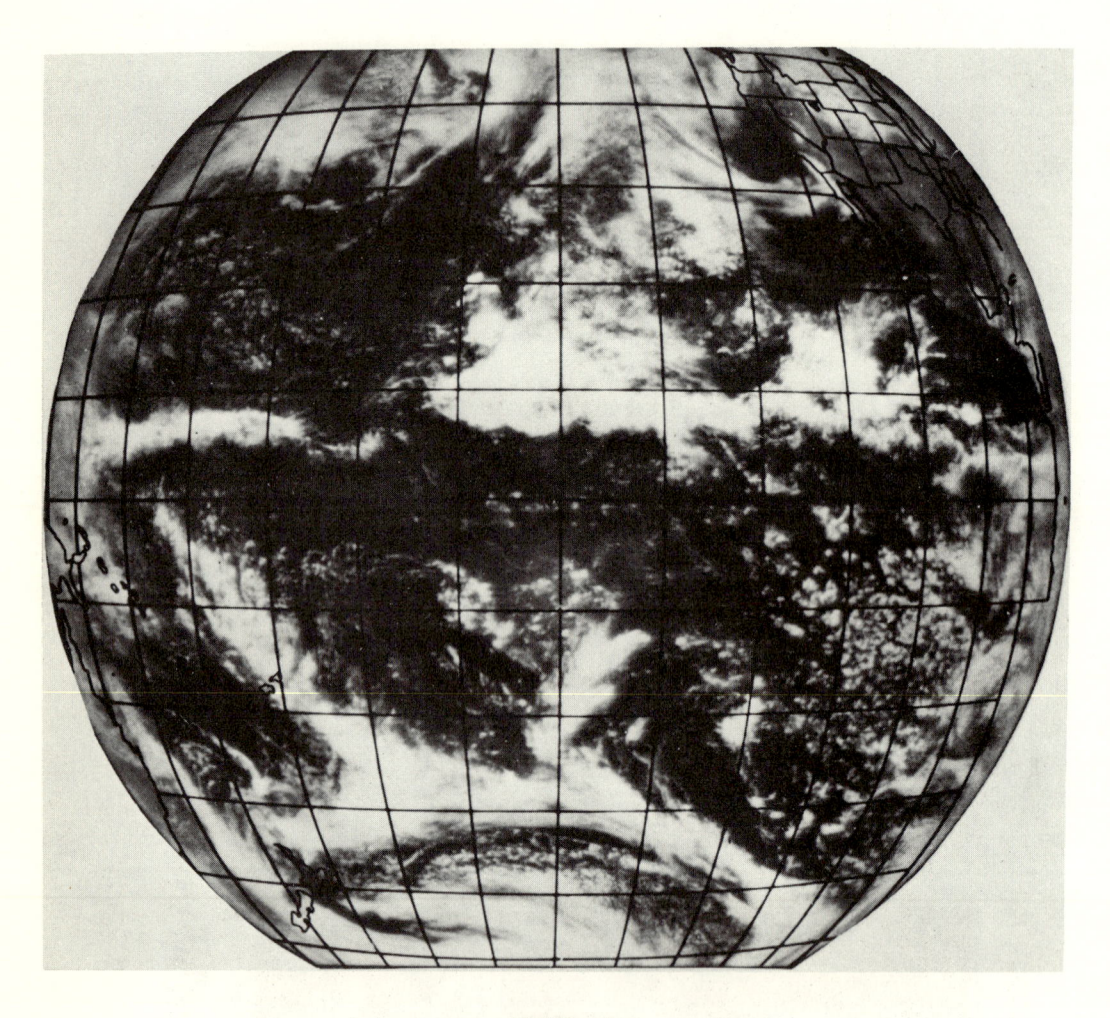

ATS–1
SSCC PICTURE, DEC, 1966

Figure 16

NASA ATS III MSSCC 19 NOV 67 161257Z SSP 49.28°W 0.12°S ALT 22242.67 SM
Figure 17

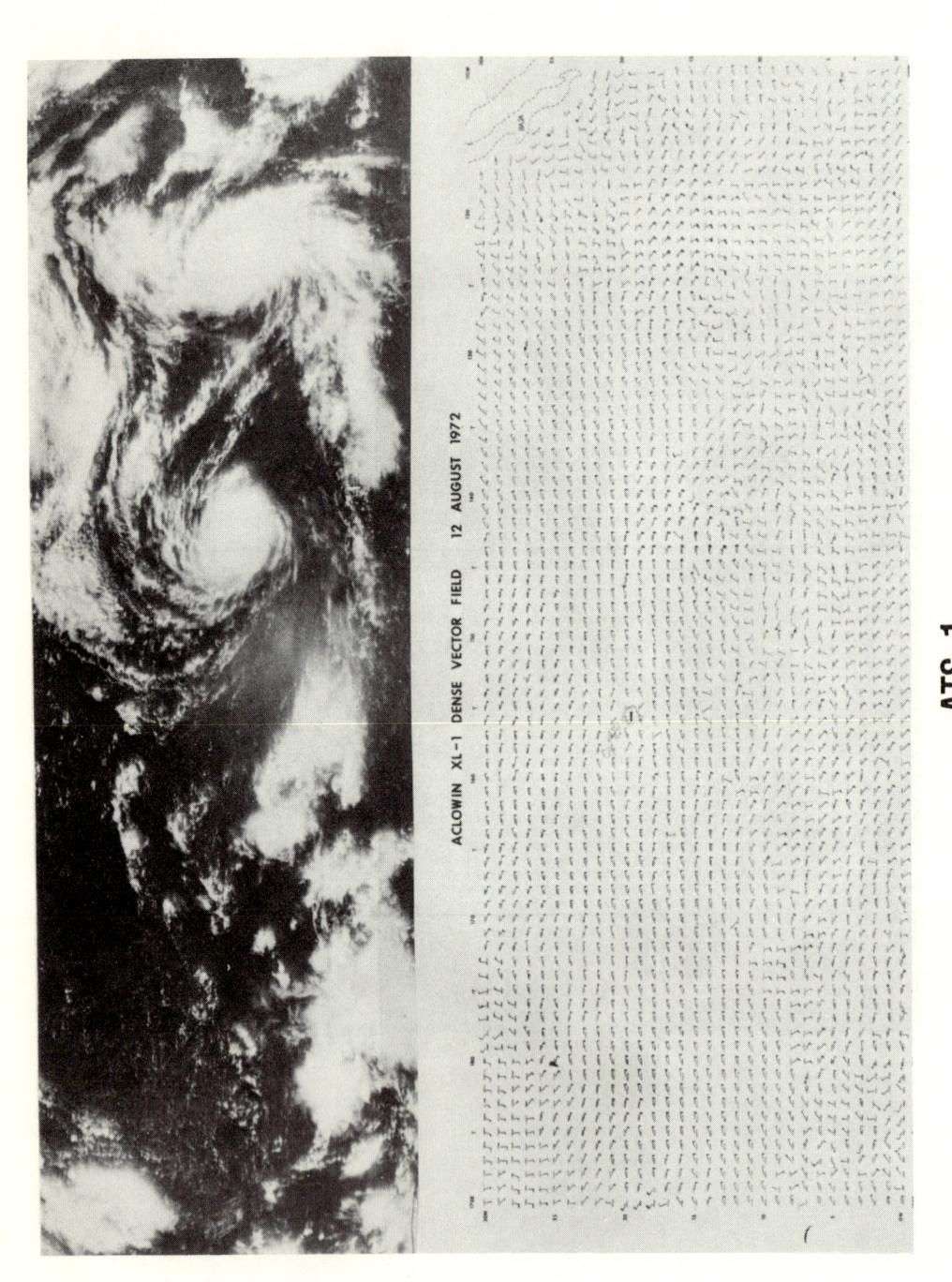

ATS-1
DERIVED WIND FIELD, 12 AUG, 1972
Figure 18

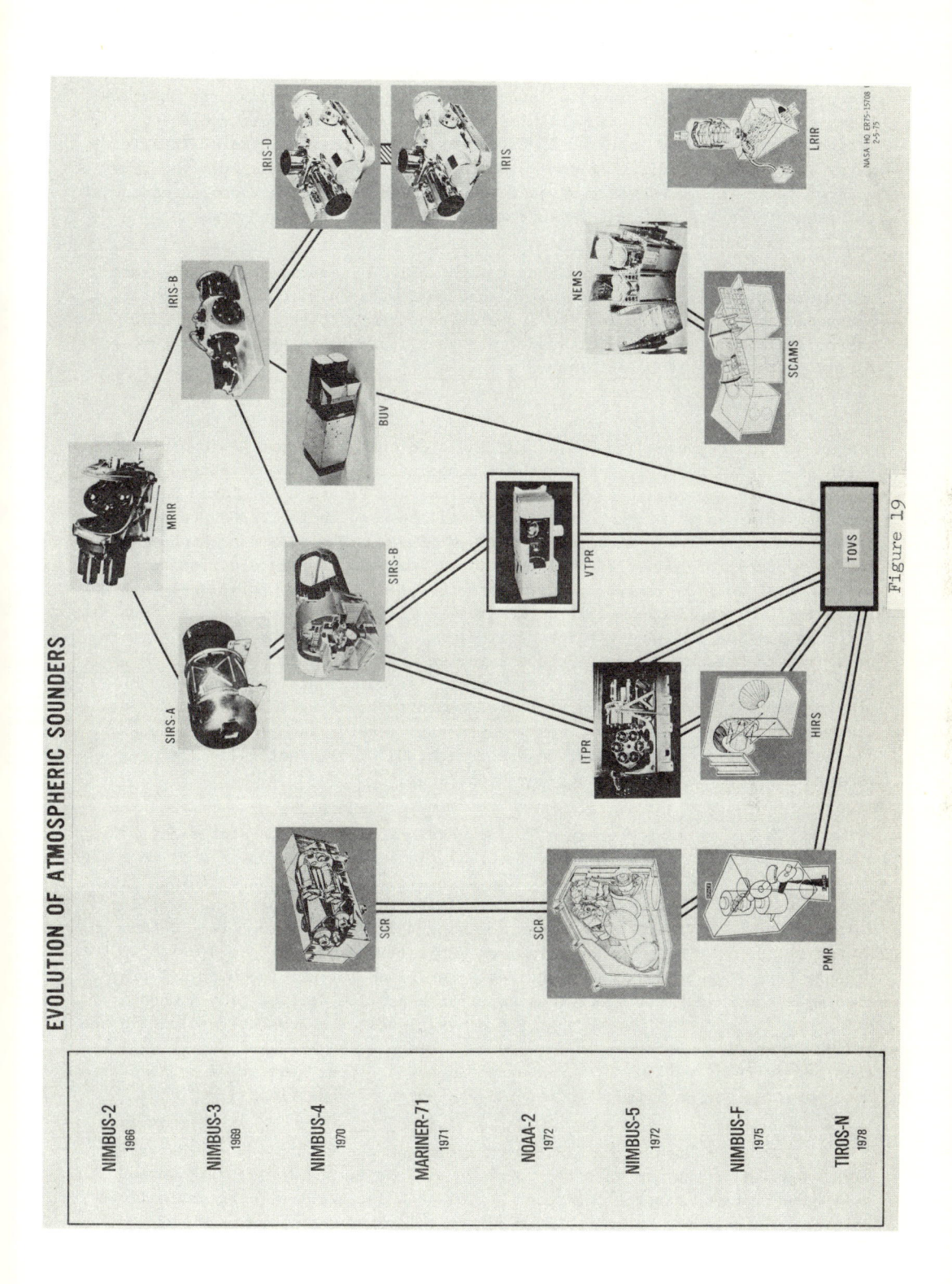

THE EVOLUTION OF ATMOSPHERIC MEASUREMENTS FROM SATELLITES

Figure 19

However, it was April 14, 1969, that is to be considered the historic date in the development of global vertical sounding techniques. On that date, Nimbus-3 carried two instruments into orbit that provided radiometric observations from which vertical temperature profiles of the atmosphere could be deduced--the Satellite Infrared Spectrometer (SIRS) and the Infrared Interferometer Spectrometer (IRIS). Examples of the temperature profiles derived from each of these instruments are given in Figure #20, where comparison is made with the standard radiosonde measurements made at the same place and time.

The Satellite Infrared Spectrometer (SIRS-A) measured eight discrete intervals in the 11-15 μm spectral region, while the Infrared Interferometer Spectrometer (IRIS-B) measured the infrared spectrum between 5 and 20 um from which, additionally, the water vapor and ozone, as well as atmospheric temperature structure, could be obtained.

Improved versions of both instruments were subsequently flown on Nimbus-4. The Nimbus-4 and 5 also carried a Selective Chopper Radiometer (SCR) which provided a measure of atmospheric temperature up to the stratopause (45 km). This instrument is the forerunner of the Pressure Modulated Radiometer (PMR) to be flown on Nimbus F (1975) and will be used in the TIROS Operational Vertical Sounder (TOVS) planned for TIROS-N (1978). The SIRS-B on Nimbus-4 (1970) was the forerunner of the operational Vertical Temperature Profile Radiometer (VTPR) on NOAA-2, 3 (1972) and the planned TIROS Operational Vertical Sounder (TOVS) on TIROS-N (1978). The NOAA-2 VTPR system provided the capability for sounding the atmosphere twice daily from 30 km down to the earth's surface, or to the top of any extensive cloud layer on a global basis. The VTPR on NOAA-3 (1973) for the first time, transmitted sounding data continuously to provide instant vertical temperature and humidity sounding information to local stations around the world that were equipped to receive and process these data. The World Meteorological Center in Washington, D. C. uses satellite VTPR temperature soundings in its regular operation.

The Infrared Temperature Profile Radiometer (ITPR) on Nimbus-5 (1972) and High Resolution Infrared Sounder (HIRS) planned for Nimbus F (1975) were also related to the SIRS-B development. The Nimbus E Microwave Sounder (NEMS) flown on Nimbus-5 (1972) has demonstrated its ability to sound temperature profiles through clouds and is related to the Scanning Microwave Spectrometer SCAMS on Nimbus F (1975). The combined data from NEMS, ITPR, and SCR have been shown to produce better soundings than either data set alone for use in numerical prediction models. Figure #21 shows the temperature profile derived from Nimbus-5 NEMS data in the clear and through the heavy clouds of a tropical cyclone in the Indian Ocean on 20 December 1972. The next Nimbus, the Nimbus-F, which will be launched in 1975, will carry a number of improved, later generation radiometers incorporating new techniques and additional spectral sensitivity.

The TOVS Sounder on TIROS-N will draw on the technology of PMR, HIRS, VTPR, and NEMS/SCAMS, so that the sounding systems at the end of this decade will have improved resolution, reach higher into the stratosphere, and will be able to penetrate practically all conditions of cloudiness, exclusive of precipitating clouds.

TEMPERATURE PROFILE INFERRED FROM RADIATION MEASUREMENTS COMPARED WITH RADIOSONDE OBSERVATIONS

SATELLITE INFRARED SPECTROMETER (SIRS)

INFRARED INTERFEROMETER SPECTROMETER (IRIS)

——— RADIOSONDE
– – – COMPUTED FROM RADIATION MEASUREMENTS

Figure 20

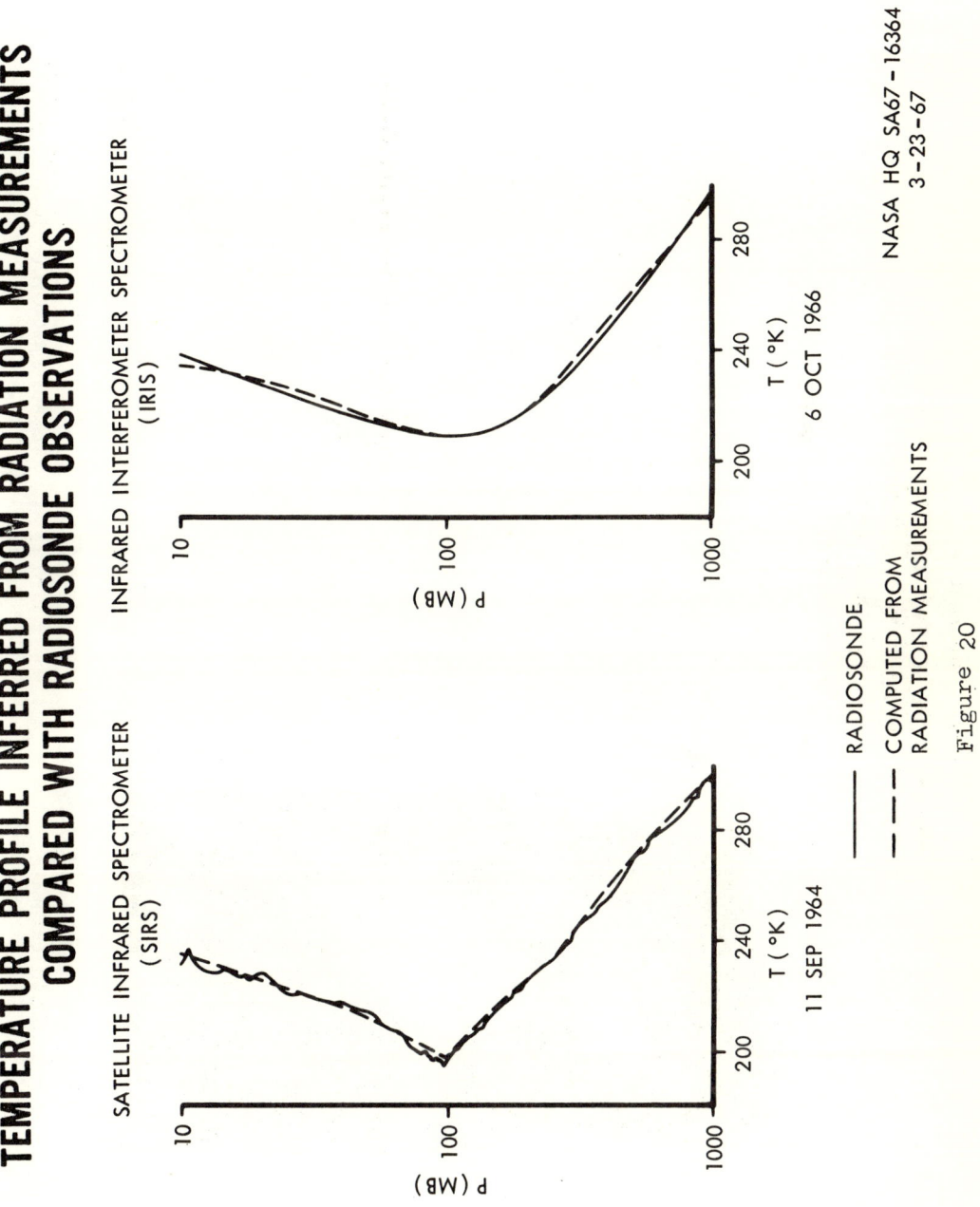

NASA HQ SA67-16364
3-23-67

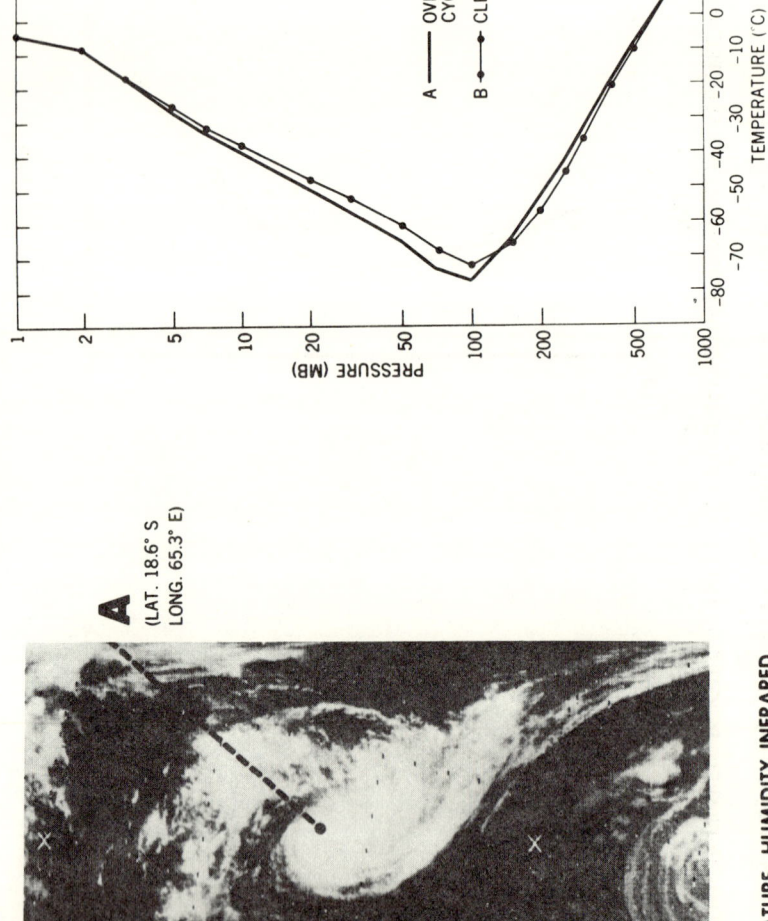

Figure 21

Finally, shown in Figure #19 is Limb Radiance Inversion Radiometer (LRIR), which utilizes the concept of sensing the atmosphere at the earth's limb. Improved vertical resolution of temperature and moisture above cloud level will be derived as well as profiles of ozone. LRIR is the first of our developments for identifying and monitoring atmospheric constituents.

IV. GARP

As we conclude this part of the discussion in describing the evolution of the meteorological satellite instrumentation and their application, we should recall our opening remarks. In those remarks, I pointed out that these developments have led to the institution of NOAA of a National Operational Meteorological Satellite System providing global data to the National Meteorological Center on a regular basis for use in weather analysis and forecasting.

Even as far back as the early 1960's, it was recognized that observations from space could become a key factor in assisting us to improve our understanding of weather and its prediction. It was this realization that led to the establishment of the international Global Atmospheric Research Program GARP .

A most important milestone that we look forward to is the implementation of the First GARP Global Experiment (FGGE) in 1978-79. This will be a truly international effort involving many countries that will be contributing a diverse number of observational systems, data analysis, and basic research capabilities. The objective of FGGE is to bring the entire global atmosphere under constant surveillance during about a two-year period to test the existing numerical prediction models and to determine to what degree our forecasting ability can be extended. Estimates range that this will be possible for as long as one to two weeks in advance, a decided improvement over our one to three day capability today.

Figure #22 is a depiction of the combination of satellites, both geostationary and polar orbiting, balloon observation systems, buoys, and augmented ground stations that will participate in the First GARP Global Experiment.

V. FUTURE THRUSTS

The scale of events to which my talk, up to now, has addressed itself is called the synoptic scale, or sometimes the medium scale, and applies to atmospheric motions and predictions covering from one day to 1-2 weeks. Figure #23 shows this scale in the center of the diagram and, as you can see, includes such things as planetary waves, extra-tropical cyclones, hurricanes, etc.

On either side of this scale are two other important scales and our research and development now is being directed towards the space applications in these scales.

On the short end of the scale are the severe local storms characterized by a short time interval and a limited aerial extent. Here, we feel that the geostationary satellite will be most useful and our future development will

Figure 22

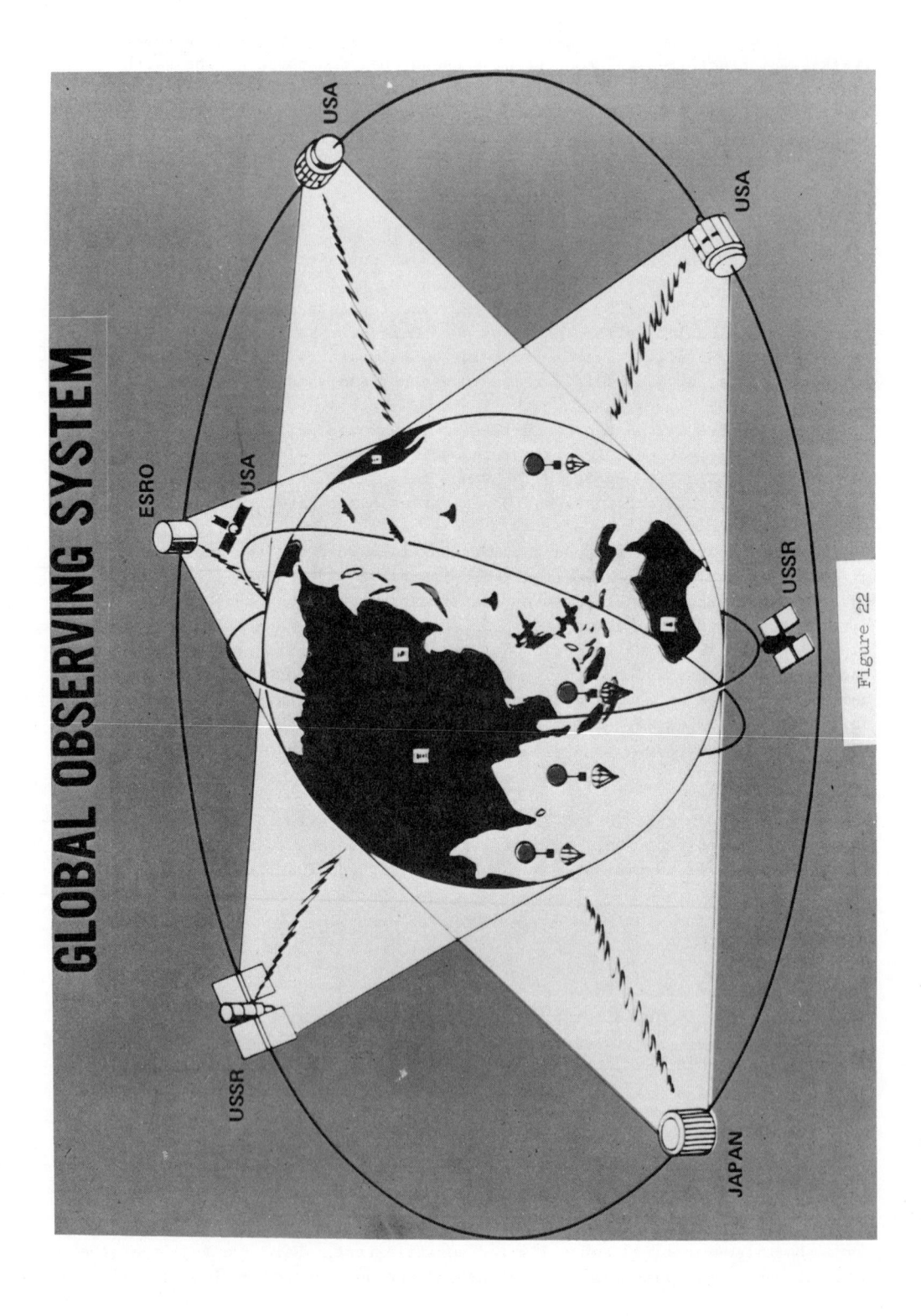

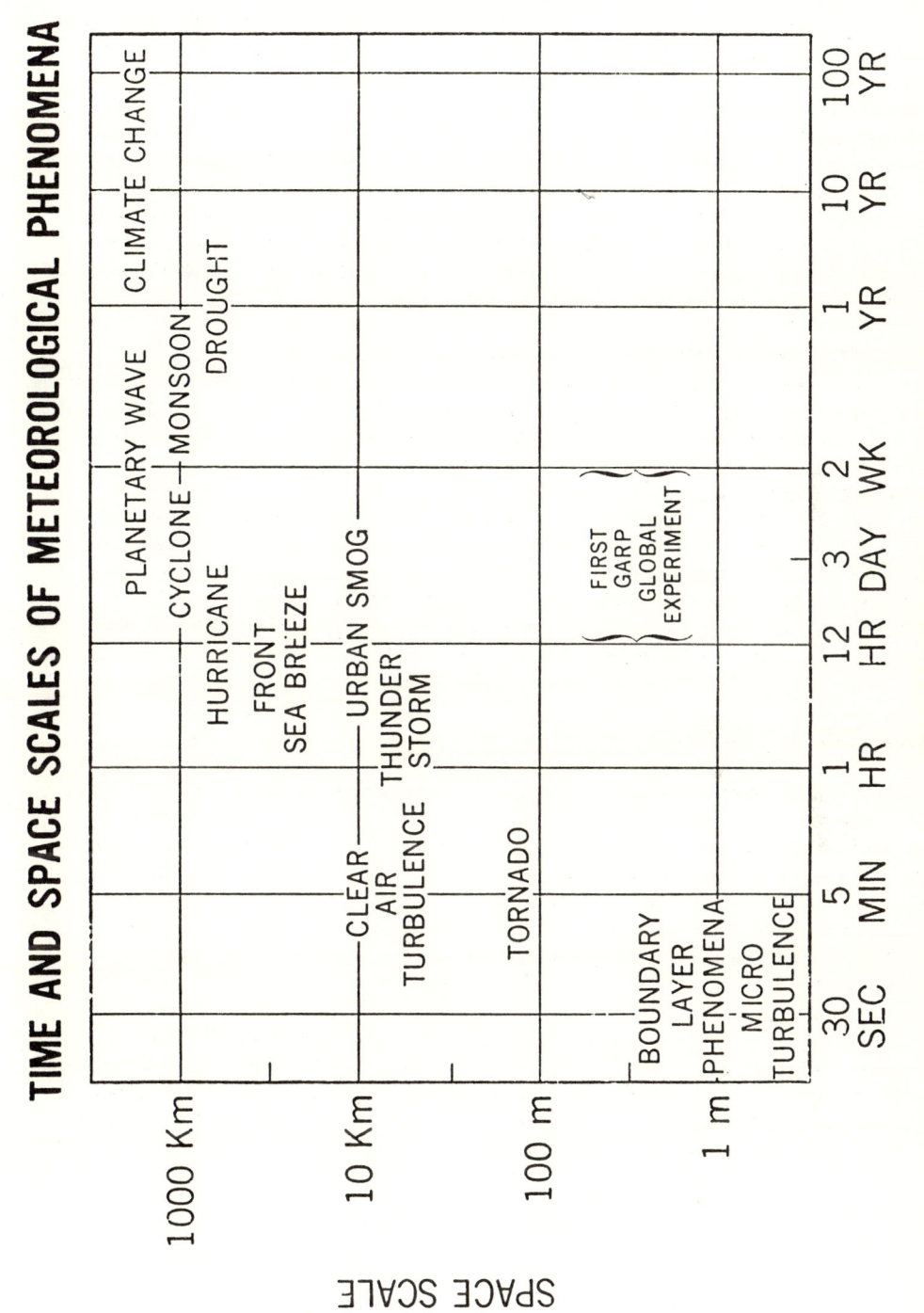

TIME AND SPACE SCALES OF METEOROLOGICAL PHENOMENA

Figure 23

consist of major improvements in the existing geostationary satellites. The first of these will be the incorporation of a sounder on board a geostationary satellite which will permit us to investigate the details of the environment in which severe local storms breed and will reveal the unique conditions that foster their development. In addition, we will concentrate on the optics of the satellite sensors in order to improve the resolution of the observations. We recognize that system developments in this area will produce a substantial amount of data to be processed so that associated with the satellite system development will be the development of a rather extensive automatic data handling system which will allow us to store, recover, and use the data in a most versatile manner. We recognize that also associated with this development will need to come capabilities for rapid communication of the information to the user in the form and time useful to him.

Finally, and on the other end of the scale, is the climate variation scale. Here we are confronted with a new set of requirements--that of monitoring those parameters that influence the climate and its variation. These include snow and ice conditions, sea conditions in the upper but sub-surface layer, variations in land use features, the measurement of the minor atmospheric constituents, measurements of the solar input solar constant , and of the earth radiation budget. Capabilities already exist with regard to some of these (such as the snow and ice conditions and surface conditions). We have instruments under development to measure the earth radiation budget and even the solar constant; we also have plans for measuring the surface conditions of the ocean. However, as yet, we are unable to probe very far beneath the surface of the ocean with remote sensing techniques.

VI. CONCLUDING REMARKS

As a final comment on the evolution of meteorological satellites, I would like to mention that they have been and will continue to be used as communications relays picking up information from unattended stationary and moving platforms and relaying this information to a central receiving station. In the case of the moving platforms, position of the platform is also available so that the velocity of the platform can be determined. If the moving platform is a balloon in the atmosphere, we are thus able to derive the local wind at balloon level.

So, as you can see, the meteorological satellite program is indeed an old and mature program and it has produced a very significant number of contributions to meteorology. However, there is also much work left to be done--in the application of space meteorology to the local severe storm problem on the one hand, and to the climate variability problem on the other.

SEASAT: A SPACECRAFT VIEWS THE MARINE ENVIRONMENT WITH MICROWAVE SENSORS

JOHN R. APEL

Director, Ocean Remote Sensing Laboratory
Atlantic Oceanographic and Meteorological Laboratories
Environmental Research Laboratories
National Oceanic and Atmospheric Administration
Miami, Florida 33149

ABSTRACT

Seasat-A is a new NASA satellite dedicated to oceanographic measurements of interest to a broad spectrum of the marine community. Its strong suit is an array of active and passive microwave instruments that give it the ability to view surface features on a day-night, near-all-weather basis. It will measure such features as wave heights, lengths, and directions; surface wind velocities; currents; temperatures; ice cover; and the marine geoid. Sensor capabilities and examples of their data output will be given, and the usefulness of these data for understanding the coastal marine environment will be discussed.

INTRODUCTION

Among the new programs initiated by the National Aeronautics and Space Administration for fiscal year 1975 is the ocean dynamics satellite, Seasat-A. This spacecraft has very real potential for enlarging the breadth and depth of scientific knowledge in several disciplines in earth science, and, in addition, promises to yield economic and social benefits of considerable magnitude, especially in the area of maritime affairs. This paper describes the objectives and background of the program, the spacecraft and its instruments, and delineates some of the scientific problems it can address.

The objectives of Seasat are (1) to develop and validate means for predicting the general ocean circulation, surface currents, and their transports of mass, heat, and nutrients, (2) to develop and validate means for synoptic monitoring and prediction of transient phenomena on the ocean surface such as wave heights and directions, surface winds, temperature, and storm surges, with an emphasis on identifying marine hazards, and (3) to make precision determinations of the marine geoid.

Because of the great length and breadth of the sea and the harsh environment it presents, the difficulties in obtaining detailed, timely information of sufficient observational density across most of its expanse have prevented an effective monitoring and forecasting system for the oceans. Thus, the prediction of wave heights depends on forecasts of the time and space histories of surface winds--the latter forecasts themselves being fraught with considerable uncertainty, as the loss of ships and oil drilling rigs at sea attests. Similarly, the locations of major ocean currents are known only approximately and the data required for shipping and fishing interests to efficiently exploit currents are lacking. The lack of sufficient wind and pressure data over the oceans has precluded an improved, longer-range weather

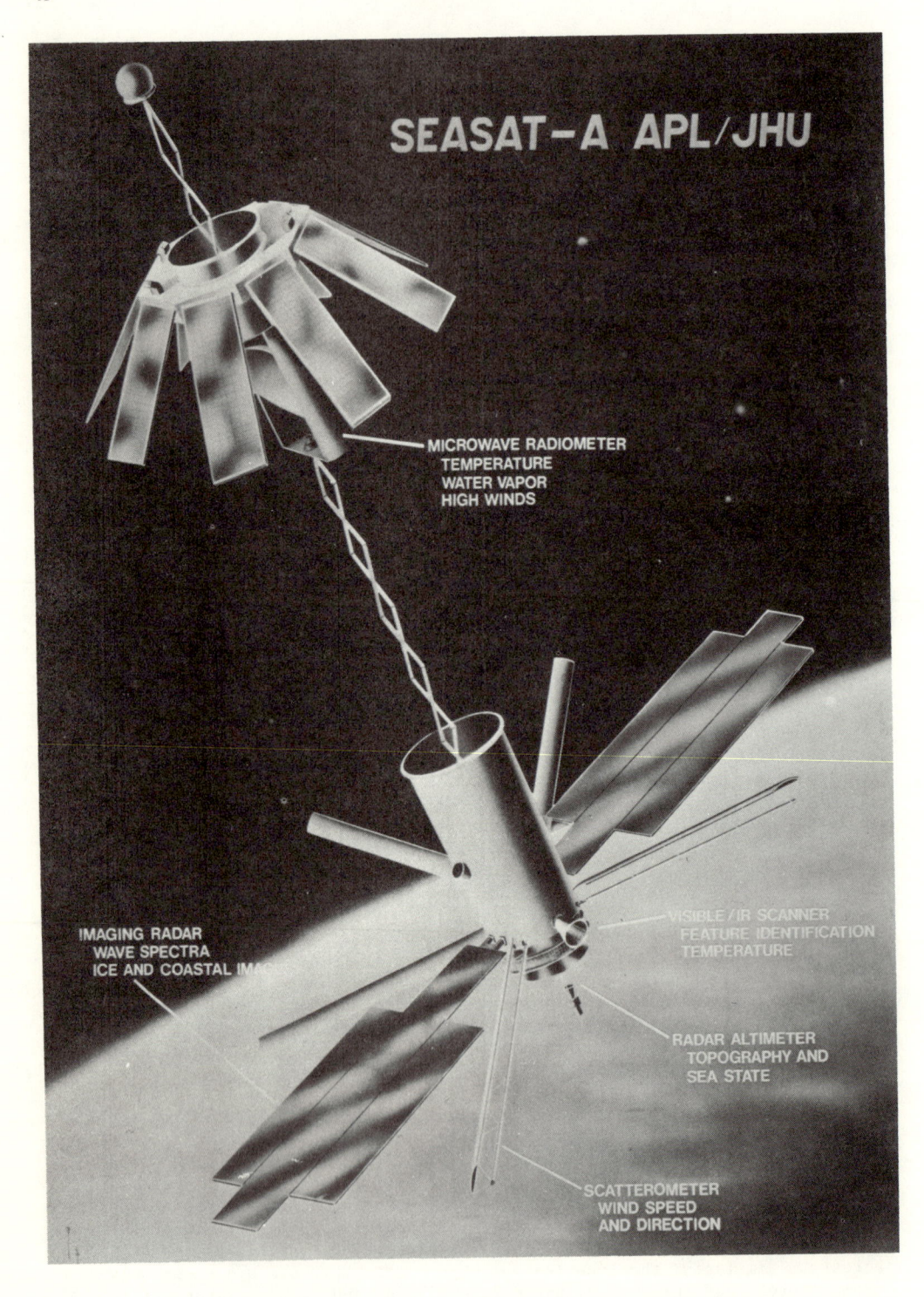

Fig. 1. - Configuration of Seasat-A, showing five sensors and their functions (JHU/APL).

forecast for continental areas. In order to achieve an effective one- to two-week forecast, observational data are needed over the oceans with about the same frequency and density as now exist in the continental United States.

SYSTEM DESCRIPTION

Basically, Seasat-A is a research-oriented program consisting of spacecraft, precision ground tracking systems, and data processing and modeling capabilities that will address both scientific and forecasting problems in ocean surface dynamics, boundary layer meteorology, and geodesy. Its strong suit is in an array of active radar and passive microwave and infrared instruments that give it the capability of observing the ocean on a day/night, near-all-weather basis. It is this group of sensors that allows Seasat-A to make quantitative measurements of oceanic, atmospheric, and geodetic parameters not only in clear weather but under wind and wave conditions perhaps approaching hurricane force, as well as over regions lying under persistent cloud cover. How this is accomplished is best appreciated after the system configuration has been laid out.

Figure 1 depicts a possible spacecraft design for Seasat-A. The prominent features on each design are the microwave antennas and, of course, the solar cell panels. The spacecraft is three-axis stabilized to point toward the vertical to within $\pm 0.5°$.

The mission profile for Seasat-A is as follows: lifetime, one year minimum; orbit, approximately 800 km altitude at an inclination of $108°$ (retrograde); eccentricity, less than 0.006, for a nearly circular orbit; period, 100 minutes, resulting in $14\frac{1}{2}$ orbits per day. This orbit is non-sunsynchronous and will precess through a day/night cycle in approximately four and one-half months. Its ground track for one day is shown on Figure 2; as can be seen, it spans almost all of the unfrozen oceans of the world from the Antarctic to the Alaskan North Slope and the Canadian archipelago. The orbit is also optimum for fine-grained mapping of the geoid over the open ocean.

INSTRUMENTS

1. The Pulsed Radar Altimeter has two distinct functions: to measure the altitude between the spacecraft and the ocean surface to a root-mean-square precision near ± 10 cm, and to determine significant wave heights along the sub-satellite path. The altitude, when blended together with accurate orbit determinations, may be used to decipher the topography of sea surface including spatial variations in the geoid and time variations due to ocean dynamics.

A current NASA estimate of the geoid in the western Atlantic, as derived from satellite tracking data and surface gravity measurements, is given in Figure 3; the figure shows the geoid--that is, the theoretical elevations and depressions of the motionless ocean surface due to gravity, with contours of constant height given in meters, relative to an elliptical earth. On land, this surface is used as a reference surface in precision surveying, and at sea, for determining surface deflections of the vertical.

The prominent ocean surface depression due to the deep ocean trench north of Puerto Rico has been observed by four different methods, the Skylab radar altimeter, S-193 being the first to give a continuous direct measurement of the sea surface topography. Figure 4 shows two altimeter traces of the overlying ocean surface and island topography. The measurement was made with the Skylab altimeter, which has a one-meter precision. The spacecraft measurements bear out both other observations and detailed calculations which indicate that the ocean surface is depressed by 15-20 m over a 100- to 150-km

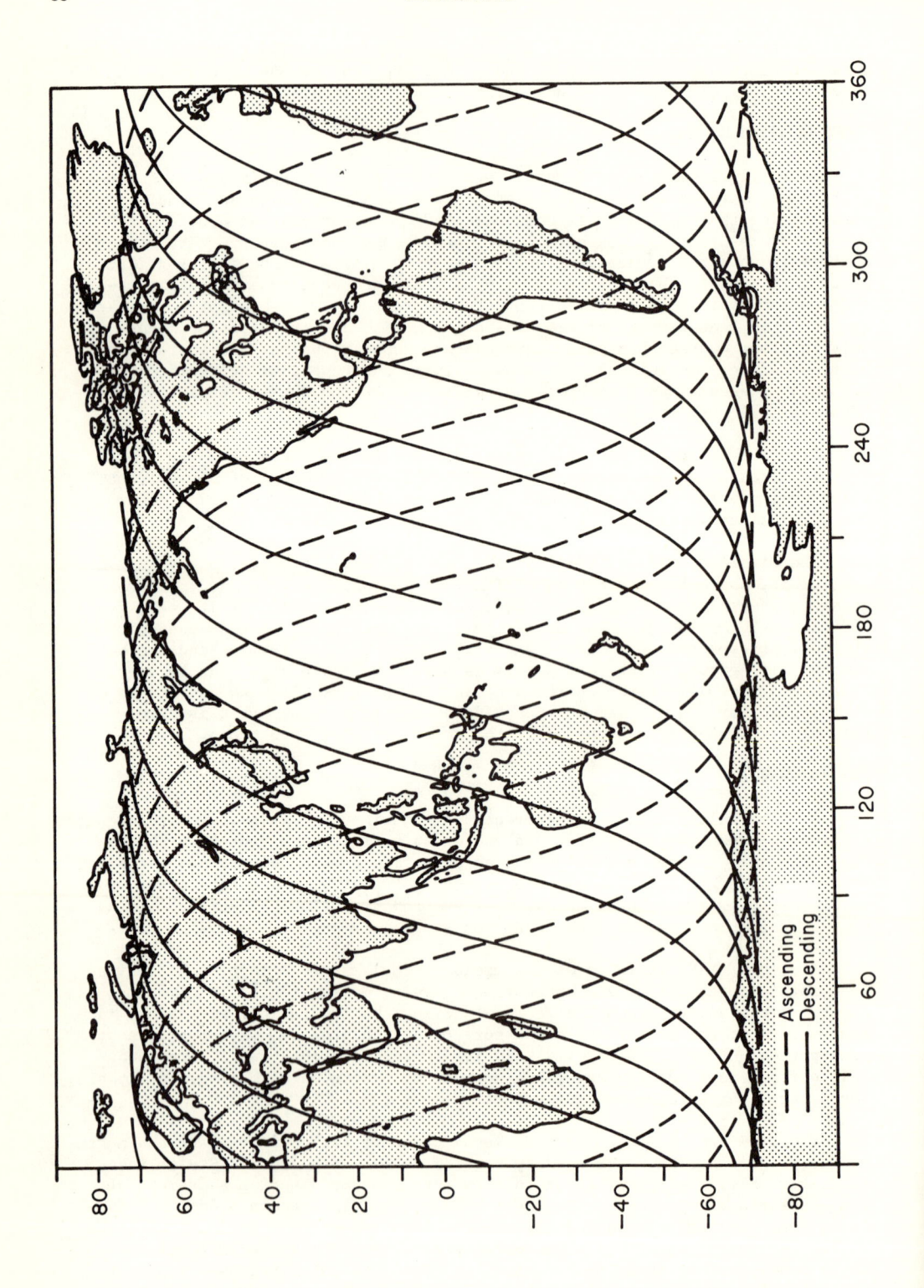

Fig. 2. - Ground track of Seasat-A for 24-hour period (JHU/APL).

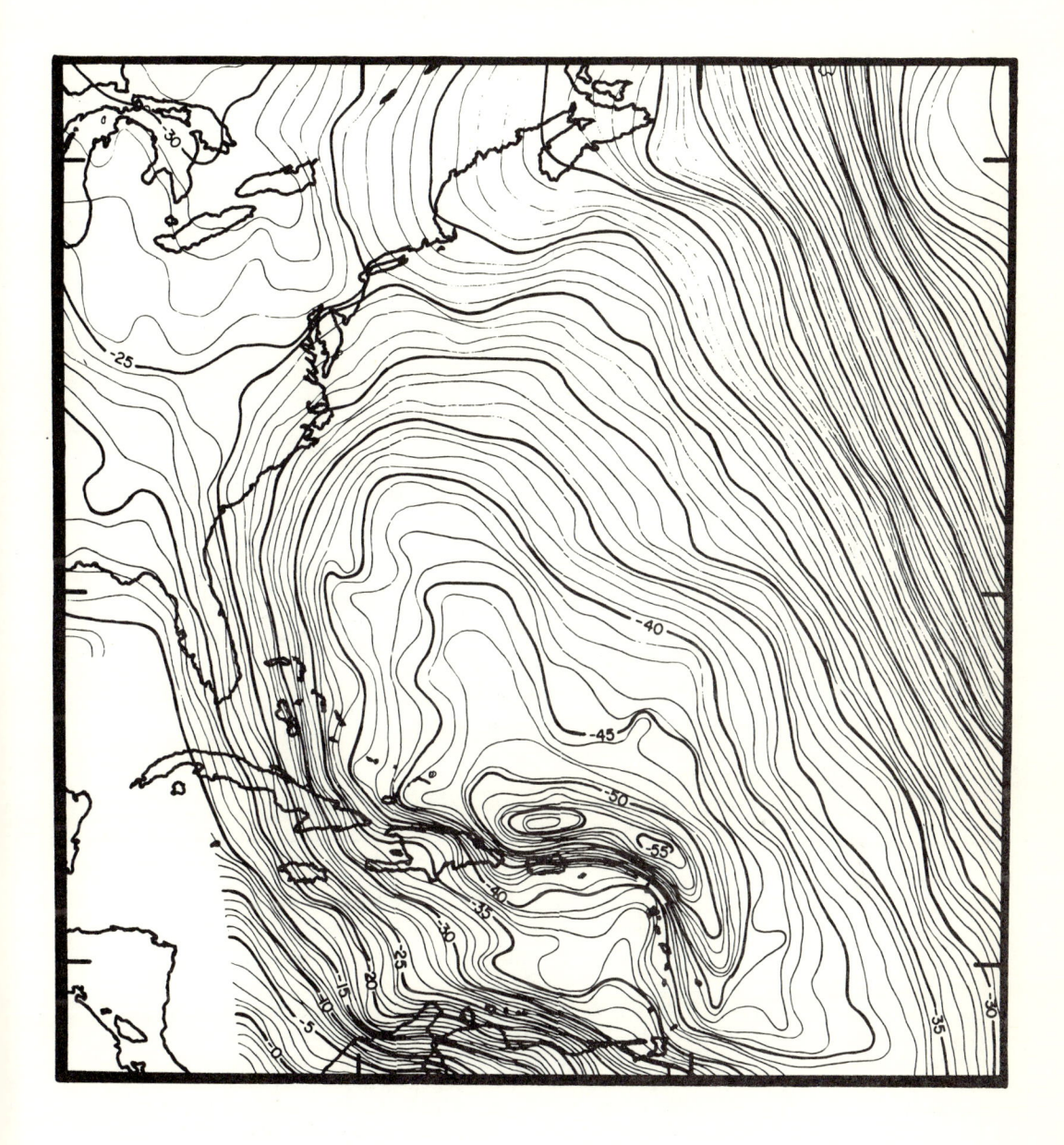

Fig. 3. - Geoid for western Atlantic as calculated from satellite orbit perturbations and marine gravity data (NASA/GSFC).

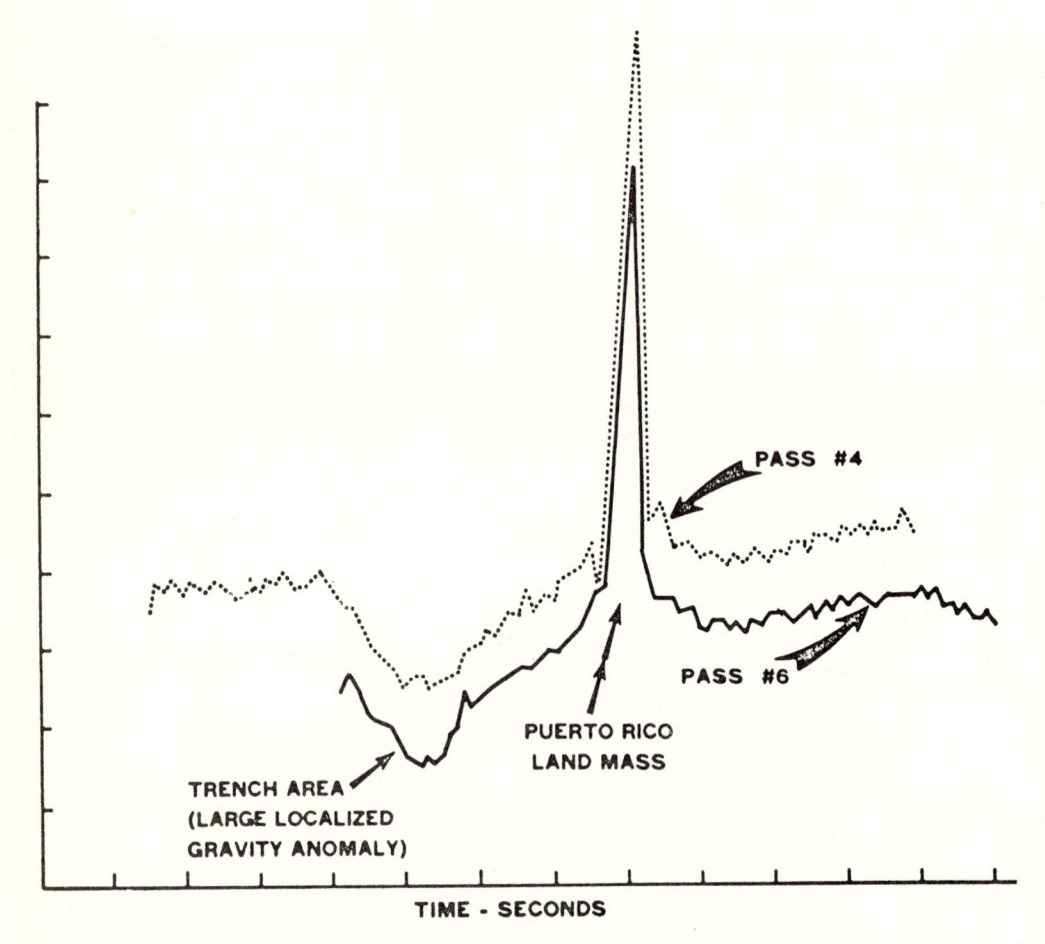

Fig. 4. - Altimeter traces from Skylab S-103 for two passes over Puerto Rico Trench. Vertical units are 10 meters (NASA/WFS).

distance north of Puerto Rico, due to the gravity anomaly associated with the trench. Anomalies which are elevations rather than depressions have also been observed from Skylab. These are due to seamounts, plateaus, and the Mid-Atlantic Ridge. Thus, the whole foundation of precision geoidal measurements via spacecraft altimetry seems to be on a reasonable theoretical and observational footing. The problems are to extend it globally and increase the precision.

There exist much smaller topographic departures from the geoid that are due to ocean dynamics effects. Such time-varying features as intense currents, tides, wind pile-up, storm surges, and tsunamis are in principle observable with an altimeter having submetric precision by way of measuring sea surface slopes relative to the geoid. For currents, the slope of the surface is proportional to the surface speed. However, the topographic variations due to even intense systems such as the Gulf Stream are quite small compared to the gravity-caused geoidal undulations, being of order 1 to 2 m at most. Another ocean dynamical feature observable at the 10-cm level of precision are waves, which may be used along with surface wind measurements to make world-wide sea state forecasts. The principle of the measurement is that a short pulse reflected from a rough sea will be broadened by the various reflecting heights caused by the waves. The broadened shape of the returned echo contains wave height information, with the rougher the sea, the longer the echo. Aircraft flights have shown this technique to work in low to moderate seas. On Seasat-A, wave heights from one to above 20 meters should be measurable along the subsatellite track on a near-all-weather basis. Other information derived from the sensor complement may be used to extend this measurement well out from the suborbital track.

2. Synthetic Aperture Imaging Radar. The required extension of wave information will be made by using an imaging radar to obtain images of the ocean on a sampled basis. Such a radar can function through clouds and moderate rain to yield wave patterns near shorelines and in storms and can see waves whose length is greater than about 50 m. It can also provide high-resolution pictures of ice, oil spills, current patterns, and similar features. Computations can be performed on the radar data to yield a quantity called the wave directional spectrum which gives the relative distribution of wave energy among different wavelengths traveling in various directions; this, together with the surface wind velocity, is the fundamental information needed in forecasting of wave conditions on the ocean.

Figure 5 illustrates two trains of waves off Kayak Island, Alaska, one of 150-m and the other of 60-m wavelength, taken from the NASA Convair 990 aircraft with the Jet Propulsion Laboratory imaging radar; the waves are being refracted and shortened by shoal water as they approach the island visible on the left-hand side. Also on the lower left and center of the figure is a directional spectrum computed for the relatively uniform part of the wave train to the right of the image. Distance from the center of the spectrum corresponds to increasing wave frequency, angle to direction of propagation, and intensity to wave energy.

The data rate from an imaging radar is high, and judicious use must be made of the device. Nevertheless, it should be possible to sample wave spectra over 10-km-square patches of ocean densely enough to obtain global data on sea conditions. Near the NASA receiving sites along the U. S. coasts, more generous quantities of imagery will be taken and studies of storm wave patterns near potential offshore nuclear power plant sites, deep-water oil ports, harbors, and breakwaters will be made. Over the Northwest Passage and the Great Lakes, a demonstration of real-time mapping of ice leads and open water will be made as an aid to navigation through those straits and inland seas.

3. Microwave Wind Scatterometer. The third radar system is a microwave scatterometer, intended to measure surface wind speed and direction by sensing the small capillary waves induced by the wind over the ocean. Previous aircraft experience and recent Skylab data taken over the Pacific hurricane Ava in June 1973 indicate this sensor is useful in winds approaching 20 m/s, yielding speeds with an error of ±2 m/s and directions to ±20°. Figure 6 illustrates Ava as taken from the environmental satellite NOAA-2 on the left; on the right are graphs from Skylab S-193 showing radar scattering, radiometric temperature, and, in the upper right-hand corner, wind speed. The peak wind of 45 knots (22 m/s) obtained from the scatterometer was observed some distance from the eyewall, which the sensor could not view because of look-angle constraints.

4. Scanning Multifrequency Microwave Radiometer (SMMR). The SMMR is a passive, nonradiating microwave device, in contrast to the three previous sensors. It simultaneously senses the microwave energy emitted by and reflected from the ocean, ice, and atmosphere. In order to separate out the

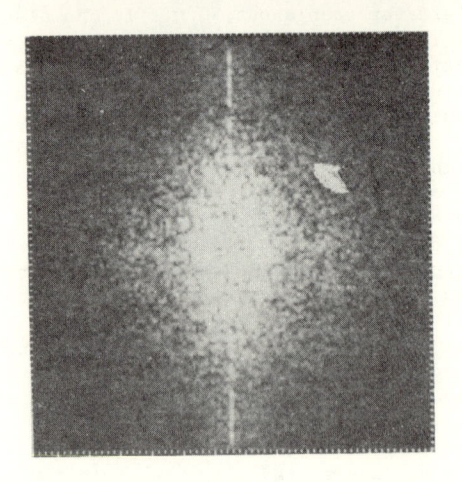

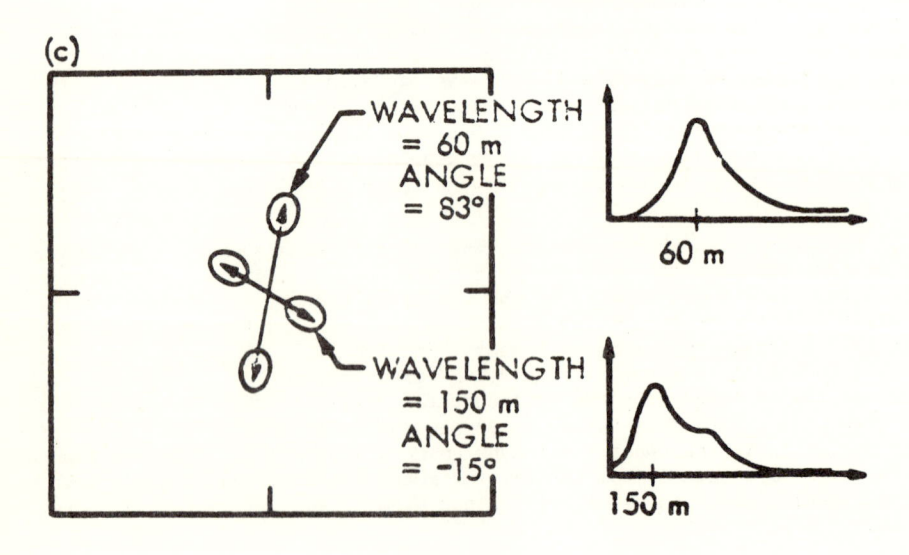

Fig. 5. - Wave data from imaging radar showing wave refraction patterns and directional spectrum (JPL/CIT).

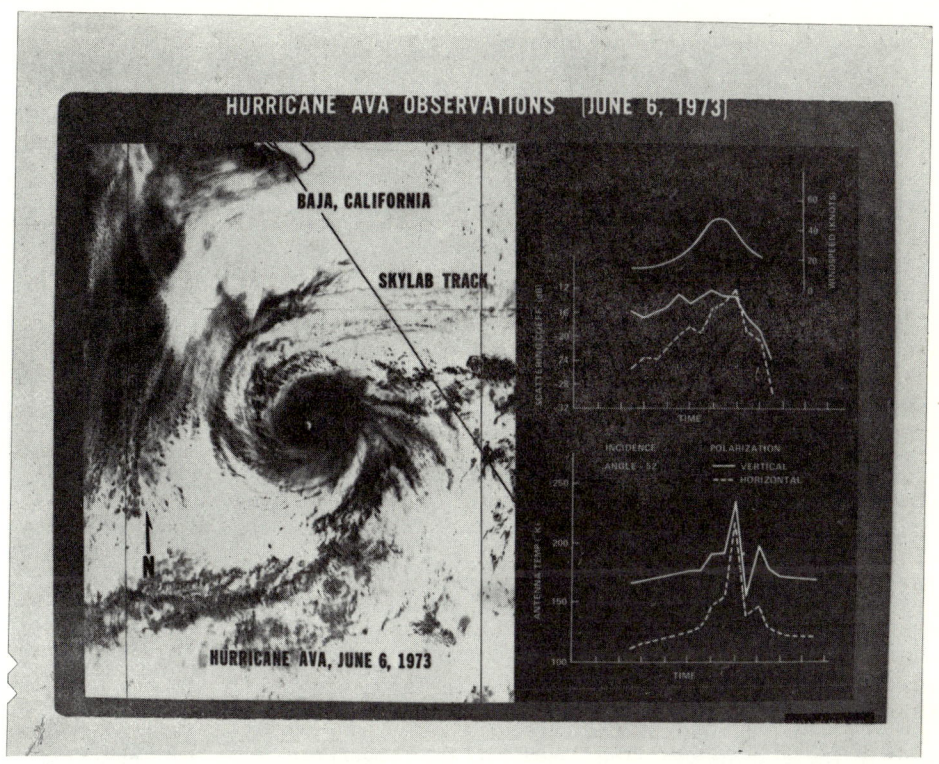

Fig. 6. - Wind scatterometer and radiometer data taken through hurricane Ava in the Pacific (NASA/HQ).

various contributions to the signal from these sources, several microwave frequencies are used with each chosen for maximum sensitivity to one of those geophysical parameters. The scanning feature will allow low-resolution images of objects along its line of sight to be constructed from the signals received.

The functions of the SMMR are severalfold. It is first a wind speed instrument that senses the increase in emitted microwave energy due to roughness, foam, and streaks on the ocean caused when higher wind speeds create wave breaking and whitecaps. The estimated observable range of speeds is from about 10 to perhaps 50 m/s, but the upper limit has yet to be firmly established. Thus, the range of speeds measurable from Seasat should be extended by SMMR from the 20-m/s limit of the scatterometer up toward hurricane-force winds. Secondly, it appears capable of measuring sea surface temperature with an accuracy of 1½-2°C, even through light clouds, where present infrared de-icers are useless. Thirdly, the other frequencies are used for determining atmospheric liquid water and water vapor content, quantities that are needed in models of oceanic and atmospheric boundary layer processes as well as for important corrections to the precision altimeter measurements. Ice fields and cover will also be observed with low resolution from SMMR. When blended with the wind data from the Microwave Wind Scatterometer, the two sensors should yield a global, quantitative determination of surface wind speed wherever it is below essentially hurricane force. The measurements will be equivalent to some 20,000 ship reports a day. When combined with available ship and buoy surface information on wind and pressure, it becomes possible to compute the atmospheric surface pressure field over the entire ocean, except perhaps near severe storms; this will also be true in the data-sparse southern hemisphere. Such results should help to improve the 24-hour weather forecasts substantially, perhaps making them extensible to two or three days. This improved predictive capability for winds implies an approximately equal improvement in forecasting waves, especially when assisted by the data on the initial state of the sea obtained from the radar altimeter and imager.

5. Infrared Radiometer. The purpose of this sensor is to provide images of thermal infrared emission from ocean, coastal, and atmospheric features, which will aid in interpreting the measurements from the other four microwave instruments. Figure 7 is an example of imagery taken from the NOAA-2 Very High Resolution Radiometer over the southeastern United States and clearly shows the Gulf Stream off the coast as a dark band of water, as well as the Gulf of Mexico Loop Current, a time-varying feature that apparently profoundly affects the fishery and the weather in that inland sea.

A word on the measurement of sea surface temperature is in order here. This seemingly inconsequential parameter is actually of considerable importance in oceanic and atmospheric processes, since it results from the absorption of that prime mover, solar energy, by the sea. For instance, the difference between active and inactive hurricane seasons may be due to just 2-3°C lower water temperature in hurricane gestation areas. Ocean temperature is a major factor determining the tone of weather and climate in many coastal regions of the world. Maps of sea surface temperature are very useful for tracing current systems such as the Gulf Stream especially in the winter months. Furthermore, open-ocean fish such as tuna tend to swim along lines of constant temperature at certain times during their excursions, and a knowledge of temperature can assist in their location. Thus, sea surface temperature offers a clue to several interesting processes in the ocean.

THE TOTAL SYSTEM

The interrelationships between data from these five sensors are suggested in Figure 8, which illustrates the complex nature of the contribution that each sensor makes to the geophysical parameters being measured. The figure indicates the importance of carrying the full sensor complement in order to achieve the measurement objectives.

Seasat-A is thus an integrated observatory addressing the objectives discussed at the beginning of this paper. Table I outlines the capabilities of the spacecraft system in meeting the requirements set forth by a community of data users who were identified by NASA at the beginning of the program.

An important element in interpreting the Seasat-A data and extending its utility will be fleshing out the information obtained from this spacecraft with the considerable data on oceans and atmosphere available from other sources. The environmental/meteorological satellites are one such obvious source for marine and weather data, as are ships, buoys, and transoceanic aircraft. In the case of ocean wave forecasts, a land-based high-frequency skywave radar that is intended for operational, detailed monitoring of wave spectra near the continental United States is expected to be in service; its fine-grained data nicely complements the necessarily coarser-space open ocean wave spectral data from Seasat-A. Similarly, research data on currents, tides, the geoid, and the other parameters of interest will be amalgamated with the Seasat-A data by individual researchers interested in specific problems.

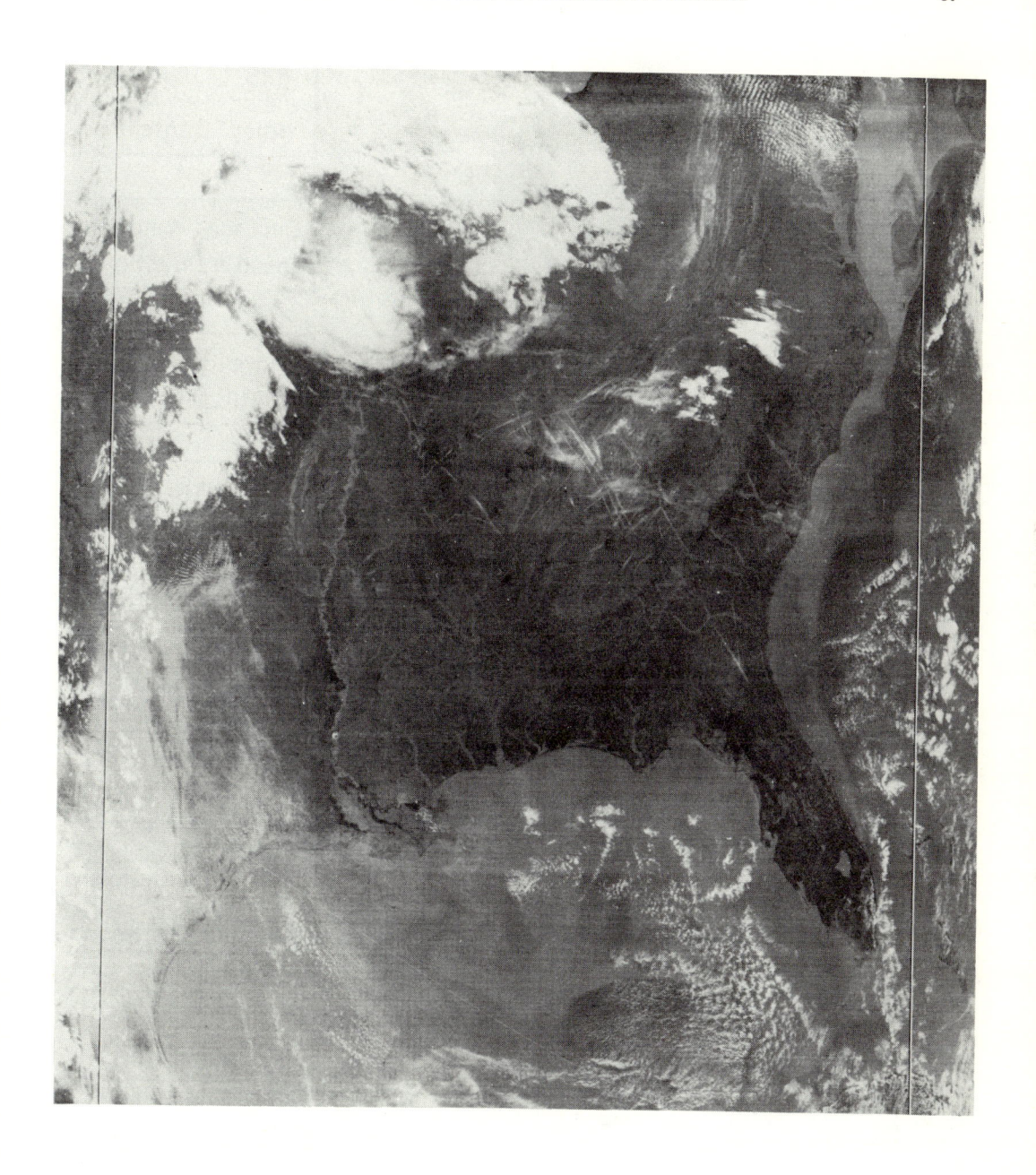

Fig. 7. - Thermal infrared imagery of eastern U. S. showing Gulf of Mexico loop current and Gulf Stream as dark water masses.

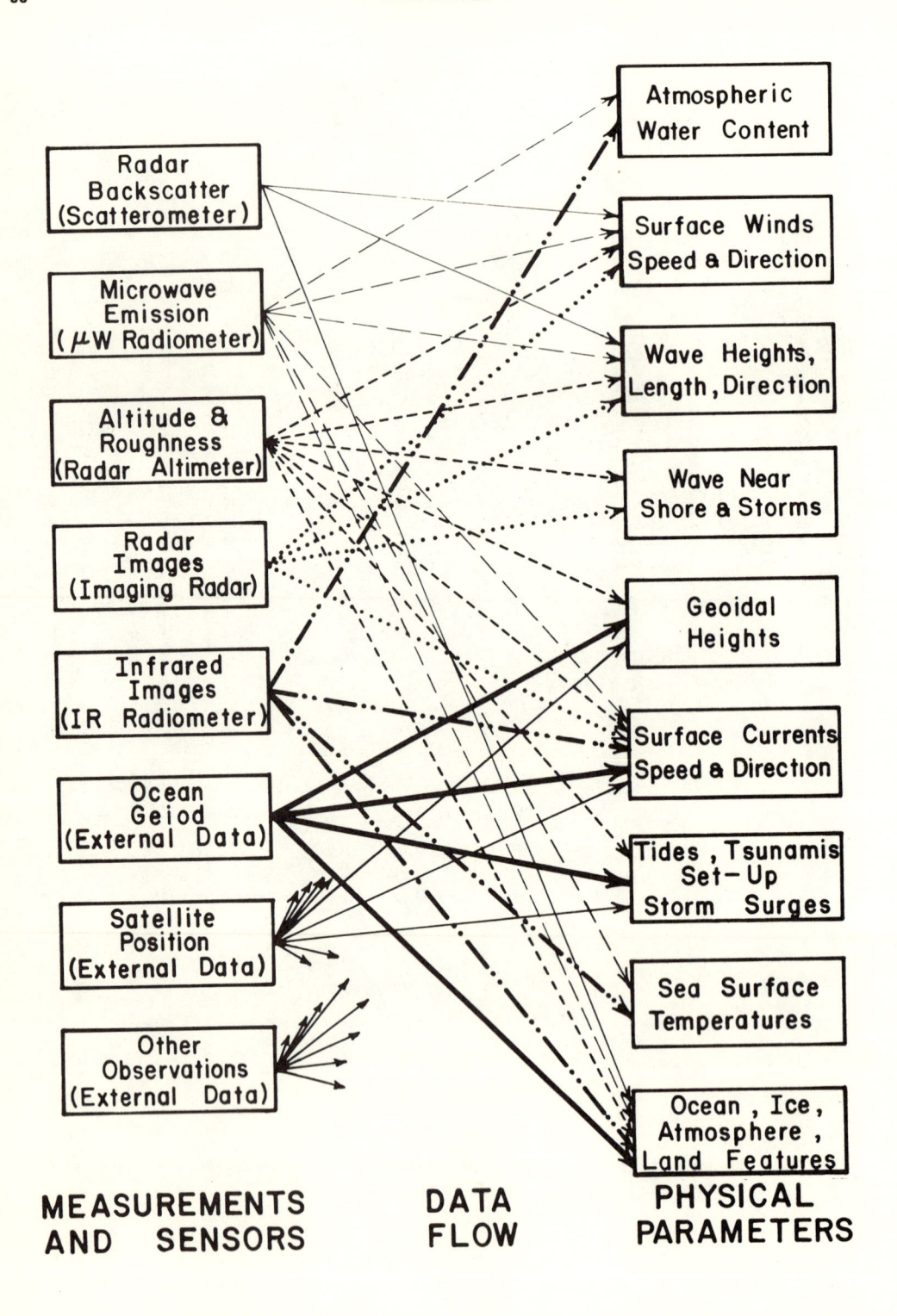

Fig. 8. - Interrelationships between Seasat-A sensors and geophysical variables to be derived.

TABLE I

CAPABILITY OF SEASAT-A IN MEETING USER REQUIREMENTS

20 February 1974

PHYSICAL PARAMETER	INSTRUMENTS	RANGE	PRECISION	RESOLUTION OR IFOV	TOTAL FOV	COMMENTS
Wave Height, $H_{1/3}$ (x,y)	Pulse Altimeter Coherent Alt.	1.0 – 20 m	±0.5 m or ±10%	2x7 km spot	2-km swath	along subsatellite track only
Directional Wave Spectrum $S(\lambda,\Theta,x,y)$	Imaging Radar (2-D transform)	S: unknown λ: 50-1000 m Θ: 0-360°	S: --- λ: ±10% Θ: ±10°	50-m resolution	20x20 km squares	global samples at 250-km intervals
	2-f Wave Spectrometer	S: unknown λ: 6-500 m Θ: 90° sector	S: --- λ: ±10% Θ: ±9°	8 x 25 km spot	300-km swath about nadir	global samples at 150-km intervals
Surface Wind Field, $\vec{U}$(x,y)	Scatterometer	U: 3-25 m/s Θ: 0-360	±2 m/s, ±10% ±20°	≤ 50 km spot	two 450-km swaths	global, 36 hrs (low speeds)
	μW Radiometer	U: 10-50 m/s Θ: unknown	±2 m/s, ±10% ---	≤ 100 km spot	900-km swath about nadir	global, 36 hrs (high speeds)
Surface Temperature Field, T(x,y)	IR Radiometer	-2° to +35°C	±¼° – 1°C	1-7 km IFOV	1500-km swath about nadir	global, 36 hrs (clear air only)
	μW Radiometer	0° to 35°C	±1.5°C	100 km spot	900-km swath about nadir	global, 36 hrs (clouds & lt. rain)
Geoidal Heights, h(x,y) (above reference ellipsoid)	Pulse Altimeter Coherent Alt.	7 cm – 200 m	±7 cm		18-km spacing along equator	sampled throughout one year
Sea Surface Topography, $\int$(x,y) (departures from geoid)	Pulse Altimeter Coherent Alt.	7 cm – 10 m	±7 cm	2x7 km spot	2-km swath	along subsatellite track only
Oceanic, Coastal, & Atmospheric Features (Patterns of waves, temp., currents, ice, oil, land clouds, atmospheric water content)	Imaging Radar	high resolution	all weather	25 or 100m	100 or 200 km	sampled direct or stored images
	IR Radiometer	high resolution	clear air	1-7 km	1500-km swath	broadly sampled images
	μW Radiometer	low resolution	all weather	15-100 km	900-km swath	global images

CONCLUSION

Seasat-A promises to be an exceptionally useful and productive program. It should have a large impact on earth science and on a community of users in the general populace.

The United States space program has produced satellite systems that have looked at the thin envelope of the atmosphere, at the browns and greens of rocks and plants, and at the bright thermonuclear fires of the sun and stars. The time has now arrived to mount a space-oriented investigation of the sea, that last remaining member of the ancient elemental quartet of air, earth, fire, and water.

REMOTE SENSING OF OCEANS USING MICROWAVE SENSORS

K. KRISHEN

Lockheed Electronics Company, Inc., Aerospace Systems Division,
Houston, Texas, U.S.A.

ABSTRACT

This paper presents a review of the results of a study of the ocean surface phenomena. The use of active and passive microwave sensors to detect ocean surface winds and waves, temperature, salinity, storm cells, oil slicks, and ice conditions is demonstrated. The aircraft- and spacecraft-acquired microwave data from the Naval Research Laboratory and the National Aeronautics and Space Administration/Lyndon B. Johnson Space Center are presented. The radar back-scattering cross section data show strong correlation between ocean surface winds/waves, storm regions, and oil slicks. A strong dependence upon these parameters has been shown in the Ku-band at a radar frequency of 13.9 GHz. The relationships between radiometric brightness temperature and ocean surface temperature, salinity, and sea state are set forth. Altimeters and imaging radars provide measurements of geoid, sea state, underwater topography, and the progress and location of storms.

NOMENCLATURE

A	illuminated area
$E(\theta)$	emissivity as a function of θ
$\lvert E_\alpha^S \rvert$	magnitude of the scattered field at the receiver with α polarization
$\lvert E_\beta \rvert$	magnitude of the incident field at the surface with β polarization
ε	complex dielectric constant of the surface
EREP	Earth Resources Experiment Package
η	main-beam efficiency
h	height from surface of water
$G(\theta,\phi)$	antenna gain in (θ,ϕ) direction
GEOS-C	geoid-sensing satellite
$H_{1/3}$	significant wave height in feet
HH	horizontal transmit, horizontal receive

HV	horizontal transmit, vertical receive
JSC	Lyndon B. Johnson Space Center
$k_1 \ldots k_n$	constants
m/sec	meters per second
NASA	National Aeronautics and Space Administration
NOAA	National Oceanic and Atmospheric Administration
NRL	Naval Research Laboratory
p, q	radian wave numbers on the ocean surface
ϕ	azimuth angle in degrees
Q_0	ratio of σ_0 at 5° to σ_0 at 45° incidence angle
$r(\theta)$	reflectivity as a function of θ
R_r	distance from radar receiver to the illuminated area
SEASAT-A	NASA satellite for sensing oceans
σ_0	backscattering radar cross section
T_A	radiometric antenna temperature in degrees K
T_B	radiometric brightness temperature of ocean surface in degrees K
T_g	physical ground temperature in degrees K
T_s	radiometric sky brightness temperature in degrees K, as viewed from surface
τ	transmittancy of the atmosphere
θ	angle of incidence at the surface
VH	vertical transmit, horizontal receive
VV	vertical transmit, vertical receive
W_1	ocean surface wind velocity

1. INTRODUCTION

Experimental studies of the radar-observed sea echo at 9.2-, 3.2-, and 1.25-cm wavelengths have been described by Kerr [1]. These studies were conducted during World War II from shore- and ship-based surveillance systems because the echo exhibited unwanted noise which could mask desired target signals. Earlier radar sea echo studies were directed to reducing the sea echo and improving radar design and performance. These studies also demonstrated the feasibility of measuring ocean surface conditions using microwave systems.

The determination of ocean surface winds, waves, salinity, oil slicks, and geoid has generated much interest during the last decade. Because microwave spaceborne sensors can provide all-weather, day-and-night, synoptic, high-resolution capabilities, they are especially suitable to these studies. Remote microwave sensors have been used by the Naval Research Laboratory (NRL), the National Aeronautics and Space Administration (NASA), and the National Oceanic and Atmospheric Administration (NOAA) to gather both active and passive microwave data over ocean scenes. The results of the measurements show several applications of microwave sensors to remote sensing of ocean surface phenomena.

Microwave scatterometers are radar devices capable of measuring the reflectivity of an object or scene in the direction of the transmitter. Such measurements are called backscattering cross sections. They exhibit a range of radar wavelengths for which backscatter is dependent primarily on the ocean surface wind velocity. The presence of oil slicks decreases the backscattering cross section significantly at higher angles of incidence. Thus, the presence of oil slicks can be detected. Furthermore, several measurement programs have demonstrated that scatterometers have the potential to categorize ice types and to distinguish water from ice.

Radar altimeters can measure accurately the distance to the object and the returned pulse shape. Recent developments have demonstrated that satellite altimeters can determine small-scale variations in the mean sea level. Although the ultimate goal in spatial and height resolution has not yet been achieved, it is anticipated that, through improvements in system and data processing, this goal will be reached within the next decade. Satellite altimetry will then produce accurate knowledge of the geoid. The positions and densities of underwater topographic features can be determined from geoid undulations and accurate satellite tracking. In addition, global knowledge of sea slopes, currents, and eddies will be attainable from waveform analysis of radar altimeter return.

Microwave radiometers provide measurements of emitted microwave energy from the ocean surface. The brightness temperatures detected by the microwave radiometer arise from several effects. These include roughness of water, salinity, physical temperature, and the presence of oil slicks, white caps, and sea foam. Thus, microwave radiometric observations can yield information concerning sea state, windspeed, and certain pollutants. At present, it does not appear that passive microwave sensors can detect ocean surface wind direction. An analysis of satellite-acquired microwave radiometric data within a season has indicated that changes in the compactness of the sea ice can be observed. Radiometers have also been able to distinguish ice types and estimate the ages of ice fields.

Imaging radars have detected the presence and extent of oil slicks on the ocean surface. Ocean swells have been imaged by airborne radar-imaging systems. The immediate capability of these systems is to map wave patterns and wave buildup during large storms. Wave data are especially significant in coastal areas. Here the wind and wave interaction is heavily influenced by coastal configuration and underwater topography. Also, the disappearance of capillary waves is a surface expression of underlying waves. The image generated by a radar system provides visible evidence of these internal waves. Furthermore, internal waves and surface currents produce oil slick patterns which can be used to detect the presence of such phenomena.

It has recently been shown that side-looking imaging radars may be used effectively in the monitoring of polar ice regions. With these systems, it is possible to identify leads, floes, and similar ice patterns. Repeated imaging can be utilized to monitor the changes and movements of these conditions.

The Skylab S-193 Microwave Radiometer/Scatterometer/Altimeter experiment was the first attempt to gather data using Earth-oriented, spaceborne, active, microwave systems. This experiment acquired nearly simultaneous radiometric brightness temperatures and radar backscatter data over land and ocean surfaces, using spaceborne microwave sensors. The experience gained from this effort and the limited analysis of the data have made noteworthy contributions to the planning of the GEOS-C and SEASAT-A programs. Definitions of sensor specifications, mission requirements, data handling, and ground truth coordination for these NASA programs were influenced directly by the data gathered by the S-193 sensors.

The complementary role of microwave sensors has resulted in the recommendation of scatterometer, altimeter, radiometer, and side-looking imaging radar as sensors for the SEASAT-A. Scatterometer-measured backscattering cross sections increase significantly with ocean surface windspeeds up to 30 knots. This increase is much less rapid for windspeeds in excess of 40 knots. The radiometer-measured brightness temperatures show rapid increases for windspeeds in excess of 30 knots. Therefore, the combined scatterometer/radiometer operation yields a better ocean surface, wind-sensing capability.

Microwave scatterometer and radiometer data gathered synoptically over the Earth's oceans will allow production of accurate worldwide wind field maps and world sea-level pressure maps. Altimeter and imaging radar measurements will be able to map the sea state globally and synoptically, thus providing data needed to plan shipping routes, ship designs, and designs of offshore structures and to assist in planning future ports. The knowledge of sea state in the form of wave spectrum can be used to measure ocean surface currents. An understanding of mass transport, heat transport, and current systems is beneficial to the economies of marine communities and global weather prediction. Satelliteborne microwave sensors can be efficiently used in monitoring ice conditions on the Great Lakes, the North Slope, and in polar regions. This knowledge will allow the forecasting of optimum ship routing in these areas. Furthermore, satellite altimetry will permit accurate measurement of the ocean geoid, and microwave sensors will provide the exciting prospect of tracking large storms.

The decade ahead calls for operational microwave sensors for remote sensing of the ocean surface phenomena, including salinity surveys and oil spill monitoring. These microwave systems must perform for extended periods of time in space. In particular, onboard data processing should be achieved for efficient dissemination of data. Improved techniques for data displays should be developed.

2. SCATTEROMETERS

Active microwave sensors have been used to measure the reflection or scattering coefficients of various objects or scenes. The ability with which a scene or object scatters incident microwave energy can be assessed by measuring the radar-scattering cross sections at various frequencies, polarizations, and incidence angles. A microwave scatterometer is a special purpose radar device which is used to quantitatively measure only the target reflectance or scattering cross section. In general, microwave scatterometers are simpler than conventional radar mechanisms because range and velocity measurement capability and the high spatial resolution (short-pulse) requirements are eliminated. Long-pulse and continuous wave scatterometers have been used to measure the scattering signatures of rough surfaces such as terrain or the ocean. The backscattering radar cross section σ_0, which is the backscattered power-per-unit area normalized for antenna gain, range loss, and transmitted power, is of great interest.

At present, vertical transmit and vertical receive (VV) backscattering cross section measurements from oceans taken by aircraft indicate: (1) there is a range of radar wavelengths for which backscatter is primarily dependent on surface windspeed and relatively insensitive to very large-scale roughness; (2) backscatter at and near nadir (0° incidence angle) monotonically decreases with increasing windspeed; and (3) backscatter from incidence angles greater than 20° off nadir monotonically increases with increasing windspeed. A representation of these three features of ocean backscatter versus windspeed and incidence angle for 2- to 3-cm wavelength radar illumination is shown in figure 1. It is noteworthy that for windspeeds below 1 meter per second (m/sec) essentially no sensible backscatter (σ_0) exists at any angle of incidence other than near-zero, and the strong radar return at the 0° incidence angle is from specular reflection (shown as a dotted line). As the windspeed increases

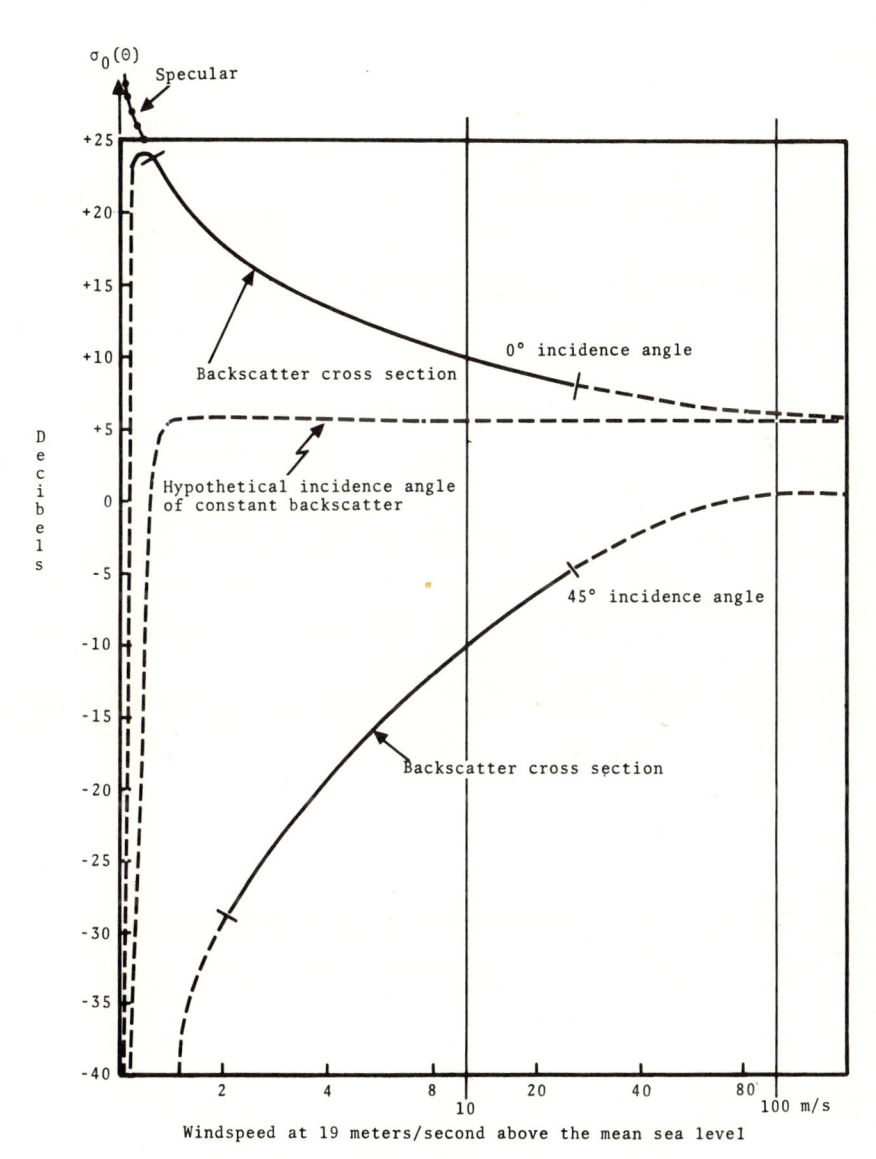

Fig. 1. — Representation of approximate magnitude of backscatter cross sections from ocean at 2- to 3-cm radar wavelengths.

toward 2 m/sec, patches of roughness start to appear on the surface where turbulent gusts at the surface exceed the threshold for wind-to-water coupling. The small capillary waves generated by these gusts produce some backscatter at all angles of incidence with very rapid rates of change in backscatter versus windspeed.

At windspeeds above about 1.5 m/sec the total surface is essentially covered with wind-driven capillary waves from which large waves will grow, and all radar return data from the ocean have the characteristics of backscatter. The approximate magnitude of backscatter cross section at 0° and 55° in figure 1 has been formed from good but sparse data taken between the extremes of about 2- to 3-m/sec windspeeds. The predicted continuation of the curves beyond 30 m/sec merely shows that the monotonic changes in backscatter cross section initiated at low windspeeds probably do not cut off abruptly but become very slight at higher windspeeds. The local slope of the curves at high windspeeds shows that accurate measurements of the backscatter must be made to sense small percentage changes in windspeeds but that this accuracy is required over only a small range of backscatter values. Alternatively, the local slopes at low windspeeds show that less accurate measurements over a large range of backscatter values sense small percentage windspeed changes.

The set of curves for backscatter versus windspeed at all angles of incidence between 0° and 55° lies between the curves shown in figure 1 (being far apart at low and close together at high windspeeds).

Since backscatter-versus-windspeed changes appear at some incidence angle between 0° and 20°, it would be desirable to discover an incidence angle for which the backscatter cross section is constant for all windspeeds. This would provide a natural reference backscatter cross section and eliminate the requirement for absolute backscatter cross section measurements if only windspeeds are required. The dashed line in figure 1 indicates this desired hypothetical incidence angle of constant backscatter lies approximately at the average level of the backscatter versus windspeed for incidence angles between 5° and 20°. Other aspects of backscatter measurements which present both opportunities and problems are: (1) the upwind, downwind, and crosswind viewing directions give different values for backscatter cross sections; (2) the sensitivity of backscatter to upwind, downwind, and crosswind viewing directions is different at each angle; and (3) the sensitivity of backscatter at each angle to upwind, downwind, and crosswind viewing also depends on windspeed. Figure 2 [2] shows this typical data. The upwind-downwind dependency was predicted and explained rather well 20 years ago by Allan Schooley [3]. Schooley's conclusions were based on specular points of the backscatter mechanism and wave tank measurements of the probability distributions of surface facets versus windspeed. These statistics included the distributions of surface slope normals and the distributions of the areas of these specular reflecting surfaces. Because he could not measure these same statistics in other directions (e.g., a long narrow tank), the crosswind dependency was not predicted. The problem created by this feature of ocean backscatter is obvious: With perfect instrumentation (i.e., no error in the backscatter measurements) but no knowledge of the viewing direction with respect to the wind direction, data spreads of many decibels will be obtained. For example, from figure 2, a 6-dB spread exists upwind to crosswind at 13.7 m/sec, and a 40° incidence angle has an average backscatter over all angles of -10 dB. At 6.5 m/sec windspeed the average is about -15.5 dB with an upwind-to-crosswind spread of 4 dB. This problem can be solved either by knowing the viewing direction with respect to wind direction or by viewing the surface from sufficient directions to determine the upwind, downwind, and crosswind curves. As input data, this curve indicates the wind direction in that resolution cell has also been measured. This resolution cell data at many incidence angles in a swath along the satellite's path is exactly what is needed to properly map the wind fields and/or the pressure fields over the oceans at the 1,000-millibar level.

With proper system design, satelliteborne scatterometry will provide a majority of the input data required for mapping the wind fields over the oceans, with new data available within 12 hours.

These global weather maps will forecast weather accurately for the first time. Improvement of these maps will provide the solid basis for weather control necessary to avoid disasters caused by inaccurate weather forecasting.

Relationships between backscattering cross sections and wind velocity, oil spills, and ice conditions. Radar observations of scattering cross sections from the ocean surfaces have been made for approximately 30 years. In most cases, only the backscattering cross sections (receiving and transmitting antennae are located at the same place) have been measured. The average scattering cross section per unit area in the far field region is given by:

$$\sigma_{0\alpha\beta} = 4\pi R_r^2 \frac{<|E_\alpha^S|^2>}{A|E_\beta|^2} \tag{1}$$

In equation 1, R_r = distance from radar receiver to the illuminated area, $|E_\beta|$ = magnitude of the incident field at the surface with β polarization, $|E_\alpha^S|$ = magnitude of the scattered field at the receiver with α polarization, and A = area illuminated by the incident wave. In equation 1, polarization α may or may not be the same as β.

The NRL has conducted comprehensive measurements of the backscattering cross sections from ocean surfaces [4 through 9]. These measurements have been conducted at several frequency and polarization combinations. Airborne and ground sensors have been used by the NRL. A summary of the significant NRL measurements is given in table 1.

NASA has flown numerous missions with spaceborne and airborne sensors over ocean surfaces [10 through 14]. NASA's Lyndon B. Johnson Space Center (NASA/JSC) has gathered airborne radar (also known as scatterometer) data at 0.4, 13.3, and 13.9 GHz and spaceborne radar data at 13.9 GHz (from the Skylab S-193 Microwave Radiometer/Scatterometer/Altimeter). Backscattering cross section data has also been collected by the Langley Research Center (NASA/LaRC) at 13.9-GHz frequency. Table 1 presents a summary of the NASA measurements.

The behavior of the average backscattering cross section σ_0 can be explained on the basis of a composite scattering model for the ocean surface. This model considers the ocean surface to be composed of a slightly rough surface (capillary waves and very high frequency gravity waves) superimposed on a larger structure (gravity waves and swells). Based on this model, it can be shown that the predominant backscattering at higher angles is because of the small structure [12]. To study the effect of the wind on the backscattering cross section, the directional spectrum of the high-frequency, gravity-capillary structure of the sea could be expressed as:

$$W(\xi) = k\xi^{-k_3}, \tag{2}$$

where k and k_3 are constants, and $\xi = (p^2 + q^2)^{1/2}$ (p and q are radian wave numbers on the ocean surface).

At higher angles of incidence, the backscattering cross section can be expressed, in part empirically and in part because of scattering from a small gravity-capillary structure, as follows [12]:

$$\sigma_{0ij}(\theta) = k_1 W_1^{k_2} |\alpha_{ij}|^2 (\cos\theta)^4 (\operatorname{cosec}\theta)^{k_3} \tag{3}$$

Table 1. — Available Radar Backscattering Cross Sections Over Water/Ocean Surfaces

Type of measurement and organization	Wavelength or frequency	Polarization combinations	Approximate range of angles of incidence	Range of winds or waves
Spaceborne experiments: NASA/JSC	13.9 GHz	VV, VH, HV, HH	0° to 53°	4 to greater than 55 knots
Airborne experiments: NRL	0.428, 1.228, 1.25, 4.425, and 8.91 GHz	VV, VH, HV, HH	0° to 89°	4 to 48 knots
NASA/JSC	13.3 GHz	VV	0° to 60°	6 to greater than 55 knots
	0.4 GHz	VV, VH, HV, HH	0° to 60°	
	13.9 GHz	VV, VH, HV, HH	0° to 50°	
NASA/LaRC	13.9 GHz	VV, VH, HV, HH	0° to 50°	6 to 40 knots
From platforms/bridges: NRL	8.6 mm, 1.25 cm, and 3.2 cm	VV	0° to 80°	0 to 25 knots
Wave tank: NRL	9.375 GHz	VV, HH	10° to 86°	Millimeter waves of 1.0 to 6 cm

In equation 3, α_{ij} is defined as:

$$\alpha_{HH} = (\varepsilon - 1) \Big/ \left[\cos\theta + \sqrt{\varepsilon - \sin^2\theta}\right]^2$$

$$\alpha_{VV} = (\varepsilon - 1)[(\varepsilon - 1)\sin^2\theta + \varepsilon] \Big/ \left[\varepsilon\cos\theta + \sqrt{\varepsilon - \sin^2\theta}\right]^2$$

$$\sigma_{VH} = \sigma_{HV} = 0 \tag{4}$$

where W_1 = ocean surface wind velocity, ε = the complex dielectric constant of the surface, θ = angle of incidence at the surface, and k_1 and k_2 are constants; k_1 is related to the constant k.

By using algorithm 178, Direct Search, from <u>Communications of the ACM</u> [15], a FORTRAN program finds the values of k_1, k_2, and k_3 and then searches for a minimum value. The value of the dielectric constant was taken as $\varepsilon = 55 + j\, 30.5$.

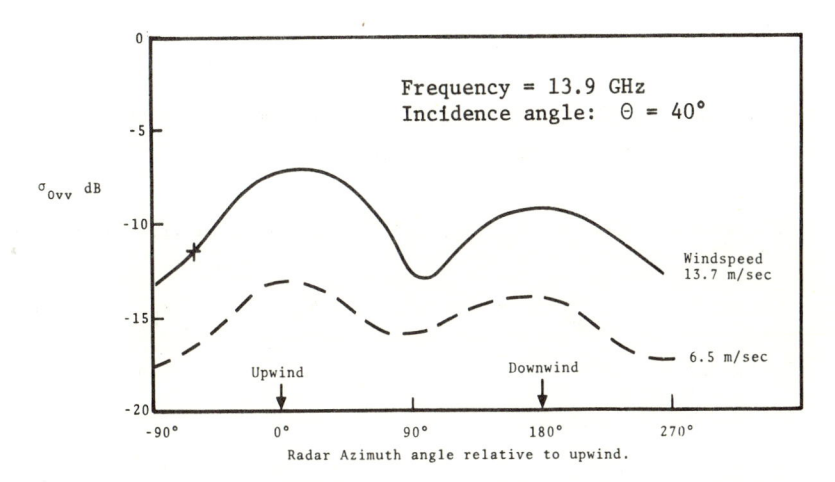

Fig. 2. — NASA/LaRC values of σ_0 versus wind heading [2].

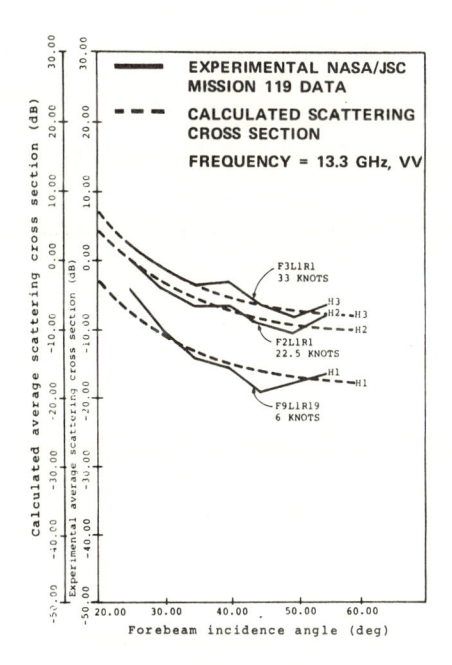

Fig. 3. — Comparison of calculated and experimental scattering cross section for Mission 119.

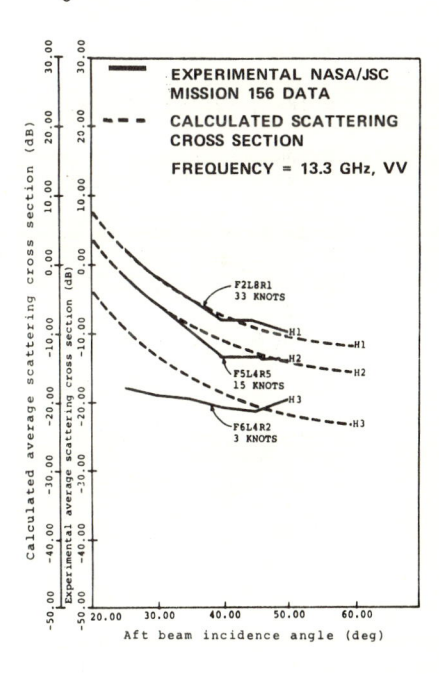

Fig. 4. — Comparison of calculated and experimental Mission 156 data.

The cosec θ form given in equation 3 can be simplified and expressed in cot θ form as follows [3]:

$$\sigma_{0ij}(\theta) = k_1 W_1^{k_2} |a_{ij}|^2 (\cot \theta)^{k_3} \qquad (5)$$

Table 2 gives the typical values of k_1, k_2, and k_3 for NASA/JSC 13.3-GHz data. Flight:line:run are designated by F:L:R.

Table 2. — Calculated Values of k_1, k_2, and k_3 for Mission 119, Using cot θ Form (VV, 13.3 GHz)

Set	Data	k_1	k_2	k_3
I	Crosswind - forebeam 25° ≤ θ ≤ 50° F5L3R1, F2L3R1, F3L3R1	0.015	1.445	5.25
II	Crosswind - aftbeam 25° ≤ θ ≤ 50° F5L3R1, F2L3R1, F3L3R1	0.0098	1.75	4.32
III	Upwind - forebeam 25° ≤ θ ≤ 50° F9L1R19, F2L1R1, F3L1R1	0.043	1.33	5.00
IV	Upwind - aftbeam F9L1R19, F2L1R1, F3L1R1	0.01	1.912	4.391
V	Downwind - aftbeam 25° ≤ θ ≤ 50° F9L1R18, F5L1R2, F2L1R2, F3L1R2	0.0184	1.785	4.26
VI	Downwind - forebeam 25° ≤ θ ≤ 50° F9L1R18, F5L1R2, F2L1R2, F3L1R2	0.0188	1.50	5.44
VII	Diagonal wind - forebeam 30° ≤ θ ≤ 55° F9L3R3, F7L3R1, F3L2R1	0.114	1.0157	4.4
VIII	Diagonal wind - aftbeam 30° ≤ θ ≤ 55° F9L3R3, F7L3R1, F3L2R1	0.061	1.07	7.07

Figure 3 shows the experimental and calculated data from JSC Mission 119 (using equation 3), for upwind 13.3-GHz data taken from aircraft. The values of constants are: $k_1 = 0.026$, $k_2 = 1.324$, and $k_3 = 5.47$.

The second set of data was chosen from Mission 156 NASA/JSC data. The data is for aftbeam 13.3-GHz VV data for upwind condition. The values of constants using the cosec θ form (equation 3) are: $k_1 = 0.0207$, $k_2 = 1.1$, and $k_3 = 6.6$.

Figure 4 shows the comparison of calculated and experimental values from data taken from aircraft for JSC Mission 156. It should be noted here that flight 6 of Mission 156 was for very calm conditions with extremely low surface wind velocity.

Table 3 is the 1.25-cm data given in Grant and Yaplee's paper [5]. The values using the cot θ form are: $k_1 = 0.00107$, $k_2 = 1.64$, and $k_3 = 5.03$.

Table 3. — Grant and Yaplee, for 1.25-cm, VV Case (No Wind Direction Reported)

Mean windspeed in knots	θ in deg, σ_0 dB			
	20°	30°	40°	50°
7.5	-14.0	-25.0	-28.75	-31.10
12.5	-12.4	-21.25	-25.00	-27.60
17.5	-9.0	-17.8	-22.5	-24.4
22.5	-7.5	-14.2	-17.5	-20.2

The four-frequency (4-FR) measurements by the NRL have been analyzed to determine dependence of σ_0 on W_1 [14, 16, 17]. The values of k_2 are somewhat smaller (ranging from 0.2 to 1.9) compared to the values for NASA data. However, the operating frequency of the NASA 13.3-GHz scatterometer is higher than for the NRL 4-FR radar. The data presented in tables 2 and 3 show that the spectrum of the high-frequency gravity-capillary structure (on top of the low-frequency gravity waves) is wind-dependent. The value of k increases as the value of k_3 decreases with wind velocity. In general, the value of the constant k_3 is also a function of the radar incident wavelength. To illustrate this point, the forebeam data for F7L3R1 gathered with the 0.4-GHz system (σ_{0VV}) for 15.5 knots was processed using equation 5. (On this run, both 13.3- and 0.4-GHz data were gathered simultaneously.) The results are as follows:

0.4-GHz data	13.3-GHz data
$k_3 = 7.3$	$k_3 = 4.7$

This comparison, which was made for the same relative conditions, shows that significant differences can result in the measured spectrum of gravity-capillary waves as a result of the incident radar wavelength, a phenomenon which is analogous to filtering. When summarized, it becomes apparent that, at 13.3-GHz, σ_0 can be related to wind velocity through the high frequency and capillary spectrum. A wind dependence of $W_1^{k_2}$, where k_2 ranges from 1.0 to 1.95, is seen. Once this is achieved, the link between sea state and wind velocity is provided through the equation

$$H_{1/3} = m W_1^n \tag{6}$$

where $H_{1/3}$ is the significant wave height in feet, and W_1 is the windspeed in knots. The values of m range from 0.004 to 0.023, while n ranges from 2.0 to 2.5 [14, 18].

The Skylab Earth Resources Experiment Package (EREP) Hurricane Ava pass provides backscattering cross sections for a wide range of wind velocities. A quick-look evaluation [19, 20, 21, 22] of the Hurricane Ava data shows a strong dependence of σ_0 on the wind velocity. The S-193 radiometer/scatterometer gathered data in crosstrack (right only) mode for nominal roll angles of

0°, 15.6°, 29.4°, 40.1°, and 48°. The actual angles attained by the S-193 antenna differed slightly from these nominal angles. The highest roll angle data (average 46.6°) was taken from approximate areas shown in figure 5. The photograph was prepared by the NOAA.

The scatterometer data corresponding to average roll angles of 31.5°, 40.61°, and 46.6° are given in figures 6, 7, and 8, respectively, as a function of time. The wind velocities in these figures were interpolated from the data given in a report prepared by Hayes, et al. [23]. During this pass the Skylab was in solar inertial mode. The data drop after GMT 18:58:17 was because of attenuation by the 0° Doppler filter. The dashed line, marked 1-dB bandwidth point, corresponds to a Doppler filter attenuation of 1 dB. No Doppler filter corrections have been made for the data given in these figures. In figure 5, locations of points 1 through 13 of figure 8 have been shown.

To study the wind dependence for data presented in figures 6, 7, and 8, a function of the form:

$$\sigma_{0HH} = k_1 W_1^{k_2} \tag{7}$$

was used. A least-mean-squares fit yields the following results:

- For the average roll angle of 31.35°, $k_2 = 0.65$.
- For the average roll angle of 40.61°, $k_2 = 0.60$.
- For the average roll angle of 46.6°, $k_2 = 1.89$.

This wind dependence has been investigated for data only to GMT 18:58:17; the wind direction was very nearly downwind for these data points. No atmospheric corrections were made to the data.

NASA/LaRC has acquired 13.9-GHz airborne scatterometer data for various windspeeds. The values of k_2 for this data range from 1.4 to 2.0 [2]. A comparison of computed and measured data is shown in figure 9 [2].

The NASA 13.3-GHz aircraft scatterometer measures σ_0 for a particular ocean cell for incidence angles from -60° to +60°. Data at 0° incidence is not processed because of the zero infrared system. As the wind velocity on the surface of an ocean increases, the scattering cross section at low angles of incidence decreases, and the cross section at higher angles of incidence increases. To exploit this phenomenon for prediction purposes, a ratio Q_0 given by

$$Q_0 = \sigma_0 \text{ at } 5°/\sigma_0 \text{ at } 45° \tag{8}$$

was introduced. The prediction curve based on this ratio for data gathered during Missions 119 and 156, over fully developed seas, using a computer fit, is given by

$$Q_0 = (6.4) \times 10^4 \, W_1^{-2} . \tag{9}$$

In order to see how well the Q_0 curve from Mission 119 can be used to predict wind velocities for Mission 156, the relationship given in equation 9 was used. Equation 9 has been graphed in figure 10, which also gives the measured values of Q_0.

The predicted value of wind velocity from each Q_0 dB value was obtained by using equation 9. The predicted and measured values are compared in table 4.

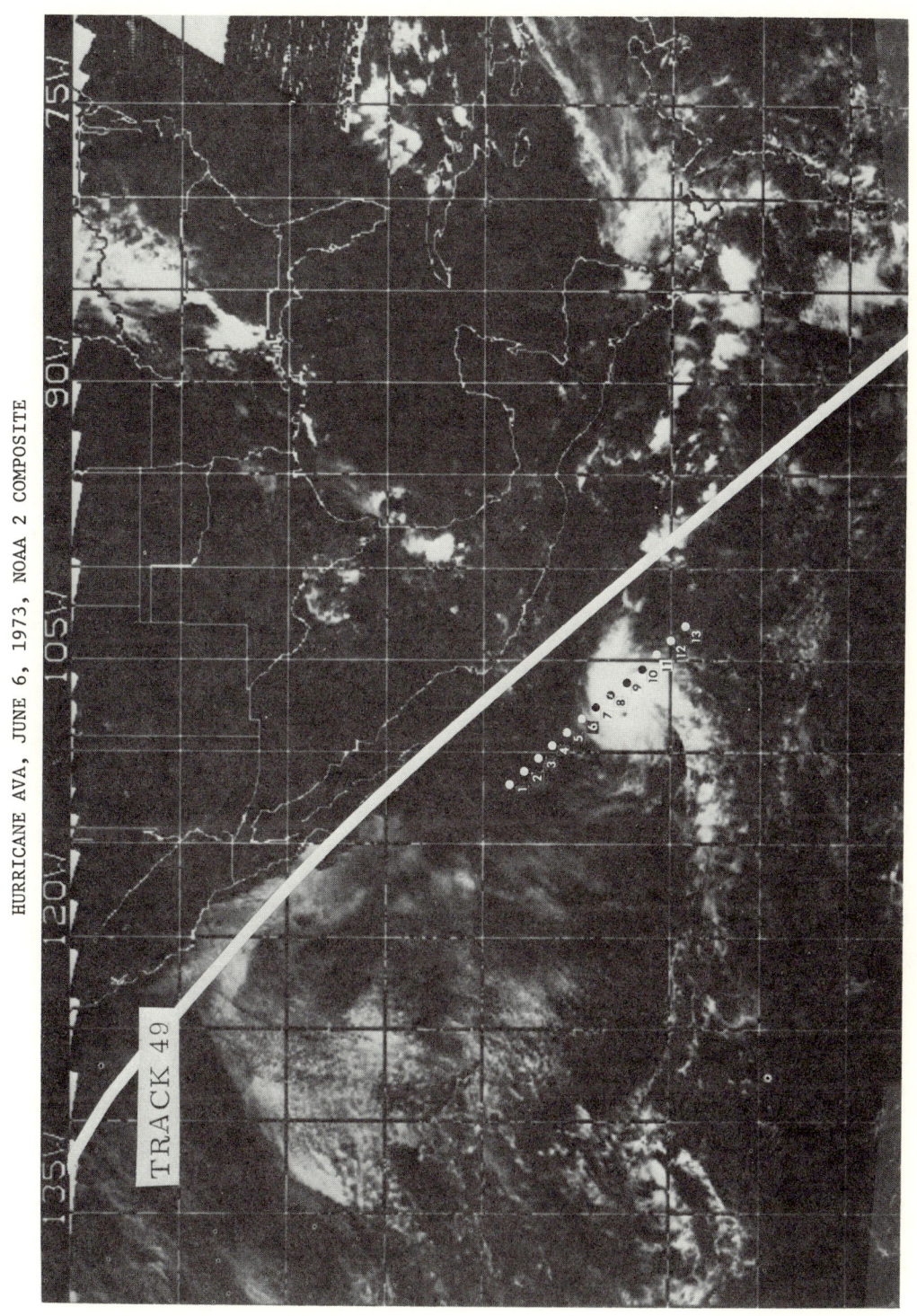

Fig. 5. — Approximate location of the 13 data points (1 through 13) from north to south.

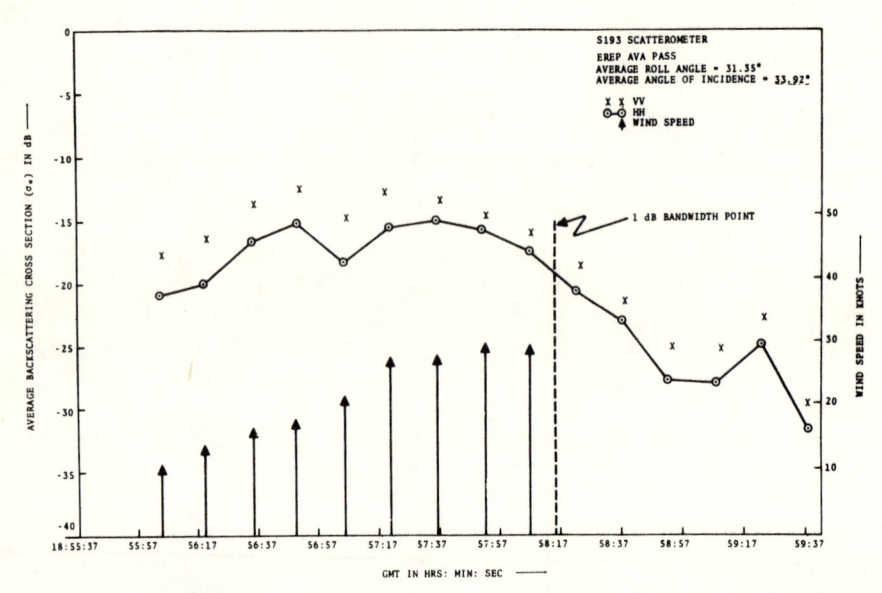

Fig. 6. — Backscattering cross section as a function of time for Hurricane Ava pass with 31.35° roll angle.

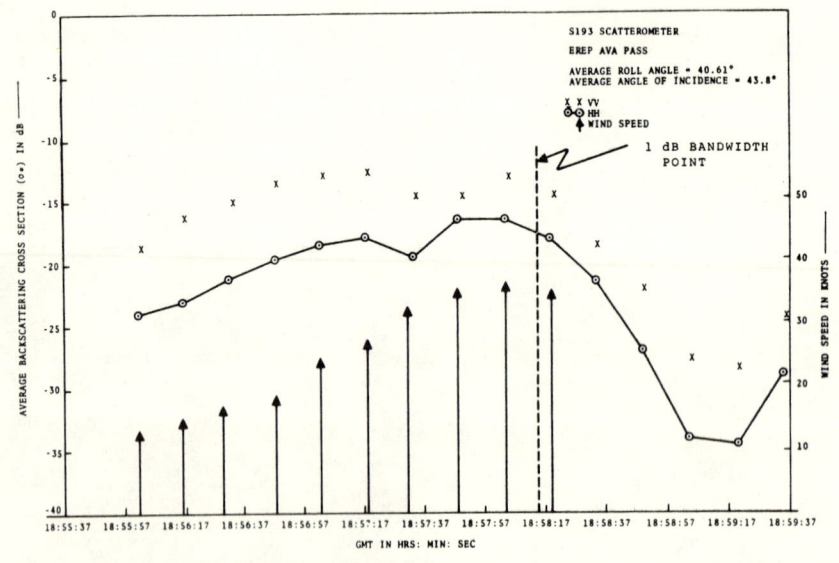

Fig. 7. — Backscattering cross section as a function of time for Hurricane Ava pass with 40.61° roll angle.

REMOTE SENSING OF OCEANS

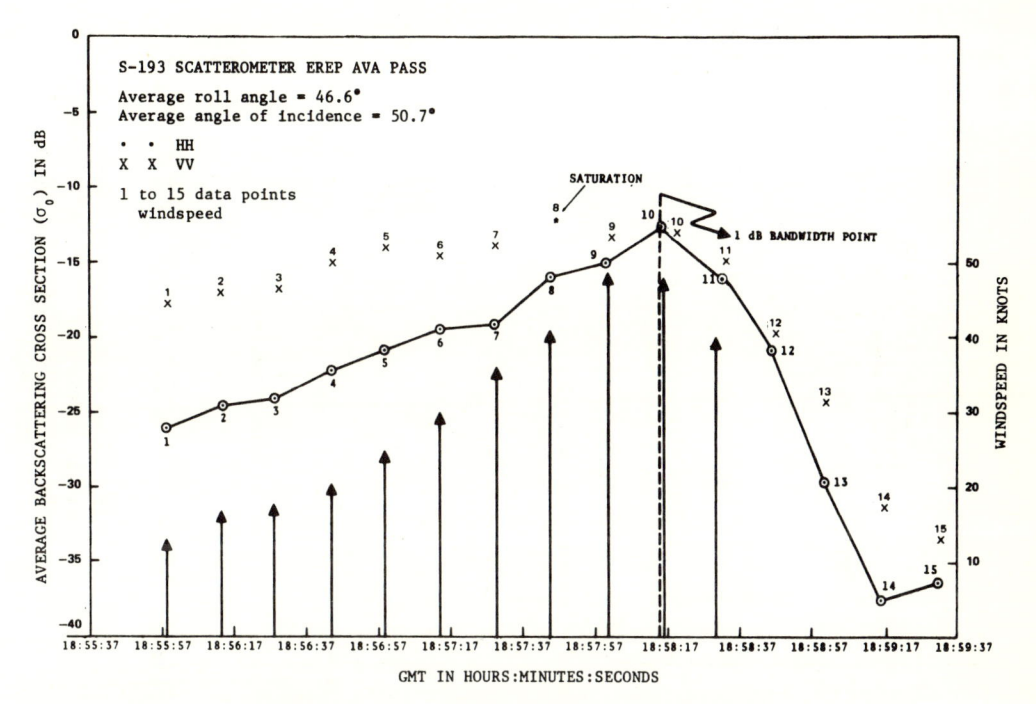

Fig. 8. — Backscattering cross section as a function of time for Hurricane Ava pass with 46.6° roll angle.

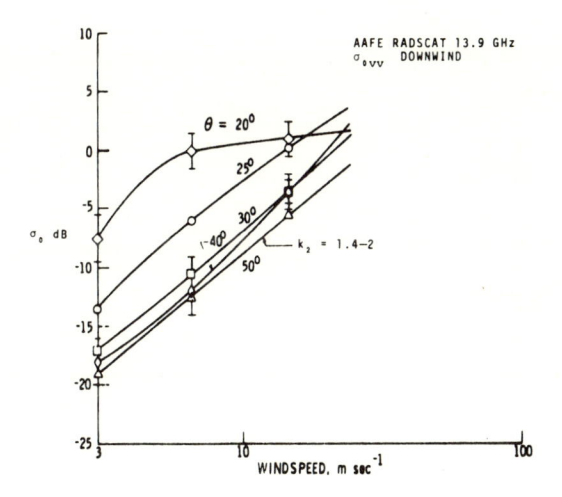

Fig. 9. — NASA/LaRC values of $\sigma°$ versus windspeed [2].

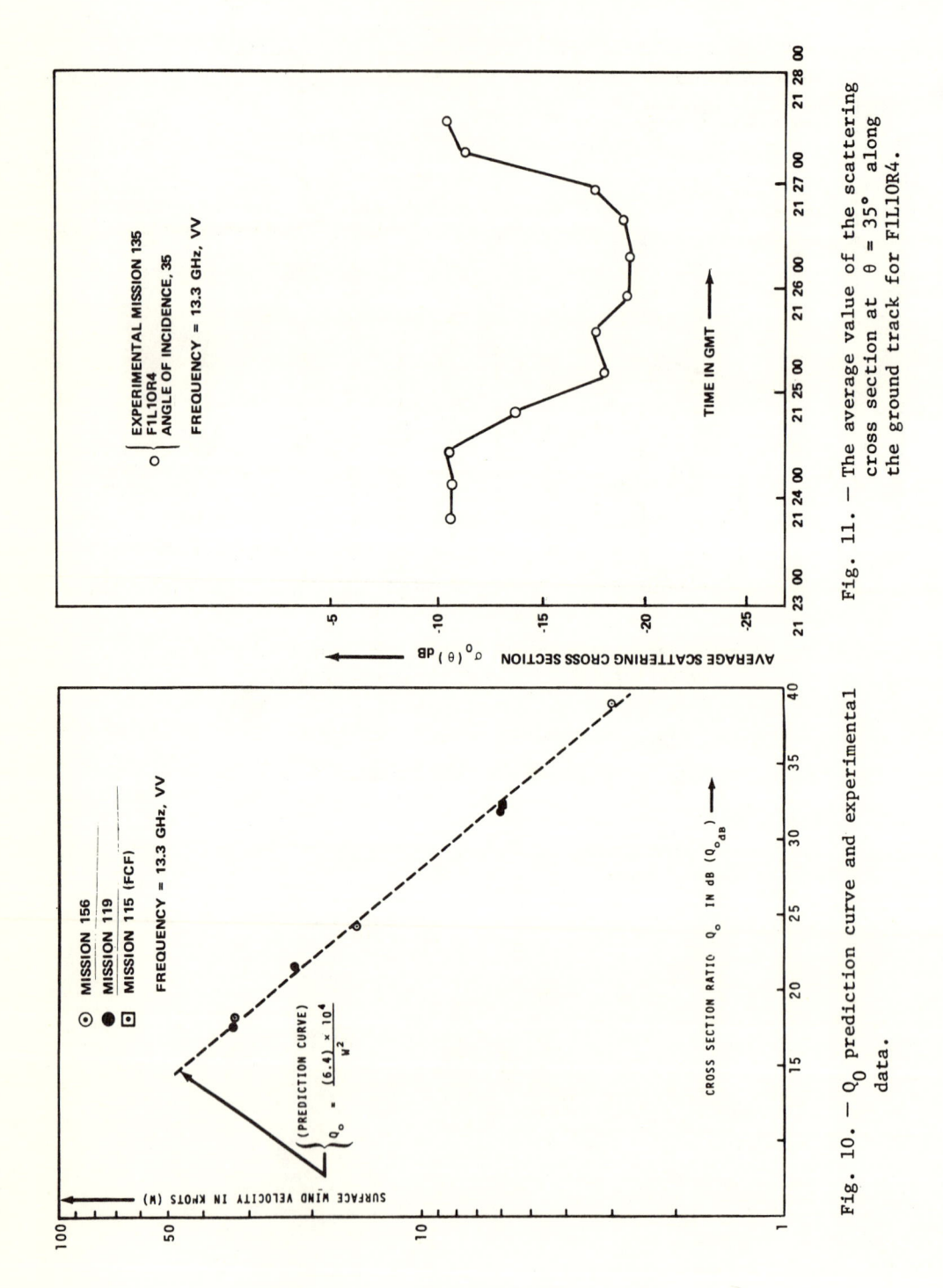

Fig. 11. — The average value of the scattering cross section at $\theta = 35°$ along the ground track for Fl110R4.

Fig. 10. — Q_0 prediction curve and experimental data.

This table demonstrates that compared to σ_0, Q_0 can be used more effectively for prediction of ocean surface winds. Fully developed seas and constant windspeeds are required to provide the accuracies shown in this table.

Table 4. — Predicted and Measured Values of Wind Using Q_0 Curve

Predicted value of wind velocity for Q_0 Mission 156 values from Mission 119 curve (W_p), knots	Actual measured value of wind velocity during Mission 156 (W_a), knots	Percentage difference $\left(\dfrac{W_p - W_a}{W_p}\right) 100$
2.8	3	7.14
15.5	15	3.22
31.0	33	6.45

Data at 13.3 GHz was collected over an oil spill in the Gulf of Mexico during NASA/JSC Mission 135. The scattering cross section at higher angles of incidence decreased by 5 to 10 dB in the presence of oil spill [24]. The behavior of σ_0 over an oil spill is shown in figure 11. The decrease in the scattering cross section is attributed to the damping of small gravity and capillary waves on the water surface.

Aircraft-acquired scatterometer data from first-year and multiyear ice are shown in figure 12, as reported by Rouse [25]. At a frequency of 13.3 GHz (VV data), older ice is shown to exhibit greater backscattering than newer ice. Similar results were obtained from the analysis of 13.3-GHz and 0.4-GHz JSC Mission 126 data. The backscattering cross section data also indicates that ice thickness can be measured from radar observations [26]. However, this application demands further investigation. The influence of salt content, roughness, thickness, physical temperature, and complex dielectric constants on the backscattering cross section must be experimentally investigated further. Open water areas can be differentiated at both 13.3-GHz and 0.4-GHz radar frequencies.

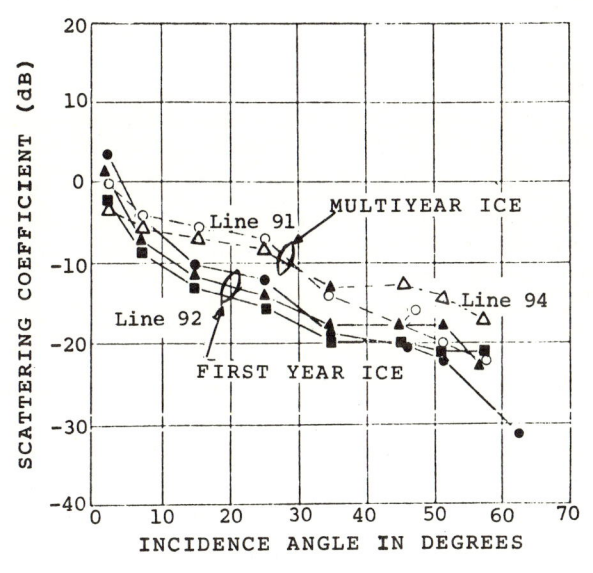

Fig. 12. — Comparison of scattering coefficient angle variations for first year and multiyear ice.

3. RADIOMETERS

Microwave radiometers measure the thermal microwave radiation emitted by matter because of its thermal nature. The intensity of microwave radiation is several orders of magnitude weaker than naturally emitted radiation in the visible and infrared parts of the electromagnetic spectrum. Microwave radiometers have essentially linear, sensitive receivers and narrow beam-width antennae. Internal calibration sources employ sources of constant microwave radiation. Thermal radiation at radio wavelengths was first detected by Jansky [27]. Dicke [28] developed the first stable microwave radiometer. Investigators have since explored the use of radiometers in remote sensing of the Earth's phenomena.

Microwave radiometer observations have shown promising uses; i.e., detection of the presence of oil on the ocean surface; reading of microwave brightness temperatures of the ocean, which are significantly dependent on the ocean surface wind field; and possible measurement of water temperature and salinity in some marine environments. Furthermore, microwave radiometers can measure liquid water content in the atmosphere. Radiometer measurements can also determine the attenuation of microwaves in the atmosphere, which can be utilized in correcting backscattering cross section data.

The brightness temperature measured from a platform at a height h can be expressed as:

$$T_B(\theta,h) = \left[E(\theta)T_g + r(\theta)T_s\right]\tau(h) + \int_0^h T_A(h)\frac{\delta\tau}{\delta h}dh \qquad (10)$$

In equation 10, T_s is the sky brightness temperature as seen from the surface; $T_A(h)$ is the brightness temperature of the atmosphere at height h; τ is the transmittancy of the atmosphere; T_g is the ground temperature; $E(\theta)$ is the emissivity of the surface; $r(\theta)$ is the reflectivity; and θ is the viewing angle of the radiometer. Equation 10 is valid for pencil beam antenna patterns, for which side lobes are negligible and uniform patterns are assumed within the main beam. For highly directional antennae with side lobes, the microwave brightness temperature can be simplified as follows [29].

$$T_B = T_{BM}\eta + T_{BS}(1-\eta) \qquad (11)$$

where $\eta = 1/4\pi \iint G(\theta,\phi)\sin\theta\, d\phi\, d\phi$; η = main lobe efficiency; T_{BM} = mean ground brightness temperature over the main lobe; and T_{BS} = mean ground brightness temperature over lobes other than the main lobe. The value of η is greater than 0.9 for modern radiometers. Thus, when the atmospheric and ground physical temperature effects are removed, it is possible to get corrected mean brightness temperatures of ground scenes.

<u>Relationship of brightness temperature with ocean surface phenomena and ice conditions</u>. The brightness temperature T_B of an ocean surface is a function of roughness and reflectivity. The T_B depends on polarization, operating wavelength of the radiometer, and angle of incidence. For calm ocean surfaces, Hyatt and Paris have modeled equation 10 in terms of water temperature, salinity, microwave frequency, polarization, and angle of incidence [30, 31]. The effects of surface roughness have been theoretically investigated by Stogryn, Ulaby and Fung, Wu and Fung, and Lynch and Wagner [32 through 35]. These theoretical studies show relationships between ocean surface parameters (salinity, windspeed, etc.) and the calculated brightness temperature.

Several NASA/JSC missions have been flown in the area of the Mississippi River as it flows into the Gulf of Mexico to study the effect of salinity changes on the brightness temperature at 1.42 GHz for vertical incidence [36, 37]. The plot of surface water salinity and temperature along the flight track (from Paris [37]) is shown in figure 13. The symbols indicate the values obtained by the surface survey conducted shortly after the aircraft flight. Symbol M indicates times of passage over the river mouth, and F represents the passage over the surface foam. In general, surface and radiometric measurements agree. Stogryn [38] has reported on a method of computing the dielectric constant of saline water. Measurements reported to date indicate that significant changes in T_B result from ocean surface salinity changes up to frequencies of 4 GHz [39, 40].

Soviet scientists have demonstrated the use of a 3- to 4-GHz microwave radiometer in mapping water surface temperatures over the Caspian Sea [41]. Investigators are determining the use of the 2.69-GHz radiometer to measure sea temperature in the open ocean [42].

The presence of an oil spill causes an increase in the brightness temperature [43, 44]. Studies by the NRL have shown that both the presence of oil and an estimate of its thickness can be detected by microwave radiometers [43].

The dependence of T_B on the ocean surface wind has been intensively investigated during the last decade [45 through 49]. Increases of 1.1° K per meter per second of horizontally polarized brightness temperatures at 19.34 GHz have been measured. Two factors cause an increase in T_B : the increase in

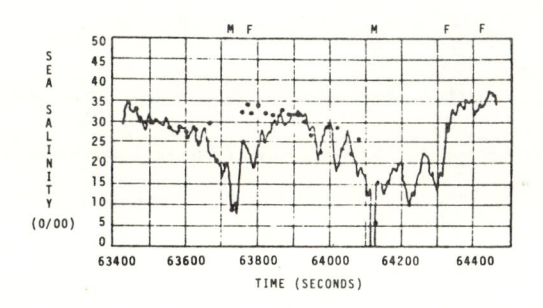

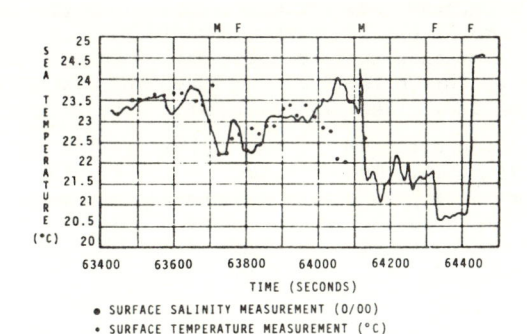

Fig. 13. — Results of JSC Mission 190, flight 3, run 8, over site 138, line 1 [37].

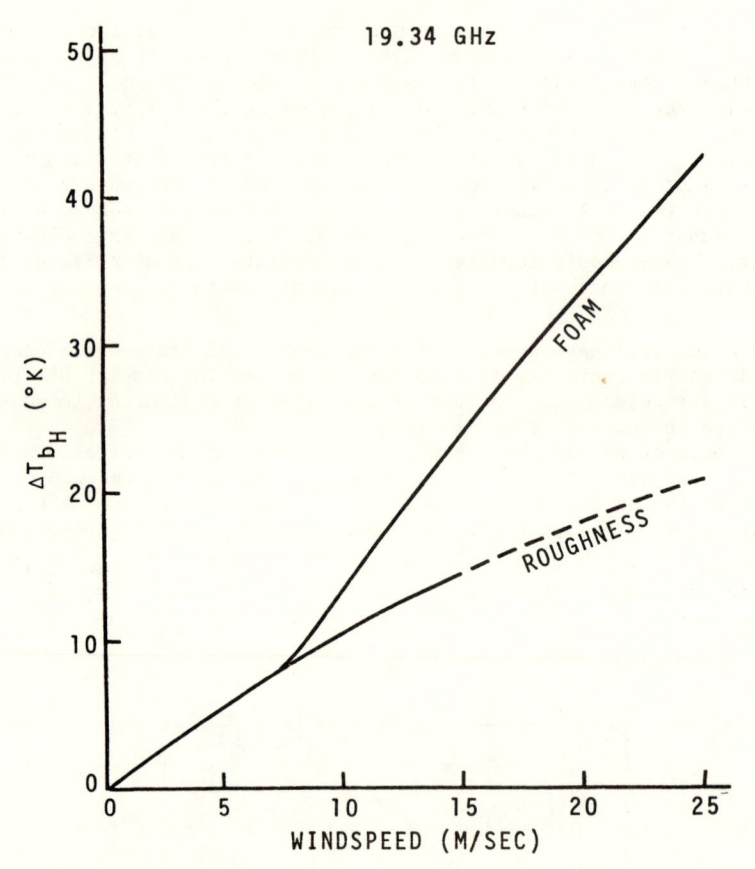

Fig. 14. — Increase in brightness temperature as a function of surface windspeed [50].

mean square slope and the increase in the extent of foam and spray. For low windspeeds as shown in figure 14 [50], the increase in T_B is attributed to the increase in sea roughness (sea state effect). At higher windspeeds, the predominant effect is the presence of spray and foam [51, 52]. Foam and spray have emissivities greater than that of calm sea. Figure 14 represents an estimate of the combined effects of surface roughness and sea foam with windspeed, as viewed from a satellite. The S-193 radiometer antenna temperature T_A (not corrected for atmospheric effects) for the highest roll angle (50.7° incidence angle) is shown in figure 15. The dependence shows a change near 30 knots of surface windspeed for the first 11 points (wind velocities are shown in fig. 8). The brightness temperature changes more rapidly for greater windspeeds. Furthermore, the difference between horizontal and vertical antenna temperatures $(T_{AH} - T_{AV})$ shows a decrease with increasing windspeed.

In summary, the microwave brightness temperatures for higher frequency radiometers are significantly dependent upon surface wind fields. This dependence encourages the use of remotely sensed surface winds of calm to 30 m/sec velocity and above to determine brightness temperatures. No dependence on wind direction has been reported to date.

Imaging radiometers have been used to generate microwave maps of the polar ice [53]. The dependence of the radiometric brightness temperature on ice

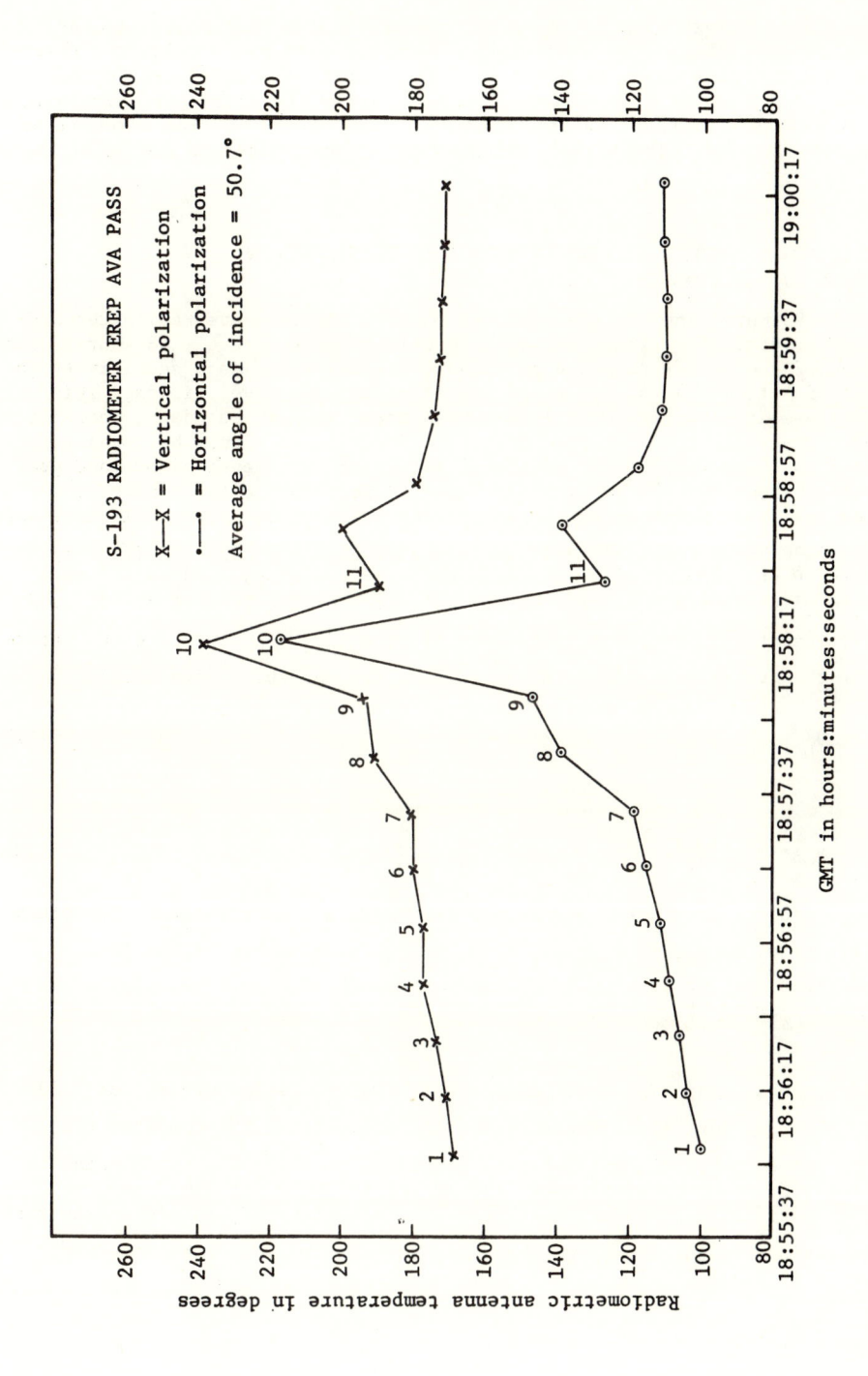

Fig. 15. — Radiometric antenna temperatures for Hurricane Ava pass.

phenomena allows the showing of ice pack boundaries, ice pack densities, and multiyear ice distribution on these maps. Repeated data-takes allow the observation of ice dynamics. The Nimbus-5 radiometer used to acquire this data operates at a wavelength of 1.55 cm. Figure 16 shows the microwave emissivities of the polar seas as a function of frequency. This figure is from a report by Gloersen, et al. [54]. The extent and location of open water in the polar regions can be determined using satelliteborne radiometers [54].

4. COMBINED RADIOMETER AND SCATTEROMETER OBSERVATION OF OCEAN SURFACE WINDS

The S-193 Hurricane Ava pass provides nearly simultaneous radiometer and scatterometer data over winds ranging from calm to approximately 48 knots. The dependence of radiometer response to winds (fig. 15) shows less sensitivity in the range of 0 to 30 knots, and rapid increases at higher velocities. This rapid increase at higher velocities has been noted in previous data, where it was due to the formation of foam and spray. However, it cannot be totally attributed to these factors for Hurricane Ava, since the dense clouds and precipitation surrounding the eye were undoubtedly responsible for some of the increases in T_A. Scatterometer data, on the other hand, shows rapid increases up to about 30 knots and less rapid increases at higher windspeeds (fig. 8). This may be partly due to microwave absorption by the clouds and rain. This dependence on windspeed suggests the investigation of the product of σ_0 and antenna temperature T_A. The dependence of $V = \sigma_0 T_A$ on windspeed was undertaken for the Ava data (σ_0 not in decibels). Two variables were used:

$$V_h = \sigma_{0HH} T_{AH} \tag{12}$$

and

$$V_v = \sigma_{0HH} T_{AV} \tag{13}$$

where V_h and V_v represent horizontal and vertical polarizations. To study the effect of windspeed (W_1) on V_v and V_h a relationship of the form:

$$V_i = k_1 W_1^{k_2}$$

(i = v or h) was used. The best computer fit yields a value of k_2 of 2.20 for V_h and 2.25 for V_v. Furthermore, the values give the computer fit, within 20 percent, up to the first nine points (fig. 17). The tenth point in figure 8 (about 47 knots of windspeed) does not fit well with the computer-determined curve. An examination of the computed values of k_2 indicates that combined radiometer and scatterometer data produce stronger dependence when compared individually.

5. ALTIMETERS

The basic ability of radar to determine the distance to targets has been known since the early 1920's. The advent of technology rapidly extended the measurement capability of radar both in terms of distance to remote targets and

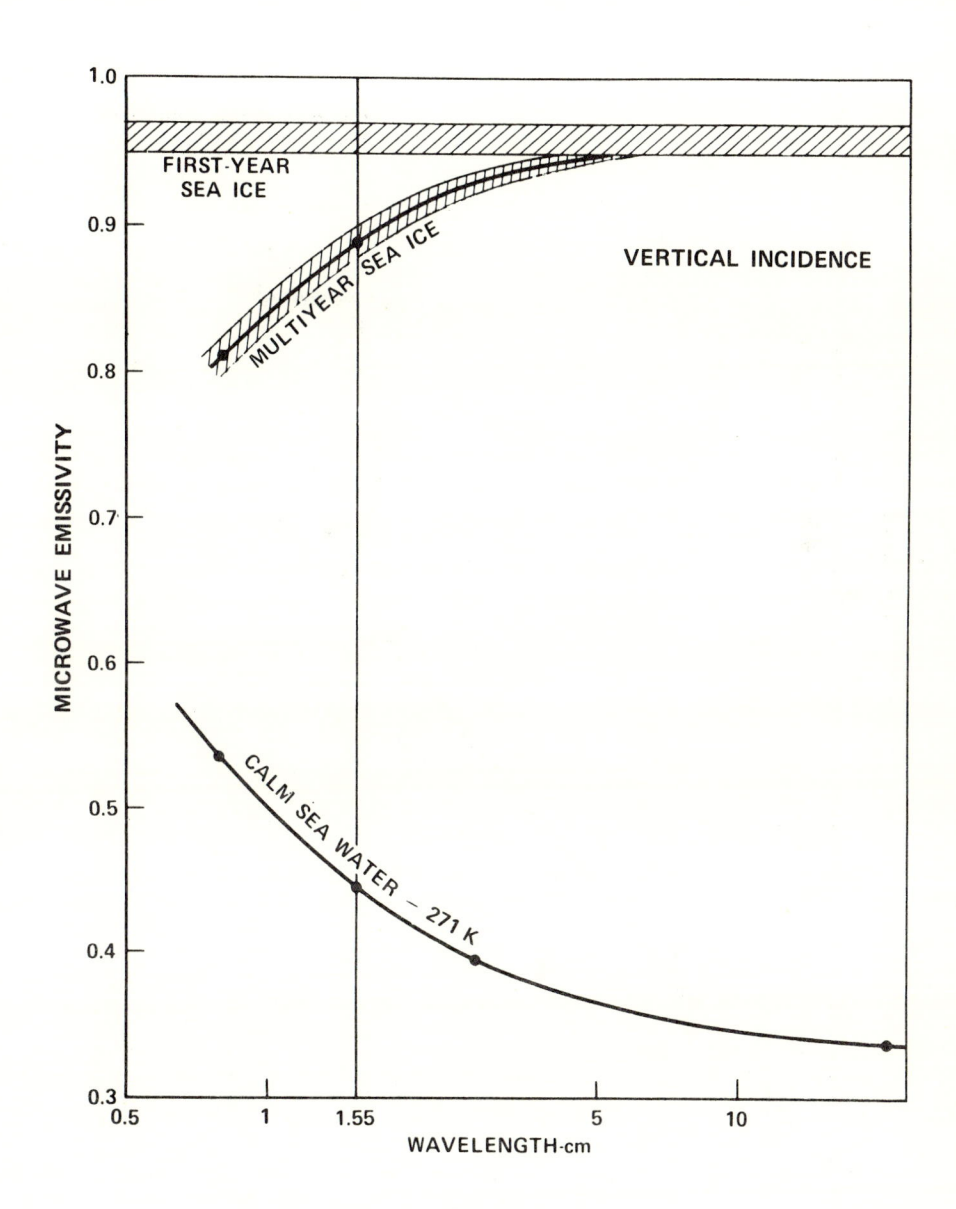

Fig. 16. — Microwave emissivities of the polar seas [54].

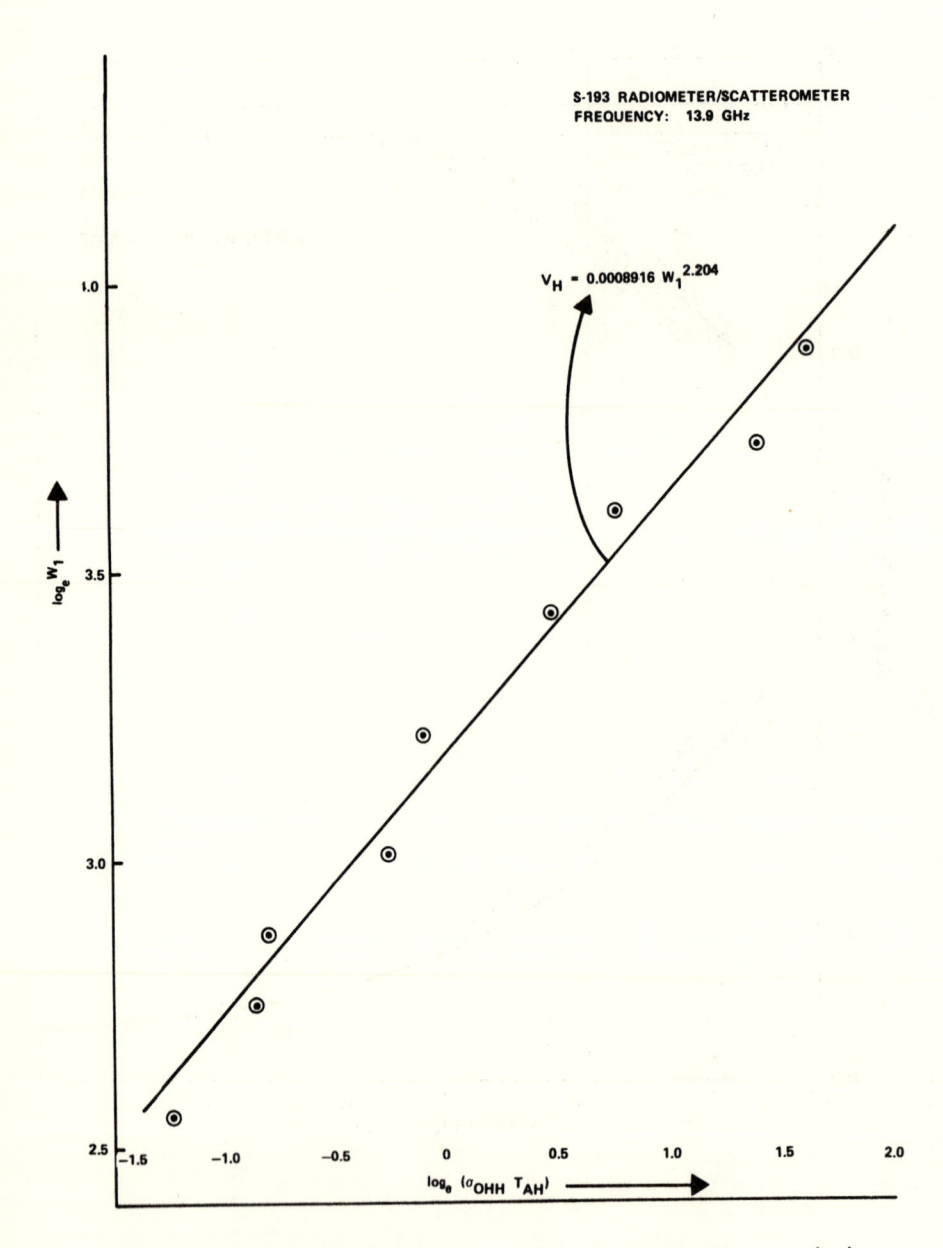

Fig. 17. — The dependence of $\sigma_{OHH} T_{AH}$ on wind velocity (W_1).

in terms of resolution of several targets within the antenna field of view. By 1960 several radar systems that could and did measure radar returns from the Moon, nearer planets, and the Sun were in operation [55, 56, 57]. The concepts of pulse-width-limited radar and extended target detection made it possible for the radar to describe the topographic characteristics of planetary surfaces.

The geometry of the lunar observations can be directly applied to radar satellite observations of the Earth at nadir, except that the relative amplitude of the curvature of the radar wavefront and the planetary surface is reversed [58].

In 1969 a 1-ns radar system designed and built by NRL was mounted on the Chesapeake Light Tower about 60 feet above the water level [59]. The system operated in beam-limited condition, producing a footprint of 60-cm diameter. The sea surface was sampled 10 times per second with 1-ns pulse, and the radar return as a function of delay was recorded on magnetic tape. The heights of the waves were sampled simultaneously with three wave poles surrounding the illuminated area. Figure 18 [59] gives a comparison of the radar and wavestaff spectra. These comparisons showed a close correspondence between radar range variations and wavestaff height amplitudes. These measurements also showed the standard deviation of the radar-derived height distribution (fig. 19) to be about 5 percent less than that computed from the wavestaff data.

Airborne altimeter data was collected by NASA/Wallops Flight Center. The NRL X-band altimeter capable of transmitting different pulse lengths was used. These aircraft observations show that excellent agreement exists between predicted and measured characteristics of the returned pulse. Furthermore, a 4-cm noise with a 3-Hz output rate is possible. The 70-cm pulse-limited footprint permitted spatial resolution of longer ocean wavelengths. Figure 20 shows an average of 30 returns of individually transmitted pulses [59]. Recent data from NASA/Wallops show that accuracy in measuring wave heights up to 15 cm or 10 percent of the wave height (whichever is greater) can be achieved using the airborne altimeter [60].

The information obtained from the S-193 altimeter experiment has considerably exceeded all expectations and has provided unique geodetic data. One of several examples of geodetic profiles obtained is presented in figure 21 [61].

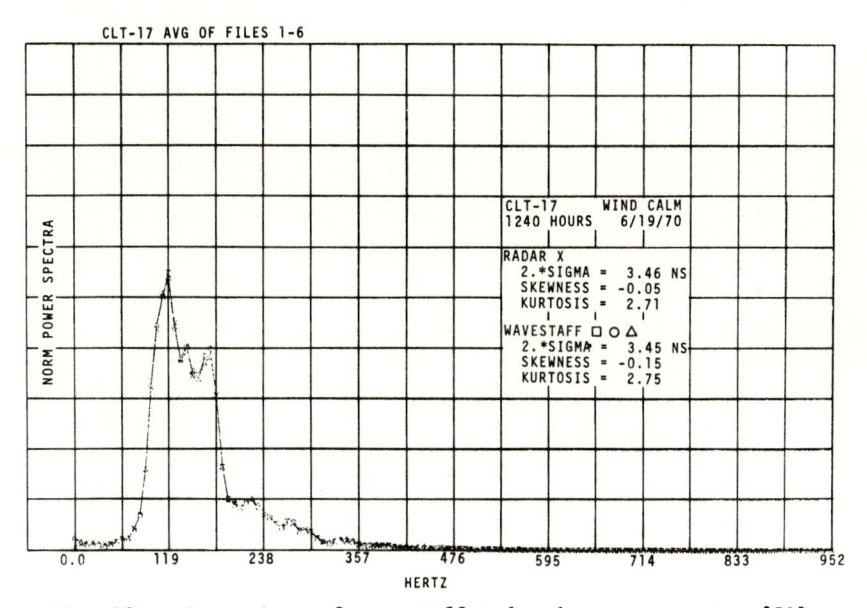

Fig. 18. — Comparison of wavestaff and radar wave spectra [59].

This example shows a comparison of the altimeter-geoidal data and the Marsh-Vincent data developed from a reference grid with dimensions of 1° latitude by 1° longitude. In the S-193 altimeter, the range in meters is the difference between the altimeter-measured geoid and the reference ellipsoid as determined by the orbital data. These differences are called "altimeter residuals." Both absolute and relative comparisons are shown. Because of a 5-meter root-mean-square uncertainty in the C-band radar Skylab tracking data, the absolute displacements between S-193 and Marsh-Vincent data may not be

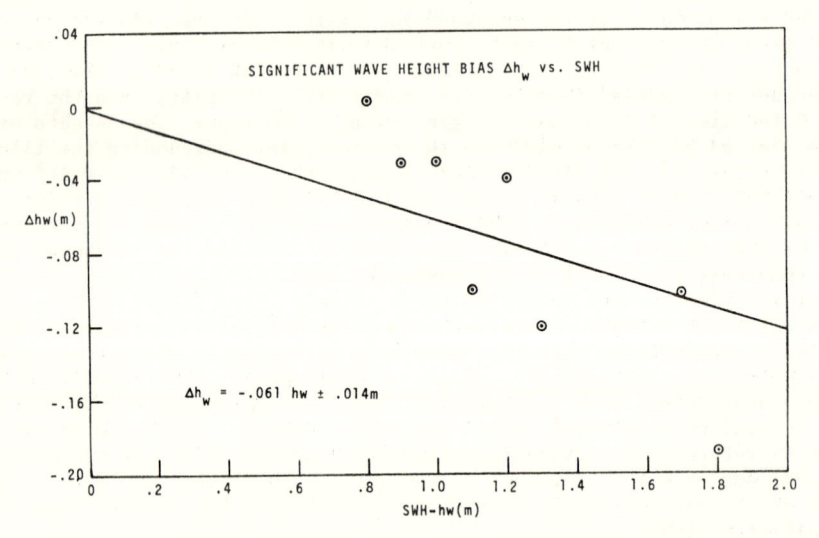

Fig. 19. — Significant wave height bias versus significant wave height (SWH) [59].

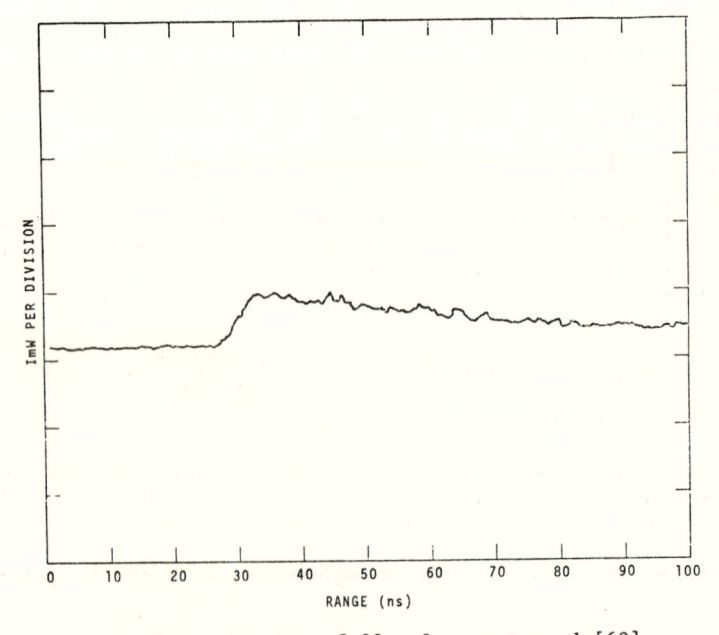

Fig. 20. — Average of 30 pulses returned [60].

meaningful, since these are largely indications of long-term errors in the orbital solutions. Data from EREP pass 9 over the vicinity of Wallops Island, Virginia, are given in figure 21. The orbital data was obtained by combining the unified S-band tracking system and the NASA Wallops Space Center FPQ-6 radar. Figure 22 shows the location of the pass 9 ground track.

Skylab EREP passes 4 and 6 were flown over a known geodetic anomaly — the Puerto Rican Trench (fig. 23) [61]. These passes have essentially the same ground track. The trench gravitational anomaly is manifested in a mean sea level depression of about 10 meters in the first 60 seconds of the elapsed time (fig. 24). The large increase in residuals given in figure 24 [61] at nearly 80 seconds of the elapsed time is due to island landscatter [61]. The S-193 data shows adequate overall agreement with the Puerto Rican Trench region data given by Vincent and even closer relative agreement with data from Von Arx [61]. The contours shown in figure 23 are derived from the Von Arx data. Geodetic surface variations with an accuracy of 5 to 20 meters on a global scale have been measured using S-193 altimeter data.

Figure 25 [61] shows the altimeter residuals beginning in the vicinity of Charleston, South Carolina, for the early portion of EREP pass 4. The ground track for pass 4 is shown in figure 22. Figure 25 also records the ocean bottom profile. It is intriguing to note the correlation between the altimeter data and the bottom features. It should be noted that at the "change in pointing angle" a programmed altimeter submode change altered the antenna pointing angle by about 0.5°. The altimeter-recorded discontinuity at this point is the result of an attitude and not an altitude change. Other examples of this correlation for S-193 altimeter data have been noted (fig. 26).

The S-193 nadir align mode gave pointing accuracy to within approximately ±0.5°. A better estimate for determining antenna pointing angle was developed using the S-193-obtained pulse shapes. These pulse shapes were compared with those calculated using altimeter antenna beam width, 100-ns pulse width, averaging time, and altitude. Using this procedure, the antenna pointing angle can be resolved to within approximately ±0.05°. It can be resolved in the range of 0.25° to 0.8° of off-nadir angles. Figure 27 [61] shows the results of this comparison. The 0.05° accuracy figure has been determined empirically to be the accuracy obtainable.

6. IMAGING RADARS

The effectiveness of imaging radars in the remote sensing of ocean surfaces has recently been recognized [62]. Ocean swells have been imaged at radar wavelengths from a few centimeters up to 2 meters. Satelliteborne imaging radars would afford global monitoring of sea states and ocean waves. The monitoring of wave patterns near coastal areas is of immediate interest for the protection of life and property. The most exciting prospect of radar is that of mapping wave patterns and wave buildup during large storms. The X-band images of waves and surf in a coastal area are shown in figure 28. The potential value of radar in detecting and monitoring oil spills has been demonstrated in experiments conducted by the NRL using imaging radars at 0.428-, 1.228-, 4.425-, and 8.91-GHz frequencies [63]. The presence of currents beneath the surface has been detected by imaging oil spill patterns on the surface of the ocean. Since these internal waves produce observable effects on small gravity and capillary waves on the ocean-surface spectrum, it is feasible that they can be monitored using imaging radars [64].

A potential application of the imaging radars is that of surveying polar ice regions. Data gathered by Goodyear using a side-looking synthetic-aperture radar [65] show that ice conditions can be determined from radar imagery (fig. 29). Radar can separate winter ice from polar ice, locate possible open channels of water, plot movement of icebergs and ice islands, and determine limits of pack ice [65].

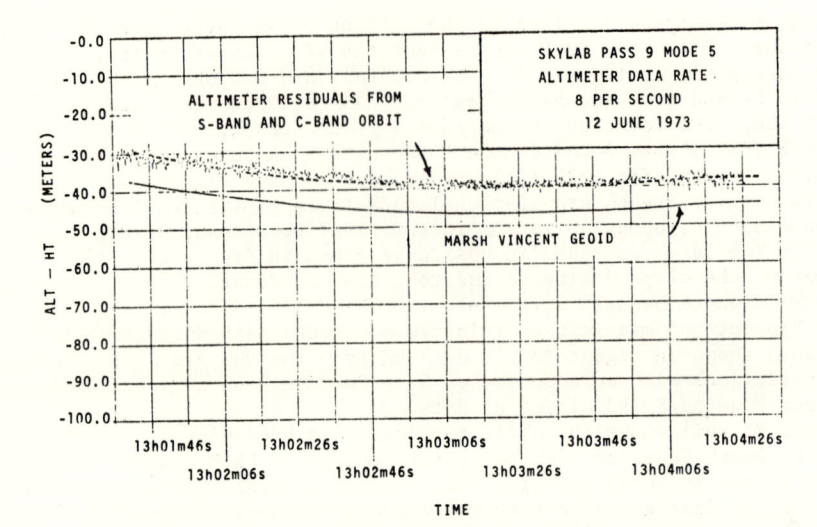

Fig. 21. — Comparison of altimeter altitude residuals (from S-band and C-band determined orbit) with the Marsh-Vincent geoid in the vicinity of Wallops Island, Virginia [61].

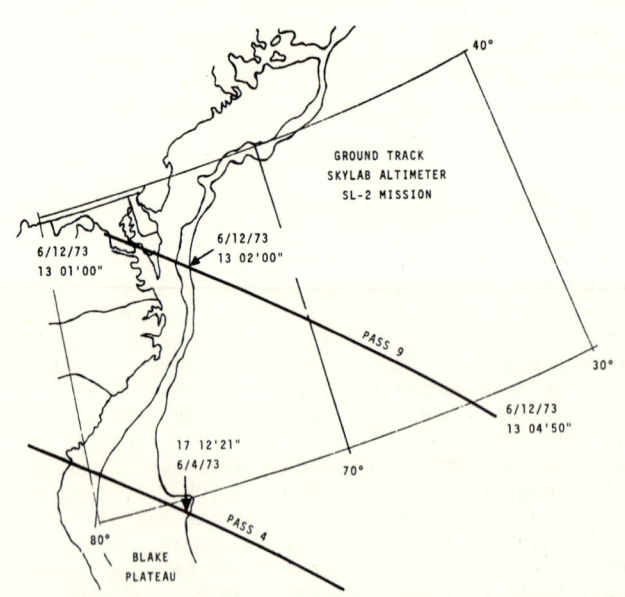

Fig. 22. — Ground tracks for passes 4 and 9 of Skylab Mission SL-2 [61].

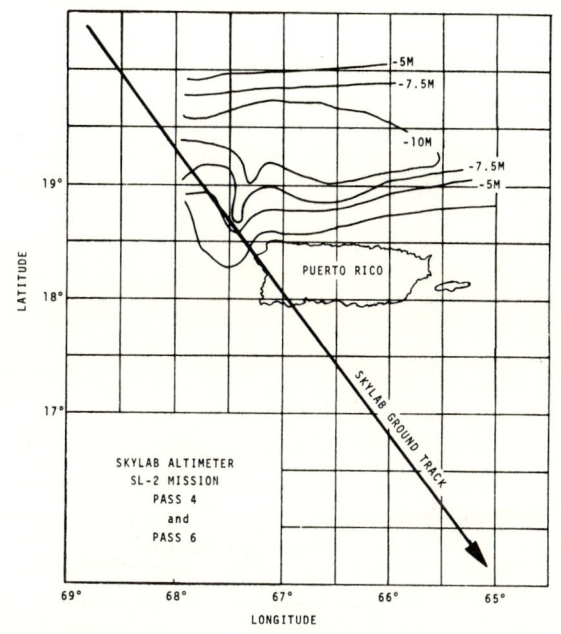

Fig. 23. — Ground track for the Puerto Rican Trench area data shown in figures 11 through 14 [61].

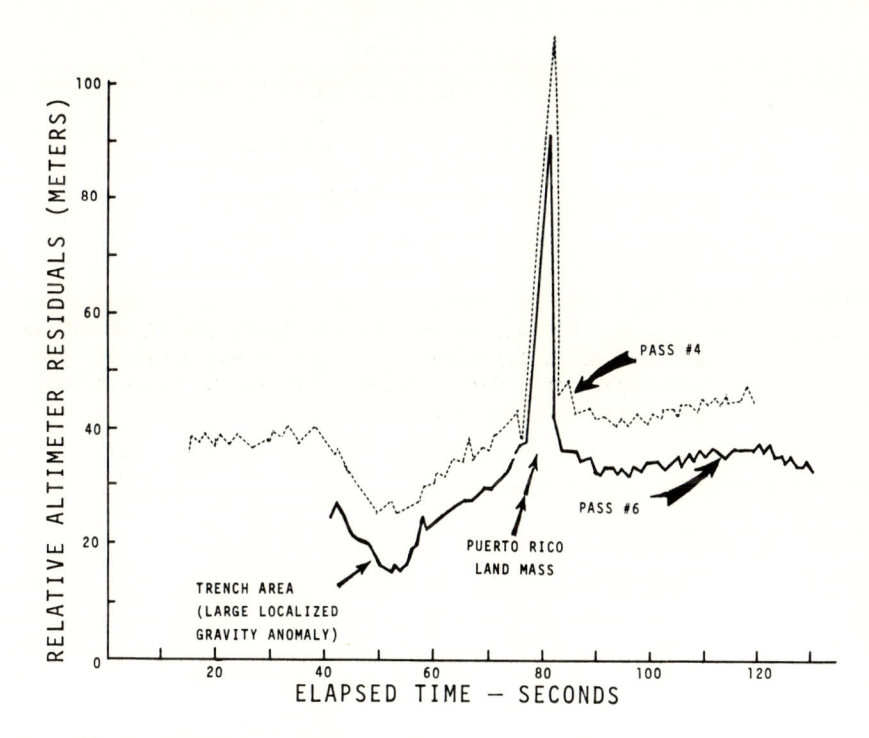

Fig. 24. — Comparison of Puerto Rican Trench area data from passes 4 and 6, Skylab Mission SL-2, on corrected time scale [61].

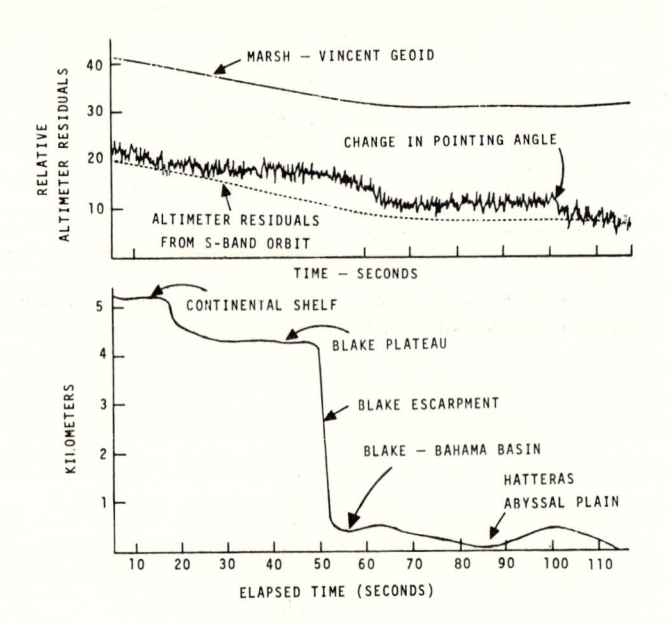

Fig. 25. — Comparison of altimeter altitude residuals (from S-band determined orbit) with the Marsh-Vincent geoid and the corresponding undersea topography (in the vicinity of the Blake Escarpment)[61].

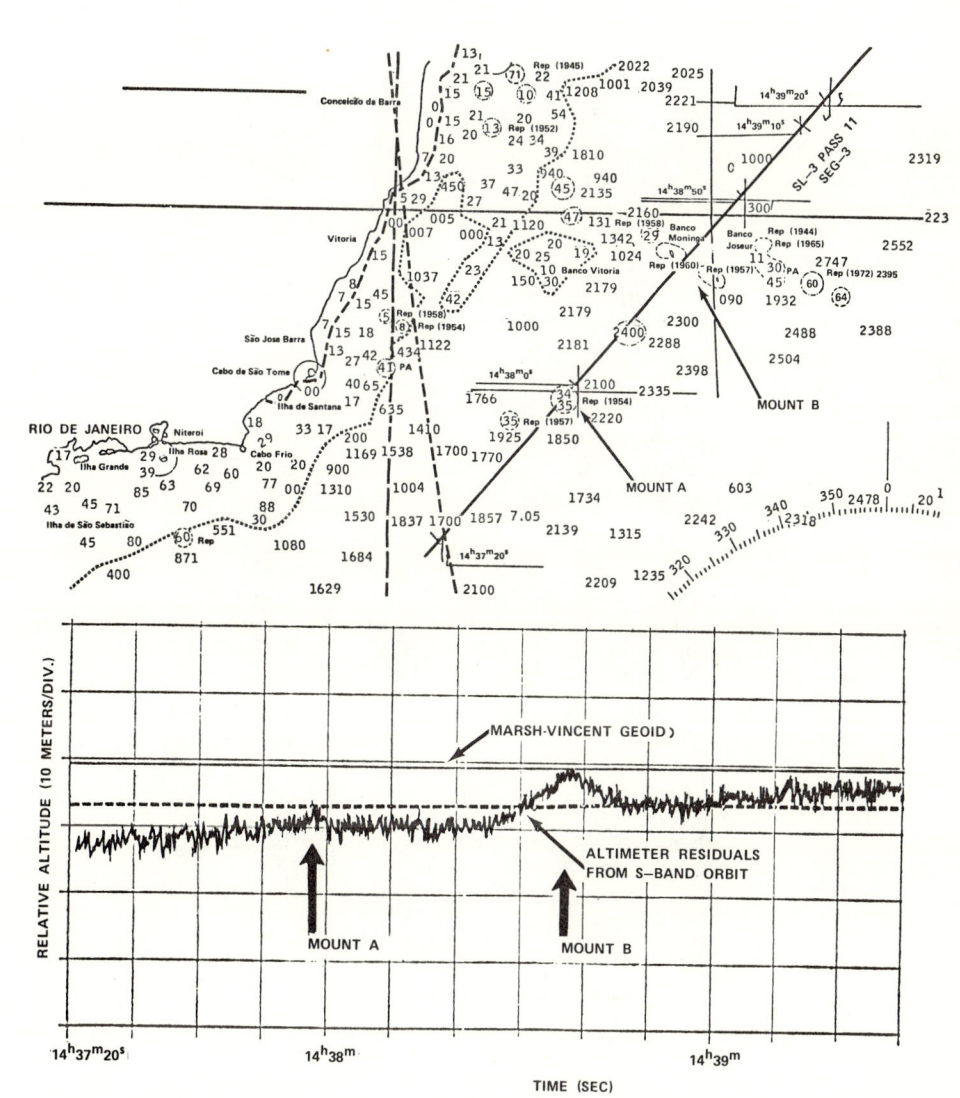

Fig. 26. — Effects of subsurface features on altimeter data; Skylab Mission SL-3 pass 11. Upper portion shows geographic area and bottom topography. Lower portion shows altimeter relative profile. (Data supplied by NASA, Wallops Flight Center.)

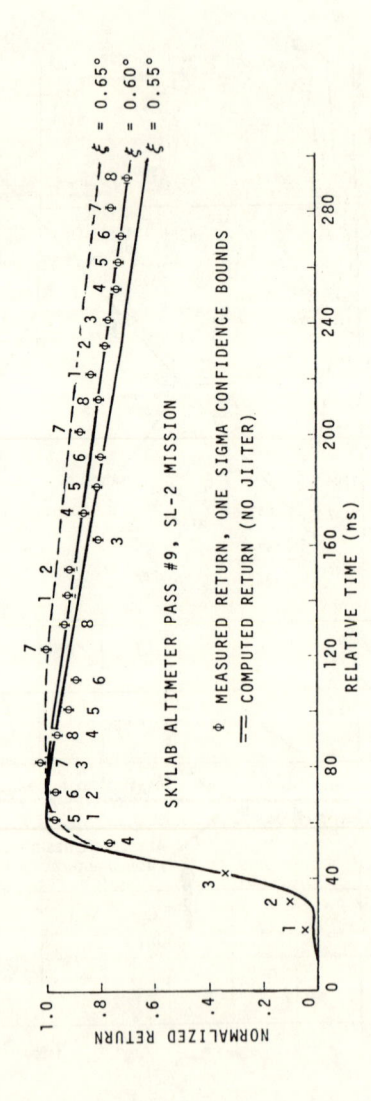

Fig. 27. — Comparison of measured and theoretical (neglecting tracking loop jitter) mean return waveforms for a 10-ns transmitted pulse width (pass 9, mode V, DAS-3) [61].

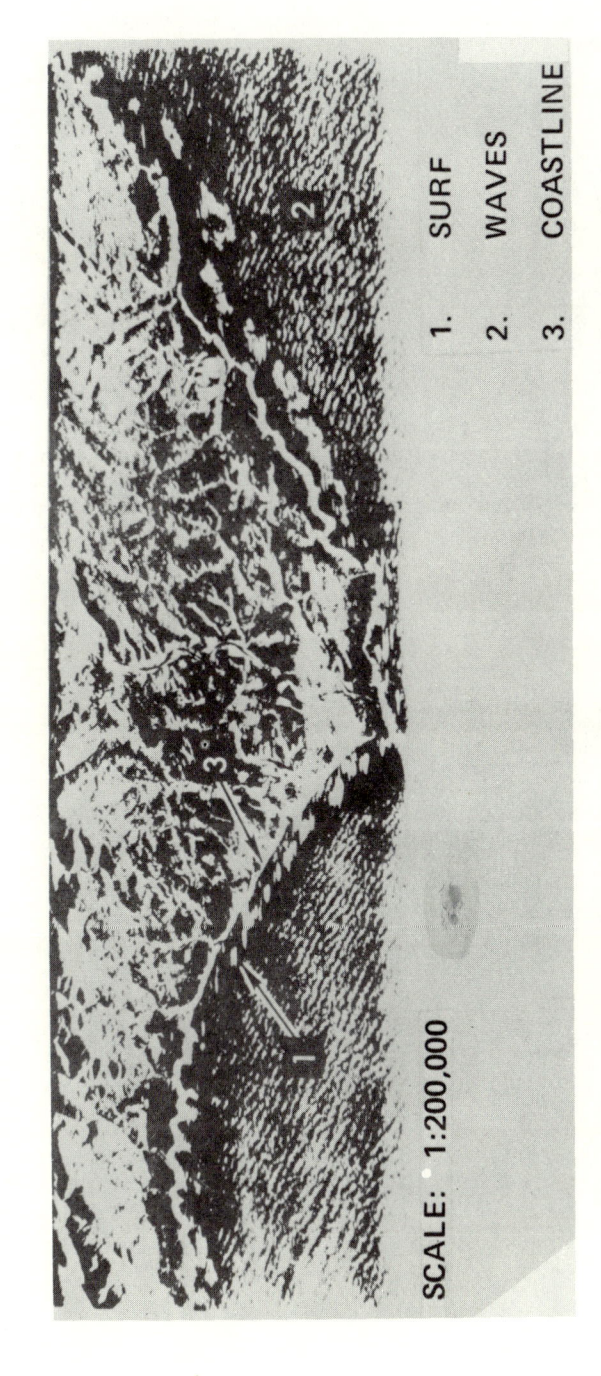

Fig. 28. — X-band radar image of waves and surf (resolution 10 to 15 meters) [64].

Scale 1:200,000   3 June 1971

Scale 1:200,000
14 December 1970

Fig. 29. — Images of the same area taken at different seasons of the year. The December image shows ice covering the water areas [65].

7. FUTURE RESEARCH AND DEVELOPMENT

The decade ahead calls for systems which can be run reliably for extended periods of time. It also calls for computational techniques which will automatically yield radiometric brightness temperature and backscattering cross section data. To accomplish this goal, the following recommendations for research and development are given.

<u>Improved radar calibration techniques</u>. A calibration period should follow each data collection period. The frequency of these calibrations will depend on the performance of the system. Some technique of external calibration should be adopted. Corner reflectors over smooth deterministic targets could be used for airborne and spaceborne scatterometers. Internal calibrations are undesirable because they do not involve all the paths through which the actual returned energy passes.

<u>Calibration data measurements</u>. Measurements of calibration data for spaceborne and airborne instruments corresponding to each subsystem are usually done in the laboratory. The amplifier, mixer, and filter gains are some examples. Internal gains change with time for instruments flown for extended periods. To alleviate this problem, automatic calibration modes should be designed to check all calibration data needed to calculate backscattering cross sections and radiometric brightness temperatures from raw data.

<u>Improved systems design</u>. Improvements are needed in the reliability of antenna pointing, switching logic, and polarization isolation. System drifts, resolution, and noise-level fluctuations must be reduced in future airborne and spaceborne systems.

<u>Data processing techniques</u>. For operational systems, highly efficient data processing techniques are needed. No reliable data display techniques have been advanced to date. Onboard processing of data should be investigated, and digitally processed data should be color coded to display ocean winds and waves for visual interpretation and use. Data processing programs should be capable of handling large volumes of emissivity, reflectivity data and displaying this data on a world map.

8. CONCLUSION

The last decade has been accelerated exploration of the applicability of microwave sensors to the remote monitoring of the ocean environment and the production of improved weather forecast models. Both theoretical and experimental studies offer conclusive evidence that a strong correlation exists between ocean surface phenomena and microwave measurements. The potential worth of earth observations using microwaves has been recognized by NASA, NOAA, NRL, and others. The SEASAT-A program is an example of this recognition. This satellite will operate a scatterometer, an altimeter, and a radiometer, as well as imaging radar. Concentrated studies should still continue to resolve unanswered questions as to the detection of wind direction and data display techniques. The suitability of microwave sensors in providing data independent of the sunlight under almost all weather conditions is vitally important to the efficient monitoring of the Earth's oceans.

9. REFERENCES

1. Kerr, D. E. (Ed.). Propagation of Short Radio Waves. MIT Radiation Laboratory Series, Vol. 13, McGraw-Hill, New York, 1951.

2. Swift, C. T.; and Jones, W. L., Jr. Satellite Radar Scatterometry. Proceedings of IEEE International Conference, March 1974.

3. Schooley, A. H. Upwind-Downwind Ratio of Radar Return Calculated from Facet Size Statistics of a Wind-Disturbed Water Surface. Proceedings of IRE, Vol. 50, April 1962.

4. MacDonald, F. C. The Correlation of Radar Sea Clutter on Vertical and Horizontal Polarization with Wave Height and Slope. IRE National Convention Record, Part I, pp. 29-32, 1956.

5. Grant, C. R.; and Yaplee, B. S. Backscattering from Water and Land at Centimeter and Millimeter Wavelengths. Proceedings of IRE, Vol. 45, pp. 972-982, July 1957.

6. Daley, J. C.; et al. Sea Clutter Measurement on Four Frequencies. NRL Report 6806, November 1968.

7. Daley, J. C.; et al. Upwind-Downwind-Crosswind Sea Clutter Measurement. NRL Report 6881, April 1969.

8. Daley, J. C.; et al. Radar Sea Return — JOSS I. NRL Report 7268, May 1971.

9. Daley, J. C.; et al. Radar Sea Return — JOSS II. NRL Report 7534, February 1973.

10. Summary of Ryan Reflectivity Program. Ryan Aeronautical Company Report 29072-11, March 1965.

11. Moore, R. K. Radar Scatterometry — An Active Remote Sensing Tool. Proceedings of 4th Symposium on Remote Sensing of Environment, pp. 339-373, April 1966.

12. Krishen, K. Correlation of Radar Backscattering Cross Sections with Ocean Wave Height and Wind Velocity. J. Geophys. Res., Vol. 76, pp. 6528-6539, September 1971.

13. Moore, R. K.; and Pierson, W. J., Jr. Worldwide Oceanic Wind and Wave Predictions Using a Satellite Radar-Radiometer. J. Hydronautics, Vol. 5, pp. 52-60, April 1971.

14. Krishen, K. Mathematical Model for the Relationship of Radar Backscattering Cross Sections with Ocean Scene and Wind Velocity. Proceedings of 7th Symposium on Remote Sensing of Environment, Vol. III, pp. 1871-1877, May 1971.

15. Hooke, R.; and Jeeves, T. A. Direct Search Solution of Numerical and Statistical Problems. J. Assoc. Computing Machinery, Vol. 8, 1961.

16. Guinard, N. W.; and Daley, J. An Experimental Study of a Sea Clutter Model. Proceedings of IEEE, 58, 543, 1970.

17. Daley, J. An Empirical Sea Clutter Model. NRL Report 2668, October 1973.

18. Moskowitz, L. Estimates of Power Spectrums for Fully Developed Seas for Wind Speeds of 20 to 40 Knots. J. Geophy. Res., Vol. 69, No. 24, 1964.

19. Krishen, K. Quick-Look Evaluation of Hurricane Ava S-193 Scatterometer Data. Lockheed Electronics Company, Inc., Technical Memorandum LEC 642-788, August 1973.

20. Moore, R. K.; et al. Simultaneous Active and Passive Microwave Response of the Earth — The Skylab RADSCAT Experiment. Proceedings of 9th International Symposium on Remote Sensing of Environment, Ann Arbor, Michigan, April 1974.

21. Ross, D.; et al. A Remote Sensing Study of Pacific Hurricane Ava. Proceedings of 9th International Symposium on Remote Sensing of Environment, Ann Arbor, Michigan, April 1974.

22. Krishen, K. Contribution to Ocean Panel Including Reports on Some Sea Return Experiments, Ocean Surface Windspeed Sensing, and Scatterometers. Lockheed Electronics Company, Inc., Technical Report LEC-3896, July 1974.

23. Hayes, J.; et al. A Preliminary Analysis of the Surface Truth Data to be Correlated with the Skylab II Data Obtained for the S-193 Microwave Investigators. City University of New York Informal Report, August 1973.

24. Krishen, K. Detection of Oil Spills Using 13.3-GHz Radar Scatterometer. J. Geophys. Res., Vol. 78, No. 12, April 1973.

25. Rouse, J. W., Jr. Arctic Ice Type Identification by Radar. Proceedings of IEEE, Vol. 57, No. 4, April 1969.

26. Parashar, S. K.; Fung, A. K.; and Moore, R. K. A Theory of Wave Scatter From an Inhomogeneous Medium With a Slightly Rough Boundary and its Application to Sea Ice. Presented at the 1974 URSI-Specialist Meeting on Microwave Scattering and Emission From the Earth, Berne, Switzerland, September 1974.

27. Jansky, K. A Note on the Source of Interstellar Interference. Proceedings of IRE, Vol. 23, pp. 1158-1163, 1935.

28 Dicke, R. The Measurement of Thermal Radiation at Microwave Frequencies. Rev. Sci. Instruments, 17, 268-275, 1946.

29. Decker, M.; and Dutton, E. Radiometric Observations of Liquid Water in Thunderstorm Cells. J. Atmos. Sci., Vol. 27, 1970.

30. Hyatt, H. A. Emission, Reflection, and Absorption of Microwaves at a Smooth Air-Water Interface. J. Quant. Spectrosc. Radiation Transfer, Vol. 10, Pergamon Press, 1970.

31. Paris, J. F. Prediction of the Response of Earth-Pointed Sensors (PREPS): Passive Microwave Sensors. Lockheed Electronics Company, Inc., LEC/HASD 640-TR-105, 1972.

32. Stogryn, A. The Apparent Temperature of the Sea at Microwave Frequencies. IEEE Trans. Antennas Propagat., Vol. AP-15, pp. 278-286, March 1967.

33. Ulaby, F. T.; and Fung, A. K. Effects of Roughness on Emissivity of Natural Surfaces in the Microwave Region. Proceedings of SW IEEE Conf., pp. 436-440, April 1970.

34. Wu, S. T.; and Fung, A. K. A Noncoherent Model for Microwave Emissions and Backscattering From the Sea Surface. J. Geophys. Res., Vol. 77, pp. 5917-5929, October 20, 1972.

35. Lynch, P. J.; and Wagner, R. J. Rough-Surface Scattering: Shadowing, Multiple Scatter, and Energy Conservation. J. Math. Phys., Vol. 11, pp. 3032-3042, October 1970.

36. Droppleman, J.; Mennella, R.; and Evans, D. An Airborne Measurement of the Salinity Variation of the Mississippi River Outflow. J. Geophys. Res., 75, 5909-5913, 1970.

37. Paris, J. F. Salinity Surveys Using an Airborne Microwave Radiometer. Proceedings of 8th International Symposium on Remote Sensing of Environment, 665-676, Ann Arbor, Michigan, 1972.

38. Stogryn, A. Equations for Calculating the Dielectric Constant of Saline Water. IEEE Trans. Microwave Theory Tech., Vol. MTT-19, August 1971.

39. Thomann, G. C. Remote Measurement of Salinity in an Estuarine Environment. Remote Sensing of Environment, Vol. 2. American Elsevier Publishing Company, Inc., 1973.

40. Lepley, L.; and Adams, W. Direct Determination of the Electromagnetic Reflection Properties of Smooth Brackish Water to the Continuous Spectrum From 10^8 to 4×10^9 Hz. Water Resources Research Center, University of Hawaii, TR No. 48, Honolulu, Hawaii, 1971.

41. Gurvich, A.; and Egorov, S. Determination of the Surface Temperature of the Sea From Its Thermal Radioemission. Izv. Atm. and Oceanic Phys., 2, 305-307, 1966.

42. Hidy, G.; et al. Development of a Satellite Microwave Radiometer to Sense the Surface Temperature of the World's Oceans. S-band Radiometer Final Report, Contract NAS 1-10106, Space Division, North American Rockwell, 1971.

43. Hindin, H. J. (Ed.). A Slick Analysis of Oil Slicks. News. Microwaves, October 1973.

44. Kennedy, J. M.; and Wermund, E. G. Oil Spills, IR and Microwaves. Photogrammetric Engineering, 1971.

45. Hollinger, J. P. Passive Microwave Measurements of the Sea Surface. J. Geophys. Res., Vol. 75, pp. 5209-5213, September 20, 1970.

46. Ross, D. B.; Cardone, V. J.; and Conaway, J. W., Jr. Laser and Microwave Observations of Sea Surface Conditions for Fetch-Limited 17- to 25-m/s Winds. IEEE Trans. Geosci. Electron., Vol. GE-8, pp. 326-336, October 1970.

47. Nordberg, W.; Conaway, J.; Ross, D. B.; and Wilheit, T. Measurements of Microwave Emission From a Foam-Covered, Wind-Driven Sea. J. Atmos. Sci., Vol. 28, p. 429, April 1971.

48. Swift, C. T. Microwave Radiometer Measurements of the Cape Cod Canal. Radio Science, 1974 (in press).

49. Williams, G. F., Jr. Microwave Radiometry of the Ocean and the Possibility of Marine Wind Velocity Determination From Satellite Observations. J. Geophys. Res., Vol. 74, pp. 4591-4594, August 20, 1969.

50. Hollinger, J. T. Passive Microwave Sensing of Marine Wind Speed. Paper communicated to NASA Active Microwave Workshop, NRL, Washington, D.C., July 1974.

51. Droppleman, J. D. Apparent Microwave Emissivity of Sea Foam. J. Geophys. Res., Vol. 75, pp. 696-698, January 20, 1970.

52. Stogryn, A. The Emissivity of Sea Foam at Microwave Frequencies. J. Geophys. Res., Vol. 77, pp. 1658-1666, March 20, 1972.

53. Gloersen, P.; et al. Microwave Maps of the Polar Ice of the Earth. Report X-652-73-269; NASA/Goddard Space Flight Center, August 1973.

54. Gloersen, P.; et al. Polar Sea Ice Observations by Means of Microwave Radiometry. Report X-642-73-341, NASA/Goddard Space Flight Center, November 1973.

55. Kerr, F. J.; and Shain, C. A. Moon Echoes and Transmission Through the Ionosphere. Proceedings of IRE, Vol. 39, pp. 230-242, March 1951.

56. Lincoln Laboratory, Haystack Facility.

57. Jet Propulsion Laboratory, Goldstone Facility.

58. Yaplee, B. S.; and Shapiro, A. Remote Sensing of Sea State Using Narrow Pulse Radar. NRL, communicated to NASA Active Microwave Workshop, July 1974.

59. Shapiro, A.; Yaplee, B. S.; and Uliana, E. A. Simulated Ocean Radar Impulse Responses From Lunar Radar Measurements. NRL Report 7050, May 7, 1970.

60. Yaplee, B. S.; Shapiro, A.; Hammond, D. L.; Au, B. D.; and Uliana, E. A. Nanosecond Radar Observations of the Ocean Surface From a Stable Platform. IEEE Trans. Geosci. Electron., Vol. GE-9, pp. 170-174, July 1971.

61. McCoogan, J.; Miller, L.; Brown, G.; and Hayne, G. The S-193 Radar Altimeter Experiment. Proceedings of the IEEE, Special Issue on Radar Technology and Applications, June 1974.

62. Apel, John R. A Hard Look at Oceans From Space. AIAA 9th Annual Meeting and Technology Display, Washington, D.C., January 1973.

63. Guinard, N. W. The Remote Sensing of Oil Slicks. Proceedings of 7th International Symposium on Remote Sensing of Environment, Vol. 2, 1971.

64. Proceedings of 1974 Active Microwave Workshop, NASA/JSC, to be published.

65. Goodyear Aerospace Corporation. Developing Earth Resources With Synthetic Aperture Radar. Report GIB-92900, Arizona, August 1973.

ACKNOWLEDGEMENT

The author wishes to thank all the Ocean Panel participants of the NASA Active Microwave Workshop held in July, 1974, for their helpful suggestions. Particular thanks are due to Mr. Godbey (General Electric) and Dr. Apel (NOAA) for their comments and reviews which were directly incorporated in this paper.

PART II

ACTIVE SENSOR APPLICATIONS

RADAR APPLICATIONS TO WEATHER ANALYSIS AND FORECASTING AND TO PRECIPITATION CLIMATOLOGY

EDWIN KESSLER

National Severe Storms Laboratory, NOAA, Norman, Oklahoma, U.S.A.

ABSTRACT

Weather radar detects radiation back scattered from hydrometeors and thereby provides means for depiction of spatial and temporal distributions of rain and snow. Radar data, through suitable processing, yields precipitation rates and amounts, and also speed and direction of motion of precipitation cells. The intensity and shape of echoing areas often indicates strong winds and hail, and intense electrical activity. In fact, radar is a good tool for assessing precipitation climatology where weather signatures are not obscured by prominent terrain features or by atmospheric conditions giving rise to anomalous propagation. Weather radar data can guide site selection and reservoir management, and provide lead time for measures to lessen storm damage.

INTRODUCTION

Meteorological applications of radar began to be developed thirty years ago, shortly after it was found that weather sometimes interfered with military objectives of radar during the Battle of Britain. Ryde [1] described basic weather radar theory and Maynard [2] illustrated some important meteorological targets such as hurricanes and squall lines. A recent general text on weather radar has been given by Battan [3]. Increased use of weather radar has followed meteorological studies and improvements in radar reliability and ease of maintenance, transition from 3-cm wavelength to less-attenuating 5- and 10-cm wavelengths, the development of clear data displays and means of rapid economical communication of quantitative data to remote points.

The great value of radar stems principally from its accurate definition of precipitation boundaries and relative intensities. These important variables are not well represented by our widely spaced conventional observing stations, nor by satellite based sensors whose vision is most effective for defining distributions of cloudiness and certain parameters measurable in the clear atmosphere.

DATA DISPLAY

The intensity of radar echoes from precipitation swings widely from pulse to pulse, the envelope of pulse averages forming a χ^2 distribution whose degrees of freedom are determined by a number of parameters involving radar electronics, the transmitted signal, and the weather itself [4,5].

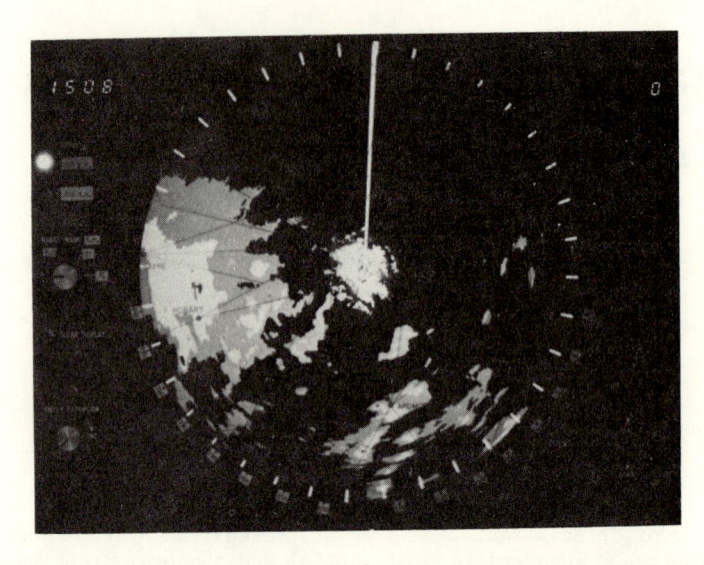

Fig. 1. – Precipitation detected by NSSL's WSR-57 radar as displayed 35 km away, 20 September 1974, 0904 CST. The radius of the display corresponds to a range of 130 km.

Capabilities in the device illustrated in Fig. 1 represent the thoughtful blending of considerations of electronic technology, signal processing, behavior of precipitating weather systems, and requirements for weather data. In Fig. 1 the data are appreciated instantly by the trained human operator, and the data can be simultaneously transmitted for computer processing [6].

DATA PROCESSING

Rain areas and locations seen in Fig. 1 can be entered manually into a desk computer for quick calculation of trends in intensity or velocity. However, most recent study has involved use of the data in large computers with range-azimuth coordinates which are non-conformal when mapped orthogonally (Fig. 2). The computer is programmed readily for calculations to identify echoes, estimate their motion, rainfall equivalence, persistence, and numerous statistical characteristics.

Rapid advances in minicomputer technology now offer capabilities at the operational level which, just a few years ago, were feasible only at substantially equipped research facilities. Nevertheless, implementation of such capabilities is not proceeding immediately because socio-technological implications of various choices need to be carefully considered and compared.

SUMMARY OF WEATHER RADAR RELATIONSHIPS

Weather Forecasting: The term "forecast" refers here to phenomena foreshadowed in advance of their development, as contrasted to description ("warning") of expected behavior of entities already formed. In this context, radar has value as a precipitation forecast monitor. In other words, the forecast "no storms" based on data other than radar data, gains confidence with radar verification, or may be subject to re-examination if radar shows developing precipitation areas. Two contrasting situations have been examined by Vlcek [7].

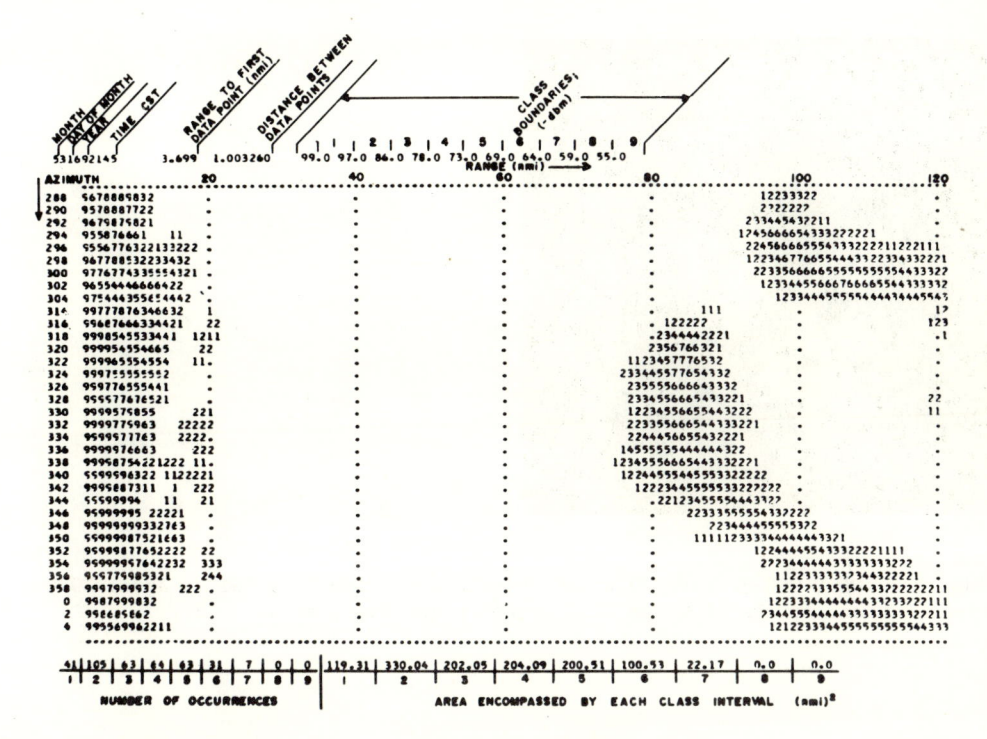

Fig. 2. - Echo intensity data is listed by computer. Azimuth varies vertically, range horizontally. Digits relate to approximate 5 db steps of echo intensity averaged in areas 2° x 1 km.

Storm Warning: Here we refer to extrapolation of trends observed. The motion of large precipitation areas can be objectively estimated by correlation of echo distributions at successive times, as discussed by Kessler and Russo [8] and even more simply by extrapolation of visual observations of the radar plan-position indicator [9]. In the case of intense isolated cells, a characteristic hook shape may identify a tornado-bearing mesocyclone [e.g. 10, p. 143] and an echo isolation algorithm is an effective storm tracker [11, 12] (Fig. 3). Association of radar echo intensity with the frequency of lightning activity is clearly shown by Kinzer [13]. Some associations between radar data parameters and severe weather manifestations including turbulence [14], are shown in Fig. 4.

Flood analysis and Warning: Since mid-1950's there has been intensive study of rainfall rate in relation to radar echo intensity [15, 16]. But the following conditions have impeded introduction of systematic quantitative techniques to the operational service: non-uniqueness of electromagnetic scattering relationships especially pronounced at short wavelengths; growth and horizontal drift of scatterers between the altitude viewed by radar and the ground; non-uniform radar calibration standards; and inadequate data processing equipment. Probably as important as these is large variability in scattering that arises from conditions of site and atmospheric structure related only remotely to rainfall intensity. Such variability has not been well revealed in research cases frequently selected for their uniformity. Effects of anomalous propagation are evident often during severe storm conditions, when intense medium-scale atmospheric variability appears to influence propagation paths strongly. Advances in the technology of data processing and communicating,

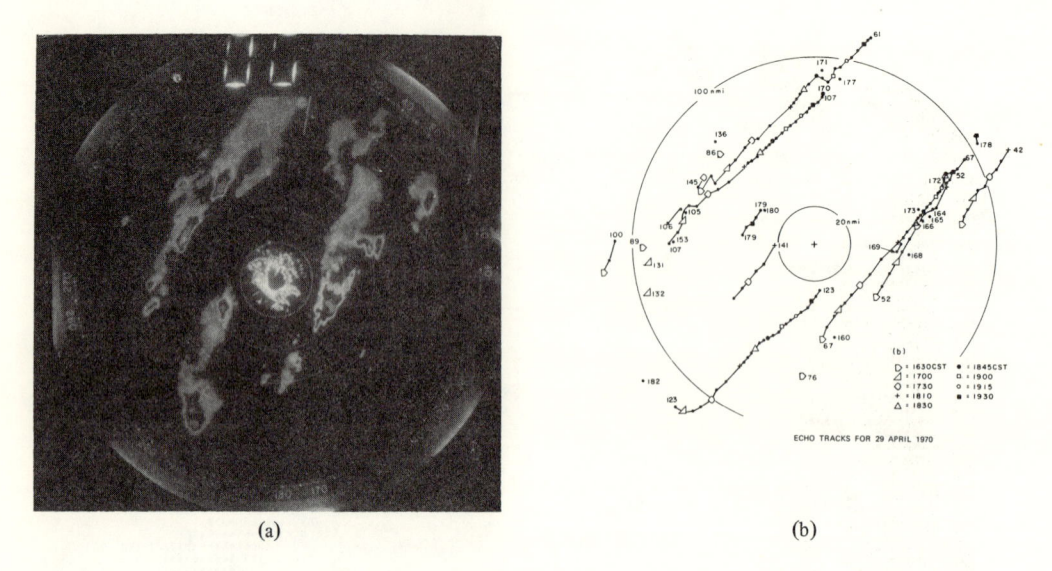

Fig. 3. - NSSL WSR-57 radar display of intense storms 1739 CST 29 April 1970, and tracks of storm centers after Blackmer, et al. [12].

Fig. 4 (*Continued*)

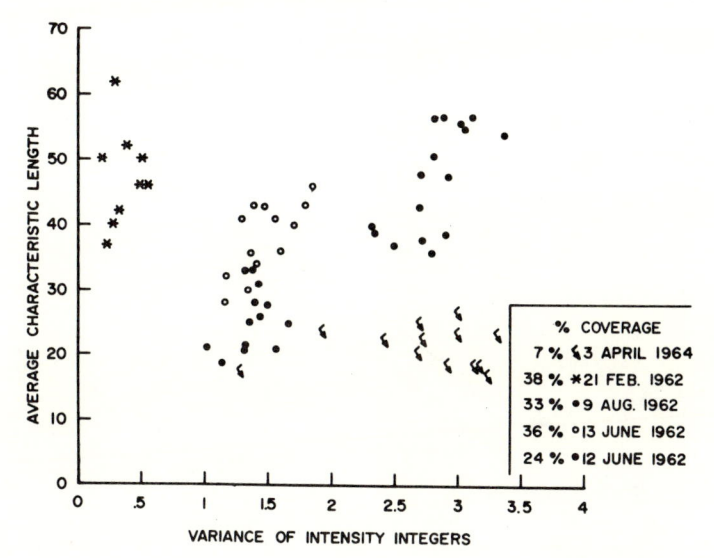

Area A of echo on the first chart, mi²	Number of cases on first chart	Percentage of cases identifiable on chart 3 hr later
$10{,}000 < A$	100	100
$5{,}000 < A \leq 10{,}000$	59	85
$3{,}000 < A \leq 5{,}000$	59	76
$2{,}000 < A \leq 3{,}000$	55	67
$1{,}000 < A \leq 2{,}000$	60	57
$A \leq 1{,}000$	34	24

* All but first row based on analysis of 240 charts, Aug–Dec 1961.

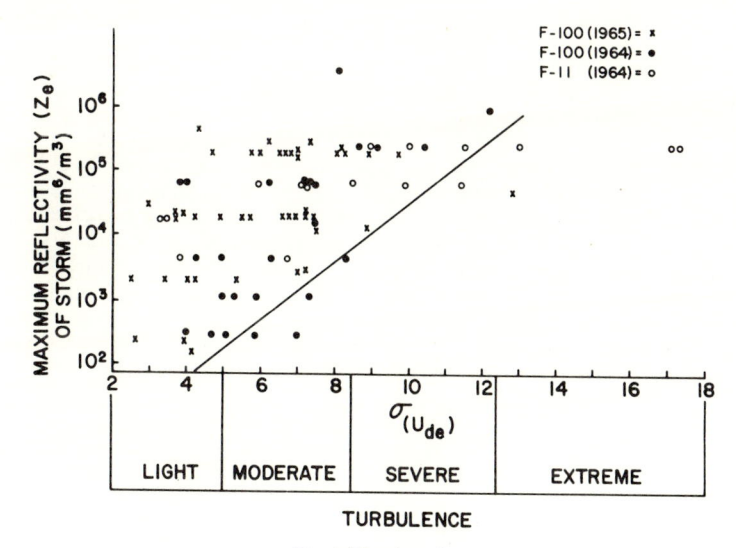

Fig. 4 (*Continued*)

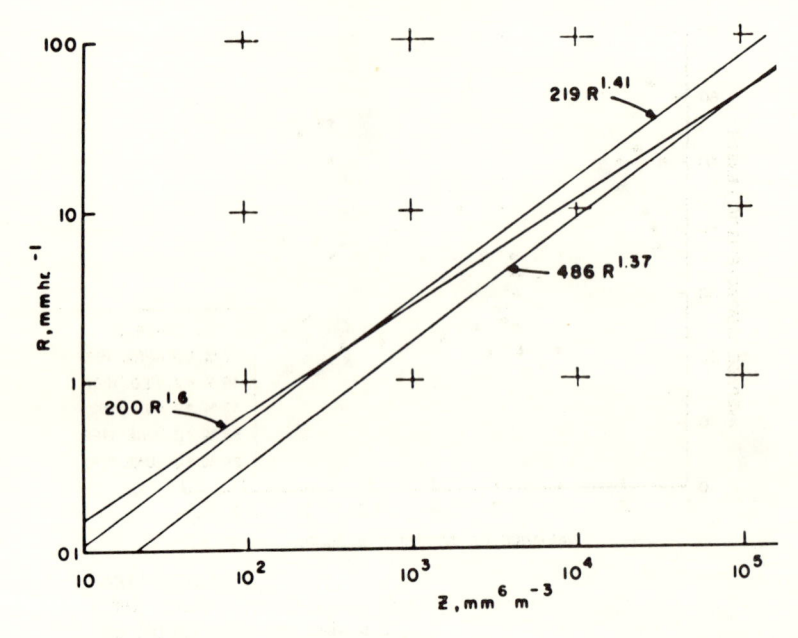

Fig. 4. – Echo characteristics and weather. Top: Thunderstorm probability increases with echo height and intensity [9]; Middle left: Illustrating persistence of echo statistical properties within storms at half hour intervals [15]; Middle right: Area persistence increases with echo size [9]; Bottom left: Envelope of turbulence encounters indicated by deviations of derived gust velocity is defined by maximum reflectivity of nearby storms. (After Lee, [13]); Bottom right: Envelope of radar rainfall relationships [15].

however, now offer practical bases for combining radar and raingage data, to realize accuracy approaching raingage point values and representativeness characteristic of the radar data continuum [17, 18, 19]. Radar-raingage combinations may soon be applied in operational rainfall analysis and flood prediction. In the meantime, with the present system, valuable flood warning services are being rendered from bases of operator experience, improved radar calibration, and manual digitizing of radar data in 40 mile squares [20].

Radar Climatology: Radar can accumulate representative statistics on the temporal and areal distribution of rainfall, more rapidly than raingages as usually spaced and, therefore, radar can be a significant aid both to site selection and also to the design of systems influenced by rainfall. Statistical or climatological studies of radar data for design of communication systems have been summarized by Drufuca [21] and Rogers [22]. A quantitative approach to relating radar data with other weather sensors has been indicated by Kessler, et al., [23]. To ensure reliability in studies of radar precipitation climatology, a few raingages should be used as aids to calibration.

THE FUTURE

Present studies foreshadow the future national weather system. With development of Doppler radar techniques for examining details of air motion in hitherto inaccessible severe storm interiors (Fig. 5), new insights into pro-

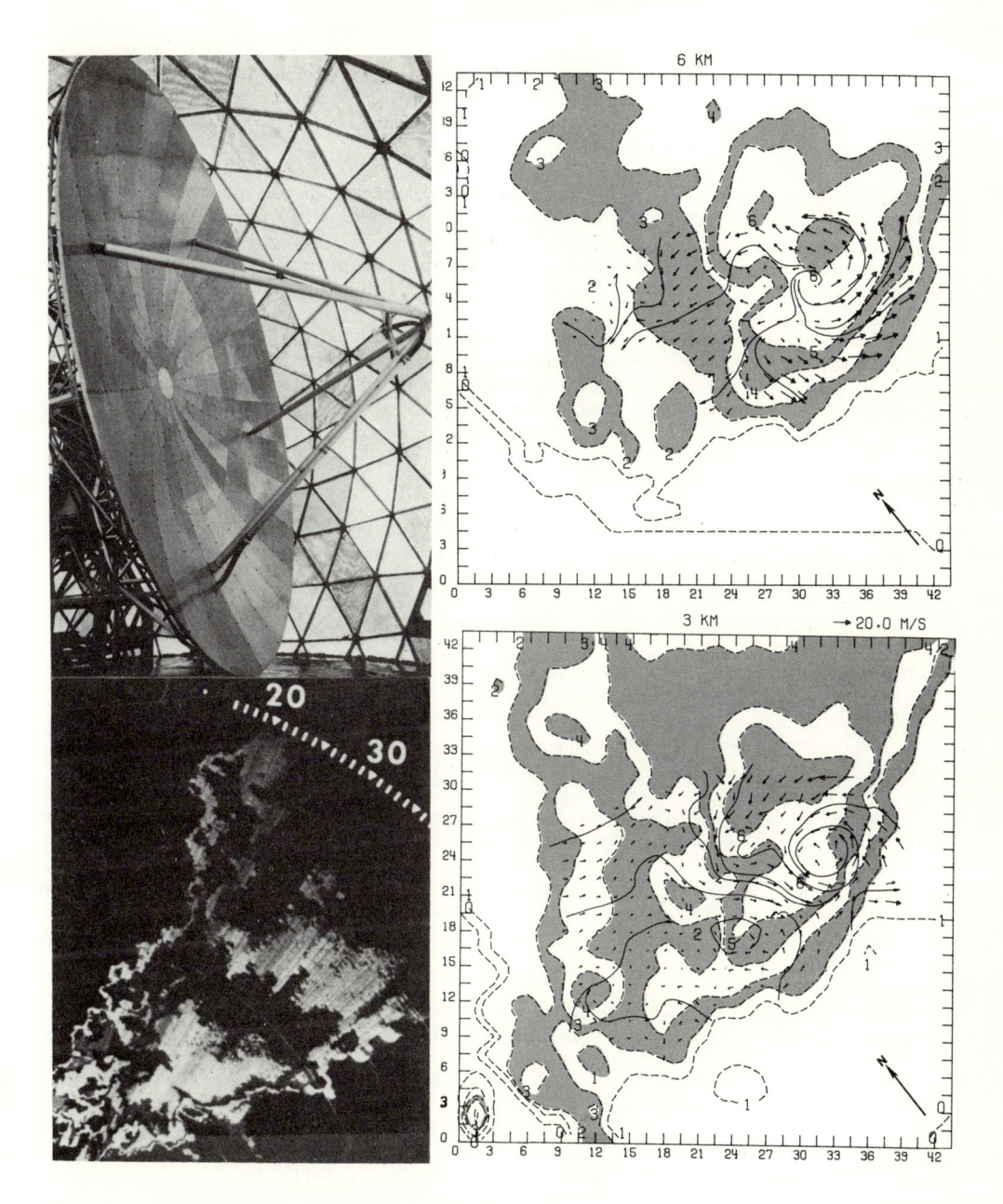

Fig. 5. – Upper left: Nine-meter antenna of 10-cm Doppler radar at NSSL Hqs.;
Lower left: Echo power at altitudes below 1 km in a tornadic storm,
20 April 1974, in plan view. Range intervals are 20 km—Norman, Okla.
is just off the lower left corner; Right: Reflectivity (alternately
shaded areas) and disturbed wind field of the tornadic storm at
altitudes of 3 and 6 km. Scale labels are in km. Data collected by
twin Doppler radars on a 40 km baseline. (Wind analyses by Peter
Ray.)

cesses responsible for storms may contribute means for both improved forecasting and more reliable hail and tornado identification. Important insights to processes of turbulence and diffusion and the scrubbing of air by precipitation may ensue also from extension of Doppler radar capabilities to routine sensing of weakly reflective clear air [24].

While radar for purely local uses may be dedicated to the protection of particularly valuable installations such as large industrial parks and power facilities, the information inherent in radar data will be processed for national use in a system of interconnected sensing stations which may follow some concepts treated by Kessler and Wilson [25] and Taylor and Browning [26].

In closing, it seems appropriate to note that radar is one of many tools and that both conventional sensors and new capabilities based on meteorological satellites are expected to find an important place in developed systems for environmental description and prediction.

REFERENCES

1. Ryde, J. W. 1945. The Attenuation of Centimetre Radio Waves and the Echo Intensities Resulting from Atmospheric Phenomena. J. Inst. Elec. Eng. XCIII, Part IVA:101-103.

2. Maynard, H. R. 1945. Radar and Weather. J. Meteor. 2:214-226.

3. Battan, L. J. 1973. Radar Observation of the Atmosphere. University of Chicago Press, 324 pp.

4. Kessler, E. 1959. The PAR-Scope: An Oscilloscope Display for Weather Radars. IRE Trans. on Aeronaut. & Nav. Elect., ANE-6, 1:31-36.

5. Sirmans, D. and Doviak, R. J. 1973. Meteorological Radar Signal Estimation. NOAA Technical Memorandum, ERL-NSSL-64. 80 pp.

6. Wilk, K. E. and Brown, R. A. 1975. Applications of Conventional and Doppler Radar Measurements in Severe Storms Research. Paper prepared for the Third Symposium on Meteorological Observations and Instrumentation, February 10-13. Washington, D. C.

7. Vlcek, C. L. 1973. Observations of Severe Storms on 26 and 28 April 1971. NOAA Technical Memorandum, ERL-NSSL-63. 18 pp.

8. Kessler, E. and Russo, J. A. 1963. Statistical Properties of Weather Radar Echoes. Proc. 10th Weather Radar Conference, Washington, D. C. pp. 25-33.

9. Wilson, J. W. and Kessler, E. 1963. Use of Radar Summary Maps of Weather Analysis and Forecasting. J. Appl. Meteor., 2, 1: 1-11.

10. Barnes, S. L., Editor 1974. Papers on Oklahoma Thunderstorms, April 29-30, 1970. NOAA Technical Memorandum, ERL-NSSL-69. 231 pp.

11. Barclay, P. A. and Wilk, K. E. 1970. Severe Thunderstorm Radar Echo Motion and Related Weather Events Hazardous to Aircraft Operations. ESSA Technical Memorandum, ERL-TM-NSSL-46. 63 pp.

12. Blackmer, R. H., Jr., Duda, R. O., and Reboh, R. 1972. Application of Pattern Recognition Techniques to Digitized Weather Radar Data. Preprints,

15th Radar Meteorology Conference, Amer. Meteor. Soc., Boston, MA. pp. 138-143.

13. Lee. J. T. 1967. Association Between Atmospheric Turbulence and Radar Echoes in Oklahoma. ESSA Technical Memorandum NSSL-32. pp. 1-10.

14. Kinzer, G. D. 1974. Cloud to Ground Lightning Versus Radar Reflectivity in Oklahoma Thunderstorms. J. Atmos. Sci., 31, 3:787-799.

15. Wilk, K. E. and Kessler, E. 1970. Quantitative Radar Measurements of Precipitation. Meteor. Monog., 11, 33:314-329.

16. Harrold, T. W., English, E. J., and Nicholass, C. A. 1974. The Accuracy of Radar Derived Rainfall Measurements in Hilly Terrain. Quar. J. Roy. Meteor. Soc., 100, 425:331-350.

17. Wilson, J. W. 1970. Integration of Radar and Raingage Data for Improved Rainfall Measurement. J. Appl. Meteor., 9, 3:489-497.

18. Brandes. E. A. 1974. Radar Rainfall Pattern Optimizing Technique. NOAA Technical Memorandum ERL-NSSL-67. 16 pp.

19. O'Brien, G. F. 1974. Digitized Radar Experiments D/RADEX Progress Report No. 44, 5 pp. (Typescript, National Weather Service, NOAA, U.S. Dept. of Commerce.)

20. Moore, P. L., Cummings, A. D., and Smith, D. L. 1974. The NWS Manually Digitized Radar Program and Some Applications. NOAA Technical Memorandum NWS-SR-75. 21 pp.

21. Drufuca, G., Chairman, 1974. Report of Working Group on the Statistics of Rainfall Rates and the Design of Communication Systems. J. de Recherches Atmosphériques, III, 1-2:473-476.

22. Rogers, R. R., Chairman 1974. Report of Working Group on the Mesoscale Structure of Precipitation and Space Diversity. J. de Recherches Atmosphériques, III, 1-2:485-490.

23. Kessler, E., Gray, K., and Dooley, J. T. 1968. Toward a Quantitative Radar Climatology. Proceedings, 13th Conference on Radar Meteorology, Amer. Meteor. Soc., Boston, MA. pp. 280-285.

24. Browning, K. A., Bryant, G. W., Starr, J. R., and Axford, D. N. 1973. Air Motion Within Kelvin-Helmholtz Billows Determined for Simultaneous Doppler Radar and Aircraft Measurements. Quar. J. Roy. Meteor. Soc., 99, 422:608-618.

25. Kessler, E. and Wilson, J. W. 1971. Radar in an Automated National Weather System. Bull. Amer. Meteor., Soc. 52, 11:1062-1069.

26. Taylor, V. C. and Browning, K. A. 1974. Towards an Automated Weather Radar Network. Weather, 29, 6:202-216.

THE USE OF LIDAR FOR ATMOSPHERIC MEASUREMENTS

M. PATRICK MC CORMICK

*NASA Langley Research Center
Hampton, Virginia 23665, U.S.A.*

INTRODUCTION

This paper discusses the application of laser radar (lidar) to atmospheric measurement, especially those related to energy related problems. The heart of a lidar system is, of course, the laser which was first conceived by Professors Schawlow and Townes in December 1958 at Columbia. It was not, however, until July 1960 that Dr. Theodore Maiman at Hughes Research Laboratory first reduced the laser principle to practice. Actually the term lidar, an acronymn for light detection and ranging, came before the advent of the laser. Middleton and Spelhous in their 1953 book on "Meteorological Instruments" first suggested the term "lidar" for light detection and ranging when discussing ceilometry. The honor of the first paper in which data were described using a lidar goes to Fiocco and Smullins for their September 1963 paper in Nature entitled "Detection of Scattering Layers in the Upper Atmosphere (60-140 km) by Optical Radar." This paper was followed by a paper by Fiocco and Grams (May 1964) in the Journal of the Atmospheric Sciences entitled "Observations of the Aerosol Layer at 20 km by Optical Radar." During this early period, papers were being presented and published in which the potential uses of lasers and lidar in meteorology were being expounded. As an example, Goyer and Watson at NCAR published a paper on "The Laser and Its Application to Meteorology" in the Bulletin of the American Meteorological Society in September 1963. In September of 1964 Myron Ligda gave a paper at the World Conference on Radio Meteorology in which he pointed out that others were also discussing the possible uses of lasers in meteorology. These included Dick Shotland, Dave Atlas, P.S. Carter, and R.B. Battelle. Also described in this presentation was the first SRI lidar (Mark I) which SRI started building in January 1963. Their first atmospheric returns were obtained in July 1963. SRI was the first, as far as I can deduce, to apply the lidar technique to meteorology in the lower troposphere.

Most measurements have utilized elastic scattering techniques but more recently have included inelastic scattering techniques and absorption studies. Presently, many groups are working with lidar in an attempt to characterize the atmosphere. For an idea of the scope of this work, see the Conference Abstracts of the Conferences on Laser Radar Studies of the Atmosphere. Comprehensive reviews of lidar measurements of the atmosphere can be found in papers given by Kent and Wright [1] and Collis [2].

In general, laser radars operate in the following manner. A Q-switched laser emits a pulse of nearly monochromatic light approximately 30 nsec in duration into the atmosphere. Molecules and suspended particulate matter (aerosols) scatter and/or absorb this radiation as the pulse propagates through the atmosphere. A small portion of this light is scattered directly back toward the laser. A receiver composed of mirrors and/or lenses collects this backscattered radiation and directs it onto a photodetector whose output is measured as a

Fig. 1. - A photograph of the 48-inch (1.22 meters) Langley Research Center laser system.

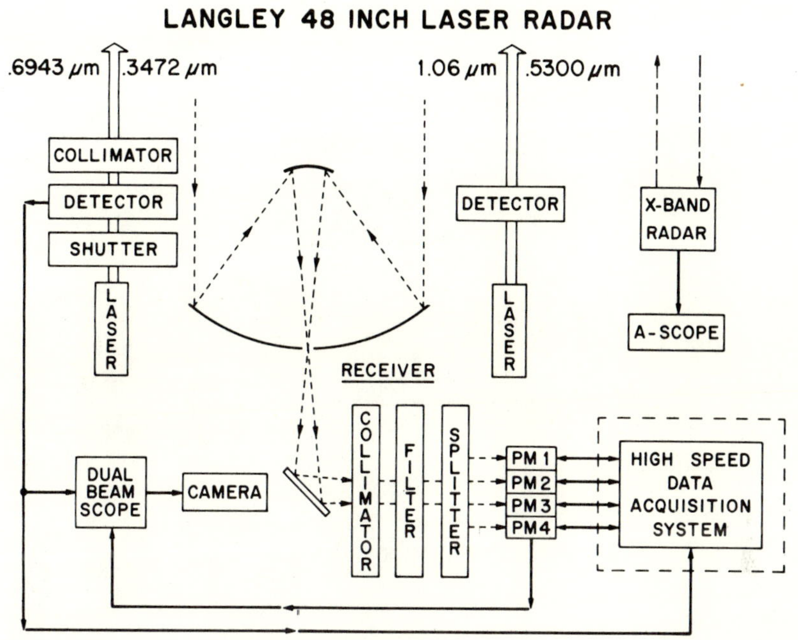

LANGLEY 48 INCH LASER RADAR

Fig. 2. - Schematic diagram of the 48-inch (1.22 meters) Langley laser radar system.

function of elapsed time after laser emission or, therefore, range. The backscattered energy incident on the photodetector is examined spectrally at or near the laser output wavelength with color filters, interference filters, or spectrometers. This enhances the signal-to-noise ratio by reducing unwanted background radiation and determines whether elastic or inelastic (Raman and fluorescence) techniques are utilized.

Shown in figure 1 is the Langley Research Center's 48" laser radar system. It consists of two temperature-controlled lasers (Ruby and Neodymium-doped glass) mounted on either side of an f/10 Cassegrainian-configured telescope consisting of a 48-inch diameter f/2 all metal primary and a 10-inch diameter secondary. A schematic of the system is shown in figure 2. The output from the detector package is recorded by a high-speed data acquisition system. Analog signals are amplified and bandwidth limited, digitized at a 5- or 10-megahertz rate with 8-bit accuracy, and recorded on magnetic tape. Pulse count data is amplified, discrimated, counted at a 200 megahertz rate, and also stored on magnetic tape. Altitude resolution is obtained by using the variable 1-, 5-, or 10-microsecond bin widths that are available. A 16K word storage computer is used to control the data acquisition system and provide data message. An X-band microwave radar, boresighted with the laser system axis, is used to ensure safe operation in the atmosphere. A rotating shutter has been recently added to reduce laser fluorescence after Q-switching.

The entire system is mobile and can scan in elevation and azimuth at a slew rate of 1° per second. As an example of the variability in lidar sizes, figure 3 shows a small lightweight lidar being built at Langley. It utilizes a dye laser, Fresnel lens receiver, transient recorder, and magnetic tape recorder.

LIDAR THEORY

After crossover of the emitted laser pulse and receiver field of view, the voltage $V(\lambda, R)$ at the photodetector output for a laser radar system is given by

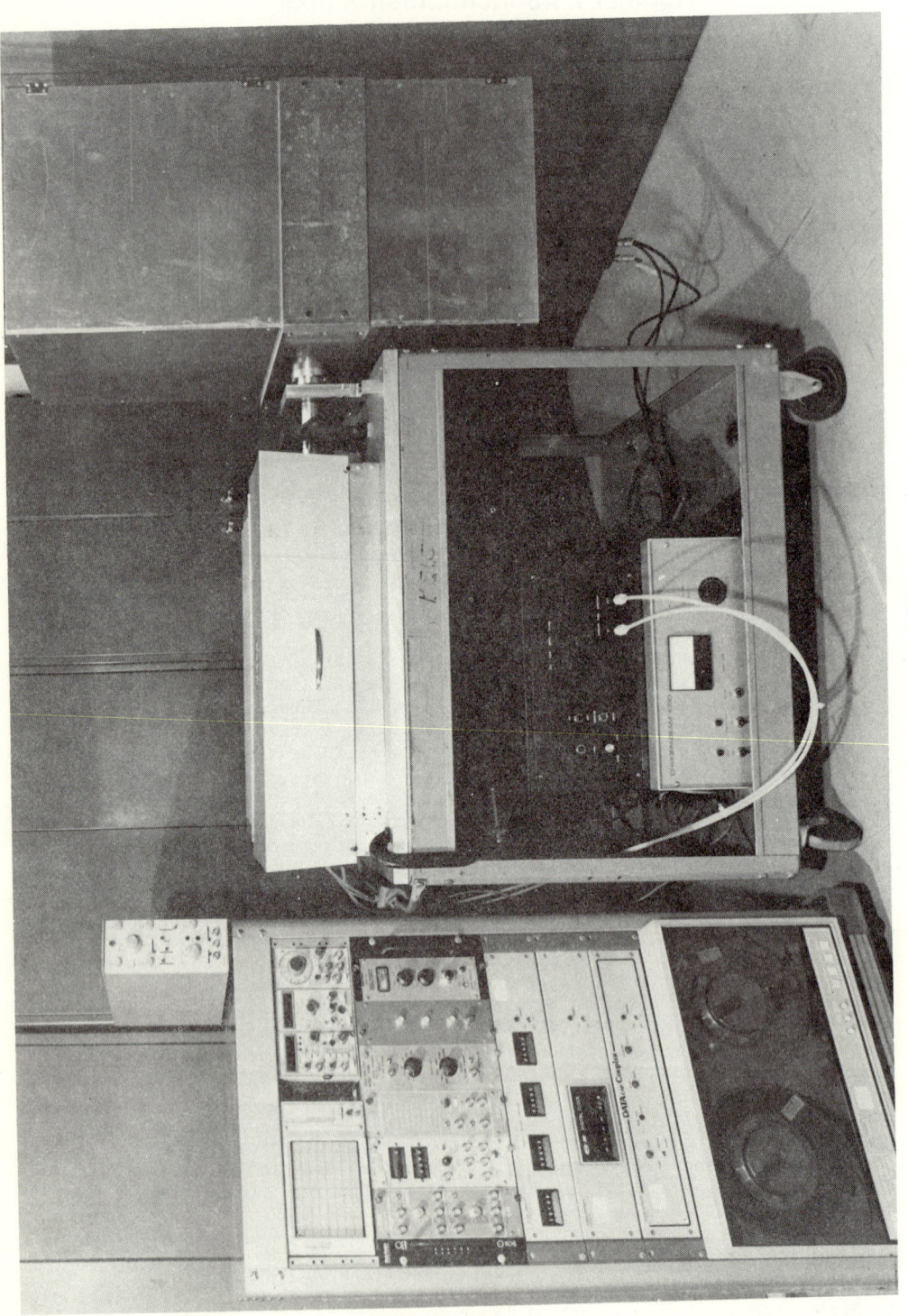

Fig. 3. – Compact lidar system.

$$V(\lambda, R) = \frac{\gamma(\lambda)E(\lambda)q_t(\lambda, \Delta R)q_s(\lambda', \Delta R)f(\lambda, R)}{R^2} \quad (1)$$

with

$$\gamma(\lambda) = \frac{cA}{2} T(\lambda)S(\lambda)L$$

where $f(\lambda, R)$ is the backscattering function at wavelength λ and range R, and $q_t(\lambda, \Delta R)$ and $q_s(\lambda', \Delta R)$ are the transmissivities from the laser radar to the scattering volume and from the scattering volume to the laser radar, respectively. The symbol λ' refers to a wavelength other than λ, L is the photodetector load resistor. $T(\lambda)$ is the system optical efficiency, $S(\lambda)$ is the photodetector spectral sensitivity, A is the area of the receiver, c is the speed of light, and $E(\lambda)$ is the laser output energy per pulse.

For elastic scattering, the emitted and return wavelengths are identical and there is no absorption. Eq. (1) reduces to

$$V(\lambda, R) = \frac{\gamma(\lambda)E(\lambda)q^2(\lambda, \Delta R)f(\lambda, R)}{R^2} \quad (2)$$

where $f(\lambda, R) = f_M(\lambda, R) + f_A(\lambda, R)$. The subscripts M and A refer to molecular and aerosol, respectively.

The Raman scattered return is shifted in wavelength from the laser output. Eq. (1) becomes

$$V(\lambda', R) = \frac{\gamma(\lambda')E(\lambda)q_t(\lambda, \Delta R)q_s(\lambda', \Delta R)f_r(\lambda', R)}{R^2} \quad (3)$$

where $f_r(\lambda', R)$ is the Raman scattering function at R and is independent of aerosol scattering. The shift in wave numbers is characteristic of the particular molecular species in the scattering volume, Raman scattering can occur for any incident wavelength but its scattering function is approximately three orders of magnitude less than molecular scattering functions.

For the case of fluorescence scattering λ' may or may not be λ with the scattering function being that of fluorescence scattering. It can only occur, however, when λ corresponds to an absorption line or band in the atmosphere and, like Raman scattering, the return radiation at λ' is characteristic of the scattering molecules. Although the cross sections for fluorescence are very large compared to Raman scattering, there are inherent problems that must be overcome in order to utilize this technique. It now appears that the differential absorption laser radar technique (DIAL) might offer the best hope for measuring pollutants with laser radar.

An atmospheric scattering model for the elastic backscattering function at two laser wavelengths is shown in figure 4. The details of its derivation are given in Ref. [3]. The molecular contribution is calculated by using the familiar Rayleigh scattering equation and the U.S. Standard Atmosphere (1962). The aerosol contribution is calculated using: Mie theory, a Junge r^{-4} size distribution from 0.125 μm to 10 μm; a 1.5 refractive index; and Rosen's [4] particle sampling data from 5 to 26 km with a sea level value of 450 particles/cm^3 and an exponential interpolation between sea level and 5 km. This model includes this interpolation since the aerosol number density in the first 5 km is highly variable.

TROPOSPHERIC MEASUREMENTS

Shown in figures 5 and 6 are results from data taken simultaneously at 0.6943 μm and 0.3472 μm on October 21, 1969 [3]. These data are compared with

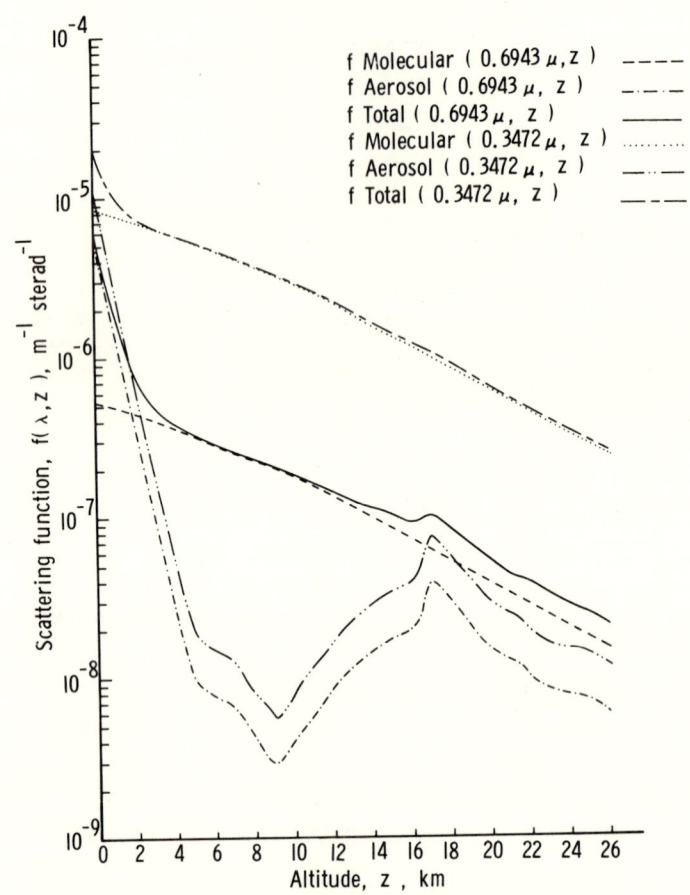

Fig. 4. - Backscattering function model of the atmosphere for a wavelength of 0.6943 μ and 0.3472 μ. The molecular contribution and their sum are shown.

radiosonde data of temperature and dewpoint taken at the same time. Note that the particulates are being "capped" by the temperature inversion, that is, vertical mixing is severely constrained by the temperature inversion. A rising parcel of air would ascend into air warmer than itself and tend to sink back to its original position. It should also be noted that the water vapor is evidently being mixed by the same mechanism. These data also show good agreement with the atmospheric scattering model shown in figure 4.

Figure 7 shows a summary of mixing layer data at 0.6943 μm from a cooperative study with Oregon State University in the Willamette Valley [5]. In the valley during summer, a temperature inversion is created by a subsidence of high-pressure air. The valley walls impede advection while the temperature inversion effectively puts a lid on the valley creating conditions of high pollutant concentration. The top of the mixing layer, as determined by laser radar and condensation nuclei counter, the altitude region of temperature inversion, and the region of decreasing water vapor are shown. Also shown are the altitude regions of significant decrease in laser radar scattering ratio and particle count.

The actual mixing height determined by lidar is defined as the average altitude during which a strong reduction in scattering ratio or aerosol concentration is observed. This strong reduction normally terminated in scattering

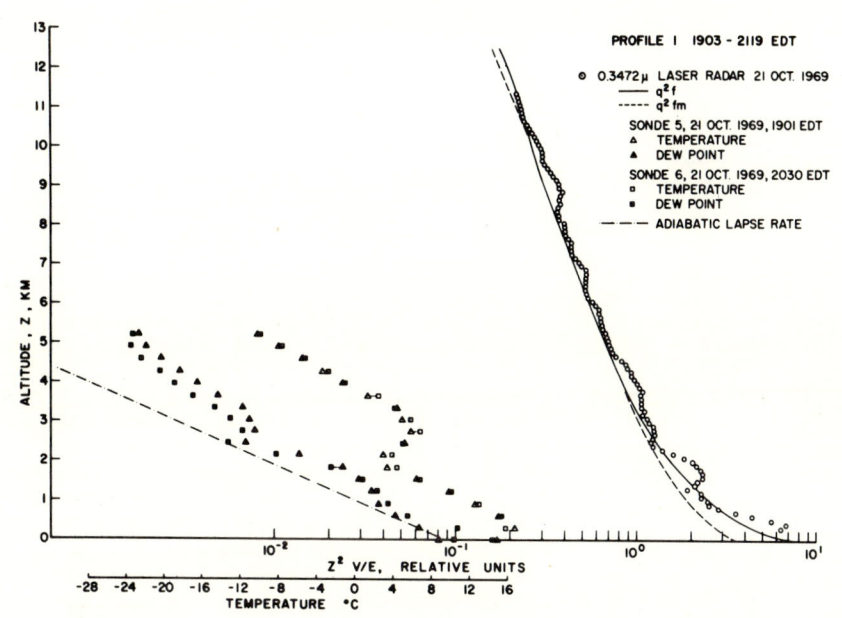

Fig. 5. — Z^2V/E versus Z laser radar data taken from 1903 to 2119 e.d.t. on October 21, 1969, for $\lambda = 0.6934$ μ. The data are normalized to the predicted profiles of q^2f. Also shown are the temperature and dew-point temperature profiles from radiosondes launched at 1901 and 2030 e.d.t. on October 21, 1969, and on adiabatic lapse rate curve.

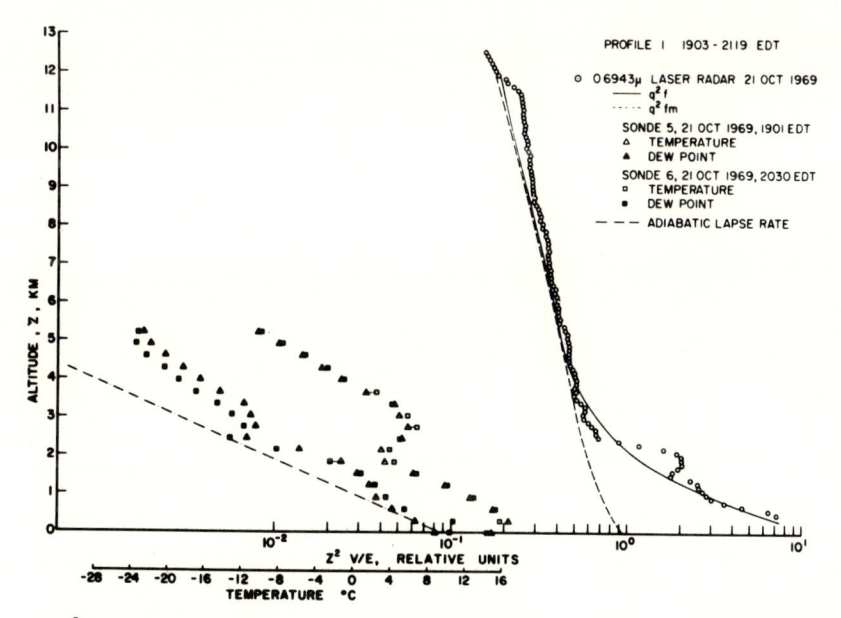

Fig. 6. — Z^2V/E versus Z laser radar data taken from 1903 to 2119 e.d.t. on October 21, 1969, for $\lambda = 0.34715$ μ. The data are normalized to the predicted profiles of q^2f. Also shown are the temperature and dew-point temperature profiles from radiosondes launched at 1901 and 2030 e.d.t. on October 21, 1969, and on adiabatic lapse rate curve.

ratio values near 1, corresponding to molecular scattering. For the aircraft-mounted condensation nuclei counter, the mixing height is arbitrarily defined as the altitude at which the concentration becomes less than 500 particles per cubic centimeter.

The laser radar scattering ratio is defined as

$$\frac{R^2 V}{q^2 d_M E} = \gamma \left[1 + \frac{f_A}{f_M} \right]$$

The ratio is normalized to 1 in a definite molecular scattering region above the top of the mixing layer. q^2 is assumed from a model atmosphere. Shown in figure 8 is an example of one such profile.

Figure 7 shows that of the nine periods, six had concurrent laser radar and particle data and of these six periods, five show the top of the mixing layer as determined by nuclei counter to lie within the band of decreasing laser radar data. For the other case (period 2), it lies within 100 meters of the laser radar data. It should be mentioned that for period 2, the laser radar did not show a well-defined mixing top and all data showed large variations. Furthermore, the top of the mixing layer, as determined by these two methods in the five cases are within 50 meters of each other, with an average difference of 22 meters. The other case shows a difference of 170 meters. In all eight periods where there are laser radar and temperature data, the laser radar data decrease lies within a temperature inversion; but for two periods, the altitude spread of the condensation nuclei data decrease does not correspond to a temperature inversion. Also, in only two of the nine periods did the water vapor decrease not include the top of the mixing layer as determined by laser radar and nuclei counter. It should be pointed out that laser radars are most sensitive to aerosols greater than approximately 0.1 μm and condensation nuclei counters are sensitive to aerosols greater than approximately 0.001 μm. Therefore, comparing these two techniques of measuring mixing height could yield differences, depending on atmospheric particle size distribution and mixing. In summary though, there was very good agreement between the two techniques. Furthermore, the actual mixing height was, in general, found to be associated with a stable layer aloft, although not uniquely with the base of this layer.

From Eq. (2) it is evident that the laser radar signal is dependent upon a combination of both aerosol and molecular scattering through the transmissivity

$$q^2 = \exp(-2\tau) = \exp\left[-2\int_0^{R'} \beta(\lambda, R)\, dR\right]. \quad (4)$$

where τ is the optical depth and β is the extinction coefficient, and through the aerosol and molecular scattering functions. The molecular scattering contribution is very easy to calculate (Rayleigh scattering) but the aerosol contribution cannot be easily (not at all uniquely) determined. Usually spherical aerosols of an assumed size distribution and refractive index is used with Mie scattering theory to calculate the aerosol contribution. A combination of Raman and elastic scattering, however, might shed some new light on the aerosol scattering functions.

The ratio of the Raman return at two altitudes Z_1 and Z_2 can be written from Eq. (3) as

$$\frac{Z_1^2 V(\lambda', Z_1)}{Z_2^2 V(\lambda', Z_2)} = \frac{q(\lambda, 0-Z_1) q(\lambda', Z_1-0) \sigma_r N(Z_1)}{q(\lambda, 0-Z_2) q(\lambda', Z_2-0) \sigma_r N(Z_2)} \quad (5)$$

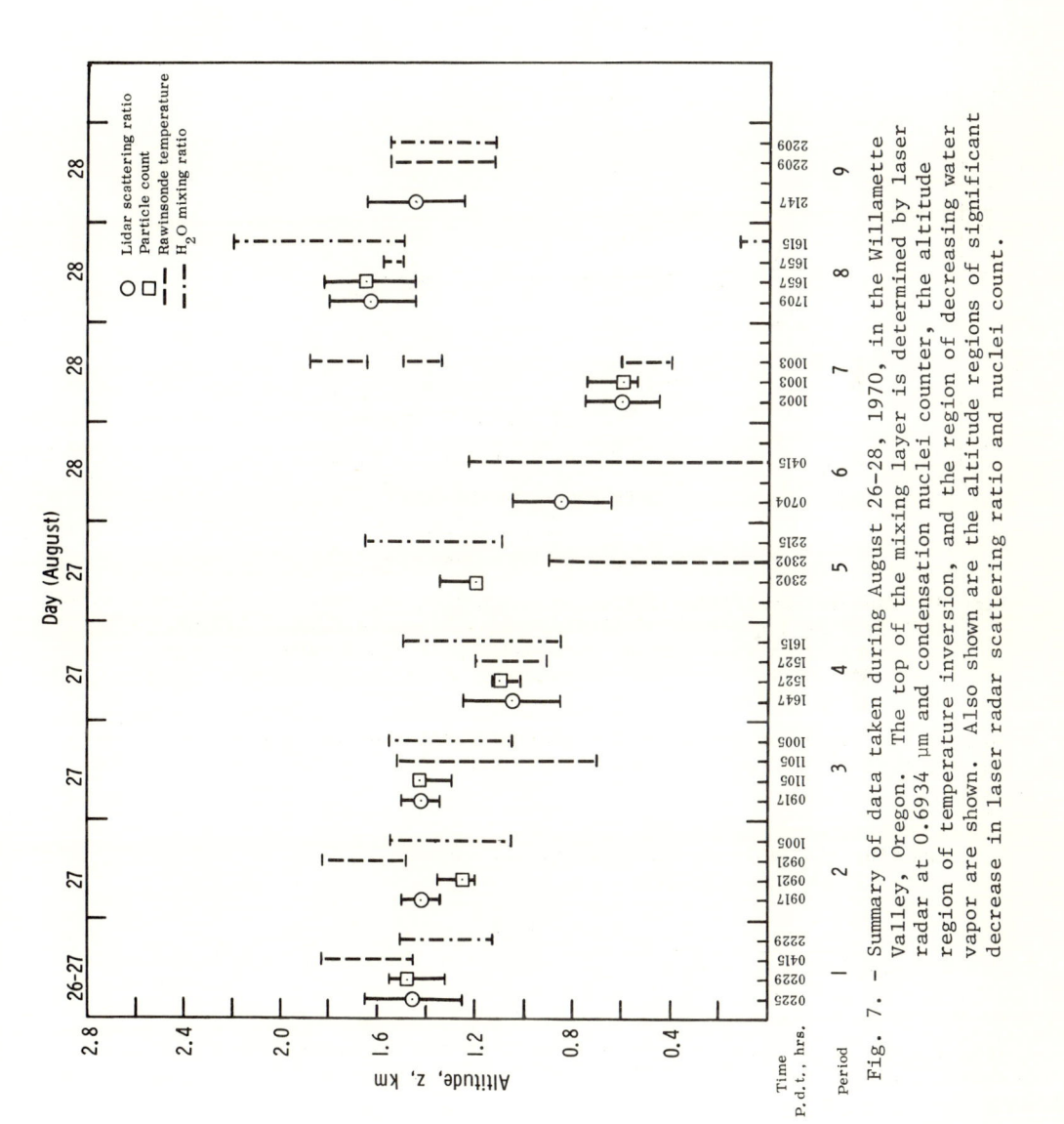

Fig. 7. — Summary of data taken during August 26-28, 1970, in the Willamette Valley, Oregon. The top of the mixing layer is determined by laser radar at 0.6934 μm and condensation nuclei counter, the altitude region of temperature inversion, and the region of decreasing water vapor are shown. Also shown are the altitude regions of significant decrease in laser radar scattering ratio and nuclei count.

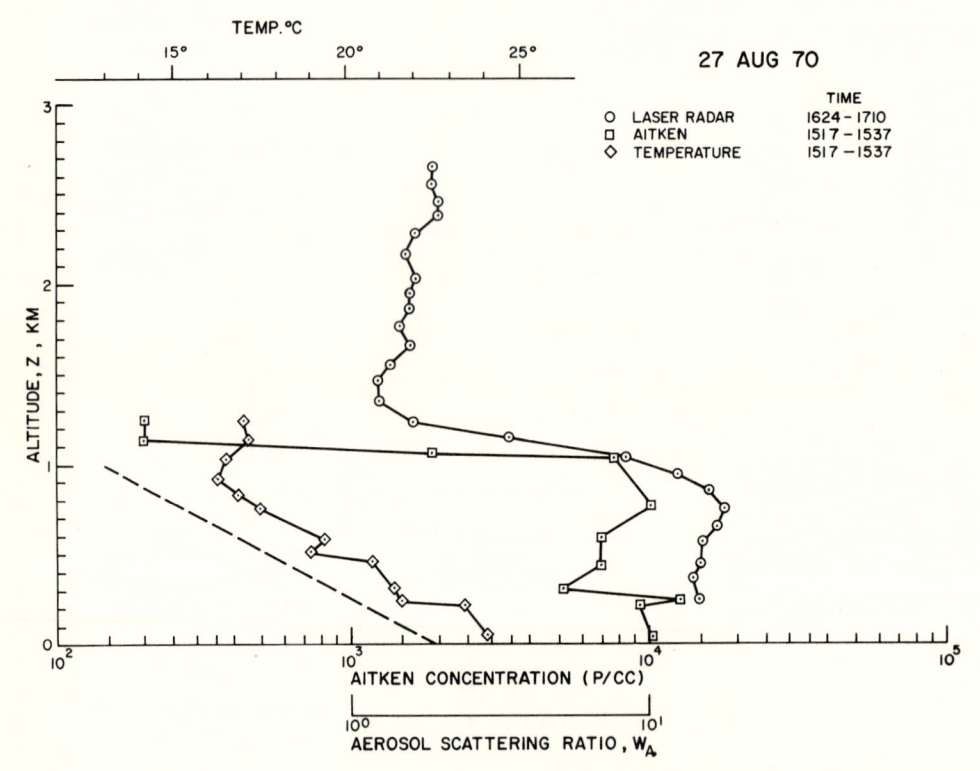

Fig. 8. – Aerosol scattering ratio, Aitken nuclei concentration and temperature data as a function of altitude in Willamette Valley, Oregon. Also shown in a dotted line is the adiabatic lapse rate.

where σ_r is the Raman cross section for the particular gas molecule of number density N and $q(\lambda, 0-Z_1)$ is the transmissivity at λ from the laser radar to altitude Z_1.

Assuming $q(\lambda, \Delta Z) \simeq q(\lambda', \Delta Z)$, Eq. (5) becomes

$$\frac{Z_1^2 V(\lambda', Z_1)}{Z_2^2 V(\lambda', Z_2)} = \frac{1}{q^2(\lambda', Z_1-Z_2)} \frac{N(Z_1)}{N(Z_2)} \qquad (6)$$

This assumption is good when there is no differential absorption or large extinction between λ and λ'. The ln of Eq. (6) gives

$$\int_{Z_1}^{Z_2} \beta(\lambda', Z) dZ = \frac{-1}{2} \ln \frac{Z_2^2 V(\lambda', Z_2) N(Z_1)}{Z_1^2 V(\lambda', Z_1) N(Z_2)} \qquad (7)$$

Assuming β constant over Z_1 to Z_2 gives

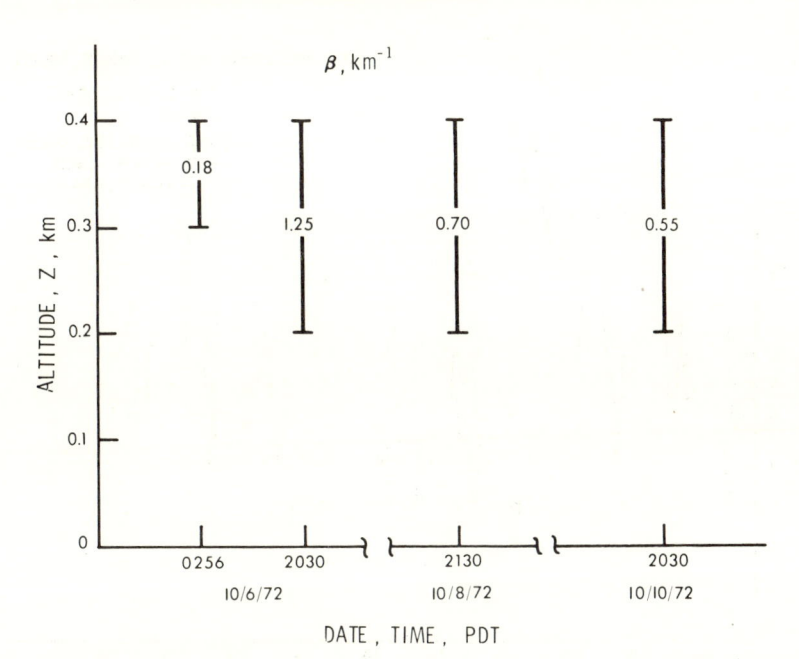

Fig. 9. - Extinction coefficient as a function of time and altitude in Azusa, California, October 1972.

$$\beta(\lambda', \bar{Z}) = \frac{-1}{2(Z_2-Z_1)} \ln \frac{Z_2^2 v(\lambda', Z_2) N(Z_1)}{Z_1^2 v(\lambda', Z_1) N(Z_2)} \qquad (8)$$

This quantity is plotted in figure 9 as a function of time and altitude in the mixing layer in the Los Angeles Basin (over Azusa, California) during October 1972 for the Raman nitrogen return using an exciting laser wavelength of 0.3472 µm. Large variations are evident. Nitrogen number densities were obtained from rawinsonde temperature and pressure data. Figure 10 compares meteorological ranges obtained from these extinction coefficients to ground visibilities in the Los Angeles Basin supplied by T.J. Howland of Meteorology Research, Incorporated, Altadena, California. The laser radar extinction coefficients were used to calculate logical ranges at altitudes and exponentially extrapolated to ground level in a manner similar to Elterman [6]. In general, there is very good agreement between the two measurements. Shown in figure 11 is aerosol extinction coefficient data obtained with an integrating nephelometer. These data fill in one of the gaps in figure 10, showing agreement with the lidar data obtained at 0256, October 6, 1972. The value of q^2 obtained in this manner can be used in Eq. (2) for elastic scattering. The Raman nitrogen data ratioed at three altitudes combined with simultaneous elastic scattering data ratioed at the same three altitudes will yield three equations and three unknowns from which the aerosol scattering function at the three altitudes can be uniquely determined. Other techniques for using laser radar data are discussed in a recent paper by McCormick [7].

STRATOSPHERIC MEASUREMENTS

Lidars have also been used extensively to study the lower stratosphere. Some of the earlier measurements are shown in figures 12 and 13, taken over

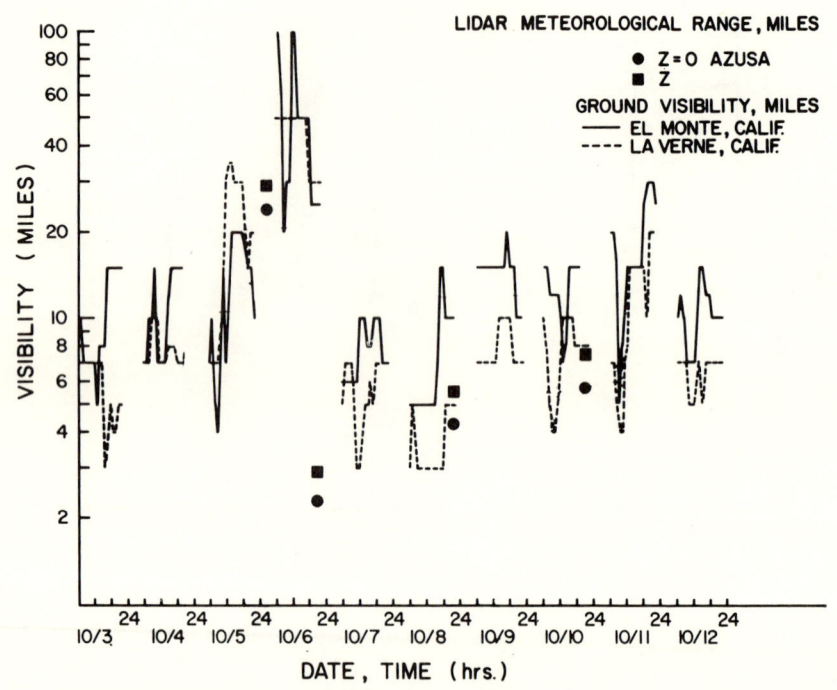

Fig. 10 - A comparison of meteorological range as determined by laser radar and ground visibility as a function of time in Azusa, California, October 1972.

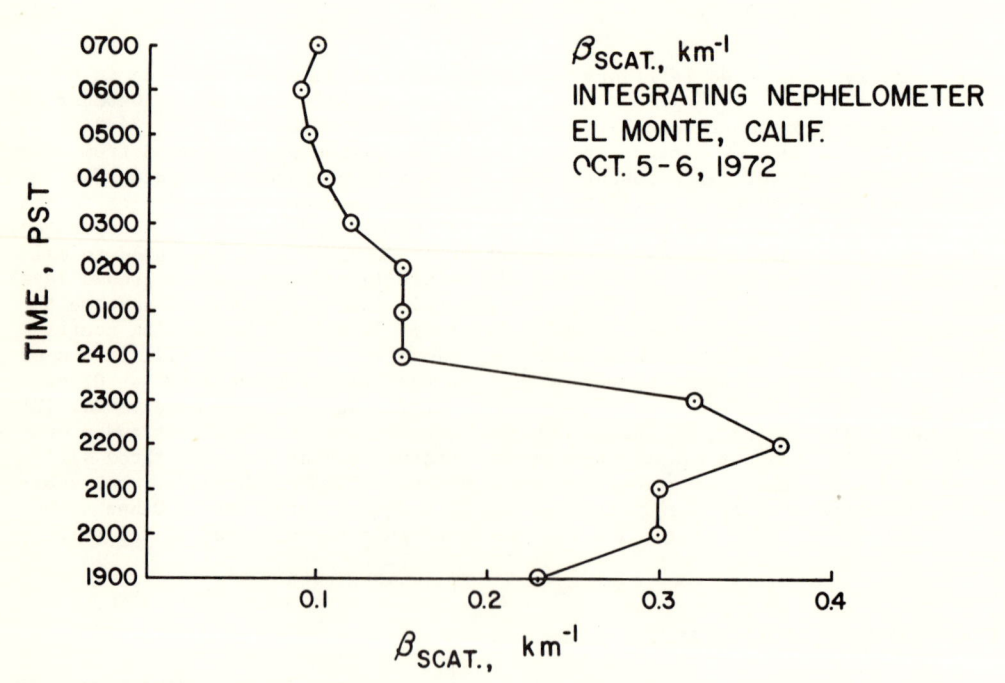

Fig. 11. - Extinction coefficient as a function of time in El Monte, California 10/5-6/72.

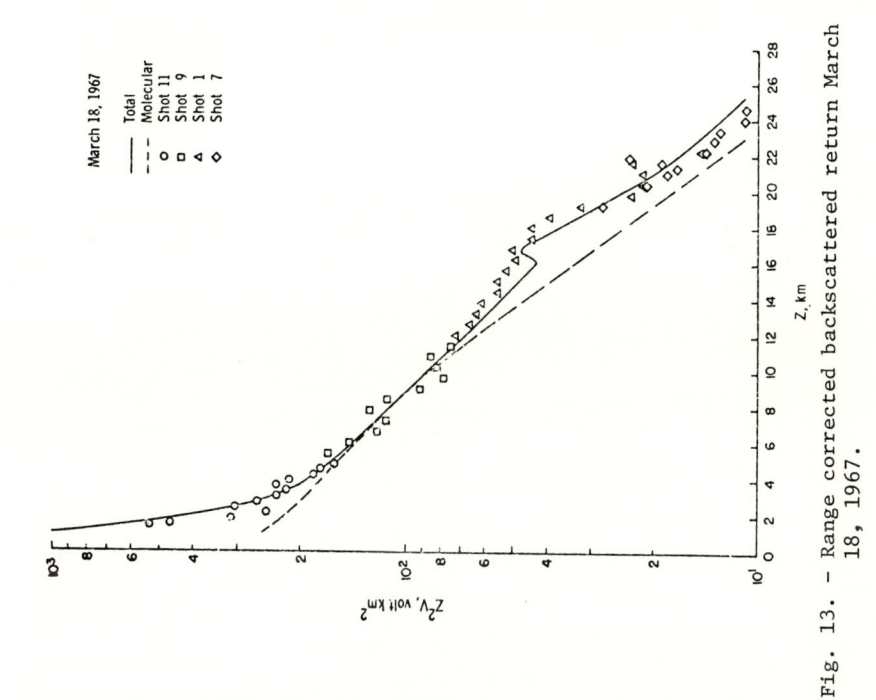

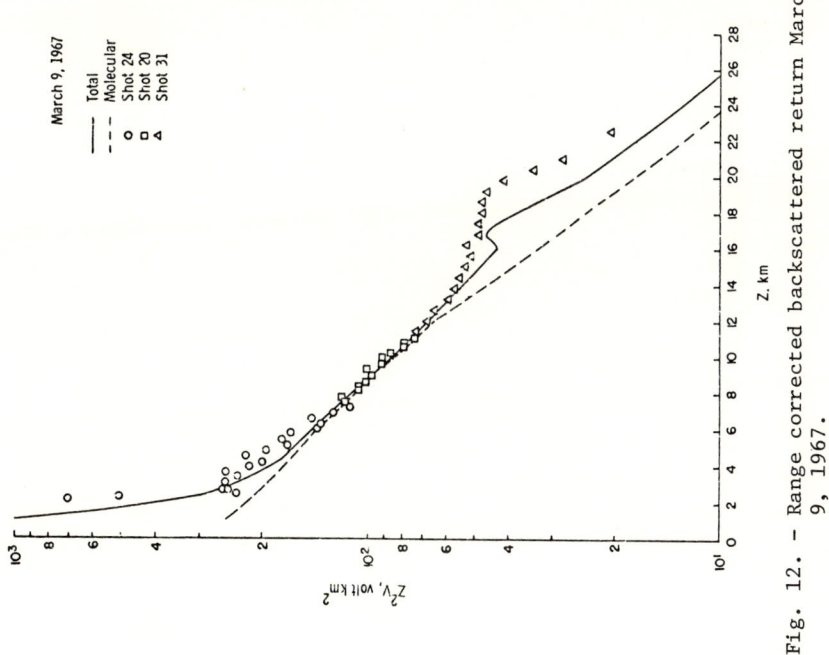

Fig. 12. - Range corrected backscattered return March 9, 1967.

Fig. 13. - Range corrected backscattered return March 18, 1967.

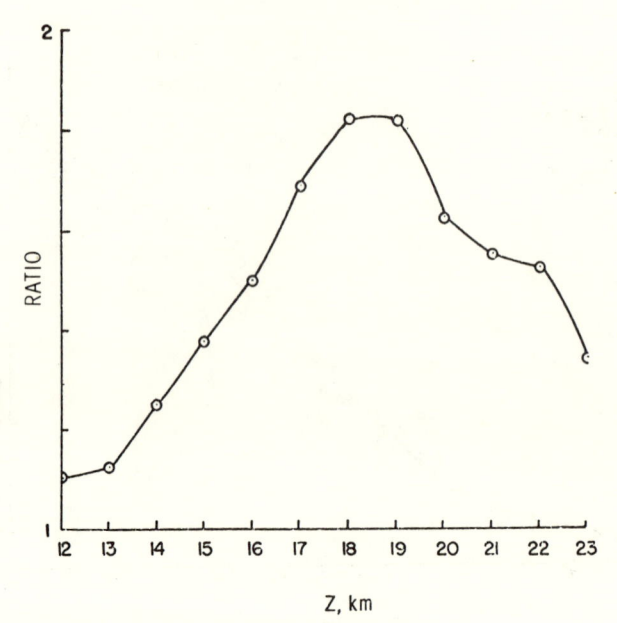

Fig. 14. - Scattering ratio for data from four nights over Williamsburg, Virginia, March 1, 9, 17, and 18, 1967.

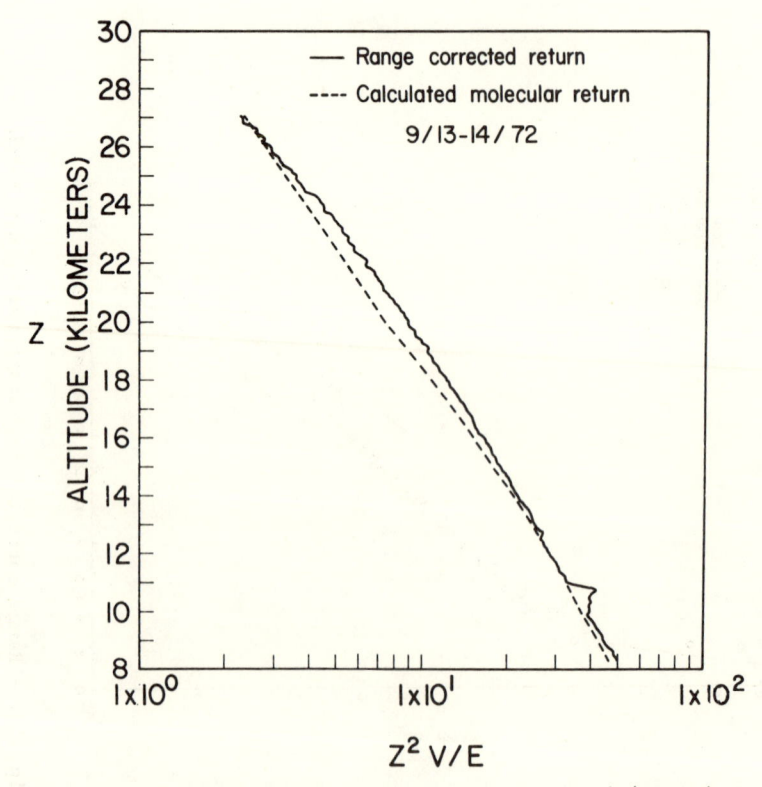

Fig. 15. - Range corrected backscattered return (9/13-14/72).

SCATTERING RATIO

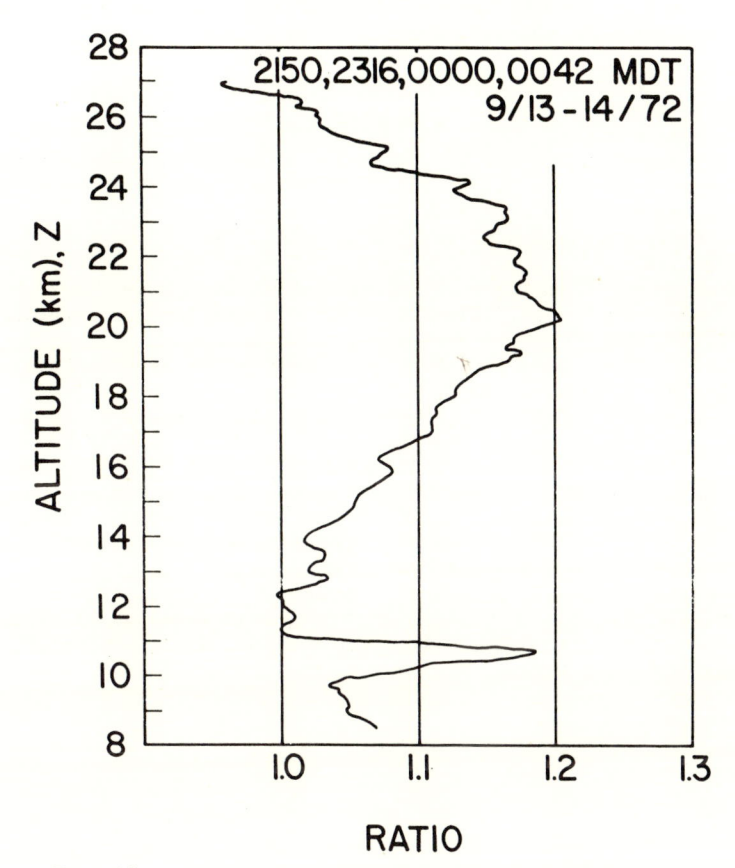

Fig. 16. - Scattering ratio for data from figure 4.

Williamsburg, Virginia, March 9 and 18, 1967 [8]. At this period of time the stratospheric aerosol layer (Junge or sulfate layer), centered at approximately 19 km, was very evident in lidar backscatter signatures representing approximately twice the scattering expected from a molecular atmosphere as shown in figure 14. More recent data [9] (figures 15 and 16) show the stratospheric aerosol concentration to have decreased significantly. The increase in scattering in this layer being only 18 percent above molecular scattering.

SUMMARY

With the increased concern over a clean environment and the effects of pollution on our climate, new measurement techniques must be developed. Lidar offers one such technique. With the present state-of-the-art it is possible to monitor: the earth's mixing layer, both the top and the spatial distribution of aerosols; cloud range, horizontal extent, and if tenuous enough, its thickness; transmission; smoke plume diffusion (if laden with enough particulates) and for limited range and concentration the SO_2 content of smoke plumes; H_2O mixing ratios in the mixing layer; the stratospheric aerosol distribution; 80-90 km sodium concentrations; and many other atmospheric parameters. Some of these are even applicable from shuttle, e.g., cloud distribution and range, and the aerosol measurements. Clearly, lidar provides a unique method of studying atmospheric structure and offers the possibility of increasing knowledge in many areas of atmospheric research and meteorology.

REFERENCES

[1] Kent, G.S. and Wright, R.W.H., "A Review of Laser Radar Measurements of Atmospheric Properties," Journal of Atmospheric and Terrestrial Physics, Vol. 32, May 1970, pp. 917-943.
[2] Collis, R.T., "Lidar," Applied Optics, Vol. 9, No. 8, Aug. 1970, pp. 1782-1788.
[3] McCormick, M.P., "Simultaneous Multiple Wavelength Laser Radar Measurements of the Lower Atmosphere," Electro-Optics International, Brighton, England, March 24-26, 1971.
[4] Rosen, J.M., "Correlation of Dust and Ozone in the Stratosphere," Nature, Vol. 209, March 1966, p. 1342.
[5] McCormick, M.P., Melfi, S.H., Olsson, L.E., Tuff, W.L., Elliott, W.P., and Egami, R., "Mixing-height Measurements by Lidar, Particle Counter, and Rawinsonde in the Willamette Valley, Oregon," NASA TN D-7103, Dec. 1972.
[6] Elterman, L., "Vertical-Attenuation Model with Eight Surface Meteorological Ranges 2 to 13 Kilometers," AFCRL-70-0200, March 1970.
[7] McCormick, M.P., and Fuller, W.H., "Lidar Techniques for Pollution Studies," AIAA Journal, 11, pp. 244-246, Feb. 1973.
[8] McCormick, M.P., "Laser Backscatter Measurements of the Lower Atmosphere," Ph.D. Thesis, College of William and Mary, Williamsburg, Virginia, 1967.
[9] Northam, G.B., Rosen, J.M., Melfi, S.H., Pepin, T.J., McCormick, M.P., Hofmann, D.J., and Fuller, W.H., Jr., "Dustsonde and Lidar Measurements of Stratospheric Aerosols: A Comparison," Applied Optics, Vol. 13, No. 110, pp. 2416-2421, Oct. 1974.

PRACTICAL CONSIDERATIONS TO THE USE OF MICROWAVE SENSING FROM SPACE PLATFORMS

R. P. EISENBERG

*General Electric Co., Valley Forge Space Center,
Philadelphia, Pennsylvania*

ABSTRACT

There exists a wide range of microwave sensing techniques (active, passive, imaging) with a varied set of characteristics (frequency, polarization, spatial resolution, etc.) that can be used in obtaining remotely sensed earth resources data from space. The recent Skylab Earth Resources Program provided a valuable opportunity to examine several of these techniques as applied to ocean phenomena and to assess their utility for future applications. The advent of the Seasat Program, as well as many Space Shuttle flight opportunities, promises the maturing of microwave remote sensing into a practical operational method of monitoring our earth and ocean environment on a global scale. This paper describes briefly some of the principles upon which microwave sensing is based, reviews the Skylab S-193 experiment and some of the results obtained, defines a possible sensor complement for Seasat and illustrates the problems and limitations of microwave remote sensing techniques from space platforms.

INTRODUCTION

The application of microwave techniques for remote sensing from space platforms has been demonstrated relatively recently but already promises to become a useful operational method for understanding and solving energy related problems. The principles involved were developed decades ago, but have been applied to remote sensing from space just within the last few years. Mariner 2, launched in 1962, carried a low resolution surface temperature radiometer [1] and Cosmos 243 and 384 flew vertical temperature profiling radiometers [2]. More recently, Nimbus 5 provided data from a 19 GHz scanning radiometer (ESMR) as well as from a 5 frequency microwave spectrometer (NEMS). It was not until December 1972 that the first active sensor, the Apollo XVII Lunar Imaging Radar, was deployed in space. The S-193 Microwave Altimeter/Radiometer/Scatterometer which was carried aboard Skylab as part of the Earth Resources Experiment Package (EREP) was the first integrated sensor to provide near simultaneous remotely sensed active and passive microwave data. In 1975 the GEOS-C Satellite will carry a high resolution radar altimeter to measure the geoid and provide ocean wave height information with a resolution approaching 50 cm. In a relatively short time the application of microwave sensing to the measurement of our terrestrial environment has matured to the point where operational systems are being considered.

Microwave sensors, when combined with and augmented by data from other remote sensors in the visible and IR region have application in areas such as energy resource exploration (e.g., geo-thermal), crop and timber surveys, soil moisture measurements, water surveys, ice coverage mapping and numerous others. In areas that are less explicitly energy resource related, microwave remote sensing from space has already demonstrated its ability to measure and return data from which ocean surface windspeed, wave height, geoid topography and sea surface temperature may be inferred. Continuously monitoring the oceans on a global scale using microwaves should provide timely and much needed data to the users of the world's oceans enabling more accurate long range weather and sea state forecasts, thereby yielding considerable savings of time, life and property and money.

Although generally limited in spatial resolution, microwave sensing techniques have a very essential advantage over visible and infrared scanners in that they can operate at night or day and under a wide range of weather and atmospheric conditions, if operating frequencies are judiciously chosen. The utility of visible sensors is greatly reduced by conditions of darkness and cloud cover while neither condition seriously compromises the performance of a microwave sensor [3]. In addition, the availability of two decades of microwave frequency spectrum enables a designer to tailor a sensor to a specific physical phenomena such as oxygen or water vapor absorption or foliage penetration [4] and even permits an extension of the concept of "color" into the microwave region.

To fully appreciate the capabilities and recognize the restrictions associated with microwave remote sensing, it is necessary to understand some of the fundamental principles upon which this technique is based. These principles are reviewed in the following sections after which a description of the Skylab S-193 microwave sensors and of the Seasat Program is provided to illustrate the problems and practical considerations involved in the use of microwave remote sensing.

MICROWAVE SENSING PRINCIPLES

There are basically two generic classes of microwave sensing - active and passive. Active sensing involves the illumination of the target of interest with a source of microwave energy and measuring the response of the target to that illumination. Altimetry, scatterometry and imaging radar techniques are illustrative of this class. Passive sensing relies on the fact that any warm body or surface with finite emissivity ϵ emits thermal radiation over all microwave frequencies which can be detected and measured in spite of it being a noisy process. Microwave radiometry is based upon this principle.

Radiometry has been well developed in the years since its foundation and is now based on a sound body of theory involving physics, black body radiation principles and the mathematics of probability, statistics and random processes [3, 5]. An object at absolute temperature T_O and with surface emissivity ϵ will emit radiation of a brightness temperature $T_B = \epsilon T_O$. The spectral radiance b is given by $b = \frac{2KTB}{\lambda^2}$ per unit volume and may be viewed as equivalent to the emission of noise with a power spectral density equal to $N_O = kT_B$ per unit bandwidth. Implicit here are the concepts of perfect impedance matching in the transfer of power from the object in question to the antenna of the sensor and that of the object's black body characteristics, i.e., it absorbs all incident radiation and re-radiates it. In a finite bandwidth B of the detection process, the power received is $P_B = N_O B = kT_B B$ - thus giving rise to the ability to distinguish temperature. Although extremely weak, the thermal noise power spectral density at microwave wavelengths is easily detected by receivers with modest sensitivity using the principles of the switching radiometer first developed by Dicke [6] and since improved upon by Hach [7] and many others. Although there are basic differences in the thermal resolutions (ability to detect two brightness temperatures ΔT apart) achievable with the different types of radiometers available, each expression derived shows resolution to be directly proportional to the total system noise temperature (antenna T_{ANT} plus receiver T_R) and inversely proportion to the square root of the product of predetection bandwidth and integration time constant (Bτ). Recent radiometer developments reported have claimed resolution of $0.1°K$, an absolute accuracy $\pm 0.2°K$ and a long term stability of better than $0.3°K$ over a one week operating period [8]. In general, such extreme accuracy and stability are achieved by very carefully controlling the temperature environment of those components in the radiometer "front end", by minimizing loss prior to the common switching element and by very accurate insertion loss measurement.

The availability of instruments as precise as that just described suggests that very accurate surface temperature measurements on the earth could easily be sensed remotely. However, although the radiometer may measure an apparent microwave temperature it is still an arduous task to relate this to surface temperature. This is due to the fact that the emissivity of a surface is a function of many variables - incidence angle at which it is viewed, polarization, frequency, roughness, complex dielectric constant and its physical temperature. For the case of the sea surface the brightness temperature is related to the physical parameters through the equation [3, 9]

$$T_B(\theta, \phi) = t_o [\epsilon(\theta, \phi) T_o + (1 - \epsilon(\theta, \phi)) T_R] + T_{ATM} \tag{1}$$

where
t_o = atmospheric transmissivity along line of sight ($0 < t_o < 1$)
$\epsilon(\theta, \phi)$ = surface emissivity ($0 \le \epsilon \le 1$)
T_o = absolute physical temperature of surface
T_R = radiation temperature incident on surface
T_{ATM} = sky brightness temperature along line of sight

The apparent microwave temperature T_A is obtained by convolution of the antenna pattern $G(\theta, \phi)$ with the brightness temperature (actually performed each time the radiometer takes a measurement) to yield

$$T_A = \int_\Omega T_B(\theta, \phi) G(\theta, \phi) d\Omega \tag{2}$$

One result of this complex relationship is that additional data must be obtained in conjunction with the radiometer to perform the inversion process and to permit extraction of the sea surface temperature from the apparent temperature sensed.

Microwave scatterometry employs an active sensor in a mode which does not measure range but instead measures the scattering coefficient (normalized radar backscattering cross section - σ^o) of the target. The measurement is accomplished by transmitting fairly long, (narrow band) microwave pulses and estimating the return signal power. If an accurate measure of the transmit power is known the ratio of P_R/P_T can be formed where
P_R = received power
P_T = transmit power
and related to σ^o by the formula

$$\sigma^o = \frac{(4\pi)^3 R^4}{\lambda^2} \cdot \frac{P_R}{P_T} L^2 \frac{1}{\int_A G_o^2(\Psi) f(\Psi) dA} \tag{3}$$

where R = range to target
λ = wavelength
L = atmospheric loss term (one way)
$G_0(\Psi)$ = antenna gain
$f(\Psi)$ = two way antenna pattern
A = illuminated area

A scatterometer can be used to measure specific areas through control of the antenna beamwidth and antenna pointing angle with discrimination in the receiver controlled by means of range gating or Doppler filtering or a combination of both. The accuracy of the measurement of σ^0 is affected by parameters such as predetection bandwidth B_{IF}, integration time for the measurement of the return signal in the presence of noise τ_{S+N} and for noise alone τ_N and the input signal to noise ration SNR_{IN} according to the relation [10]

$$K_p \cong \left[\frac{1}{B_{IF} \tau_{S+N}} \left(1 + \frac{1}{SNR_{IN}}\right)^2 + \frac{1}{B_{IF} \tau_N} \left(\frac{1}{SNR_{IN}}\right)^2 \right]^{1/2} \quad (4)$$

where the assumptions of $B_{IF} \tau_N \gg 1$, $B_{IF} \tau_{S+N} \gg 1$ have been made and where K_p is a normalized standard deviation of the estimate of received power P_R. Using this relationship and by exercising the antenna parameters, scatterometers may be designed which achieve very accurate ($K_p \leq 5\%$) measurements of σ^0 over a wide range of input signal to noise ratios and with a large variety in ground footprint resolutions (IFOV).

However, as in the case of radiometry, a good determination of σ^0 does not necessarily imply the measurement of the phenomena under examination in geophysical units. Scatterometry from space has its highest value when applied to measurements of the ocean surface in order to infer information about the local winds. Briefly, the ocean's surface responds almost immediately to winds on the order of 2m/sec or larger by generating capillary waves whose wavelength is on the order of a few centimeters. These capillary waves increase in amplitude as the windspeed increases and result in a larger radar backscattering cross section. With microwave wavelengths from X- to Ku-band the cross section is a maximum due to the wavelength relationship between the incident energy and the physical measurements. When viewed at angles from vertical incidence to near grazing the cross section for a target under the influence of a constant windspeed varies as shown in Figure 1. It can also be seen that the radar cross section increases measurably at incidence angles between approximately 25° and 55° as the windspeed increases [11]. Other effects noted to date, either by aircraft observations or spaceborne measurements include a larger σ^0 response when viewing the surface in the upwind direction, decrease in σ^0 measured at nadir as sea state increases and better sensitivity to horizontal polarization at large incidence angles. Although the response of the ocean surface (σ^0) is a very complex function of frequency, complex dielectric constant, polarization, incidence angle and surface roughness, there is still overwhelming evidence that scatterometry will ultimately provide better remotely sensed windspeed data than previously thought possible.

The use of radar altimetry from space for geodesy, sea surface topography and ocean dynamics measurements was suggested several years ago [12]. A radar altimeter utilizes very short (wideband) pulses and a well controlled nadir oriented antenna beam. By very accurately measuring the total round trip delay time from time of transmission to when the pulse is received an estimate of the spacecraft altitude can be made. By carefully sampling and recording the return pulse so that it may be reconstructed and analyzed additional data such as waveheight may be obtained. Factors which affect the

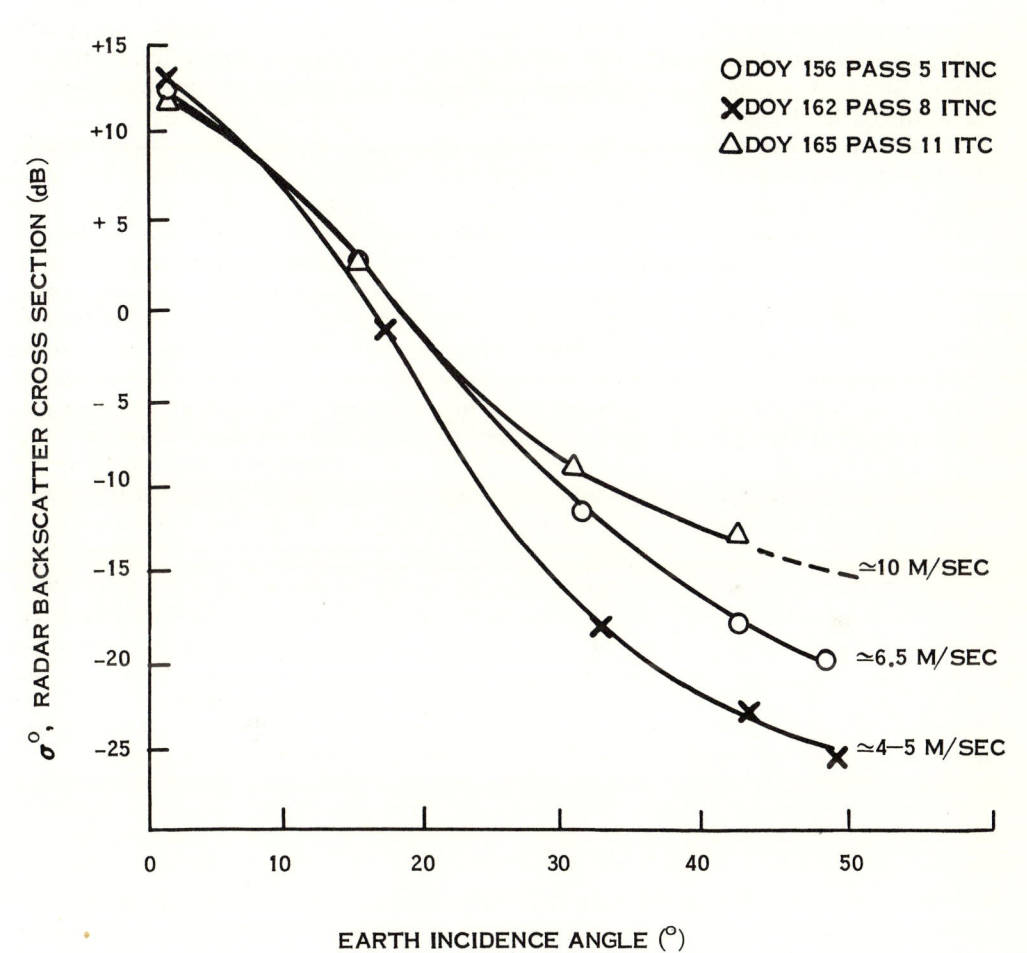

Figure 1. σ^o vs Incidence Angle

design and performance of spaceborne altimeters are antenna beamwidth, antenna pointing accuracy (and associated spacecraft attitude), transmit pulse width and ocean surface condition. All of these influence the shape and duration of the return pulse which in turn determines the response of the pulse tracker in the altimeter. In general, a very narrow pulse is required to enable accurate tracking and measurement of waveheight. As in the two techniques discussed previously satellite radar altimetry will not yield direct geophysical measurements unless supported by associated data - in this case very precise orbital tracking information. However, if such data is available, topography and geodesy information can be obtained with a precision of less than 1 meter. The altimeter flown aboard Skylab as part of the S-193 Experiment verified the concept and provided proof of the utility of this technique.

Obtaining images from radar has been in practice for several years. More recently the principles of coherent synthetic apertures have been employed which allow resolutions comparable to visible and IR scanners to be achieved. This is accomplished in the following manner: A single antenna of horizontal aperture length L is used to transmit and receive signals to the desired target. For each return the phase and amplitude of the reflected signal is preserved. The antenna is then displaced a distance X (in this case through spacecraft motion and control of repetition frequency) and the new return data also stored. After a predetermined distance has been covered all of the signals in storage are then processed with the effect of having synthesized an array of length $\frac{R\lambda}{L}$ where R is the range to the target. The azimuth resolution achieved in this manner can approach a value of $L/2$ if each measurement is focused through proper phasing. When the synthetic aperture radar is combined with an optical processor images can be generated which compare favorably with conventional photographic techniques. Thus restrictions of daylight and cloud free targets disappear in the presence of this type of microwave sensor.

At present it is hoped that imaging radar can be further developed to the point where it is able to accomplish several of the remote sensing tasks now done by scatterometers, altimeters and visible scanners. There is however, a large complexity factor associated with a coherent synthetic aperture imaging radar system and an attendant increase in data rate and processing involvement. Power requirements for such a system are also larger than those of conventional sensors and its application in the immediate future for remote earth sensing will be limited to demonstrational and less than full time missions.

THE SKYLAB S-193 EXPERIMENT

The Skylab Program provided the first opportunity to place a variety of remote sensors in earth orbit together and enabled significant advances in earth resources sensing technology to be achieved in a relatively short period. Five sensors were included in the Earth Resources Experiment Package (EREP) spanning the discipline spectrum from photographic through thermal/IR scanners to microwave. The S-193 Experiment was the first integrated, active and passive system ever deployed in space and consisted of a radar altimeter and a radiometer-scatterometer (RADSCAT) instrument which enabled nearly simultaneous backscatter and emission measurements to be made. The experiment was located outside of the habitable portion of the Skylab (see Figure 2) on the truss assembly on the underside of the Multiple Docking Adapter (MDA) and was completely self contained except for power, data recording and crew command interfaces. There have been several recent papers [2, 13, 14, 15] describing characteristics of the entire S-193 Experiment, including the radiometer-scatterometer and altimeter portions.

The broad range objective of the S-193 Experiment was to assess the performance of a microwave sensor in measuring earth resource phenomena in order to provide

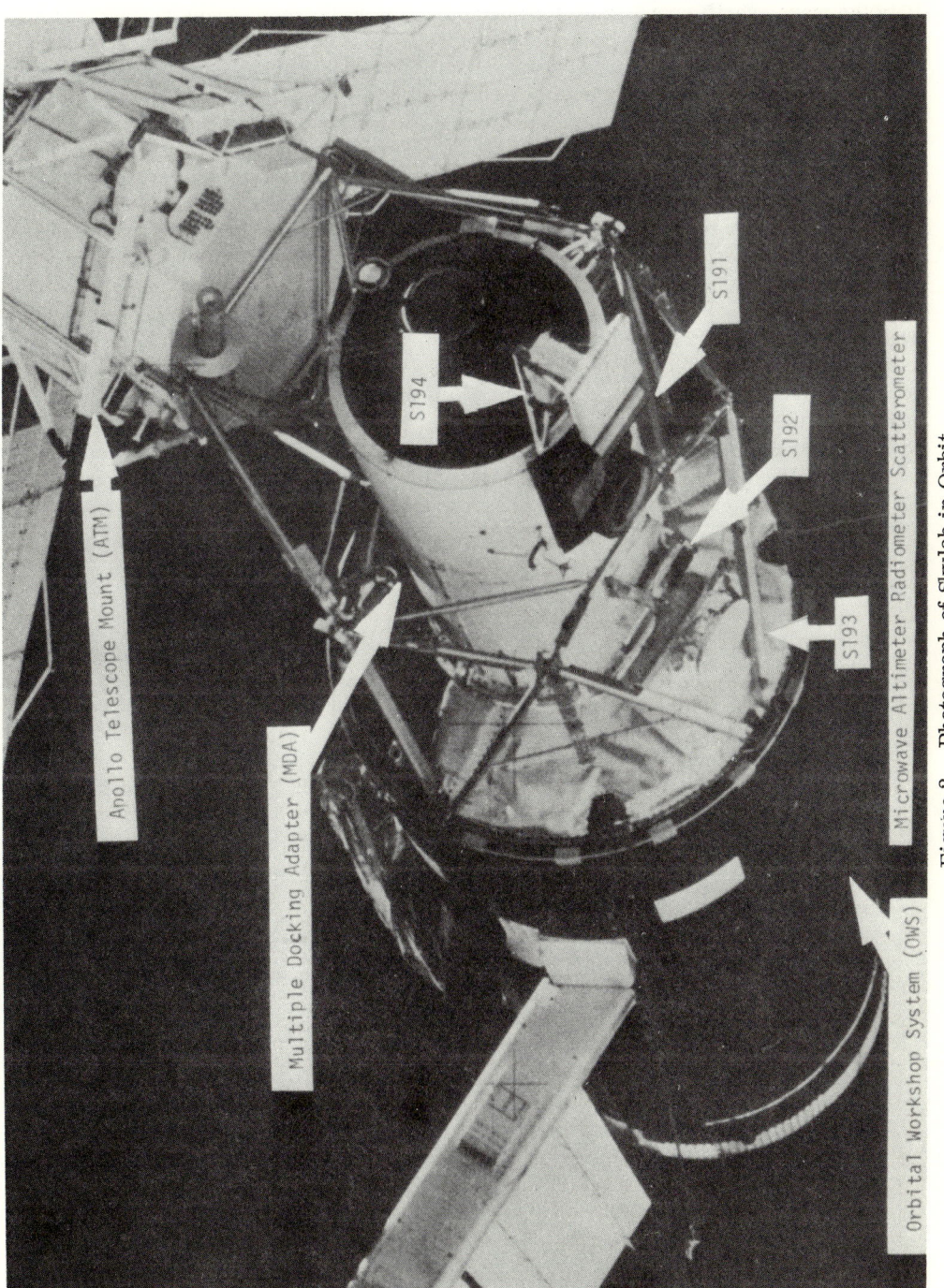

Figure 2. Photograph of Skylab in Orbit

remote sensors with optimum performance tailored to a specific application. Table 1 lists the more specific performance objectives of the radiometer-scatterometer and altimeter portions. The instrument was launched aboard Skylab I in May 1973 and was operated intermittently over a nine (9) month period during more than 100 earth resources passes. Each crew received training in operating and assessing the status of the experiment on the Command and Display Panel inside the Multiple Docking Adapter. The performance during the first manned mission was nominal, however, a failure in the antenna scanning mechanism occurred near the end of the second visit which threatened the loss of the experiment. By analyzing instrument data it was determined that a short circuit had occurred in the feedback potentiometer in either the pitch or roll gimbal assembly and tools and procedures were developed to allow the astronauts to attempt repairs. An EVA was conducted at the beginning of the last mission which confirmed the short circuit, (probably caused by conductive debris) pinned the pitch gimbal at $0°$ and enabled continued use of the experiment in a restricted capacity for the remainder of Skylab IV. Subsequent to the completion of the last mission it was learned that two additional anomalies had occurred during the final visit, a further failure in the scanning mechanism (roll axis) and loss of the reflector cup from the antenna feed. Despite these problems, data collected during Skylab IV will prove valuable with proper adjustments in the data processing.

Table 1. S-193 Experiment Objectives

Radiometer/ Scatterometer	• Correlate relationship between scat and rad returns using nearly simultaneous measurements of nearly identical earth surface cells - from orbital altitude and through seasonal changes. • Assess relative atmospheric effects on rad/scat return data. • Catalog and characterize $\sigma°$ and T_{ANT} data over a wide range of land and sea areas under varying atmospheric conditions. • Acquire and assess data to determine feasibility of rad/scat techniques for sensing: wind velocity and direction, sea surface roughness and wave patterns, sea and lake ice characteristics, seasonal changes, flooding and rainfall over large inaccessible regions, and soil moisture. • Compare orbital data to "ground truth" data gathered from aircraft sensors and ground stations. • Advance state of art in microwave transmitter and receiver technology.
Altimeter	• First in a 10-year series toward a 10 Cm resolution altimeter including S-193, GEOS-C and Seasat. • Provide knowledge of ocean scattering properties so that future altimeters may: measure geoid characteristics to great precision, provide all weather detection of ocean currents, measure ocean wave height, sea slope and detect gravitational anomalies, detect storm surges. • Demonstrate feasibility of an automatic nadir seeking system for alignment to the local vertical to optimize results. • Advance state of the art in high resolution altimeter design (PRF vs peak power, pulse compression, double pulses).

Figure 3. S-193 Block Diagram

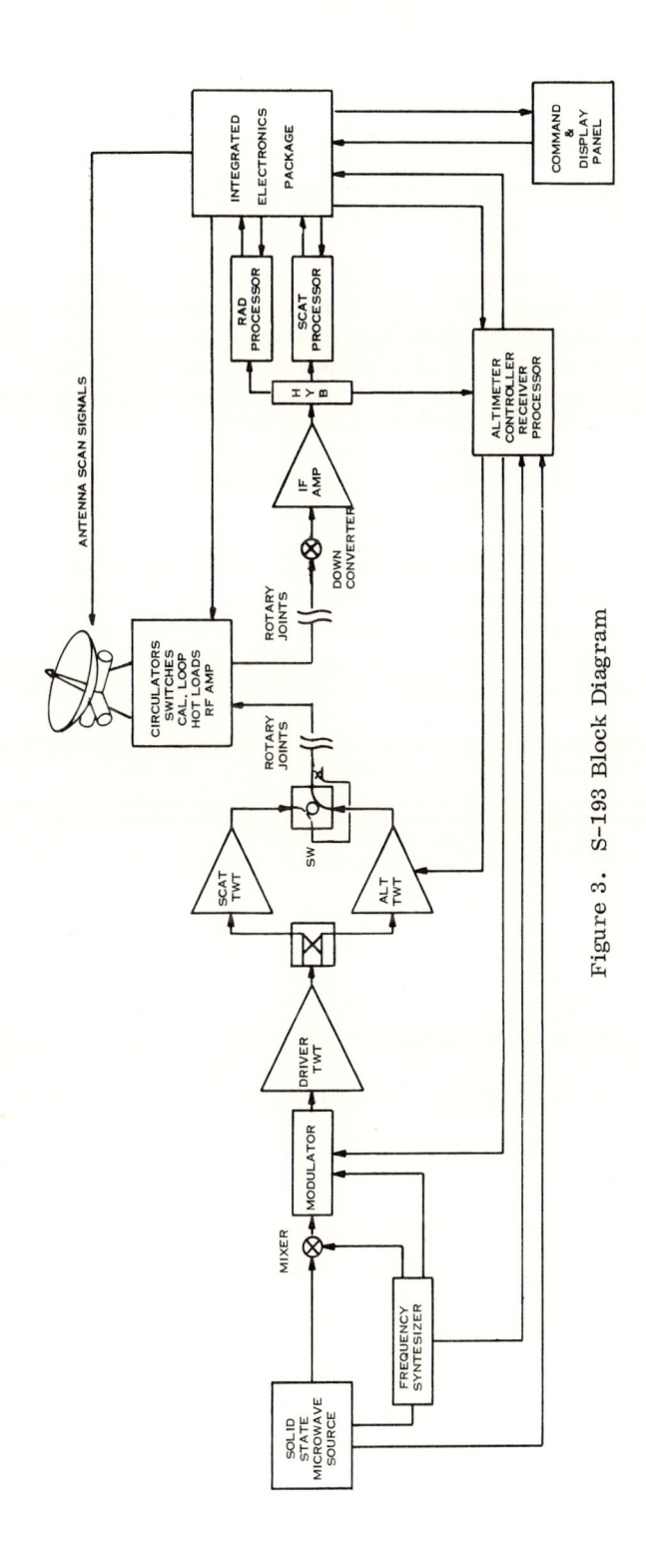

Because of its experimental nature the instrument created for Skylab was much more complex than would normally be required for a specific remote sensing application. It used common microwave components wherever possible for the RADSCAT and altimeter and employed a 1.15 meter parabolic reflector which could be automatically scanned in pitch and roll. Several operating modes were available for data collection depending on the position of the desired target and whether altimetry or RADSCAT data was required. Table 2 lists the overall performance specifications and characteristics of the S-193 while Figure 3 shows a system block diagram. The crew would select the desired operating configuration, having control over scan mode, polarization, offset angle and altimeter or RADSCAT operation and turn the instrument on. All operation from this point was automatically controlled within the experiment by digital logic contained in the integrated digital controller or in the altimeter. Tables 3 and 4 summarize the available operating modes. Figure 4 illustrates the half power ground footprints and the four basic scan modes. The scan cycle time for the ITNC mode was designed to allow viewing the earth cell from 5 successive incidence angles while the ITC cycle time provided continuous radiometer data over all incidence angles and scatterometer data at the same nominal incidence angles as in ITNC. Cross track continguous scanning (CTC) was designed with a small (1°) pitch motion correction to achieve a true orthogonal (to flight path) scan on the ground while still maintaining continuous measurements across track. Figure 5 shows two views of the experiment hardware prior to final enclosure in the aluminized Mylar thermal insulation blanket. Figure 6 is a view of the rear surface of the antenna showing the receiver "front end" components, the scanning gimbal and the microwave rotary joint.

Altimeter operations were designed to be independent of the RADSCAT portion since peak power, duty cycle and PRF were significantly different and precluded simultaneous operation. Its operation was also fully automatic and included a mode which performed an alignment of the antenna electrical axis with the local vertical to compensate for variations in the spacecraft attitude. It utilized a split gate tracker and would automatically perform an internal calibration at the end of a prescribed operating mode or in the event of an abort due to loss of track lock. A waveform sampler obtained 8 video pulse samples at a spacing of 10 or 25 nsec for the first and third pulse transmitted during each subframe equivalent to 832 pulse samples every 1.04 seconds. When processed, these samples enable reconstruction of the return pulse waveform and lead to a determination of wave height and pointing attitude [15].

Although the Skylab Program ended in February of 1974 with the recovery of the last crew, efforts are still continuing to reduce, process and analyze the large volume of S-193 data returned. These efforts will probably continue over several years but preliminary results have already been reported [2, 13, 15]. Figure 1 is an actual plot of average radar backscatter cross section versus earth incidence angle obtained form S-193 ITNC data over the Gulf of Mexico during several Skylab II passes. It clearly shows the measurable increase in σ^o as windspeed increases for a specified incidence angle.

The unexpected advent of Hurricane Ava on June 6, 1973 in the Pacific Ocean off the coast of Mexico provided the first opportunity to assess the feasibility of microwave sensing to observe large weather disturbances at sea. Several recent papers [2, 13, 16, 17] have presented measurements of this storm gathered from the Skylab microwave sensors and from other satellite and aircraft observations. Figure 7(a) shows the scatterometer VV and HH response at 50° incidence angle as the S-193 Experiment viewed the hurricane. In Figure 7(b) the radiometer brightness temperature for vertical and horizontal polarization is presented while Figure 7(c) illustrates the geographic relationship between the ground cells viewed and the eye of the storm. The value of complementary simultaneous radiometer and scatterometer measurements is well illustrated in the fact that the σ^o HH measurements follow the windspeed build up very closely while the antenna

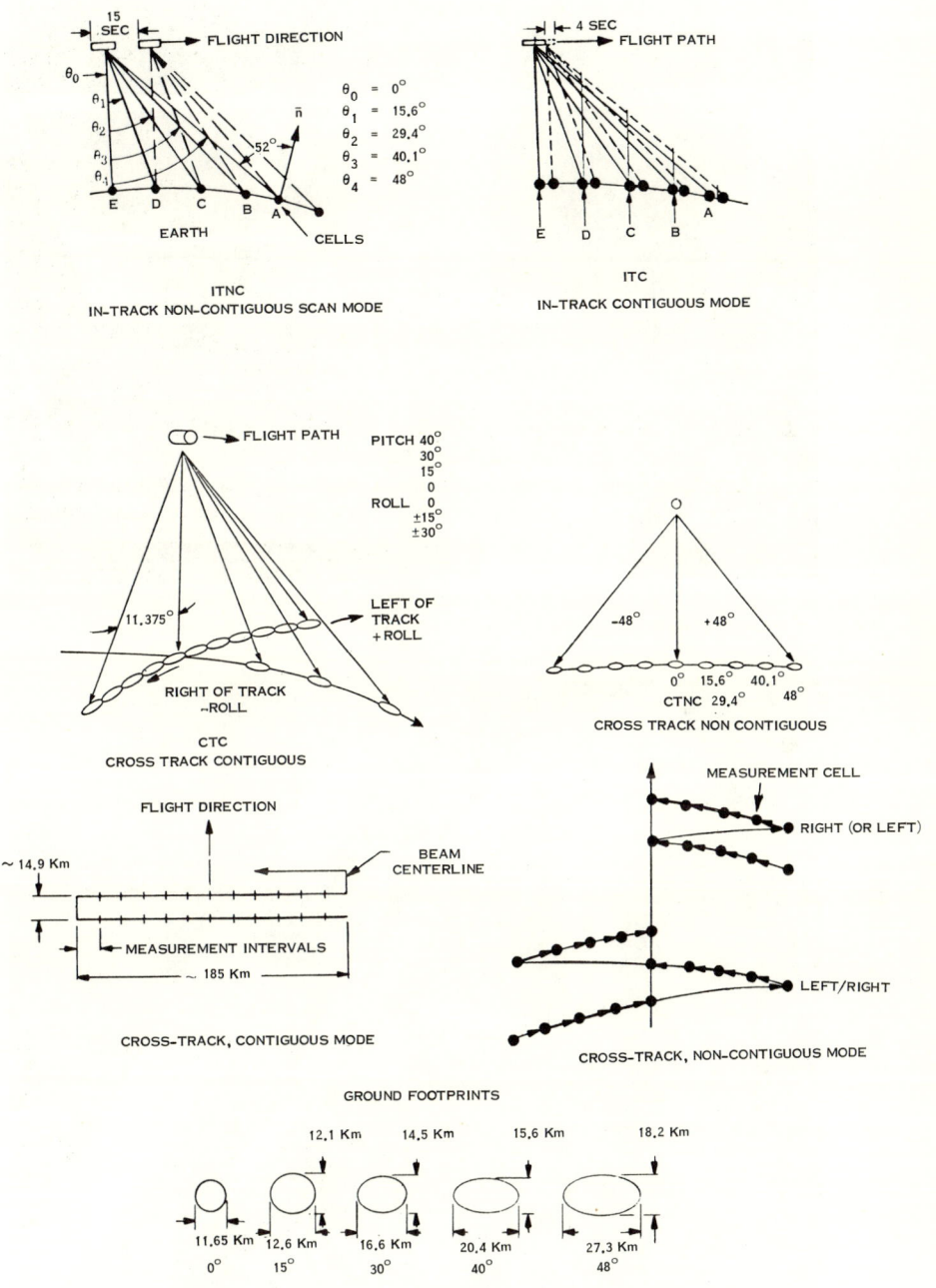

PRACTICAL CONSIDERATIONS TO THE USE OF MICROWAVE SENSING 139

Figure 4. S-193 Scan Modes and Ground Footprints

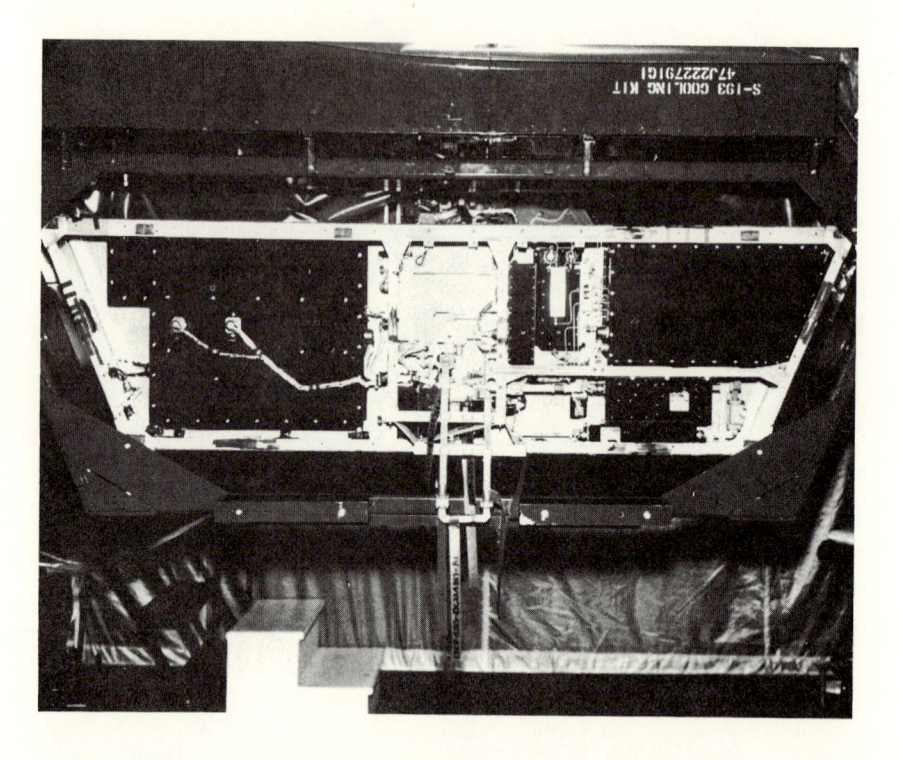

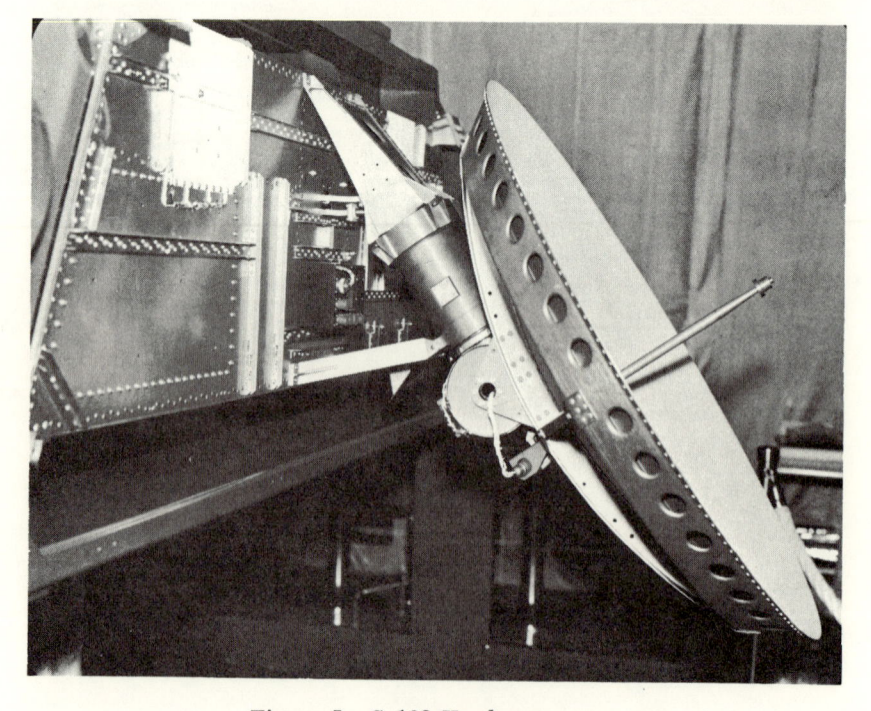

Figure 5. S-193 Hardware

Figure 6. Antenna Mounted Components

Table 2. S-193 Experiment Performance Specifications and Characteristics

Antenna and Scanning System	
Gain	41.0 dB, min
Beamwidth	1.5 ± .15°, nom
Sidelobe Level	>20 dB, down
Beam Efficiency	>85%
Polarization	Dual, Linear, V or H
Scan Capability	0° to 48° pitch (in track) +48° to -48° roll (cross track)
Scatterometer	
Transmitter	
Frequency	13.9 GHz
Output Power	20 W (pk from TWT)
Power to Antenna	12.7 W pk
Pulsewidth	5.0 m sec All Angles
Pulse Repetition Frequency	125 pps All Modes
Pulse Shape	Rectangular, 100 μsec max rise and fall time
Receiver	
Center Frequency	13.9 GHz
First IF Frequency	500 MHz
Second IF Frequency	50 MHz
System Noise Temperature	< 1200°K (7.0 dB Noise Figure)
Front End Amplifier	TDA, 5 dB max noise figure
Second IF Bandwidths	Function of antenna pitch angle (51.6 to 72.5 KHz)
Signal Plus Noise Integration Times (τ_{S+N})	8, 18 to 170 m sec
Noise Integration Times (τ_N)	6, 7 to 125 m sec
Detection	Square law device
Dynamic Range	55 dB, -80.8 dBm to -135.8 dBm at antenna
Measurement Precision	3 to 7% for $\sigma° = -30$ dB in non contiguous modes, 12% max
Radiometer	
Center Frequency	13.9 GHz
System Noise Temperature	<1200°K
Temperature Resolution	1°K
Dynamic Range	<50°K to 350°K
Calibration	Automatic, in flight adjustment of receiver gain to ΔT of hot loads
Integration Times	4, (256, 128, 58, 32 m sec)
Altimeter	
Transmitter	
TWT Output Power	2 KW, min
Frequency	13.9 GHz
Pulsewidth	20, 100 and 130 nsec
Pulse Mode	Single or Dual
Pulse Repetition Frequency	250 pps
Receiver	
Center Frequency	13.9 GHz
IF Center Frequency	350 MHz
Bandwidth	10 and 100 MHz, selectable
Pulse Compression	
Type	Binary Phase Code
Pulsewidth (Uncompressed)	130 nsec
Pulsewidth (Compressed)	10 nsec
Code	13 bit-Barker code
Signal Processing	
Altitude Tracking Loop	Digital, 200 MHz logic
Loop Bandwidth	2 Hz
Altitude Output	32 pulse avg. of 2 way delay
Altitude Granularity	1.25 ft
Acquisition Time	< 6 sec
Waveform Sampling	
Number of Sample and Hold Gates	8
Sample Date Width	10 or 25 nsec
Sample Gate Spacing	10 or 25 nsec
General	
Experiment Data Rate	4.33 Kbps Rad/Scat
	10 Kbps Altimeter
DC Power Consumption	Rad/Scat 210 W avg, 300 W pk
	Altimeter 250 W avg, 350 W pk
Weight	280 lbs.

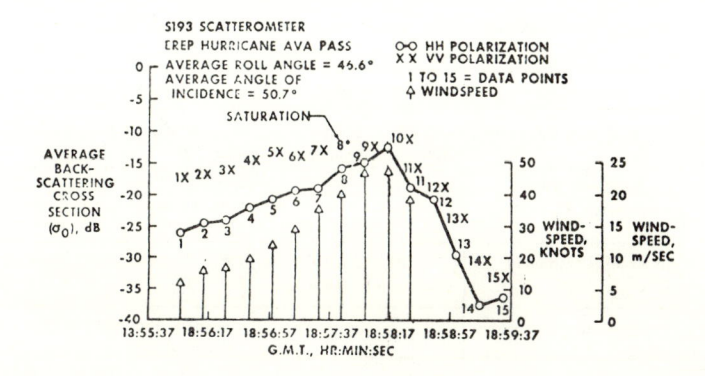

PRACTICAL CONSIDERATIONS TO THE USE OF MICROWAVE SENSING 143

Figure 7(a). Scatterometer Response, σ_{VV} and σ_{HH} (from [13])

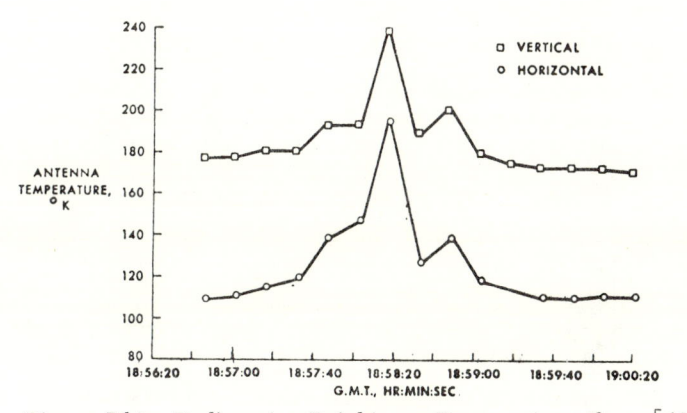

Figure 7(b). Radiometer Brightness Temperature (from [13])

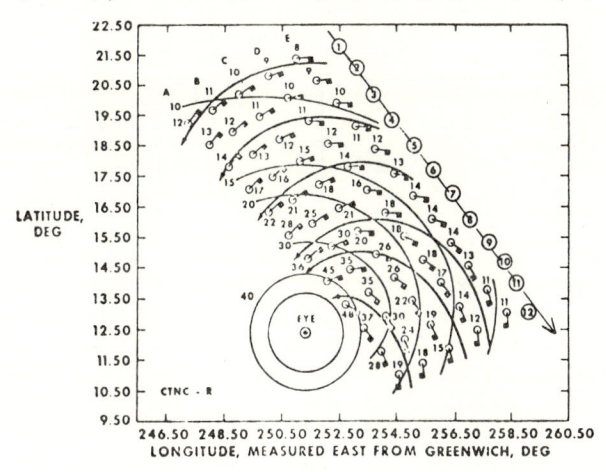

Figure 7(c). Ground Cells and Preliminary Wind Field Map (from [2])

Figure 7. Hurricane Ava Results

Table 3. Radiometer/Scatterometer Operating Modes

Operation	Scan Mode	Polarization	Antenna Scan Angles
RAD/SCAT	In Track Non Contiguous (ITNC)	VV, HH, VH, HV or Automatic Sequencing	$48°, 40.1°, 29.4°, 15.6°, 0°$ in Pitch; Roll = $0°$
	Cross Track Non Contiguous (CTNC)	Same as for ITNC	$48°, 40.1°, 29.4°, 15.6°, 0°$ in Roll; Left, Right or Left and Right of Track; Pitch = $0°$
	In Track Contiguous (ITC)	VV, HH, VH, HV - One only - No Sequencing	$48° - 0°$ in Smooth Continuous Scan; Roll = $0°$
RAD Only	Cross Track Contiguous (CTC)	Automatic POL Sequencing Between V & H	$\pm 11.375°$ Roll SCAN About Nominal Starting Angle
SCAT Only			Nominal Starting Angles:
RAD/SCAT		VV or HH Only	Pitch / Roll: $0°/0°$, $15.6°/0°$, $29.4°/0°$, $40.1°/0°$, $0°/+15.6°$, $0°/-15.6°$, $0°/+29.4°$, $0°/-29.4°$

Table 4. Altimeter Operating Modes

Mode	Antenna Angle	Pulsewidth (nsec)	Sample and Hold Gate Width (nsec)	IF Bandwidth (MHz)
I Pulse Shape	Nadir; +0.5 Pitch	100 nsec	25	10 and 100
II Backscatter Cross Section	Nadir, $+0.4°, +1.3°, +2.7°$ $+7.6°, +15.6°$ Pitch	100 nsec	25 and 100	10 and 100
III Time Correlation	Nadir	20 / 100 (Double Pulse - Variable Spacing)	25 and 10	10 and 100
V Pulse Compression	Nadir	20, 100, 130	25 and 10	10 and 100
VI Nadir Align	Nadir Seeking Scan	100	None	10

temperatures observed are relatively insensitive to windspeeds below 15 m/sec (30 knots) but increase dramatically as the winds grow from 15 m/sec to 25 m/sec (50 knots). The radiometric data is also used to correct the backscatter observations because of its sensitivity to rainfall. The combination of both techniques appears to provide a wider range of measurements than would be possible with either one alone.

EREP Pass 5 from Skylab II over the Great Salt Lake Desert provided another illustration of the value of microwave remote sensing techniques. Figure 8(a) illustrates the ground track for this pass with the crosshatched area representing the data presented in Figures 8(b) and 8(c). The instrument was operated in the CTC RADSCAT scan mode with a $0°$ pitch angle offset. Figure 8(b) shows the scattering cross section as a function of incidence angle and is more specular in nature than was expected for a target as

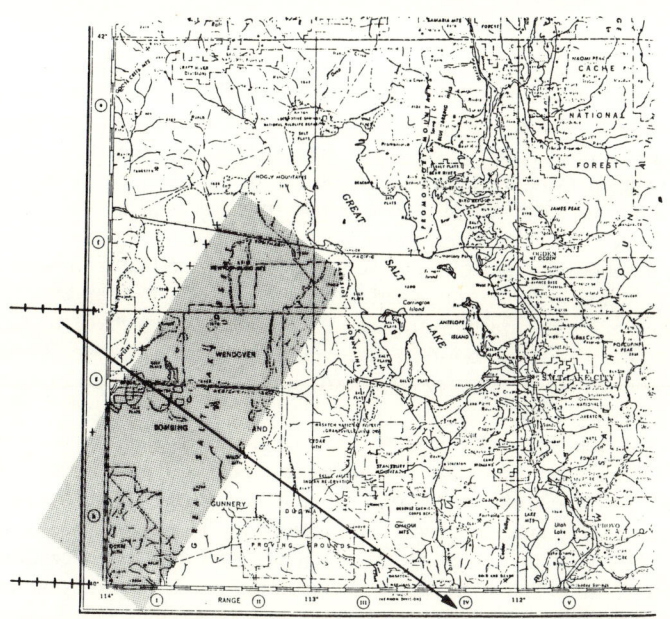

PRACTICAL CONSIDERATIONS TO THE USE OF MICROWAVE SENSING

Figure 8(a). Skylab II, Pass 5 Ground Track and S-193 (CTC-R) 5 Ground Cell Coverage

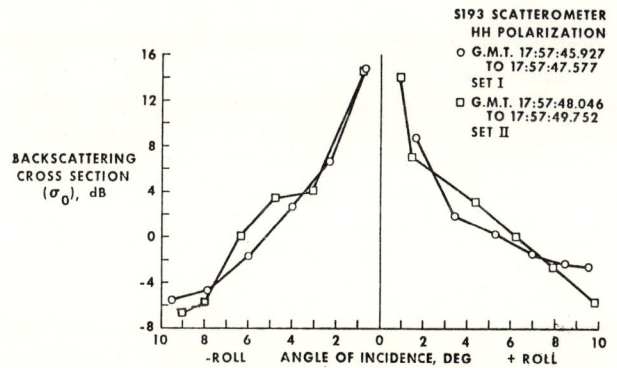

Figure 8(b). Scattering Cross Section as Function of Incidence Angle (from [13])

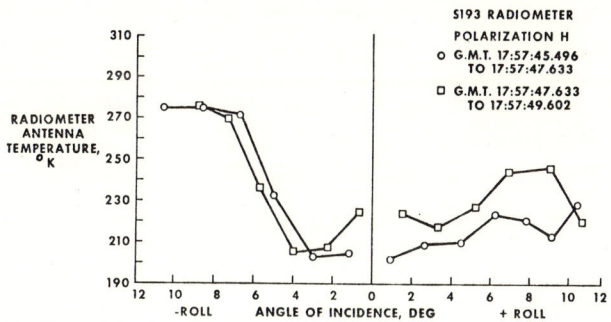

Figure 8(c). Radiometric Brightness as Function of Incidence Angle (from [13])

Figure 8. Great Salt Lake Desert Results

homogeneous as the Great Salt Lake Desert. Fortunately, the radiometric response shown in Figure 8(c) provided the explanation. In general, radiometric brightness temperatures of about $270^\circ K$ would be expected from a target such as this but as can be seen they are much lower in the areas where the σ° data are large. The lower temperatures measured can be expalined by significant amounts of moisture in the area. Increase in moisture content tends to lower the surface emissivity (and thus the radiometric temperature) and increase the backscatter cross section.

Figure 9 is a composite windfield map for a portion of the South Atlantic Ocean off the coast of Rio de Janeiro, Brazil obtained from S-193 data. The instrument was operated in the CTC-R/S mode at a pitch offset angle of $+30^\circ$. Cloud cover photographs and meteorological charts were used to aid in determining the atmospheric corrections and wind direction biases. This format for remotely sensed ocean data could be very useful in long range forecasts.

Several other interesting portions of S-193 radiometer and scatterometer data have been processed and analyzed but will not be presented here. It is felt that the possibility of using microwave sensing techniques to measure ocean surface phenomena has been well proven by this experiment. Figure 9 [18] shows the results of the correlation of many meteorologically measured wind fields plotted against those determined from S-193 RADSCAT measurements. The results are encouraging and illustrate that we may begin considering operational uses of microwave remote sensing.

The altimeter portion of the S-193 has also provided some very encouraging results [15, 19]. Figure 11 illustrates the capability of a satellite borne microwave altimeter to measure sea surface topography. Two data sets from Skylab II representing passes over the Puerto Rican Trench on June 4 and 9, 1973 show excellent agreement in relative value. The localized gravitational anomaly in the area produces a depression in sea surface elevation of more than 10 meters over a distance of approximately 150 km. Additional Skylab S-193 Altimeter data has provided updates to the present geoid model and has even shown that many large ocean floor features such as sea mounts can be detected on the surface through altimetry. The results of altimeter data analysis to extract waveheight information have also been encouraging and indicate that future altimeters with finer resolutions will be a valuable tool in monitoring our seas.

THE SEASAT PROGRAM

Building upon the experience and achievements of the Skylab and previous Earth Resources Programs, NASA has established the SEASAT Program as the next step in its larger Earth and Ocean Physics Applications Program (EOPAP). SEASAT-A, to be launched early in 1978, is a satellite devoted wholly to ocean phenomena observations. As presently conceived its payload sensor complement will be all microwave except for a visible and IR scanner. The microwave sensors to be flown include a nadir looking radar altimeter, a fan beam scatterometer and a coherent synthetic aperture imaging radar [20]. The goals and objectives of the program are listed in Table 5.

Preliminary work on the SEASAT Program began in early 1973 with the formation of two consultant groups. The User Working Group consisted of representatives of government agencies involved in maritime work, research organizations and commercial users of the seas such as shippers and fishing fleet representatives. The second body, known as the Instrument Working Group was composed of experts in the fields of remote sensing from various NASA centers and other research organizations. These two groups met together and individually several times in effort to define the measurement and instrument requirements for SEASAT-A. Table 6 lists the results of the work performed by the User Community in determining its needs for oceanographic measurements, while Table 7 and 8

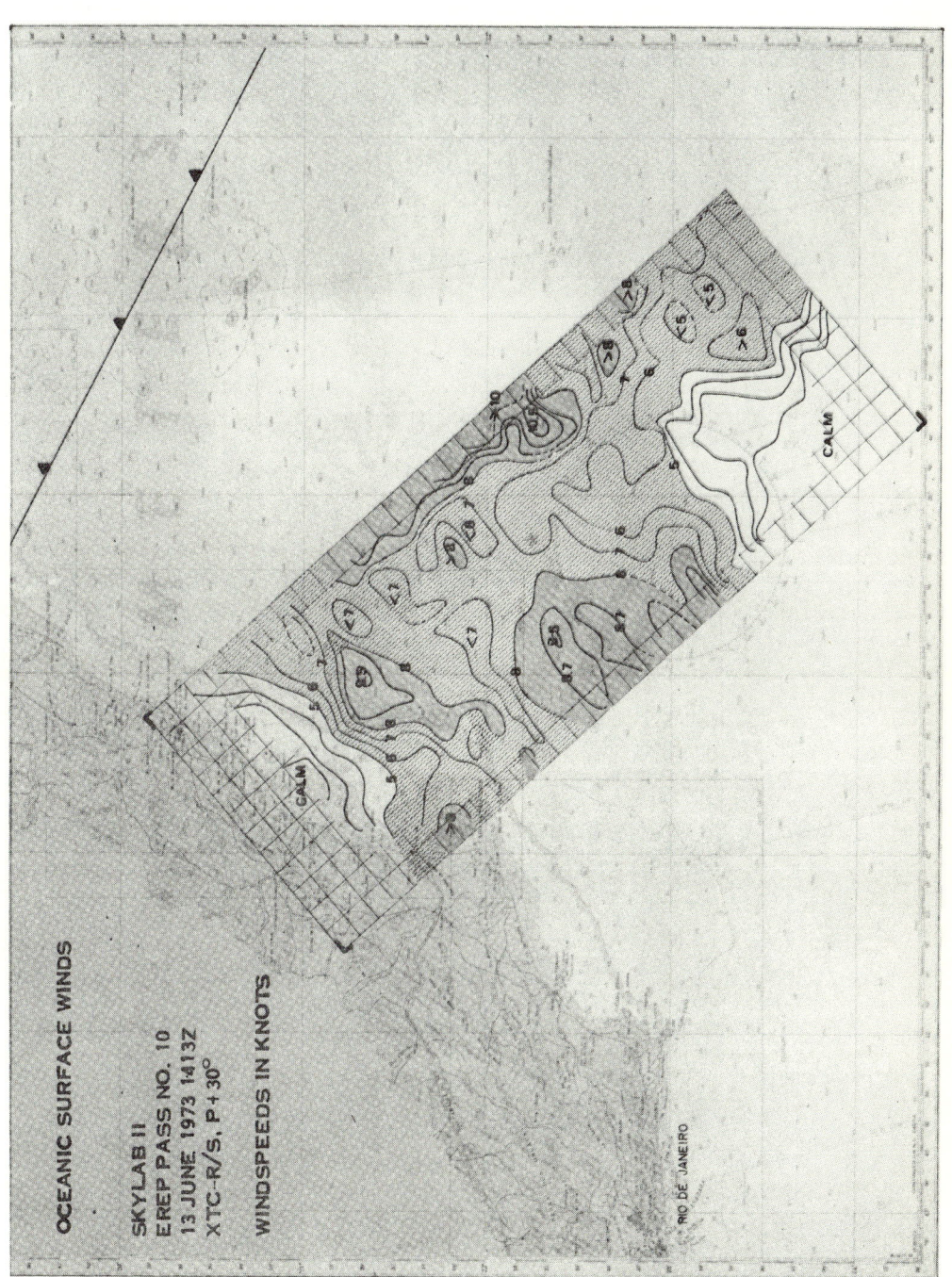

Figure 9. Windfield Map from S-193 Data

Table 5. Goals and Objectives of Seasat-A

- Demonstrate Satellite System Capability for
    1. Global monitoring of wave height and directional spectrum, surface wind, ocean surface temperature and current patterns.
    2. Measuring precise sea-surface topography, locating mid ocean currents, measuring amplitude of tides, storm surges and other transients.
    3. Charting ice fields and coastal processes.
    4. Mapping global ocean geoid.
- Provide Prototype Data to Users for Applications Involving
    1. Predictions of wave height, directional spectrum and wind fields for ship routing, storm damage avoidance and disaster warning.
    2. Maps of current patterns and temperatures for fishing fleets, pollution dispersion forecasts and iceberg hazard avoidance.
    3. Ocean geoid maps.
    4. Charts of ice fields for navigation.
- Demonstrate Operational Potential
    1. Near real time data for users.
    2. Command capability for obtaining user specified data.

Table 6. Oceanographic Measurement Requirements in Geophysical Terms (From [20])

Measurement			Range	Precision/Accuracy	Resolution	Spatial Grid	Temporal Grid
Topography	Geoid		5 cm - 200 m	$< \pm 10$ cm	< 10 km	< 20 km	Weekly to Monthly
	Currents, Surges, etc.		10 cm - 10 m 5 - 500 cm/sec	$< \pm 10$ cm ± 5 cm sec	10 - 1000 m	< 10 km	Twice a day to Weekly
Surface Winds	Amplitude	Open Ocean	3 - 50 m/s	± 2 m/s OR ± 10 - 25%	10 - 50 km	50 - 100 km	2 - 8/day
		Closed Sea			5 - 25 km	25 km	
		Coastal			1 - 5 km	5 km	Hourly
	Direction		0 - 360°	$\pm 20°$			
Gravity Waves	Height		0.5 - 20 m	± 0.5 m OR ± 10 - 25%	< 20 km	< 50 km	2 - 8/day
	Length		6 - 1000 m	± 10 - 25%	3 - 50 km		2 - 4/day
	Direction		0 - 360°	± 10 - 30°			
Surface Temperature	Open Ocean		-2 - 35°C	0.1 - 2° Relative 0.5 - 2° Absolute	25 - 100 km	100 km	Daily to Weekly with Spectrum of Times of Day and Times of year
	Closed Sea				5 - 25 km	25 km	
	Coastal				0.1 - 5 km	5 km	
Sea Ice	Extent and Age				1 - 5 km	1 - 5 km	1 - 5 km	Weekly
	Leads		> 50 m	25 m	25 m	25 m	
	Icebergs		> 10 m	1 - 50 m	1 - 50 m		2 - 4 Day
Ocean Features	Open Ocean			50 - 500 m			Twice Daily to Daily
	Coastal			10 - 100 m			
Salinity			0 - 30 ppt	± 0.1 - 1 ppt	1 - 10 km	100 km	Weekly
Surface Pressure			930 - 1030 mb	± 2 - 4 mb	1 - 10 km	1 - 10 km	Hourly

list the experimental areas of investigation and the proposed sensor complement to achieve them.

The data to be returned from Seasat will be handled by existing ground stations to the fullest extent possible in order to keep costs at a minimum. At this point in the program

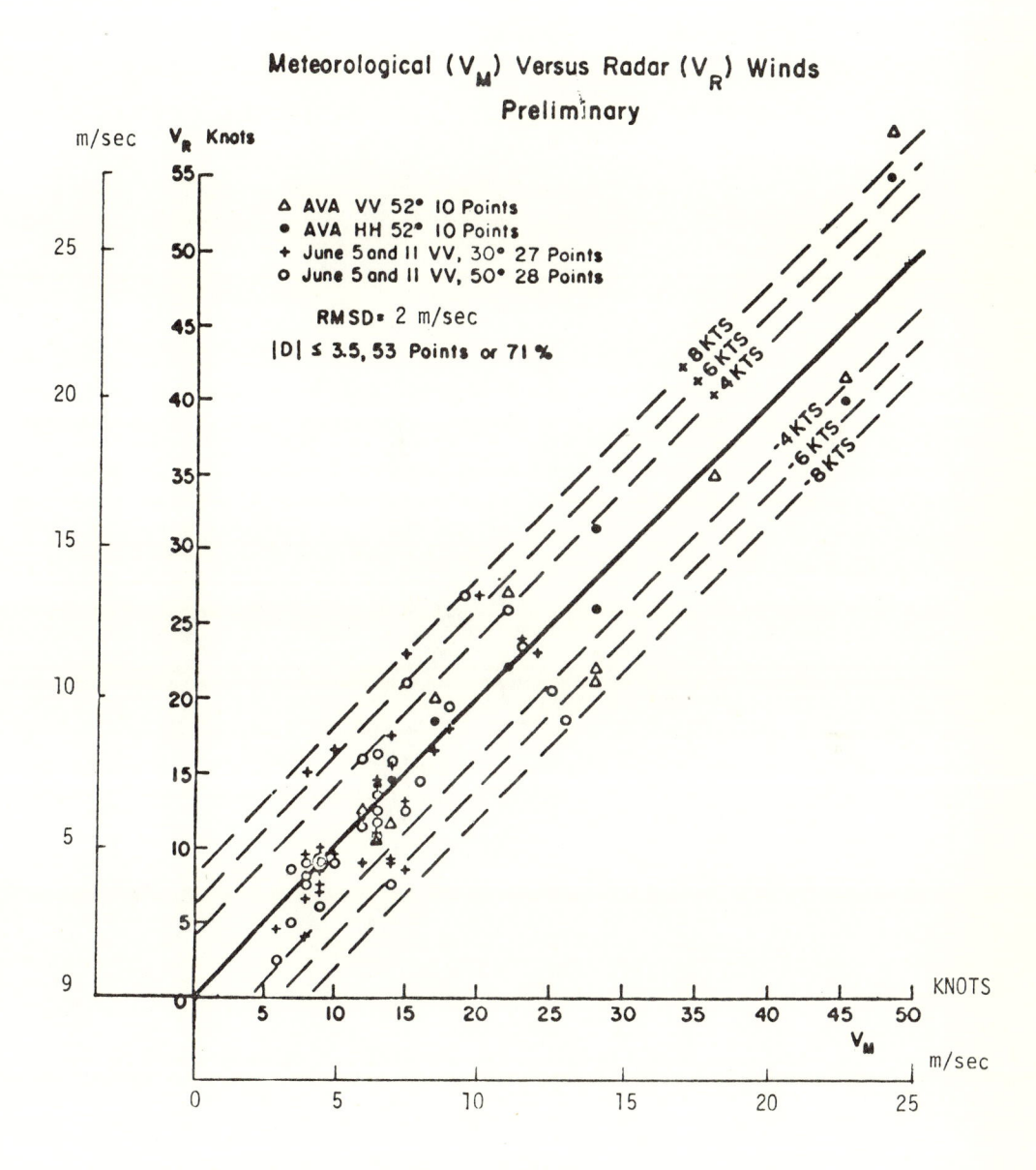

Figure 10. Comparison of Meteorologically Measured Windspeed Against Radar Determined Windspeed (from [18])

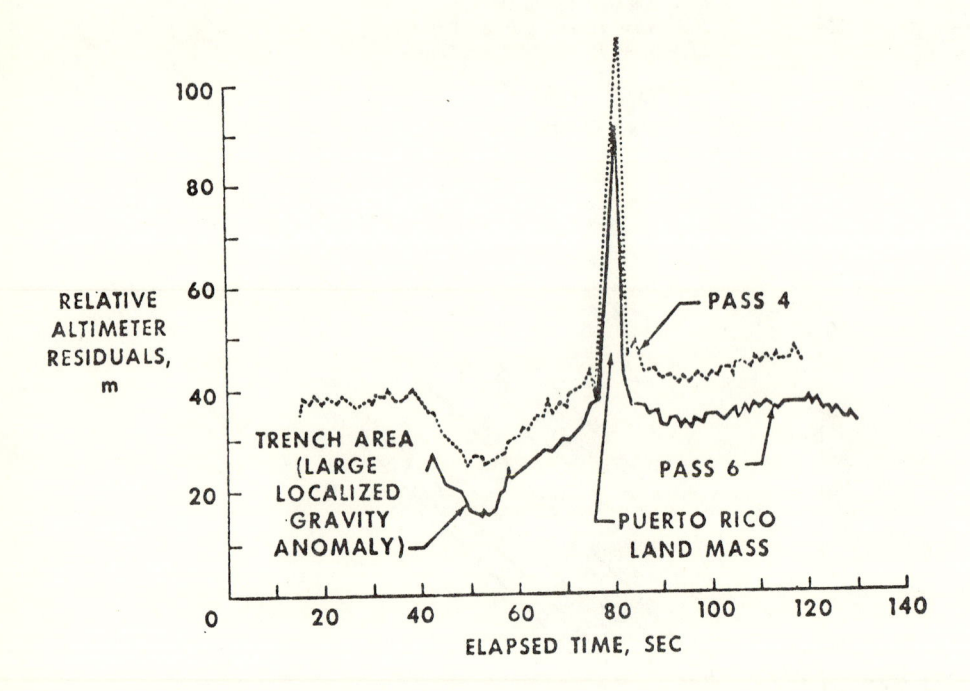

Figure 11. S-193 Altimeter Data Over the Puerto Rican Trench (from [19])

Table 7. Physical Oceanography Experiments (from [20])

Discipline Area	Experiment Area	ALT	SCAT	SMMR	SR	SAR
Geodesy	Equipotential Surface	X				
	Slope of Mean Sea Level	X				
	Fine Geoid Features	X				
Wind and Wave	Large Gravity Wave Generation	X	X			
	Wave Propagation Near Storms	X				X
	Wave Propagation Near Shore	X				X
	Wave Propagation at Continental Shelf	X				X
	Internal Wave Propagation					X
	Regular Verification of Forecasts	X				X
	Storm Wind/Cloud Relation		X	X	X	
	Wind/Rain Interactions		X	X		X
Dynamic Ocean Features	Location/Dynamics of Ocean Currents					X
	Transport of Chemicals, Pollutants,				X	X
	Salinity and Nutrients				X	
	Upwelling	X				
	Tidal Behaviors	X				
	Tsunami Propagation (Fortuitous)					X
Coastal Processes	Water Pile Up From Storms	X	X			X
	Fresh Water Influx					X
	Shoal and Shore Line Dynamics					X
	Kelp Extent					X
Ice Processes	Ice Distribution/Extent	X		X	X	X
	Ice Leads for Navigation					X
	Ice Formation/Ridging/Breakup					X
	Ice Transport					X
Thermal Processes	Surface Temperature			X	X	
	Poleward Transport of Heat			X	X	X

Table 8. Seasat Sensor Complement (from [20])

Active			Passive	
Altimeter	Scatterometer	Synthetic Aperture (Imaging) Radar	Scanning Multifrequency Microwave Radiometer	Scanning (Visible and Infrared) Radiometer
Global Ocean Topography	Global Wind Speed and Direction	Wave Length Spectra	Global All-Weather Temperature	Global Clear-Weather Temperature
Global Wave Height		Local High Resolution Images	Global Wind Amplitude	Global Feature Identification
				0.52 - 0.73 μm
				10.5 - 12.5 μm
13.0 GHz	13.9 OR 14.5 GHz	1.37 GHz	6.6, 10.69, 18, 22.235, 37 GHz	5 m (IFOV)
1 m Parabola	5 - 2.7 m Stick Array	10 x 1.2 m Microstrip Array	1.25 m Offset Parabola $\pm 35°$ Cross Track Scan	
2.5 kW Peak	125 W Peak	400 W Peak		
125 W	107 W	200 - 205 W	50 W	10 W
7, 0.42 kb/s	400 b/s	15.06 mb/s	4 kb/s	12 kb/s

the final configuration for the total data system (from sensor-to spacecraft-to ground station-to processor-to user) has not been defined but will be developed at the same time the hardware is being built in order to avoid any delays in processed data once the satellite is launched.

PRACTICAL CONSIDERATIONS

After learning about the capabilities of microwave remote sensing techniques and reviewing the results obtained from some early experiments it is easy to conclude that extensive potential exists for this application. However, there are some very real, physical limitations imposed upon this method as well as other practical considerations which serve to illustrate the problems associated with microwave sensing from space platforms.

Spatial resolution or the amount of surface area sensed for each measurement always appears in the requirements for earth observations from satellites and is a good example of the physical limitations of microwave sensors. For an active sensor control over the area illuminated is exercised through control of the antenna beamwidth or the transmitted pulse length (i.e., beamwidth limited or pulse width limited), while control over the size of the illuminated area actually measured can be maintained through the techniques of range gating or Doppler filtering. Exercising these parameters in the design of a microwave sensor system usually will result in a set of parameters tailored to the application in question, but may also produce severe restrictions on other spacecraft subsystems or components. For example, to minimize the area illuminated (within the half power beamwidth) it is desired to reduce the antenna beamwidth to as small a value as possible. Since the beamwidth of a microwave parabolic antenna is inversely proportional to the product FD where F is the frequency and D is the aperture diameter, small beamwidths require large antennas and high operating frequencies. However, it is not always possible to arbitrarily increase the operating frequency due to the interference of atmospheric absorption and attenuation effects which could obscure the desired measurement. Also, large antennas become difficult to realize for spacecraft due to stowage and deployment considerations and because the higher frequencies also dictate tighter dimensional tolerance which are hard to achieve over large surfaces. Assuming a reasonable tradeoff between frequency, antenna size and beamwidth has resulted in a practical antenna design and that use has been made of range and Dopper discrimination further complications can result if care has not been taken in specifying attitude control tolerances. For example, small errors in spacecraft pitch and roll can produce large range errors when the antenna is operated at angles off nadir. For fixed range gate timing this results in a measurement taken at a portion of the return pulse different than designed for and could mean an error of several hundred kilometers. In addition, spacecraft yaw errors rotate the ground footprint and will introduce different Doppler shifts on the return than originally anticipated and could also result in measurements of areas vastly different from those originally intended. Each of these considerations complicate the realization of microwave sensors and must be carefully analyzed in the system design.

The accuracy of microwave sensor measurements, particularly scatterometry is very tightly coupled to the final signal to noise ratio and measurement time achieved. The former is in turn related to the power transmitted, receiver noise figure, bandwidth and background clutter levels, while the latter is influenced by the permissible "smear" of an individual measurement and the required spacing between successive measurements. It is possible, using state-of-the-art microwave technology to achieve receiver noise figures as low as 4 dB (total, including RF amplifier and front end losses) and output devices exist which can provide hundreds of even thousands of watts of output power. However, the efficiencies and permissible duty cycles of these output devices (usually traveling wave tubes) are limited and their use imposes penalties in terms of high average DC power consumption for the sensor as a whole. In addition, generation of high power microwave energy in vacuum devices requires extreme high voltages (4 KV to >12 KV) which complicates the design of power supplies and imposes additional reliability considerations when required to work over extended durations in a space environment. Solid state microwave technology in the area of output devices is progressing and promises future relief for some of these problems but is not yet sufficiently mature in all respects to be considered a direct substitute.

There are circumstances in which further improvements in input signal to noise ratio will not serve to improve measurement accuracy. For example, it is conceivable that an RF amplifier noise temperature of $150°K$ can be achieved using parametric amplifier techniques. However, in the case of a radiometer, resolution is a function of total system noise temperature which includes the RF amplfier, system front end losses and antenna temperature. For sea surface observations the average antenna temperature may (vertical polarization) be as high as $150°K$ and further improvements in RF amplifier performance would tend to be masked by the antenna and front end contributions. A similar example exists in the case of altimetry in terms of the altitude resolution achievable using conventional split gate pulse tracking techniques. A recent study [21] has shown that tracker resolution (normalized standard deviation of successive time delay measurements) reaches an asymptotic value which is an order of magnitude coarser for split gate techniques as compared to adaptive processing (maximum likelihood tracking) for a given signal to noise ratio. Thus increasing signal to noise ratio past the asymptotic value would not improve measurement accuracy for a given tracker. However, savings in total system power could be realized by trading front end receiver performance for output power to achieve the same signal to noise ratio.

One final, very important practical consideration centers around the utility and timeliness of data returned. As mentioned previously, microwave sensors are capable of very accurate measurements of parameters such as radar backscattering cross section or radiometric brightness temperature, but not of ocean phenomena directly in geophysical terms. It is a long step from sensor data to geophysical units and requires extensive processing and the involvement of physical oceanographers and atmospheric physicists as well as specialists in electromagnetic scattering and other disciplines. Any satellite sensor system must be built as part of an extensive end to end data system which insures that the desired sensor data is transferred to the spacecraft data system quickly and efficiently, stored (for later transmission) or relayed to the ground, processed either into geophysical units or into an intermediate format for interpretation by specialists and finally, disseminated in time to be of value to the ultimate user. Only when microwave sensors are employed in this fashion will they realize their full potential for helping to solve energy related problems.

CONCLUSIONS

The S-193 Experiment has demonstrated the feasibility and value of microwave remote sensing from space. The application of these techniques on the GEOS-C Satellite and SEASAT Program suggest that this method is gaining acceptance as a viable means for monitoring the earth's environment and managing its resources. Limitations and practical considerations exist which somewhat confine the application of microwave remote sensing, but do not restrict its use when properly accounted for in the design of overall systems.

ACKNOWLEDGEMENTS

The author appreciates the suggestions and assistance of Dr. F. C. Jackson, Dr. K. Tomiyasu, Mr. M. Berkowitz and Mr. R. Bianchi. Acknowledgement is also given to the many excellent sources of information and data relative to the S-193 Experiment and microwave sensors in general, which are cited in the references.

REFERENCES

1. F. T. Barath, A. H. Barrett, J. Copeland, D. E. Jones and A. E. Lilley, "Mariner 2, Preliminary Reports on Measurements of Venus", Science, Vol. 139, pp. 908-909, 1963.

2. R. K. Moore, et al, "Simultaneous Active and Passive Microwave Response of the Earth - The Skylab Radscat Experiment", Presented at 9th International Conference on Remote Sensing of Environment (Ann Arbor, Mich.), April 15, 1974.

3. K. Tomiyasu, "Remote Sensing of the Earth by Microwaves", Proceedings of the IEEE, Vol. 67, No. 1, pp. 86-92, Jan. 1974.

4. R. K. Moore, "A Consultant Report Submitted to the European Space Research Organization", The University of Kansas, August 1973.

5. Hugh P. Taylor, "The Radiometer Equation", The Microwave Journal, pp. 39-42, May 1967.

6. R. H. Dicke, "The Measurement of Thermal Radiation at Microwave Frequencies", Rev. Sci. Instr., Vol. 17, pp. 268-275, July 1946.

7. J. P. Hach, "A Very Sensitive Airborne Microwave Radiometer Using Two Reference Temperatures", IEEE Transactions Microwave Theory and Techniques, Vol. MTT-16, pp. 639-636, September 1968.

8. W. N. Hardy, K. W. Gray, A. W. Love, "An S-Band Radiometer Design with High Absolute Precision, IEEE Transactions Microwave Theory and Techniques, Vol. MTT-22, pp. 382-39, April 1974.

9. W. J. Webster, T. C. Chang, P. Gloersen, T. J. Schmugge, T. T. Wilheit, "Satellite Microwave Radiometry", Proceedings IEEE Intercon 74, Session 34 Earth and Ocean Physics Application Program, March 1974.

10. R. E. Fischer, "Standard Deviation of Scatterometer Measurements from Space", IEEE Transactions and Geoscience Electronics, Vol. GE-10, No. 2, pp. 106-113, April 1972.

11. C. T. Swift, W. L. Jones, Jr., "Satellite Radar Scatterometer", Proceedings IEEE Intercon 74, Session 34, Earth and Ocean Physics Application Program, March 1974.

12. J. R. Apel, Ed., "Sea Surface Topography from Space", Nat. Oceanic and Atmospheric Admin. Tech. Rep. ERL-228-AOML-7, Vols. 1 and 2, Oct. 1971.

13. J. G. Zarcaro, "Significance of the Skylab Earth Resources Experiment Results to the NASA Earth Resources Program", Proceedings XXVth Congress, International Astronautical Federation (IAF) October 1974, Amsterdam.

14. L. S. Miller, D. L. Hammond, "Objectives and Capabilities of the Skylab S-193 Altimeter Experiment", IEEE Transactions Geoscience Electronics, Vol. GE-10, pp. 73-79, January 1972.

15. J. T. McGoogan, et al, "The S-193 Radar Altimeter Experiment", Proceedings of the IEEE, Vol. 62, No. 6, June 1974.

16. V. J. Cardone, R. K. Moore, W. J. Pierson, Jr. and J. O. Young, "Preliminary Report on Skylab S-193 Radscat Measurements of Hurricane Ava, CUNY UIO Report #23, KU RSC Report 254-1, October 1973.

17. D. Ross, B. Au, W. Brown, J. McFadden, "A Remote Sensing Study of Pacific Hurricane Ava", Presented at 9th International Conference on Remote Sensing of Environment, (Ann Arbor, Mich.), April 15, 1974.

18. R. K. Moore, W. J. Pierson, et al, "A Preliminary Difference Analysis of the Meteorological and Radar Wind Vector Magnitudes for the Skylab II", A Joint Meteorological, Oceanographic and Sensor Evaluation Program for Experiment S-193 on Skylab, EPN550, Contract NAS 9-13642, May 20, 1974.

19. J. T. McGoogan, "Precision Satellite Altimetry", Proceedings IEEE Intercon 74, Session 34, Earth and Ocean Physics Application Program, March 1974.

20. R. G. Nagler, A. A. Loomis, "SEASAT-A Monitoring the Oceans from Space", Presented at IEEE Electronics and Aerospace System Convention, EASCON 74, (Washington, D.C.), October 9, 1974.

21. R. P. Dooley, F. E. Nathanson, L. W. Brooks, "Study of Radar Pulse Compression for High Resolution Satellite Altimetry", Final Report Contract No. NAS 6-2241, TSC-WO-111, May 1973.

LASER MEASURE OF SEA SALINITY, TEMPERATURE AND TURBIDITY IN DEPTH

JOSEPH G. HIRSCHBERG, ALAIN W. WOUTERS and JAMES D. BYRNE

Laboratory for Optics and Astrophysics of the Department of Physics, and the Department of Mechanical Engineering, University of Miami Coral Gables, Florida, U.S.A.

ABSTRACT

A method is described in which a pulsed laser is used to probe the sea. Backscattered light is analyzed in time, intensity and wavelength. Tyndall, Raman and Brillouin scattering are used to obtain the backscatter turbidity, sound velocity, salinity, and the temperature as a function of depth.

THE PROBLEM

It is becoming clearer every year that as man's domination over the land areas of the world becomes nearly complete, the oceans are emerging as the new frontier for research and exploitation. Compared to our knowledge of islands and continents, our quest for information about the waters of the earth has virtually only begun. Although gross descriptions of major ocean currents and water masses are known, there has been until very recently virtually no continuous source of information about even such elementary quantities as sound velocity, temperature, salinity and turbidity. Without a knowledge of these basic building blocks of any theory, we shall find it difficult to proceed in our conquest of the last frontier on earth, which comprises more than three fourths of its surface.

Very recently, it has become possible through the methods of remote sensing to obtain some the needed data. By measuring the infrared emission from the water, the surface temperature may be continuously determined.

The infrared portion of the electromagnetic spectrum lies between the visible and microwave regions (0.78 and 1000 micrometers). The existence of this "invisible" radiation was discovered in 1800 by Sir Frederick William Herschel, who used a thermometer to detect the energy beyond the red portion of the visible spectrum produced by a prism. In 1861, Richard Bunsen and Gustav Kirchoff established the principles underlying infrared spectroscopy. After a century of further progress, the uses of infrared technology have been extended to a host of applications. Among them is the use of infrared sensors to measure the surface temperature of a distant object. This is done passively; that is, unlike radar no energy is sent to the object being investigated. Such measurements are based upon the fact that all bodies above absolute zero radiate electromagnetic energy to their surroundings: the higher their temperature, the more they will radiate. The most efficient radiator at a given temperature is the so-called "black body", which is also a perfect absorber. The radiation of a black body is described by the Planck Radiation Law:

$$I_{\lambda T} = \frac{2hc^2}{\lambda^5} \cdot \frac{\Delta\lambda}{e^{hc/k\lambda T}-1}$$

Here, h is Planck's constant, c is the velocity of light, λ is the wavelength, $\Delta\lambda$ is the wavelength range, k is Boltzman's constant and e is the base of natural logarithms. This formula gives the energy emitted per second from a square centimeter of surface at absolute temperature T at wavelength λ in a wavelength band of width $\Delta\lambda$ into an unit solid angle. Putting in values for a 1 micrometer band:

$$I = \frac{1.192 \times 10^{-16}}{\lambda^5 (e^{1.44/\lambda T} - 1)} \text{ watts/cm}^2 \text{ ster. } \mu. \tag{2}$$

At the sun's surface, for example, if the black body temperature is 5500°K at λ = 6000 A (0.6µ), I = 6000 watts/cm² ster. micrometer.

The emission from a black body versus wavelength is plotted in Figure 1 with temperature as a parameter. One can see from this figure that the emission maxima occur toward longer wavelengths with decreasing temperature. The emission

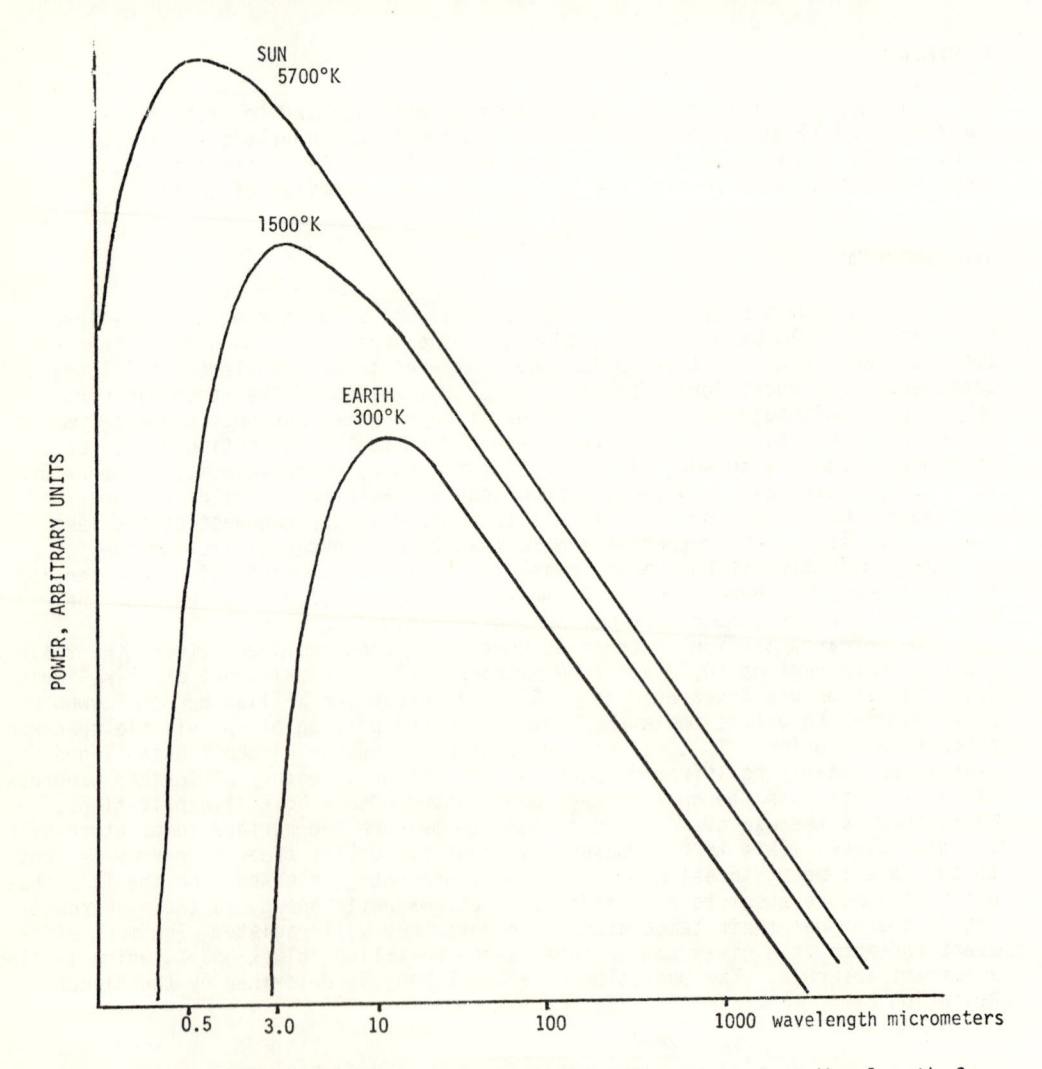

Fig. 1. - The Energy Emission of a Black Body Plotted against Wavelength for Various Temperatures.

maximum from the sun with a surface temperature about 5500°K is near 0.5 micrometer (in the green-yellow region of the visible spectrum). At the earth's ambient temperature (approximately 300°K), the maximum is at about 10 micrometers. Therefore, the region of most interest in the passive measurement of surface temperatures is in the region 8-14 micrometers, the so-called thermal infrared region.

Looking at Figure 1, one might conclude that the sun's own light would tend to drown out the emission from the earth's surface, but this is not so. This is because the sun is at a great distance, and the curves in Figure 1 are calculated for the hot surfaces themselves. At the distance of the earth, the sun subtends about 1/2° or about 7.6×10^{-5} steradian. The maximum irradiance arriving at the top of the atmosphere from the sun is therefore reduced by a factor of 7.6×10^{-5}. The sun's radiation is compared to the earth's emission at 50°F (300°K) in Figure 2. At 10 micrometers, the arriving sunlight is about 100 times less intense than the emission from the earth. Therefore, ordinary remote sensing thermal measurements are made in the infrared in a band centered near 10 micrometers. Luckily it turns out that the atmosphere is particularly clear at that wavelength so that corrections due to atmospheric absorption are small.

Although the infrared method is useful and has been applied extensively, both from satellites and aircraft, it has an important limitation: measurements by infrared methods can only give the surface temperature.

Water is very opaque to the infrared in the neighborhood of 10 micrometers. In fact such radiation is only transmitted appreciably through water for a fraction of a millimeter. This means that the temperatures measured by infrared methods come from less than the upper half-millimeter of the sea; no direct measurement of deeper temperatures can be made by the infrared method.

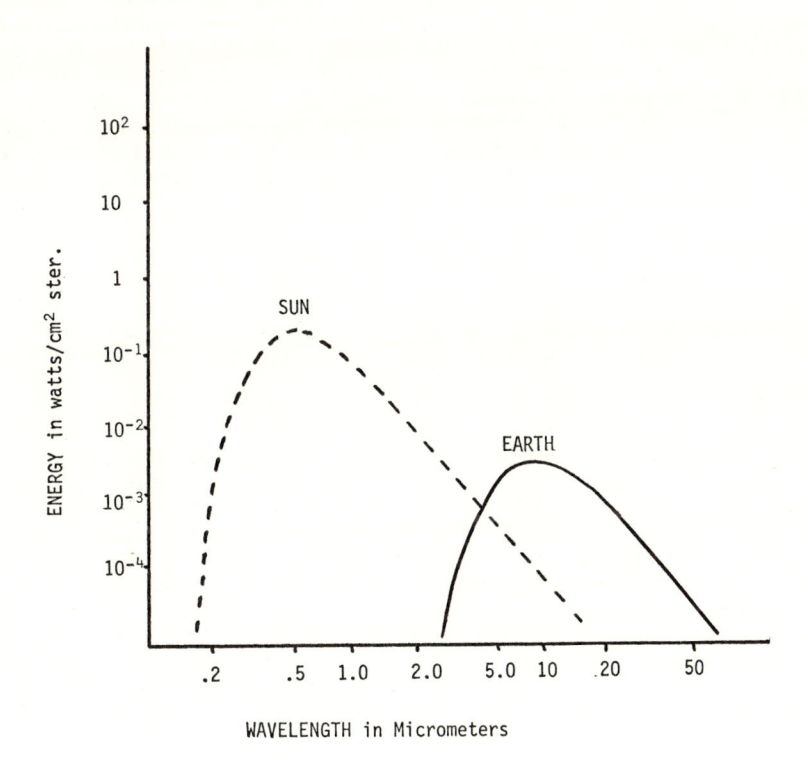

Fig. 2. - Energy Emitted from a Black Body at the Average Temperature of the Earth's Surface Compared with Reflected Sunlight.

Many problems in oceanography require data not only from the surface but from deeper layers in the ocean. Among the most vital of these problems is the question of survival of the biota in the neighborhood of a thermal power plant cooling canal outfall. Here, for example, if the deeper waters remain tolerably cool, the damage is minimized, while if the sea becomes hot all the way to the bottom the result is catastrophic. Another important piece of data is the existence (or absence) of a layer of rapid change in temperature (thermocline) in the water beneath the surface. Such layers are common in the sea and cause problems in acoustic transmission, since they bend or reflect sound waves.

At present, the only way to measure the subsurface temperature at depth is to introduce thermal measuring devices directly into the water to the desired levels. Such methods are expensive, time-consuming, and do not provide the continuous picture that remote sensing can provide of the surface. It would be very valuable, therefore, to be able to provide remote sensing of the ocean parameters at depth. The methods proposed here will provide continuous information on temperature, salinity and backscatter turbidity by remote sensing without directly introducing sensing devices into the ocean. Ultimately the remote sensing will be carried out from an aircraft, or even a satellite. Initially, however, it will be done from a ship.

A LASER PROBE

To obtain information from layers beneath the surface, it is necessary to use radiation for which the sea is transparent. The band of electromagnetic radiation which penetrates best into seawater lies between 0.48 and 0.55 micrometers, or between 4800 and 5500 Angstroms*. It is, therefore, proposed to project a beam of light of this wavelength into the sea and to use it to probe the deeper layers for information on oceanographic parameters, particularly temperature, salinity and turbidity. For this purpose, rapid modulation of a continuous light or short pulses of light (of the order of a few nanoseconds) are necessary in order to provide meaningful depth information. A powerful light source is also required, so that measurements may be made to as great a depth as possible. Finally, the temperature and salinity information is contained in small changes in wavelength of the light returned from the sea, so that the light source used must have an extremely narrow wavelength spread. All these requirements can be satisfied by the use of a laser. The laser will be mounted on a ship, aircraft or satellite and will project its light downward into the sea. Returning light scattered in a backward direction by the sea will be detected and analyzed with respect to time, intensity, polarization and wavelength. This information will in turn be recorded and processed to yield the desired parameters as a function of depth. The system is shown schematically in Figure 3.

SCATTERING

When a monochromatic beam of light penetrates into water, four principal kinds of scattering generally occur. In two of these, the wavelength centroid of the scattered light is unchanged (except by mass motions of the water, which we will ignore in this preliminary treatment), and in two it is shifted.

a) Tyndall scattering: Here the wavelength of the incident light is unchanged. The intensity is, in general, proportional to the density of particles suspended in the water. Tyndall scattering gives rise to backscatter, which is the main cause of the turbidity that interferes with the clarity of the water as observed with backlighting from daylight or artificial illumination. Tyndall scattering includes backward Mie scattering as well, which arises from smaller particles suspended in the water and has much the same obscuring effect.

---

* One Angstrom equals 10^{-10} meter.

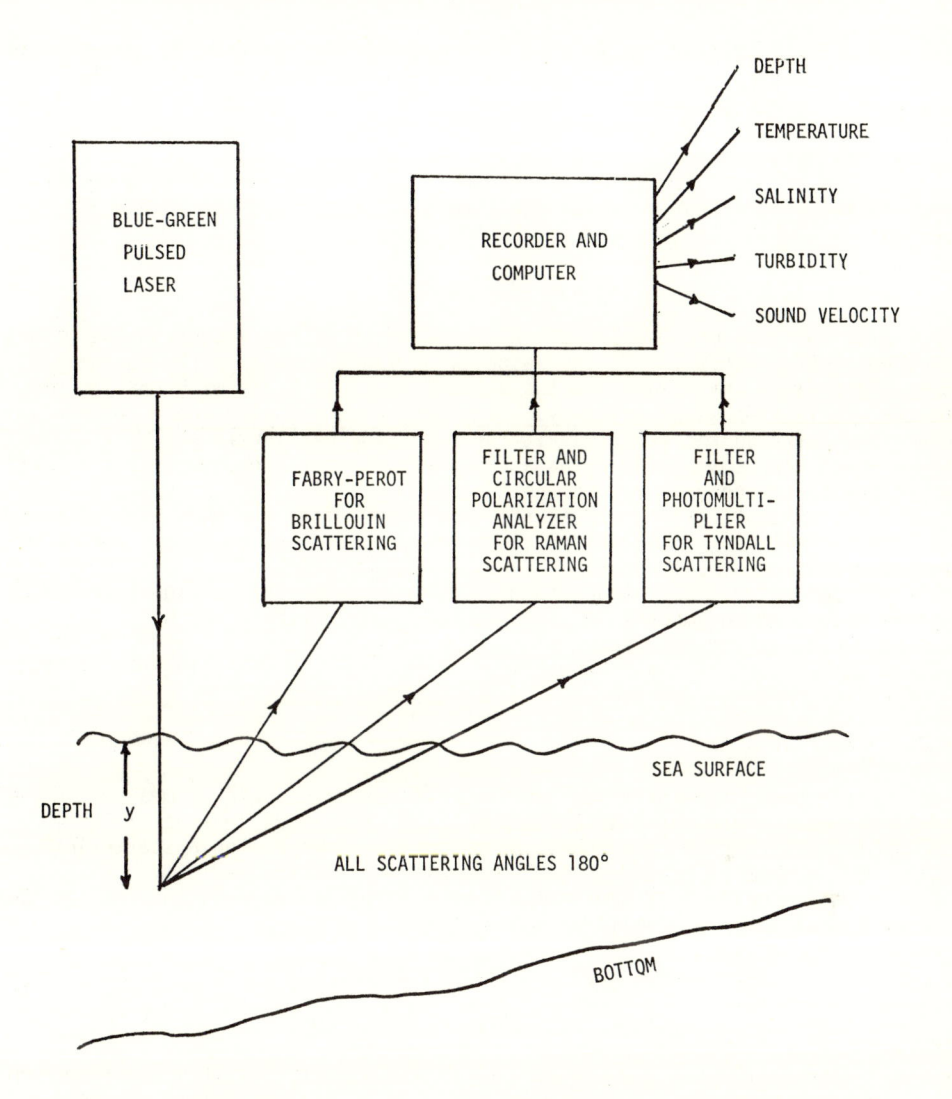

Fig. 3. - Schematic Plan of a Laser Remote Sensing Probe of the Sea.

    b) Rayleigh scattering: This occurs in clear water and is analogous to the molecular scattering in the atmosphere that gives rise to the blue color of the sky. The intensity of Rayleigh scattering is usually much less than Tyndall scattering; since its intensity is a known function of temperature, its intensity can be predicted.
    c) Raman scattering: Here the wavelength is considerably altered by the interaction of the light with energy levels of the water molecule. Since the population of these levels is temperature dependent, the characteristics of the Raman spectrum depend on the water temperature and the salinity. It turns out that both the wavelength and the polarization of the Raman light are temperature and salinity dependent. A recent study [1] by Chang and Young has shown that, because seawater may absorb certain colors in unpredictable quantities, the wavelength shifts of Raman spectra are not useful in measuring temperature and salinity. However, the study further shows that, if the exiting light is circularly polarized, the Raman-scattered light will show depolarization which is

dependent on salinity and temperature. Unfortunately, the circular depolarization**, ρ_c, cannot yield the salinity or the temperature separately. The additional information needed is supplied by Brillouin scattering as described below.

d) Brillouin scattering: Photons in the incident light react with phonons in the liquid to produce wavelength-shifted light, where the shift is proportional to the velocity of sound in the liquid. The shifted wavelength [2] is given by:

$$\lambda = \lambda_0 (1 \pm 2n \frac{v_s}{c} \sin \frac{\theta}{2}) \tag{3}$$

where λ_0 is the incident wavelength, n is the index of refraction of the water, v_s is the sound velocity, c is the velocity of light, and θ is the scattering angle. This means that there are two shifted spectral lines with spacing proportional to v_s, the sound velocity. For backscattering, the angle between incident and scattered radiation is π. The resulting wavelength shifts are about 1/50 Å, which can be measured using a Fabry-Perot interferometer. This measurement yields the sound velocity in the sea, v_s.

The Fabry-Perot is a spectroscopic instrument with extremely high resolving power and luminosity. In addition, it can be prepared [3] to have a number of channels operating simultaneously. Thus, the measurement of the Brillouin peaks can be carried out rapidly enough to preserve the depth information provided by the time delay, which amounts to 6 nanoseconds per meter of depth.

The depth, y, for each value of v_s, can thus be found by measuring the time delay in the arrival of the backscattered light from the sea using the arrival of reflection from the surface.

MEASUREMENT OF ρ_c and v_s

As described in the previous section, two parameters can be measured directly from the wavelength-shifted scattering: the depolarization coefficient, ρ_c, from Raman scattering and the sound velocity, v_s, from Brillouin scattering*. Both ρ_c and v_s depend jointly on the temperature T and the salinity s.

a) Sound velocity v_s: Many authors have suggested expressions for the sound velocity in the sea. A simplified expression [4] is:

$$v_s = 1449 + 4.6T - 0.055T^2 + 0.003T^3 \\ + (1.39 - 0.012T)(s-35) + 0.017y \tag{4}$$

where T is temperature in degrees Celsius, s is salinity in parts per thousand, and y is depth in meters. A family of velocity versus temperature curves for different values of s is plotted [5] in Figure 4. Since the depth, y, can be found by measuring the time delay, the unknowns are the temperature T and the salinity s.

b) Depolarization ratio ρ_c: Chang and Young [1] measured ρ_c for various conditions of temperature and salinity. Using a laser at 4600 Å and observing the Raman light at 5470 Å, a strong dependence on both temperature and salinity was measured and is plotted in Figure 5.

These two methods both yield information on T and s together. The question is, can T and s be evaluated separately? It turns out that this is indeed possible under certain conditions that will be discussed below.

** Defined as the ratio of the intensity in the reverse circular polarization to the intensity in the normal circular polarization during backscattering.

*Since the parameter measured is actually nv_s, a small correction has to be included for variations of n with temperature and salinity.

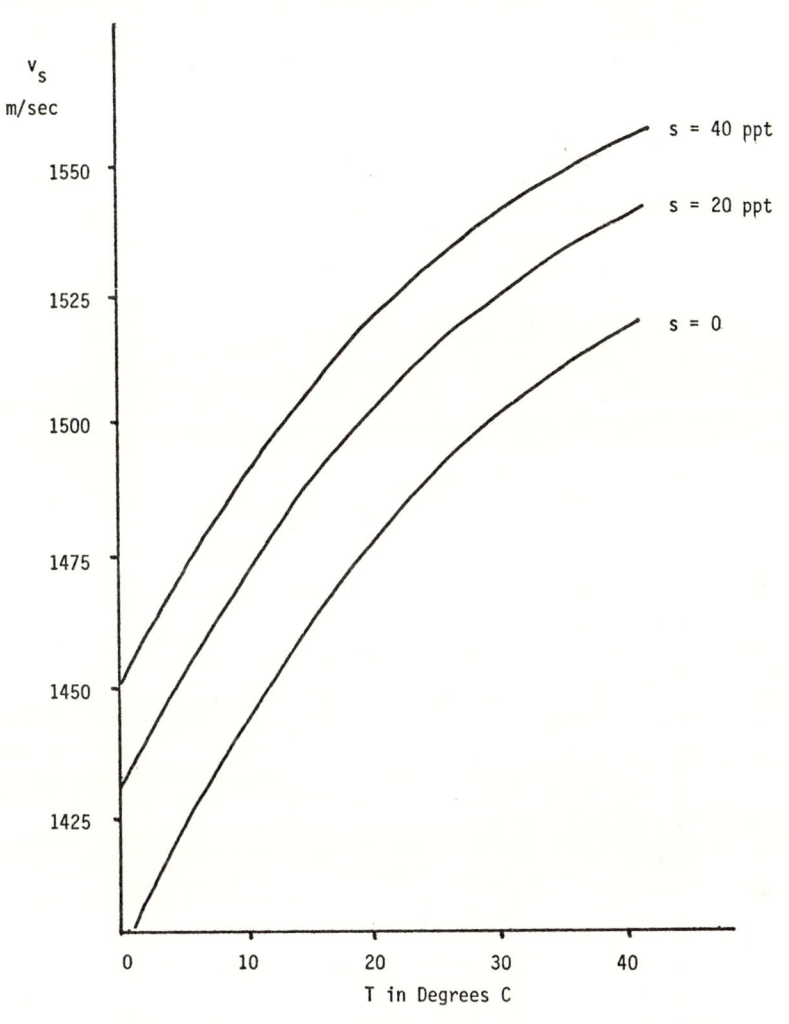

Fig. 4. - Sound Velocity v_s Plotted against Temperature T for Various Values of Salinity (Redrawn from R.J. Urick, Principles of Underwater Sound for Engineers).

DETERMINATION OF T AND s

In preceeding sections it has been shown how both v_S, the velocity of sound in water, and ρ_c, the circular depolarization of Raman scattering, can be evaluated. Each of these quantities depends on both the temperature, T, and the salinity, s, of the water.

For each measured value of ρ_c, there is a relationship between T_R and s_R, where T_R is the temperature and s_R is the salinity deduced by the Raman method. Similarly, for each measured value of v_s, there is another relationship between T_B and s_B, where T_B is the temperature and s_B is the salinity deduced by the Brillouin method. These relationships can be obtained from the curves shown in Figures 4 and 5 and assuming linearity are:

$$T_R = A + \frac{dT_R}{ds_R} s_R \tag{5}$$

$$T_B = B + \frac{dT_B}{ds_B} s_B .$$

Since the salinity and temperature of a given sample of sea must be the same measured by any method, it is required that $T_R = T_B$ and $s_R = s_B$. Solving equations (5) simultaneously:

$$s = \frac{A - B}{\frac{dT_B}{ds_B} - \frac{dT_R}{ds_R}} \qquad (6)$$

which has a solution if:

$$\frac{dT_B}{ds_B} - \frac{dT_R}{ds_R} \neq 0. \qquad (7)$$

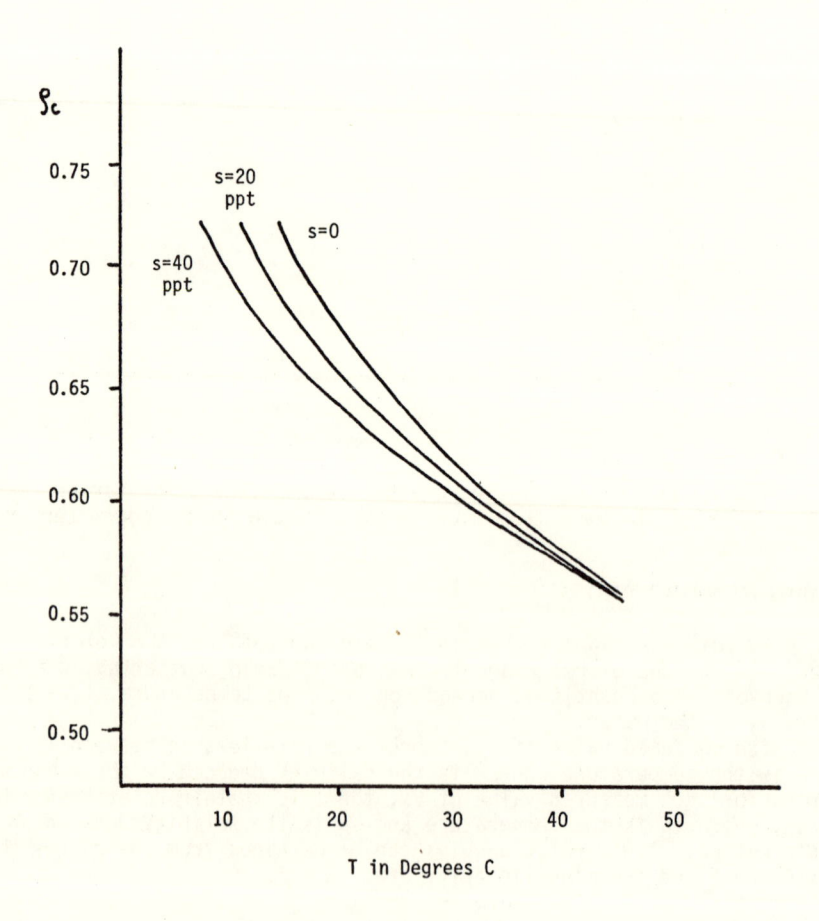

Fig. 5. - Circular Depolarization Ratio, ρ_c, Plotted Versus Temperature T, and with Salinity, s.

In Figure 6 $\frac{dT_B}{ds_B}$ and $\frac{dT_R}{ds_R}$ have been plotted together with their difference. For temperatures of interest in temperate and tropical seas, the difference is not zero, and accordingly s and T can be determined from a knowledge of ρ_c and v_s.

TURBIDITY

There are many definitions of turbidity, depending on experimental conditions and upon the use to which the parameter is to be put. Laser backscatter turbidity (LBT) can be defined as the ratio of intensity of backscattered light per unit depth to the intensity of the downward traveling incident laser light. This definition of turbidity corresponds well to the common phenomenon where upwelling scattered light interferes with the observation of detail below the region where the scattering occurs. Since different layers can be determined by timing the return signal, the LBT will be obtained as a function of depth. Most of the backscatter is caused by Tyndall scattering from particulate matter in the sea, so these measurements will reveal layers of particles, or even schools of fish, and give their density and depth.

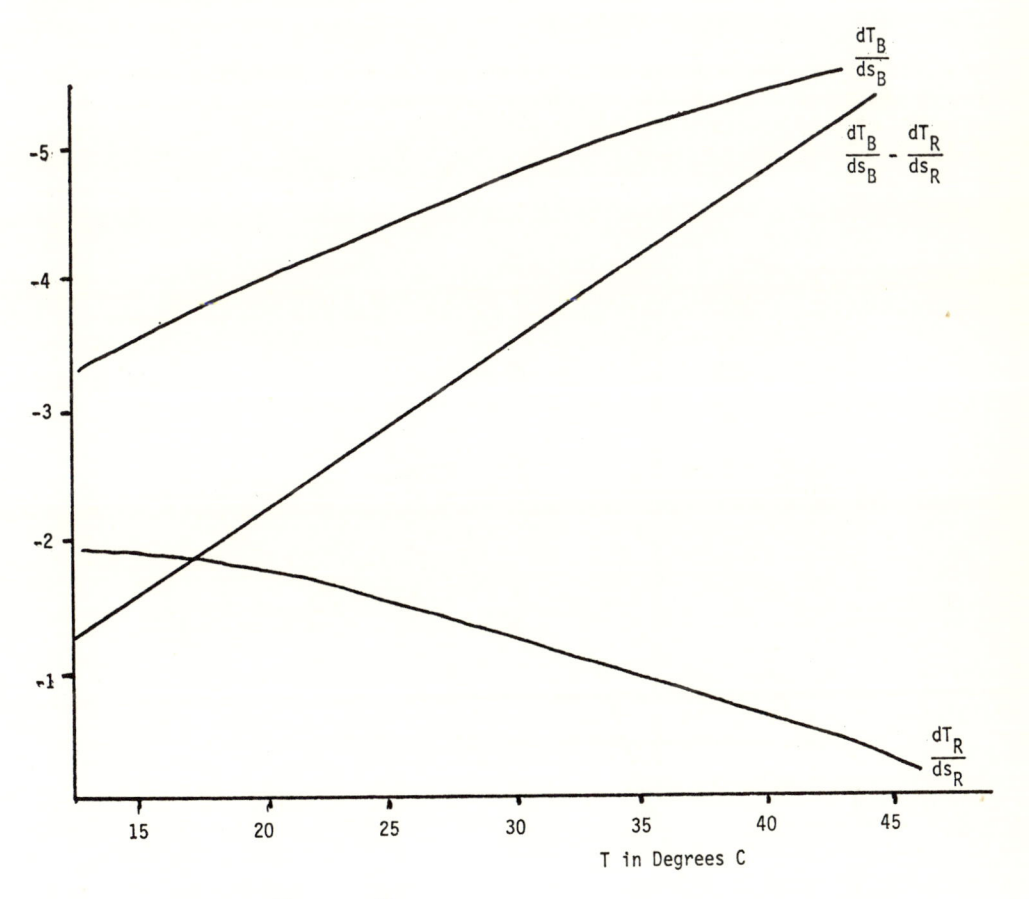

Fig. 6. - Values of $\frac{dT_B}{ds_B}$ and $\frac{dT_R}{ds_R}$ Are Plotted for the Temperature Range of Interest in Temperate and Tropical Waters. Their Difference Is Also Shown.

The intensity of the Brillouin scattering corresponding to a known incident intensity is a known function [6] of the water parameters and will give a direct measure of the intensity of the laser light at a particular layer of the sea. In Figure 7 the incident light I_0 is shown falling on an increment of water at depth dy. The Brillouin scattering intensity coefficient is b and the intensity of the Brillouin scattering is bI_0. This is measured in the course of determining v_s as mentioned above. (Alternatively, the Raman scattering intensity can be used, but, since it is at a different wavelength, the results are impaired if the water shows color absorption.) Finally, the non-wavelength-shifted Tyndall and Rayleigh scattering is measured. The Rayleigh scattering coefficient for water, r, is known [6] as a function of wavelength. Therefore:

$$R(y) = \frac{(t + r) I_0}{b I_0} \qquad (8)$$

where R(y) is the measured intensity ratio between the non-wavelength-shifted components and the Brillouin shifted components. Since r and b are known, t, the Tyndall scattering coefficient for backscatter, can be directly calculated:

$$t = R(y)b - r. \qquad (9)$$

The amount of Brillouin and Rayleigh scattered light is negligible compared to Tyndall scattered light in the presence of appreciable turbidity and can be neglected without noticeable error in the evaluation of turbidity. Therefore, LBT can be identified directly with the Tyndall coefficient t, so that:

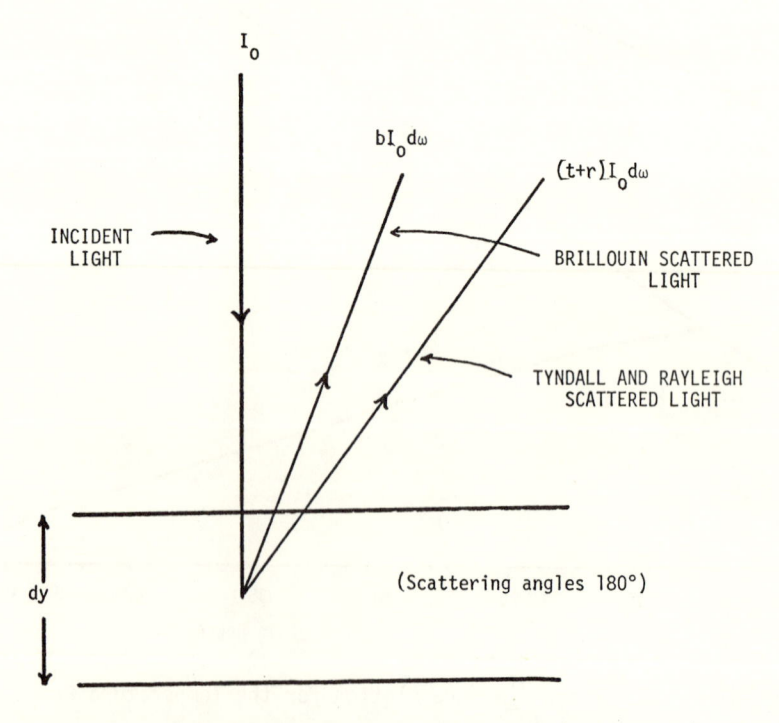

$$LBT = R(y)b \qquad (10)$$

Fig. 7. - Brillouin, Tyndall and Rayleigh Scattered Light from a Layer of Water of Depth, dy.

Since R(y) can be measured and b can be calculated as a function of depth, the Laser Backscatter Turbidity can be determined as a function of depth.

CONCLUSION

We have described a method of determining the sound velocity, turbidity, temperature and salinity beneath the sea as a function of depth by active remote sensing. This is done by projecting a pulsed laser beam in the blue green region of the spectrum into the sea and evaluating the resulting backscatter. Tyndall, Raman and Brillouin scattering are separated spectroscopically, and measured as a function of time. Spectral position, intensity and polarization state are all utilized.

ACKNOWLEDGEMENTS

This work is supported by the National Aeronautical and Space Administration of the United States of America under Contract NAS10-8600.

REFERENCES

1. Chang, C.H., and Young, L.A. AVCO Everett Research Laboratory Research Note 960. Everett, Mass., 1974.

2. Gross, E. Naturwissenschaften $\underline{1}$, 718 (1930).

3. Hirschberg, J.G., and Platz, P. Applied Optics $\underline{4}$, 1375 (1965).

4. Fairbridge, R.W. Encyclopaedia of Oceanography (New York: Reinhold, 1966).

5. Redrawn from R.J. Urick. Principles of Underwater Sound for Engineers (New York: McGraw-Hill, 1967).

6. Fabelinskii, I.L. Molecular Scattering of Light (New York: Plenum Press, 1968).

PART III

LAND USE MONITORING

REMOTE SENSING FOR WESTERN COAL
AND OIL SHALE DEVELOPMENT PLANNING
AND ENVIRONMENTAL ANALYSIS

H. DENNISON PARKER, *Manager*
Ecosystems Analysis and Remote Sensing Applications
Western Scientific Services, Inc.,
328 Airpark Drive
Fort Collins, Colorado 80521 U.S.A.

ABSTRACT

There are two broad categories of application of remote sensing technology in the development of fossil fuel resources in the Western U.S. The first includes pre-construction site evaluations, land use and usability mapping, and environmental baseline data acquisition. The second category involves long-term environmental monitoring. The geographic magnitude of these developments, particularly when multiple mines or processing plants are considered on a regional scale, precludes the use of conventional ground-based analysis techniques. The time frame in which these resources must be developed also limits the utility of conventional methods. This paper will discuss both categories of remote sensing applications and the overall role that remote sensing can play in furthering the national goal of major dependency on internal sources of energy.

BACKGROUND

In discussing applications of remote sensing to the coal and oil shale industries of the western United States, it is important that one have an appreciation of the magnitude of these developments. It is also important to realize the extent of the dependency of this country on coal and oil shale resources. Therefore, I would like to first describe briefly the types of energy developments that are now in various stages of design or construction, the fossil fuel resources on which the developments depend, and the importance of these resources to our national energy supply.

OIL SHALE

The oil shale region of the western United States encompasses portions of three states, Colorado, Wyoming and Utah (Figure 1). Although other areas of oil shale deposits are known in the U.S., none are as rich in recoverable shale oil as those of the Green River formation, in these three states. The Green River oil shales occur under approximately 16 million acres of land in these states, of which approximately 11 million acres contain oil shales of potential commercial importance, [1].

The Colorado oil shale area is the smallest in geographic distribution, but it contains the richest and thickest oil shale deposits of all three states. The oil recoverable from Colorado oil shale deposits has been estimated by the Department of the Interior at 480 billion barrels. As a comparison, the total United States domestic production of oil in 1973 was approxi-

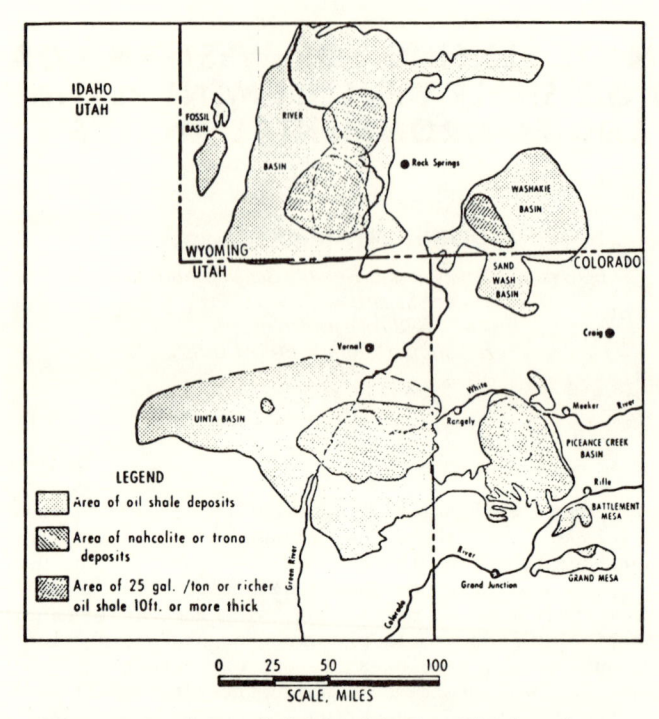

Figure 1. Distribution of oil shales of the Green River formation.

mately 4 billion barrels. A few calculations will show that the shale oil from Colorado alone could supply this level of production for over 100 years.

The largest area of the oil-bearing Green River formation occurs in Utah. However, these shales are not as rich as the Colorado shales - the Utah deposits are estimated at 80 billion barrels. Wyoming has the smallest amount of rich oil shale deposits, approximately 30 billion barrels.

The presence of oil-bearing rocks in Colorado is not a new discovery. In fact, some oil was produced from shale prior to the 1859 discovery of natural petroleum in the United States. However, shale oil was not seriously considered until about 1920, and only last year did the Department of the Interior successfully implement a program aimed at development of oil shale resources on public land, the Prototype Oil Shale Leasing Program.

Much of the Colorado oil shale area is privately owned, mostly by major oil companies (Figure 2). However, approximately half the area is government land, administered by the Bureau of Land Management. The entire area has been termed "semi-wilderness," because it is relatively undisturbed. The Northern portion, known as the Piceance Basin, is semi-arid country capable of supporting only sparse vegetation. It is low in elevation, ranging down to about 6,000 feet. However, the basin is of critical importance as a winter range area for mule deer.

The Southern part of the region is the higher elevation, Roan Plateau, which rises to 8,500 feet and higher. This area contains some very spectacular topographic features, including canyons up to 2,000 feet deep, and interesting rock outcrops and formations. Ecologically, the plateau area is quite

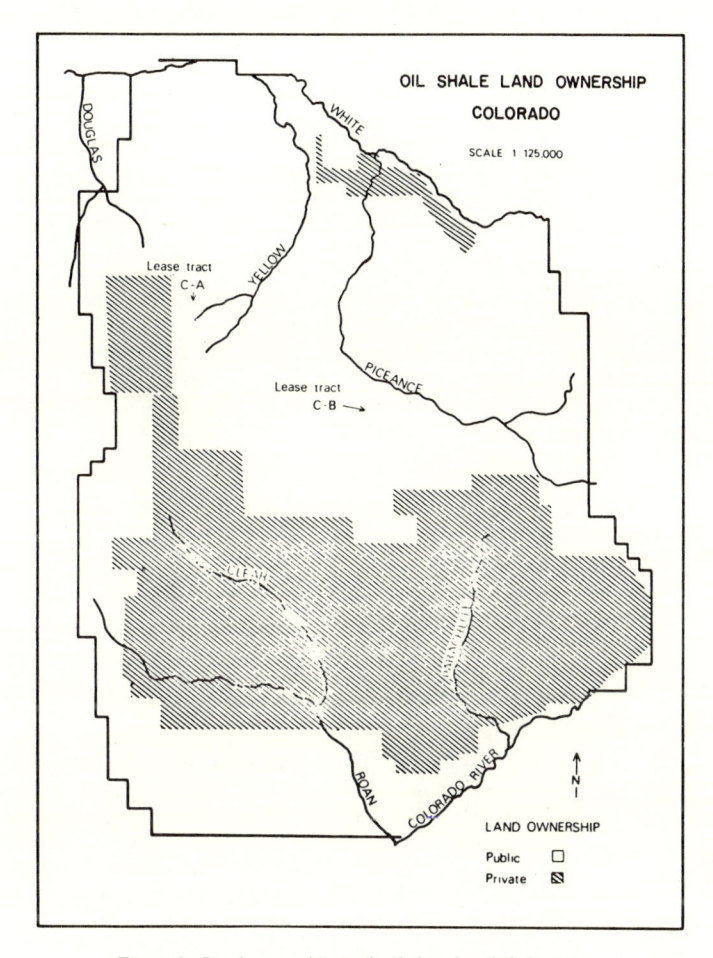

Figure 2. Land ownership in the Colorado oil shale region.

different from the Piceance Basin, basically as a result of greater precipitation. Forests of aspen (Populus tremuloides) and douglas fir (Pseudotsuga menziesii) are widespread and interspersed with sagebrush (Artemisia tridentata) and other mountain shrub communities.

The technology for extracting oil from shale has been explored on a developmental scale by several companies, and by the U.S. Bureau of Mines. The most advanced overall technology to date has been developed by a consortium known as Colony Development Operation, managed by the Atlantic Richfield Company. This group has conducted major research and development activity in oil shale processing and the associated processed shale disposal problem. Government research has been limited to a program conducted in the 1950's in Western Colorado which concentrated only on development of a retorting process for oil recovery.

Public apprehension over possible environmental damage has been a significant factor in oil shale development for many years. The greatest environmental problem associated with production of shale oil is unquestionably, disposal of the waste material created in the shale processing.

In order to obtain the hydrocarbons bound up in the shale, the rock must be mined, broken and crushed, in preparation for heating to temperatures in excess of 900 degrees Fahrenheit, at which the oil, more accurately a compound known as kerogen, is released. The waste product produced in this operation, known as "spent shale," is a salty, black substance with physical characteristics dependent largely on the retorting process used. Rich oil shales are generally defined as those which yield 25 gallons of oil for each ton of shale processed. For a plant which produces 50,000 barrels of oil per day, the amount of waste material, or spent shale, will total over a half million cubic feet per day.

The problem of what to do with this much waste material has been the subject of considerable research by the Colony Operation. Their approach, and that generally acknowledged as the most practical, is to deposit the material into one or more of the natural canyons in the oil shale region, leach it to remove excess salts, compact it, and establish a vegetation cover to stabilize the surface. The technology for accomplishing the disposal task is still under development.

Most of the country's major oil companies are involved in oil shale development in one way or another. The Federal, Prototype Oil Shale Leasing Program was initiated late last year on two tracts each in Colorado, Utah, and Wyoming. The first two Colorado tracts drew competitive bids from such companies as Shell Oil Company, Exon, Atlantic Richfield, Sun Oil, and Marathon Oil, among others. The first tract leased was won by a joint bid between Standard Oil Company of Indiana and Gulf Oil Corporation of over $210,000,000. The other Colorado tract and both Utah tract bids were even higher per barrel of oil. Neither Wyoming tract drew any bids.

I mention these figures to convey an impression of the industrial and governmental interest in the development of an oil shale industry, and the potential magnitude of the industry. However, that interest has only increased the public concern for environmental protection measures during the past few years. As a result, the lease agreement signed by the winning bidders on the two tracts in Colorado, and on the two Utah tracts, contained very comprehensive environmental protection stipulations. These lease stipulations document most of the comprehensive environmental protection concepts which were basic to the National Environmental Policy Act of 1969 (NEPA). However, the difference is that NEPA does not define the details of environmental analysis and protection measures which must be accomplished on projects which come under its jurisdiction. The oil shale lease stipulations are quite specific by comparison, even though some portions of them are poorly defined in ecological and scientific terms.

Among other things, the stipulations require comprehensive, detailed, baseline ecological inventories, long-term environmental monitoring, and active measures for the prevention of environmental degradation for the 25 to 30 year life of the leases.

There are two basic characteristics of oil shale development which suggest that remote sensing technology offers a very practical means by which to accomplish the necessary environmental analyses and monitoring tasks. They are the geographic scope of the analyses required, and the time in which the analyses must be completed.

Each of the four oil shale tracts leased in Colorado and Utah is approximately 5,000 acres in size, but the total area which must be ecologically inventoried and monitored is closer to 20,000 acres per tract. The government leasing program requires submission of each lessee's detailed development plan, following completion of the baseline environmental data acquisition, in 3 years. Considering the integration which must occur between engineering design and environmental protection plans to meet the lease environmental stipulations, 3 years is a very short time period.

COAL

The coal resources of the western United States are easily of greater significance to the nation's future energy supply than oil shale, coal requires much less processing to produce a usable fuel, and on a pure energy basis, our coal resources are far greater than our oil shale. Coal as a fuel for electric power generating plants will be the basic fuel on which this nation will depend for at least the next quarter of a century. The conversion of coal to a synthetic gas, coal gasification, will also be of increasing importance as our natural gas supplies dwindle. But, there are some problems with coal.

Many of you are no doubt acquainted with the controversy over the use of low or high sulfur coal for electric power generation, and the effects of these fuels on air quality. The primary contaminant which is produced as a result of burning coal with high sulfur content is sulfur dioxide, a compound which has caused great concern because of its deleterious effects on humans and on plant life. In evaluating the extent and utility of our remaining coal resources, we must therefore take into consideration the sulfur content of the coal to be used as fuel.

There is a possibility, indeed a strong probability that some existing air quality regulations will be relaxed. In my opinion, some of the State and Federal standards are unnecessarily stringent. But, it would be a tragic mistake to eliminate them completely in the name of an energy crisis or for any other reason. Therefore, I expect to see a continuing requirement for low-sulfur coal which can be burned without producing the comparatively large amounts of sulfur oxides which have contributed significantly to the air pollution problems in industrialized areas throughout the world.

In general, the bituminous coals of the eastern United States are of high sulfur content (Figure 3). However, in the west, the sub-bituminous and lignite coals are of low sulfur content, and are therefore much more desirable from the standpoint of reducing air pollution from sulfur oxides. Some Western coals are also more economical to mine, since they are in comparatively shallow deposits which can be surface mined. Of nearly 128 million acres underlain by coal in the western United States, about 1½ million acres could be surface mined using current methods, [2].

Now, how much low-sulfur coal is there in the western United States? The Department of the Interior has estimated that the low-sulfur coal and lignite resources of the United States are in excess of 1 trillion tons. As a comparison, the United States consumption of coal of all types for 1971 was slightly over 500,000 tons, a figure which must be almost tripled by the year 2000 of meet the projected energy demands of the Country, [3]. Projecting the life of this low-sulfur western coal at an annual consumption rate of 1½ million tons, the estimate for the year 2000, we find that our western reserves would last almost seven centuries.

The coal area of the Rocky Mountain and Northern Great Plains coal provinces totals approximately 193,345 square miles. The area is vast, scenic, and by eastern standards, very sparsely populated. The region of most sub-bituminous coal mining activity during recent years and that which is destined

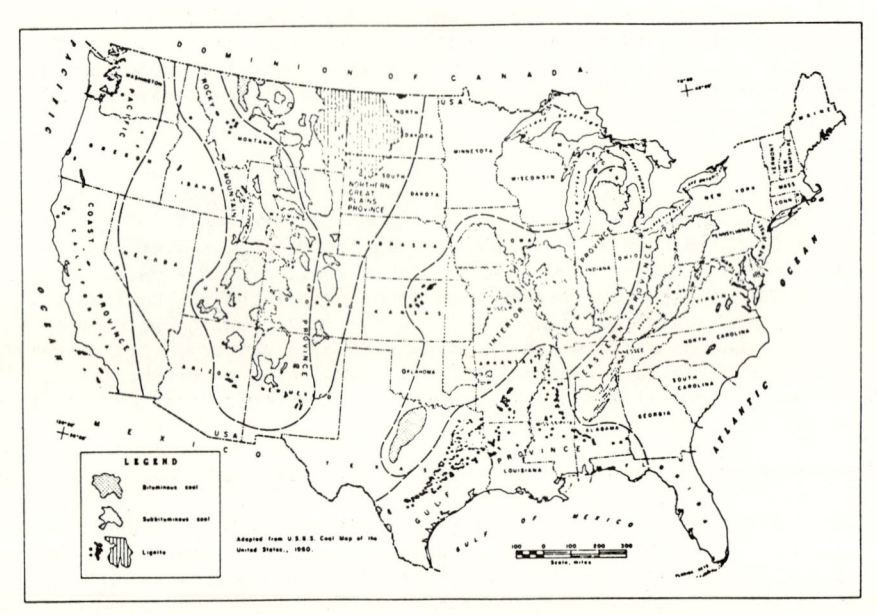

Figure 3. Distribution of coal in the continental United States.

to receive the greatest industrial attention in the near future is the Powder River Basin of Montana and Wyoming. The Powder River Basin includes portions of eight counties in Montana and Wyoming, and exceeds 33,000 square miles in area. This is roughly equal to the areas of New Hampshire, Vermont, Massachusetts, Connecticut, Rhode Island, and Delaware, all combined. These six eastern states have a total population of about 11½ million. The Powder River region's population is about 75,000. Obviously, the environmental impact both socially and ecologically will be far greater than that which would occur under similar development circumstances in a more populated region (Figure 4).

In excess of thirty oil, mining, and power companies are actively engaged in various stages of planning or construction of coal production or other coal-related energy facilities in the Powder River region. Since most of the coal is federally owned, the developing companies must adhere to strict environmental stipulations attached to Federal leases, which are similar to those imposed in the Prototype Oil Shale Leasing Program. These stipulations, as with oil shale, are in addition to the requirements of the National Environmental Policy Act and other State and Federal legislation. Incidentally, some of the state environmental regulations are considerably tougher than the Federal regulations.

Although I haven't the time to go into detail on the specific regulations for environmental planning and protection which are applicable in the various states, the impact of them is to impose requirements for very detailed inventory and analysis of the environment, comprehensive planning for environmental protection, and long-term monitoring of environmental quality.

As with oil shale, when you consider the land areas involved, the detail of ecological and other environmental analyses which are necessary, and perhaps most importantly, the time period during which these resources must be developed, the role of remote sensing technology becomes clear. There is no other means by which to gather the types of information which are required as rapidly and as accurately as remote sensing techniques will allow.

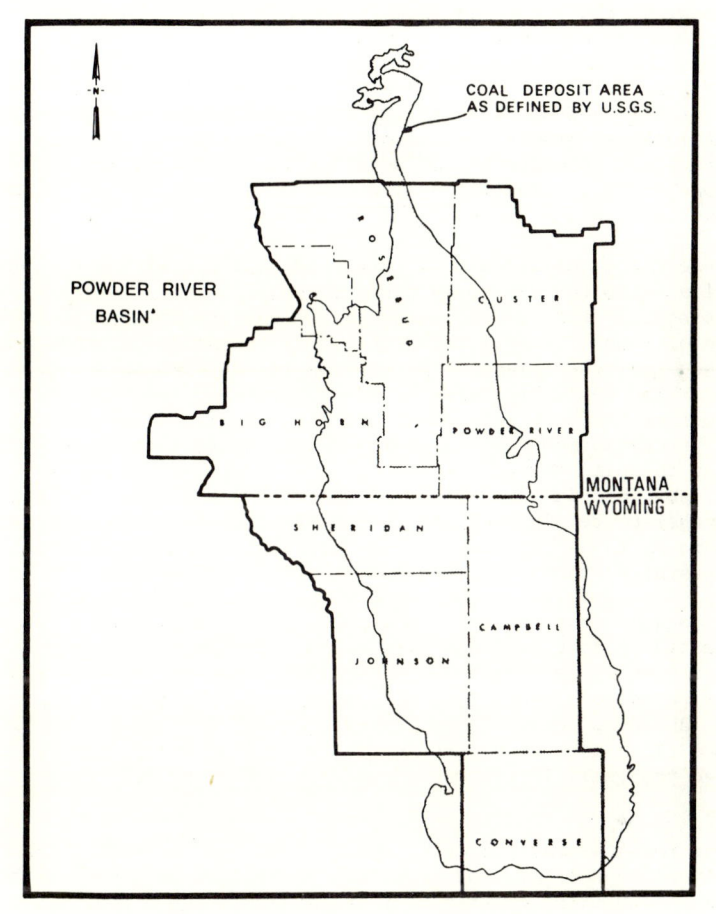

Figure 4. The Powder River Basin region of Montana and Wyoming.

THE UTILITY OF REMOTE SENSING

In remote sensing research activity, we are generally concerned only that a particular technique performs as required in a given application, from a technical standpoint. However, demonstrated technical performance is not sufficient to assure the utility of a technique in routine application.

When we consider applications to coal and oil shale development, we must recognize that there are many other considerations of equal importance to the technical sophistication of the sensor or analytical technology. Having worked closely with private industry on the numerous real and potential problems associated with oil shale and coal development, I think I can suggest a few points which are important in assessing the real applicability of remote sensing to these industries.

INDUSTRIAL REQUIREMENTS

I have already mentioned two characteristics of Western fossil fuel developments which at once severely limit the utility of conventional ground techniques, and offer for the first time an opportunity to bring into routine application many of the remote sensing technologies which have thus far been limited to applications trials.

First, we are talking about data which must be collected in a relatively short period of time - most coal and oil shale developments must await detailed environmental analyses before commencement of construction activity. Second, the geographic areas involved are of such magnitude as to preclude the acquisition of comprehensive, scientifically procured information (not data) in the necessary time interval by conventional methods.

Both these characteristics tend to support the use of remote sensing. We are all familiar with the extremely rapid geographic coverage capability of most sensor systems. But, alone, these advantages are insufficient to justify the selection of remote sensing techniques over conventional ground methods. In private industry, the client is purchasing remote sensing services only incidentally. What he really needs is the information remote sensing can yield. In general, the oil and coal industries are not interested in paying for research or development of new ecological analysis techniques. Their objectives are to build the facilities which will provide the fuels on which this Country runs, and in the process to make the reasonable profit which is required for them to remain a viable business. Therefore, in order for remote sensing techniques to be applied in these developments, they must be able to compete successfully, on an economic basis, with other methods.

Another consideration of extreme importance is the accuracy and validity of the information produced. Just because an experimental procedure has worked once, does not mean it is proven and dependable. Remote sensing techniques, as with other technical procedures, must produce repeatable results before they can be classed as proven technologies. Again, if remote sensing is to take its place alongside other techniques for application to the nation's energy-related problems, it must be dependable and practical in operational application.

I believe today we have numerous examples of remote sensing techniques which are proven, dependable, and certainly economic. But, we also have perhaps even more examples of experimental techniques which are not yet ready for application. It is of basic and critical importance that we recognize the difference.

INFORMATION QUALITY

In the abstract of this paper I have identified two broad categories of information requirements to which remote sensing technology is applicable. The first is in the acquisition of baseline information, both for engineering planning and design, and as an environmental reference point. The second category involves long term environmental quality monitoring, which is important not only because it is required by law, but also because it must become a routine practice, the costs of which are born by society in general, if we are to maintain the quality of life which is desired by society in general.

In both baseline and monitoring information acquisition, it is critically important that the information obtained is real information, and not just an array of raw data followed by unjustified conclusions and "results." Too often, the ecological analyses which have been conducted for major power, mining, or other energy developments have been obtained merely with the objective of filling the pages of an enormous environmental impact statement. The environmental impact statement, required by Section 102 of the National Environmental Policy Act (NEPA), cannot alone accomplish the real objectives of the act.

It has been historically demonstrated that NEPA is only as effective as the courts will allow it to be. With all due respect for our judicial system, our legal experts are not also scientific experts. Too frequently, the "technical experts" who have had the opportunity to directly or indirectly pass judgement on the adequacy of environmental analyses are either incompetent in the field in which they profess to have knowledge, or they have their own subjective axes to grind. As a result, we have these enormous environmental impact statements which to the uninformed, may appear quite authoritative just as a function of their size. On the other hand, scientific authorities may know the document and the work which generated it are inadequate and misleading. But, typically, they cannot even communicate their objections in a believable manner. More often, such objections are labeled as environmentally radical and are ignored.

In the long run, the companies which pay for such superficial environmental analyses are the ones who are being short-changed, and they usually don't even realize it until a suit is filed. In the process, the company frequently, and quite innocently, is thrust into the role of the villain, the greedy, profiteering industrial giant, which has no concern for the environment. In my experience dealing with many of the large energy companies, I have found the reverse to be generally the case. Most companies are willing, indeed desirous of avoiding environmental damage where possible or mitigating damage where it cannot be avoided. However, most do not have the expertise and experience in dealing with these types of problems which will allow them to foresee impending damages or even to assess the evaluations of others.

When we speak of good baseline environmental information, we are talking of facts and figures which should, indeed must, be incorporated into overall development planning if all engineering, ecological, social, and economic information is to be integrated into a mutually compatible, smoothly functioning operation. Therefore, it is a disservice to those who must depend on the information to obtain less than the highest quality, scientific, quantitative environmental baseline information. Just as the company must be able to depend on its engineering design data, it must also be able to depend on its environmental design data.

The rub comes when one realizes that the data acquisition and analytical sophistication required to determine, for example "ecological inter-relationships," as required by the oil shale environmental stipulations, are far more complex and difficult to accomplish than are the design input requirements for the engineering aspects of such a project. Ecology is not an exact science. There are those who believe that the structure and function of ecological systems will never be subject to deterministic definition. However, we can produce very strong statistical inferences on the likelihood for ecological damage of various sorts if the basic data acquisition and subsequent analyses are carried out in a scientific manner with the constant objective of producing dependable, accurate information.

In field ecological studies, it is generally not possible to intensively study every square foot of soil, each individual plant, or each member of all wildlife populations in the are to be studied. Therefore, these and other ecological phenomena must be sampled in a manner such that the sample parameters measured will be representative of the entire population, whether the population be animals, plants, soils, or air parcels. The point is that in some way, one must be relatively certain that the samples obtained are representative of the whole, or else the whole will remain either undefined or even worse, mis-defined.

Using remote sensing methods, we have extremely valuable techniques for viewing the whole, regardless of how large it may be. If our objective is general information over very large areas, we have at our disposal the unique data products produced by earth-orbiting satellites. At the other extreme, we can observe various phenomena just a few inches in size with very large scale aerial photography. Again, it is of extreme importance that we apply the right technique to the job.

It is also important that we recognize the limitations of our remote sensing tools. I said previously that with remote sensing techniques, we can view the whole, and so we can. But, we cannot necessarily analyze the whole in detail. With very few exceptions, some amount of ground truth information is required in primarily aerial surveys and analyses.

Keeping in mind the need for sampling, one of the greatest advantages of remote sensing techniques is the ability to extend very limited ground-acquired information over very large areas. Of course, there are some types of information obtainable by remote sensing which cannot be reasonably obtained on the ground at all. But, of more importance from an operational standpoint, are those numerous types of applications which can be used to extend ground information more accurately, economically, and much faster than additional ground studies. The U.S. Forest Service, in an unpublished study, found that remote sensing techniques offered a 5 to 1 time advantage in certain types of forest surveys, and a 10 to 1 advantage in the analysis of rangeland, [4]. While such comparisons are convincing and attractive, we must be careful that they are accurate and scientifically derived, as the one cited above, lest we be guilty of "oversell," a trap which is easily fallen into in remote sensing technology.

INTEGRATED ENERGY DEVELOPMENT PLANNING

In planning for a commercial shale oil extraction plant, coal gasification or liquefaction plant, electric power plant, or other major energy facility, there is a tremendous array of economic, engineering, and environmental questions which must be answered. Particularly in the Western coal and oil shale regions where previous development is virtually non-existent, such basic considerations as access, transportation, utilities and labor supply pose serious pre-development problems.

For example, a single coal gasification plant in New Mexico will require over 5,000 gallons per minute of water, 57,000 kilowatt hours of electricity, new roads for an estimated 200 vehicles per day, railroad and truck loading facilities, etc. The water and power are not available by merely tapping onto a nearby pipeline or transmission line. Dams, reservoirs, pipelines and pumping stations must be constructed; electricity must be generated and distributed; roads, railroads and pipelines must all be built in an area where the greatest construction activity previously was of a magnitude on the order of grading an unsurfaced road across a nearby Indian reservation.

On top of these more or less traditional process-related development requirements, we see increasing examples of more restrictive environmental constraints, imposed at all governmental levels. It is my contention that most such environmental restrictions are foresighted and in the best interest of the Nation in general. I do not agree with those who insist that our need for new energy supplies is of overwhelming importance and should take precedence over the maintenance of a quality future environment. At the same time however, environmental constraints must not be allowed to stifle the development of new energy sources, for our future energy supply picture is truly bleak.

Therefore, our goal is obvious, and in my opinion, quite attainable – we must develop our energy resources as necessary, while at the same time, providing for the quality environment we owe to future generations.

Regardless of one's own personal persuasion in the energy/environment question the immediate situation legally mandates consideration of the environmental aspects of development along with the engineering and economic questions discussed previously. Therefore, it is to the benefit of all concerned that we pursue both the engineering and environmental objectives of development in an integrated and comprehensive fashion rather than as separate, antagonistic activities.

The most basic prerequisite to such a comprehensive approach is information, and remote sensing techniques can frequently provide it in the most efficient manner, regardless of the information's intended use. Much of the early development planning information obtained by remote sensing will be of value both from the standpoint of engineering and construction of roads, pipelines and other structures, and as baseline environmental information. One of the basic and major advantages of a complete and detailed information base prior to development, is that opportunities for integrating environmental and engineering planning and design into one comprehensive development planning effort will be much more apparent. The advantages of such integration lie in the increased probability for insuring the environmental compatibility of all aspects of the development both at the outset and for many years into the future.

I would like to cite a few examples to illustrate this very important point. In the Prototype Oil Shale Leasing Program, the lessee is specifically restricted from interrupting or disturbing natural stream courses, or other surface water drainages and channels, including areas of natural runoff. In addition, the developer must maintain 200 foot buffer strips along all waterways in their natural condition. To meet such requirements, not only the developing company, but also its road building contractor and his employees must be aware of these limitations, and work according to plans specifically designed to meet the environmental stipulations. But, merely to be aware of such environmental constraints is not sufficient. In order for engineering plans to be drafted in compliance with the law, all areas of natural erosion must be known and mapped in detail. Therefore, the traditional approach of merely mapping the topography, and analyzing slopes from the civil engineering standpoint is not sufficient. A watershed scientist, probably in consultation with a geologist must have access to the appropriate data with which to map and render opinions on the locations and condition of any areas which are unusually susceptible to erosion, earth movement, or other watershed damage.

Another example of the desirability for coordinated engineering and environmental planning concerns land reclamation. Both in coal and oil shale development, developers are specifically required to provide for reclamation of all disturbed areas following mining. The term reclamation (as opposed to restoration) infers the establishment and maintenance in the long-term, of vegetation capable of stabilizing the landscape and providing for subsequent beneficial land uses. Establishment of such vegetation will require suitable soils for plant growth which may not be available on-site. Perhaps some of the overburden can be used, if it is stored separately from the less desirable subsoils. Or, there may be areas of suitable soils close by which can be used specifically for revegetation purposes. In either case, the surface soils throughout the area should be mapped in detail, not just as structural support for construction but also as plant growth media.

A more difficult problem in oil shale development is the requirement that the lessee either not disturb wildlife populations and migration routes, or else provide for the mitigation of any disturbance which occurs. Again, in designing the facility to meet these stipulations, the lessee must know the types and distribution of all wildlife populations which might be affected, and particularly for mule deer, their seasonal movement patterns.

Many other examples could be cited as strong legal or merely logical arguments that construction site selection and engineering of all development components both on and off-site go hand-in-hand with detailed, quantitative, environmental planning.

SPECIFIC REMOTE SENSING APPLICATIONS

BASELINE DATA ACQUISITION

In the category of baseline environmental remote sensing applications, we find similar information requirements to those which apply on developments in other parts of the country for the usual municipal projects, industrial construction, etc. Frequently, environmental analyses in more studied regions amounts to little more than literature reviews. However, the information currently available in much of the coal and oil shale region is so sparse as to preclude merely the compilation of existing information. Although this approach has been attempted in some instances, it meets neither the spirit nor the letter of some legal requirements and will almost inevitably lead to expensive court proceedings and delays for the developing company. In addition, none of the advantages of integrated engineering and environmental planning as discussed earlier can be realized.

Therefore, we must literally start from a zero data base in so far as specific, quantitative ecological information on any given site is concerned. Although general information on the occurrence of soils, vegetation, and wildlife can be found in the literature for all ecosystems of the coal and oil shale regions, it is not specific enough to provide the developing company or concerned governmental agencies with the kind of planning information needed. Neither is it specific enough to be of value as references from which to measure the long-term ecological effects, or more generally, the environmental impact of the development.

Keeping in mind opportunities for broadening the utility of field data obtained, by planning for its combined application to engineering and environmental analyses, I would now like to discuss some of the specific types of environmental data, the sensors which, in my opinion, are best suited for their acquisition, and the analytical procedures which will reduce the data to useable information.

Most of the environmental applications can be discussed under the general heading of ecological applications of remote sensing. This terminology is not intended to connote a strictly environmental utility or objective, however. When one realizes that the existing ecosystem will be modified extensively by any development of the type we are discussing, it is important to attempt definition of all initial ecological parameters and relationships, to accurately predict what new ecological conditions, including potential damage or detrimental modification will ensue as a result of development.

Therefore, the first step must be an inventory, or ecological classification scheme which results in the quantitative definition of all ecosystem components. Assuming the geology and geomorphology of the area has already been documented, the most basic step in ecosystem classification is a delineation of native vegetation types and distribution. From the subdivision of existing vegetation into well-defined, quantitative units, much additional ecological information can be extrapolated or deduced.

Although vegetation type-mapping by aerial photography has been employed for many years by such agencies as the Soil Conservation Service, the U.S. Forest Service, and other agencies, fine discrimination into more detailed vegetation units has only recently been developed to the point where it can be routinely applied. The most useful type of remote sensing data for this purpose, from the standpoint of practicality in operational applications, is color infrared aerial imagery. Multispectral scanners and the associated computer processing of digital data has been intensively researched and in some cases offers more rapid and equally accurate identifications to those obtainable by analysis of color infrared film. However, these systems are neither generally available, nor economically feasible for application on site-specific projects at the present time. Their employment in the private sector for routine application is therefore severely limited.

On the other hand, the acquisition of color infrared aerial photography, although more exacting than conventional color or black and white films, has become more or less routine with most firms and institutions capable of obtaining high quality, aerial photography. Therefore, acquisition of the raw data is no problem. In fact, even the more conservative Federal land management agencies are tending to obtain color infrared imagery on missions which formerly were flown with black and white, even though the immediate intended use may not require the infrared imagery. In most cases, the additional cost of the color infrared film is more than justified by the tremendous increase in information content of the imagery.

The analysis of color infrared imagery remains largely an art, but an art which must be accomplished by scientists. Do not be mislead by the superficial paradox of this statement. It takes no scientific training to draw lines on an overlay between a sharply defined forest and an adjacent mountain shrub community. But, to merely label the forest as forest, and the mountain shrub community as such, is not sufficient for an in-depth ecological analyses. Neither does it represent extraction of all the information the imagery contains. If this basic information is to be useful in a broad ecological approach, we must extract the answers to much more detailed questions. For example, what tree species occur in the forest? What is their condition? Is there any harvestable timber which should be considered in assessing the utility of the land area? In the mountain shrub areas, similar questions occur. Is the area important for wildlife? What are the component shrub species, and what types of wildlife do they support? How much of the area is eroded, and does this reflect detrimental use of the area?

Acquisition of these types of information requires detailed analysis of each type or site as delineated previously. Since very few stands of natural vegetation are homogenous in space, the next step following delineation of the major vegetation types is to define each type's individual plant communities in terms of species composition, cover, and the frequency of dominant or otherwise important species. This information is very important from the standpoint of assessing the area's value as wildlife habitats. It is also important to establish the likelihood for occurrence of rare or endangered plant or wildlife species, for establishing reference points with which to measure ecological change, whether natural or man-caused, and perhaps as an inference to the types of soils on which the vegetation is growing. If it is planned that the area be reclaimed to its natural state, then it is necessary to know quantitatively what the natural state was prior to development. Even more basically, such detailed information can contribute substantially to an assessment of reclamation potential.

A whole additional class of information which can be derived from color infrared imagery concerns ecosystem quality. Differential reflectance in the infrared wavelengths between vigorous and diseased vegetation is a well known and documented phenomenon. In the case of oil shale or coal development, diseased trees may at first appear to be unimportant - timber harvest is not the objective of the development. But, a few years after the mine or processing plant has begun operation, when the public for the first time notices large areas of dead trees, a good baseline inventory of the disease's prior distribution under natural conditions would be invaluable evidence that the forest's demise was not caused by the development. In cases where environmental damage of various types is specifically prohibited and regulated, as in the oil shale program, it is even more important to document any evidence of natural environmental degradation before development.

Another category of ecological application in which remote sensing techniques can be used very effectively is the definition and analysis of watershed characteristics. Most basic, of course, is the mapping of drainage patterns, streambeds, canyons and other topographic features. More subtle watershed information which can be derived from color infrared imagery includes erosion history, for example. What is the distribution of gully and sheet erosion on the site and surrounding area? Of even greater importance is an assessment of erosion susceptibility. This information can be inferred by examination of the imagery for erosion history, in combination with data on the amount of vegetation cover, the area's topography, immediate drainage area, and other factors, most of which can be obtained directly from the imagery.

Of great concern in oil shale and coal developments is the effect of mining operations on the quality of both surface and ground water. Although to my knowledge a dependable technique for mapping ground water by remote sensing has not been found, it is almost a trivial exercise to map surface water in great detail. Where ephemeral springs and seeps must be inventoried, as in the oil shale program, the presence of such areas even though dry during the overflight, can be inferred very dependably from the existing vegetation species and its condition.

I have discussed the more important and general types of baseline ecological information which are both required by the coal and oil shale industries, and which are obtainable by remote sensing methods. There are other more detailed types of baseline information on surficial characteristics which are also of importance on specific projects. Most are actually sub-categories of the general ecological applications discussed previously.

ENVIRONMENTAL MONITORING

Environmental monitoring involves basically the observation of change over time of many of the same parameters which were quantified originally in the baseline program. Some, such as disease or accelerated erosion are subject to rapid change. Others may require many years observation before any change is detected.

But, regardless of the relative dynamism of the parameters of interest, aerial surveys at regular intervals offer the most efficient techniques for detecting change or disturbance. Remote sensing techniques may not provide the reasons for change, such as diagnosing the plant disease or measuring the quantity of sheet erosion. But, they can alert one to the fact that a change has occurred, and that ground inspection of the area is required.

To perform this type of monitoring function, the original baseline information and subsequent years' data must be of comparable quality and format. One of the most critical considerations in detecting change is that the data medium have a precision of finer definition than the minimum amount of change which is to be detected. For example, the most direct way of documenting the extent of a plant community, eroded area or other phenomena with quantifiable spatial distribution is to map it. But, the map generated must be sufficiently precise that minor changes in the distribution of the phenomenon of interest can be noted on subsequent surveys of the same area.

Therefore, an important aspect of both baseline and subsequent monitoring data acquisition programs is the accuracy with which surface features are mapped. Corollary to this consideration is the requirement that each succeeding data acquisition program generate maps, overlays or other graphic products of the same scale and format. This provision will allow overlays or maps of the same features from different missions to be registered for detection of changes which may have occurred in the interim. An alternative approach which may be more desirable for larger areas is to digitize either the original film or image overlays in film density classes, for computer storage.

Regardless of the data documentation and storage procedure used, the geometric precision of the stored data will be no better than that of the original information extracted from the film image. Therefore, it is important that any major image distortions be removed or compensated in the information extraction process. In the case of stereo coverage at large or intermediate scales, it is usually necessary to transform all spatial information to a planimetric base by conventional stereoplotting procedures. If the information is to be computer-stored, such transformation is necessary prior to image scanning and conversion to density classes.

OTHER APPLICATIONS

Any discussion of remote sensing applications would be incomplete without addressing the very important category of land use. The main thrust of Western fossil fuels development is, of course, a private interprise effort. However, such regional considerations as land use planning are not necessarily excluded from consideration by the various energy companies. Although commonly considered to be primarily a government responsibility, land use planning is being viewed by a few companies involved in larger developments as a prerequisite to optimal use of all the natural resources under their ownership or control. Therefore, coupled with economic and energy technological considerations, information on current and potential land uses constitutes an important basic input to overall project planning.

Of more importance than current land uses is potential land use or land "usability." The information required to estimate the land's potential falls naturally into the category of ecological suitability, since most contemplated land uses other than mining will be based on the capability of the land to support them. Grazing, farming, timber harvest, and even wildland recreation, uses formerly associated primarily with public lands, are under serious consideration by companies holding major blocks of land in the region. If such land uses are to be profitable for the developing company, then they must be engaged in with full attention to the soils, climate, renewable natural resources, and other natural attributes of the land on which they depend.

Another category of land usability of concern to all mining companies is the potential for reclamation following mining. Although much of the region cannot be restored completely to pre-development conditions, it can probably be reclaimed to a useful and environmentally desirable state. But, planning for such reclamation requires very detailed information and analysis of pre-development ecological systems. Without massive, unreasonable manipulation of the landscape for an indefinite period into the future, they cannot be improved on, they can only be emulated.

Other more specialized applications of remote sensing technology include geothermal prospecting and mapping by airborne thermal infrared scanning, pipeline and transmission line monitoring, and detection of environmental damage of various types.

SENSOR CONSIDERATIONS

I have emphasized the importance of film data products over digital electronic data primarily because multispectral scanners and other electronic data acquisition systems remain generally unavailable for widespread, routine application. In addition, the digital data produced by these systems requires sophisticated and costly computer hardware and software for processing.

There are some more basic problems with electronic systems, however, including the sometimes extreme variations in scene radiance which can occur over time and space, even from the same target surface. Variables like cloud shadow, solar angles, time of day and season, and topography can cause great difficulty with computerized automatic pattern recognition routines. A human image analyst, especially if he is also a trained ecologist can compensate for and overcome most of these limitations in either manual or machine analysis of film imagery. But, to accomplish the same compensation in computer processing requires advanced interactive systems of which very few even exist. They are certainly unavailable for general use at the present time. An obvious exception is the use of single-band thermal infrared scanners for geothermal prospecting, thermal pollution studies, specialized wildlife detection problems, and other applications which depend on radiation in the thermal infrared wavelengths.

The use of color infrared film imagery does not necessarily imply strictly manual image analysis. The employment of modern film image density analyzers, scanners and associated digital processing may offer real advantages in time savings without sacrificing the subjective human advantage of familiarity with the scene characteristics, especially in the analysis of large areas in small scale aircraft or satellite imagery. However, caution must be exercised that the machine only supports and does not dominate the analysis process. As with scanner data, various natural perturbations of otherwise pure spectral signatures can result in serious errors in machine classifications.

ERTS satellite data constitutes a specialized product which offers some very significant advantages in certain applications. Although of insufficient spatial resolution for most of the detailed applications discussed previously, ERTS data can provide an extremely valuable regional-scale view of major surface features. This capability is particularly important in regions where existing information is sparse. As an initial means of mapping drainage patterns, major vegetation types, and large-scale land development activity, satellite data is the most efficient and least costly medium available.

SUMMARY

In closing, I would like to emphasize the importance of care and conservatism in assessing the utility of remote sensing techniques for any application, including those related to energy development. I believe remote sensing has been oversold in some respects, and those of us in the scientific community must accept some of the blame. The unfortunate aspect of this oversell is that is is unnecessary.

We have in remote sensing technology a uniquely valuable tool which can and should stand on its own merits as demonstrated by numerous proven and dependable applications. It is our responsibility to insure that remote sensing applications are developed and applied with maximum attention to scientific integrity and quality of results, and that we do not promote the technologies beyond their capability, merely to enhance our own "visibility" or public stature.

We all are aware of the dwindling research support which has restricted our research efforts in recent years. Perhaps this is merely characteristic of our times. But, I believe we will continue to be so restricted until such time as remote sensing is "brought down to earth," and makes a fresh start at aiding in the solution of the real problems our society faces today. I am not advocating applied over basic research - certainly both must continue. I am advocating adherence to the highest levels of scientific professionalism, and more attention to the quality of the results of our work. Until such time as the public and political eyes of the nation view remote sensing as a proven, useful technology, which in its own right can make a valuable contribution to our energy, food, and other major world problems, we will continue to be hampered by a lack of confidence and support of our efforts.

In Western coal and oil shale development, the environmental and technological problems are numerous and substantial. Anyone who denies their substance is, indeed, poorly informed. For the most part, they are not insoluble. But, the national need for the energy these resources can provide will not allow for enough time to approach them in the conventional manner. We must mount a concerted and comprehensive effort at their solution if we are to maintain the standard of living to which we have all grown accustomed. I am very firmly of the opinion that remote sensing has an important role to play in development of our fossil fuel resources, especially in solution of the associated environmental problems. But, we will not even be allowed on stage unless we can demonstrate our capability to perform.

LITERATURE CITED

1. U.S. Department of the Interior. 1973. Final environmental statement for the Prototype Oil Shale Leasing Program. Superintendent of Documents, U.S. Government Printing Office, Washington, D.C. 3200 pp.

2. U.S. Senate, Congressional Record. October 8, 1973.

3. Federal Energy Administration. 1974. Background Paper, Washington energy conference. February 11-12.

4. Private communication. R.S. Driscoll. U.S. Forest Service, Rocky Mountain Forest and Range Experiment Station, Fort Collins, Colorado.

EXPLORATION FOR FOSSIL AND NUCLEAR FUELS FROM ORBITAL ALTITUDES

NICHOLAS M. SHORT
*Earth Resources Branch, NASA/Goddard Space Flight Center
Greenbelt, Maryland 20771*

ABSTRACT

Studies of ERTS-1 and Skylab-EREP data have defined both the advantages and limitations of space platforms as a new "tool" in mineral exploration. Generally, useful information can be extracted from synoptic imagery and/or the direct measurements of surface reflectances from which small-scale reconnaissance geologic maps are produced, previous maps are edited and refined, landform types are better evaluated in context, and structural deformations and fracture trends are determined on a regional basis. Information about rock composition, stratigraphic ages, and "telltale" chemical alterations as guides to subsurface deposits are less reliable.

Remote sensing data from satellites are being applied to hydrocarbon exploration by many oil companies, although very few have reported "open file" results to date. One ERTS investigation in the Anadarko Basin of Oklahoma has demonstrated a remarkably high correlation between several types of anomalies recognized in the imagery and the locations of known oil and gas fields. These exceptional surface features include: 1) surface expression of underlying structures, 2) circular drainage patterns, 3) linear controls (faults and fractures), 4) tonal anomalies (nature unknown), and 5) "Hazy" anomalies (chemical alteration; geobotanical anomalies; man's activities?). When used during the exploration phase, ERTS data serve to localize areas for more intensive study by field mapping, geophysical and geochemical methods, and subsurface drilling. By reducing large areas to prime "targets" of maximum likelihood, the ERTS overview approach should save a considerable fraction of the exploration costs, and in some instances should assist significantly in making new discoveries.

In addition to supporting several ERTS follow-on investigations in petroleum exploration, NASA has approved a broad in-house study at Goddard Space Flight Center designed to verify the general applicability of the initial Anadarko Basin results. Using both conventional photogeologic methods and special computer processing, imagery taken over such oil-producing provinces as the Williston Basin, North Dakota, the Green River and Wind River Basins of Wyoming, the Denver-Julesburg Basin, several west Texas fields, and Gulf Coast salt domes is being subjected to detailed analysis in search of definitive recognition criteria.

Goddard is also engaged in a smaller-scoped study of ore guides, such as "iron stain" surface alteration and delineation of fracture traces, associated with the sedimentary "roll" deposits of uranium minerals in Wyoming.

INTRODUCTION

A pressing need to find new methods and approaches in exploring for oil and gas and other sources of fuels no longer requires demonstration or proof. Any American who experienced interminable waits in long gas lines in the 1973 winter of discontent, or reacts to his ever-rising monthly utility bill, or simply reads the depressing reports of still-impending energy crises in his local newspaper will not require any convincing to become alarmed about the outlook for the future. Many responses to these problems have been advanced—conservation, pressures on our foreign supplies, and accelerated development of solar, aeolian, geothermal, and fusion energy sources lead the list.

An acceleration in exploration is another obvious response. Tried and true techniques will be applied with increased vigor. But, innovative and even unconventional techniques must also be devised, tested, and put into operation as soon as they are declared to be practical and productive.

One such novel technique has moved across the horizon of possibility and looms now as one of the more promising approaches in an expanding search for fuel sources and other raw materials relevant to a rising worldwide demand for more energy. The approach is simply to use orbital space platforms—both automated satellites and manned space vehicles—from which to gain a new perspective of the Earth's surface by means of standard and/or specialized remote sensing techniques. This approach is actually an outgrowth of many years of aerial photography as applied to geologic mapping and mineral exploration. Its important new (although not unique) advantages over the commonly used conventional aerial systems are:

(1) the greatly expanded synoptic view provided by individual images owing to a ten to hundredfold increase in operating altitudes coupled with high resolution imaging sensors,

(2) the high frequency of (repetitive) coverage resulting from the numerous orbital passes available during spacecraft missions of months (manned) to years (satellites) duration,

(3) the use of a variety of sensors extending through several regions of the electromagnetic spectrum, including multispectral imagers, that acquire coherent data simultaneously, and

(4) the ease with which orbital remote sensing data can be transmitted or later converted into a digital mode, allowing their further treatment by a wide range of computer-processing techniques involving enhancements, selective information extraction, comparison of repetitive scenes, etc. to be made on a large volume of information.

Practical remote sensing from space platforms began almost with the launch of the first rockets that carried recoverable photographic equipment. Photographic data recording wide areas of the Earth's surface continued to be gathered on nearly all manned missions from Mercury and Gemini through Apollo. Weather satellite systems, such as TIROS and Nimbus, returned useful data on the land and sea surfaces in addition to the atmospheric data for which they are designed. Drawing upon years of experience in remote sensing from aircraft, the designers of orbital remote sensing platforms placed the first "sophisticated" multispectral sensor system on ERTS-1, the Earth Resources Technology Satellite launched in mid-1972. Soon thereafter the Skylab manned laboratory began to acquire an even broader range of remotely-sensed data, using film cameras, multispectral scanners, an infrared spectrometer, and several radiometers. Future systems now on the "drawing boards" will incorporate this growing wealth of experience with a new generation of varied sensors in space. Thus will emerge ever more versatile research satellites and, in all likelihood, on-line or operational satellites (such as the EOS series) and spacelabs (including shuttle-serviced stations).

In keeping with the author's current base of experience, this paper will confine its topics almost exclusively to results from the ERTS program pertinent to exploration for oil and gas and, more briefly, for uranium deposits. The remainder of the paper is divided into four parts: 1) a review of achievements to date in relevant aspects of general geologic studies from ERTS, 2) a survey of reported accomplishments oriented specifically towards exploration for energy sources (with emphasis on petroleum), 3) an evaluation of the prospects and limitations of the space platform approach to fuel exploration, and 4) an examination of continuing programs now funded or planned that are designed to "prove out" the use of ERTS and other space systems in exploring for fuel resources.

Before proceeding through these four sections, it is appropriate to summarize as a "preview" the present status of exploration for fossil and nuclear fuels from space. Simply synopsized, it must be stated that, as far as NASA officials can determine, no previously unknown petroleum or uranium deposits have yet been discovered directly (and probably indirectly) from interpretation of images or other forms of remotely-sensed data obtained either from satellites or from visual and/or instrument observations made by astronauts. However, "rumors" have reached the author and others of considerable interest in this approach by oil companies and other exploration-minded enterprises. It is in the nature of the petroleum industry to be secretive about successes with new techniques that score in finding oil and gas — witness the delay of several years in making seismic prospecting generally available throughout the industry after a few companies had verified its value. What has been brought to light by ERTS in particular which provides invaluable adjunct or supplementary data to the exploration scientists are these accomplishments: 1) updating and refining rock unit boundaries and distributions on small-scale maps, 2) recognition of hitherto unsuspected lineations — especially regional straight to circular or arcuate fracture systems — that could control or influence the location of oil traps or uranium concentrations, 3) better definition of often subtly expressed geomorphic "anomalies" that bear some relation to subsurface structures, and 4) an apparent surface manifestation of alterations of rock, soil, or vegetation tied to escaping hydrocarbons or to redistribution of elements associated with shallow uranium deposits. According to the growing convictions of those geologists now working with NASA's Earth Resources programs, it is just a matter of time before an oil field or a nuclear ore deposit is found through the use of significant data obtained from ERTS or some other spaceborne sensing system as an integral part of the exploration program.

PART 1: PROGRAM RESULTS*

A. General View: The ERTS-1 spacecraft was launched on July 23, 1972 from NASA's Western Test Facility in California. Both the Return Beam Vidicon (RBV) and the Multispectral Scanner (MSS) began to transmit image data on July 26, 1972. Owing to a switching circuit problem, the RBV has been shut down since early August of 1972. As of October 1, 1974, the MSS has imaged more than 100,000 scenes covering greater than 85% of the Earth's land surface. About 30%, on average, of these scenes are largely cloud-free and well-illuminated. Coverage of all of North America has been continuous but coverage of other continents became severely limited by tape recorder difficulties since March 1973.

The reader is urged to consult the three volume Proceedings of the Second (March 5-9, 1973) and Third (December 10-14, 1973) Symposia on Significant Results from the Earth Resources Technology Satellite** for detailed treatment by the investigators, program managers, agency representatives, and invited speakers of the major findings of ERTS-1 for the various discipline groups comprising the Earth Resources program.

B. Value of Synoptic Coverage: ERTS provides a remarkable sequence of uniformly illuminated, essentially planimetric vertical views of the Earth's surface that cover very large areas (~12,500 square miles) in an image obtained from the 570 mile orbital altitude. Like the Gemini-Apollo pictures, these images are invaluable because of their synoptic aspects—that is, a single scene (approximately 115 miles on a side) surveys a wide variety of terrain, geology, and land use types under near-optimum viewing conditions that emphasize the contextual relations of these surface features. Unlike the Gemini and Apollo pictures, or those from Skylab, the ERTS images meet the necessary conditions for being readily joined together in mosaics. This greatly increases the synoptic character of this imagery and extends the assessment of contextual relationships to regional and even subcontinental proportions. The resulting mosaics are surprisingly close to being orthographic, as can be determined by comparing both individual images and composite mosaics to such projections as Albers Equal Area or Lambert Conformal.

---

*Most of Part 1 has been extracted from the NASA-Goddard Space Flight Center X- Document X-650-73-316 entitled Earth Observations from Space: Outlook for the Geologic Sciences, by Nicholas M. Short and Paul D. Lowman, Jr., October, 1973.

**Volume I of each Symposium Proceedings is available from the Government Printing Office; inquiries about availability of Volumes II and III can be made there.

ERTS-1 has now acquired enough cloud-free imagery to allow assembly of vast areas of Earth in mosaics. Some spectacular examples have already been made public. A black and white (red band 5) mosaic of the entire state of Oregon is reproduced here (Figure 1) and many other states individually as well as the entire continental United States have been similarly mosaicked. Color mosaics are completed for the eastern United States from Maine to Florida, the entire western coastal United States, Florida, Louisiana, Michigan, Wyoming, and Montana (among others), and for several regional sections. Foreign areas available in color now include Italy, parts of west Africa, all of Iran, the Red Sea area, and Yemen.

Beyond the esthetic and technical achievements associated with such mosaics, the scientific merts alone—particularly in geology—are a sufficient justification for their production. Comparison of small-scale geologic or physiographic maps of large areas with ERTS mosaics reveals at once the remarkable utility of the latter for presenting regional interrelationships among major structural or land-form units. Such mosaics are especially suited to lineaments analysis where uniform lighting serves to highlight trends of continuous, often deep-seated fracture zones.

C. <u>Other Special Advantages of ERTS</u>: Geologists experienced in using radar imagery look forward to the day when an active microwave system is flown on a spacecraft. One property of radar, that of cloud penetration, offers a distinct advantage in regions which are habitually overcast. Another important property, at least of some of the microwave bands now used, is the "apparent penetration" of heavy foliage which allows radar to "see" the terrain beneath an extensive tree canopy. It is surprising to note that under some conditions, the images made from the ERTS infrared (IR) channels (Bands 6 and 7) will provide a rendition that has a "quasi-radar" aspect. This is graphically displayed in Figure 2 where the thick jungle in southern Venezuela appears almost uniformly black in the green and red bands (4 and 5) but seemingly is "stripped" to bare ground in the IR bands. In reality, the close association of canopy profile with ground topography is being revealed when the foliage is examined in the highly reflective infrared. The crystalline rock terrain can be broadly defined and differentiated in the IR band image.

Some stereo capability exists for ERTS images where sufficient sidelap is maintained. However, most of each scene cannot be viewed in this way. Using the multispectral scanner data stored on the digital tapes, members of the U.S. Geological Survey remote sensing facility at Flagstaff, Arizona have developed a reprocessing procedure that produces a pseudo-relief or "3-D" effect. One phase of that procedure involves a change in the contrast ratios of grey levels in a given band—this selective darkening and lightening acts much like air-brush shading used to give an appearance of relief to a topographic sketch map of mountainous terrain. An example of the final product is shown in Figure 3.

D. <u>Specific Geologic Applications</u>: It should be kept in mind that ERTS is basically an extension of aerial photography to large area mapping. Aerial photographs in the past have been used chiefly as aids to the geologist in preparing or revising various kinds of geologic maps. The techniques and approaches of photo-interpretation developed prior to availability of multispectral imagery from space remain the major tools by which geologists extract information from ERTS data. Because of the decrease in resolution (by factors ranging from 5 to 100), certain types of information found in aerial photos are inherently unrecoverable from ERTS but such deficiencies are offset by the synoptic overviews that provide hitherto unobtainable information. Thus, some tasks will still be done better from aircraft but others may be done best—and even exclusively—from space platforms.

One measure of success of ERTS in geology is the extent to which new information has been acquired. While it is still premature to assign a dollar value or cost-effectiveness rating to the results reported so far, it is clear that significant benefits are gradually accruing from ERTS data to the specific applications outlined in Table 1. In time, accomplishments in each of these fields will be translated into economic payoffs as new mineral deposits and oil prospects are discovered and engineering projects are undertaken because of these scientific and technological advances flowing from ERTS and other programs that couple space-acquired data with conventional ground exploration methods. The outlook is especially promising for certain parts of the world where ERTS images represent the first detailed surface coverage of regions never before surveyed or mapped beyond a reconnaissance level.

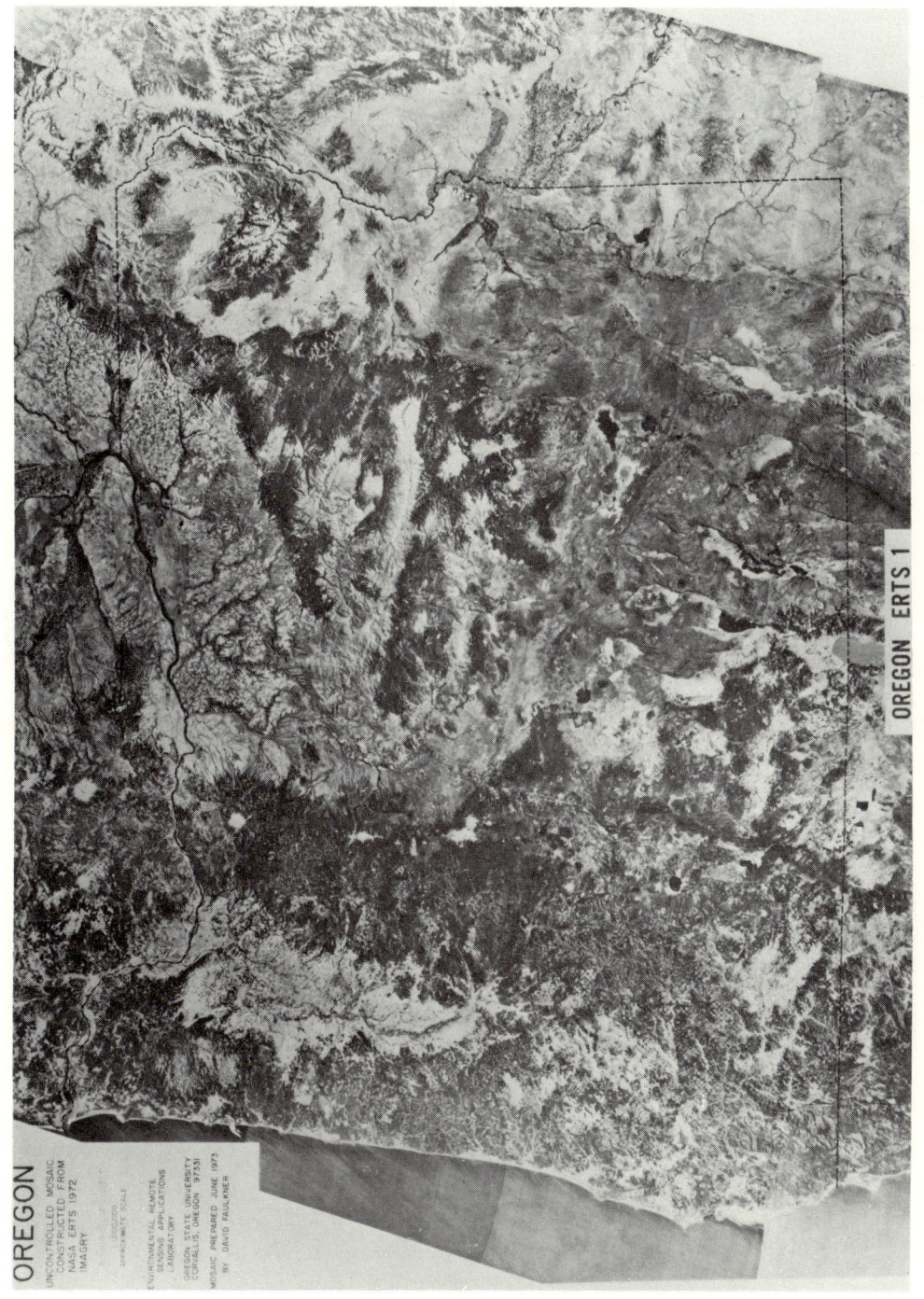

EXPLORATION FOR FOSSIL AND NUCLEAR FUELS FROM ORBITAL ALTITUDES 193

Figure 1. Uncontrolled photomosaic made from ERTS images acquired in 1972 showing the entire state of Oregon and parts of the surrounding states of California, Idaho, and Washington. All images were produced from red band 5 (courtesy Oregon State University).

Figure 2a. Red band 5 ERTS image taken on March 9, 1973 over the Orinoco River basin; the river divides eastern Columbia from southern Venezuela. The Llanos, a grass-covered plains, appears on the west side; a thick jungle-like forest covers the terrain on the east side.

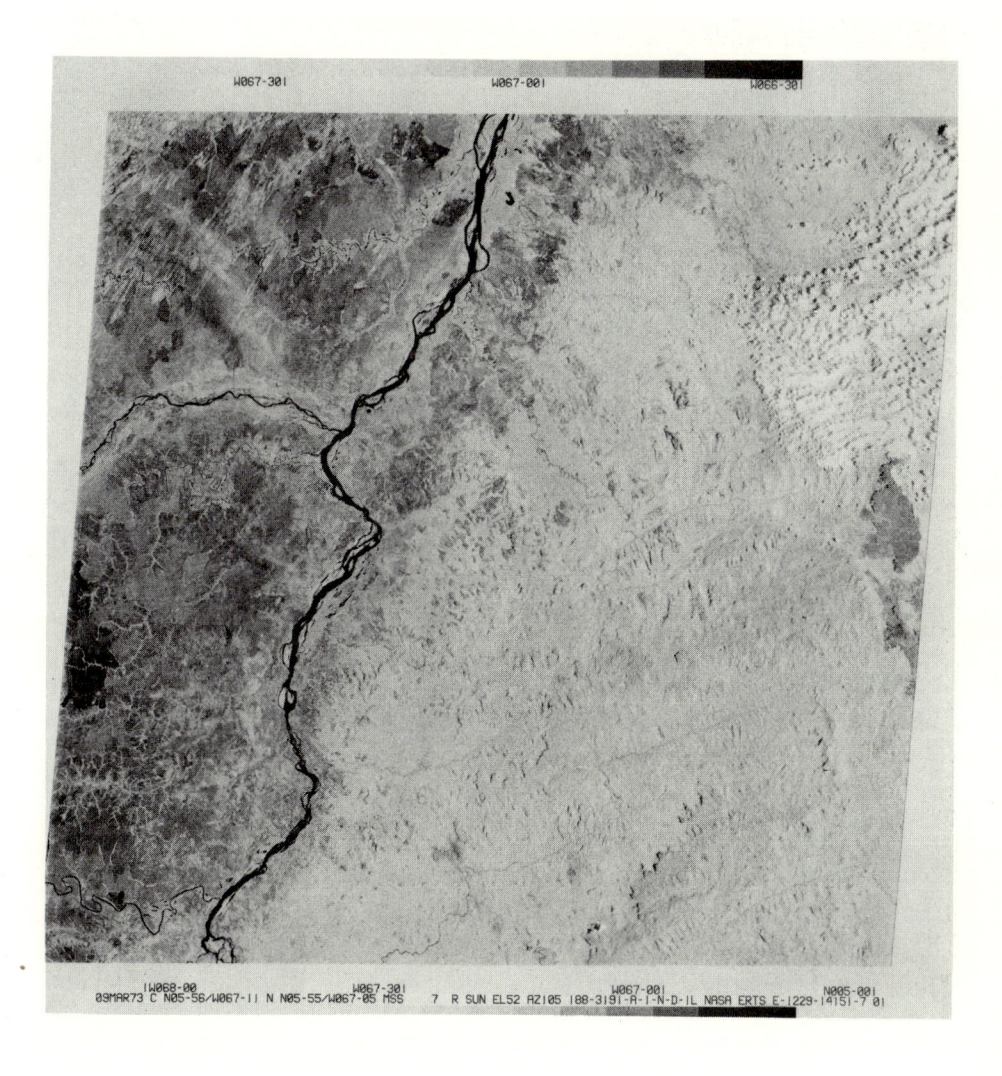

Figure 2b. IR band 7 ERTS image of the same scene shown on the opposite page. Some of the details in the Llanos are emphasized by the stronger contrast. The dark jungle cover seen in the band 5 image now acts as though it has been "penetrated", revealing the underlying terrain, here consisting of Precambrain igneous and metamorphic rocks.

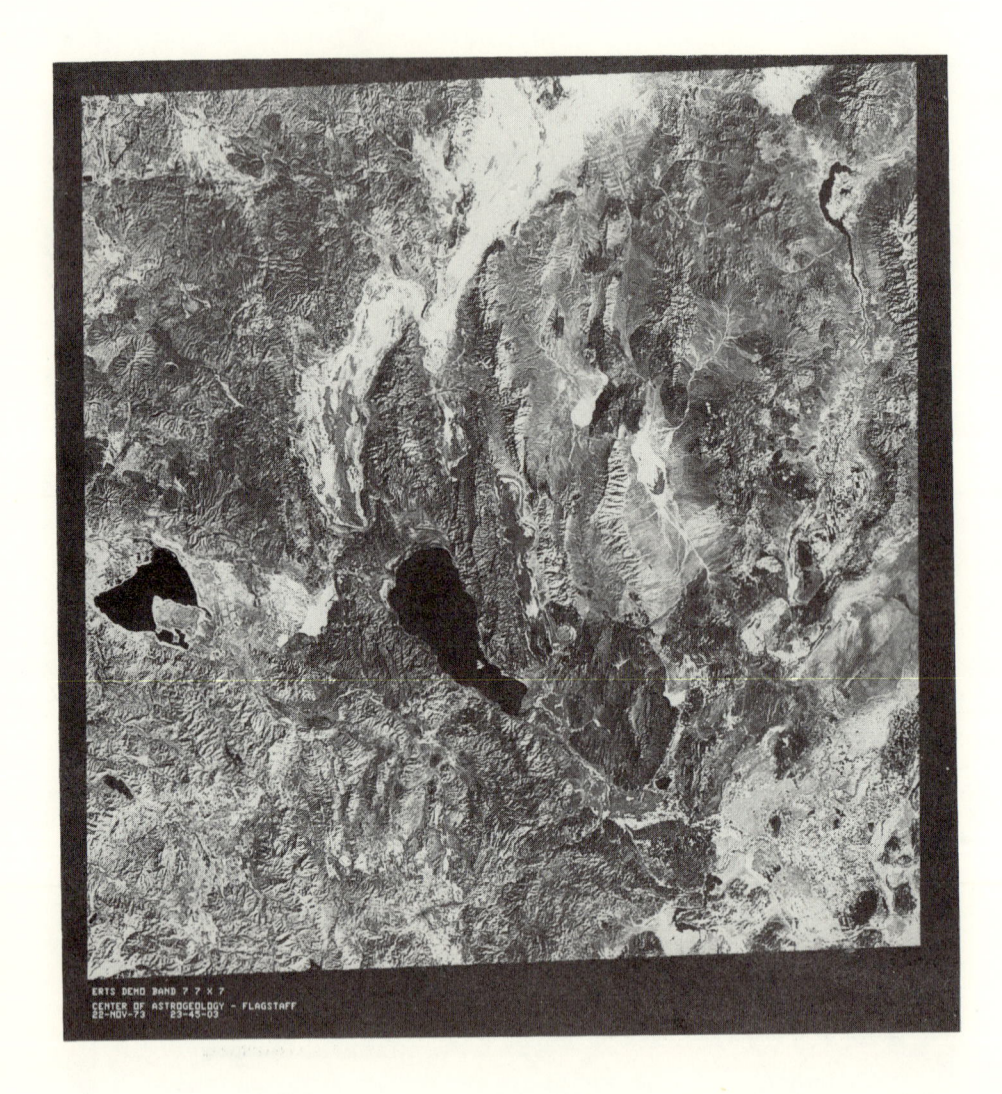

Figure 3. Computer-reprocessed rendition of the band 7 ERTS image taken on November 22, 1973 by the multispectral scanner as the spacecraft passed over western Nevada (Reno in lower left; Pyramid Lake in center; Black Rock desert at top). The pseudo-relief effect is the result of contrast stretching (courtesy U.S. Geological Survey).

Table 1

Applications of ERTS to Geology

Map Editing: • Boundary and Contact Location   • Stratigraphic and/or "Remote Sensing" Unit Discrimination   • Scale-change Corrections   • Computer-processed "Materials" Units Maps
Landforms Analysis: • Regional or Synoptic Classification and Mapping   • Thematic Geomorphology (e.g., Desert, Glacial, Volcanic Terrains)
Structural Geology: • Synoptic Overviews of Tectonic Elements   • Appraisal of Structural Styles   • Lineaments (and "Linears") Detection and Mapping   • Metamorphic and Instrusion Patterns   • Recognition of Circular Features
Lithologic Identification: • Color-Brightness (Spectral Reflectance) Classification   • Ratio Techniques   • Photogeologic Approach
Mineral-Exploration: • Reconnaissance Geologic Mapping   • Lineaments Trends (especially Intersections)   • Surface Coloration ("Blooms" and "Gossans")   • Band Ratio Color Renditions
Engineering and Environmental Geology: • Dynamic Geologic Processes (Sedimentation and Coastal Processes; Sea Ice; Active Glaciers; Permaforst Effects; Landslides and Mass Wasting; Shifting Sand Seas; Land Erosion)   • Strip Mining; Surface Fractures – Mine Safety   • Construction Materials

SOME OF THESE APPLICATIONS ARE REVIEWED IN THE FOLLOWING SECTIONS:

(1) <u>Map Editing</u>: To some extent, ERTS data can be used to make new maps but these would not be equivalent to those produced from aerial photographs. In order to construct a standard geologic map, it is normally necessary to recognize stratigraphic units and sequences at the <u>formation</u> or even <u>member</u> level and, to a lesser extent, to discriminate among the major lithologies in the area. For mapping from aerial photos, this requires recognition of unit boundaries and definition of differences among units on the basis of rock color and/or surface weathering effects, topographic and/or geomorphic expression, soil associations, and characteristic vegetative cover, among other criteria. Most units depicted on large-scale maps range in thickness from a few tens to a few hundred feet at most. However, because of resolution limitations, most stratigraphic units (defined from ground studies by criteria that usually require close-up or even hand specimen examination) cannot be recognized and separated in ERTS images along the same boundaries selected for mapping purposes. Several ground-distinguishable units with similar reflectance properties (termed remote sensing units) might group or blend into a single discernible unit in an ERTS image that may or may not have a meaningful stratigraphic and/or lithologic significance. When examined in the field, some remote sensing units actually correspond to single stratigraphic units but others are comprised of several stratigraphic units having similar reflectance.

Nevertheless, under suitable conditions geologic maps of considerable usefulness have already been produced from ERTS imagery. The example from Wyoming shown in Figure 4, when compared with the published map of the same area, indicates that new units having a field-checked reality were defined from the ERTS images even though the contacts among these or previously known units may not be as precisely located as those in the ground-based map version. In some instances mapping from ERTS can be made more accurate by referral to coverage from several seasons, as effectively illustrated in Figure 5. This makes use of the repetitive (18-day cycle) aspect of ERTS coverage from its near-polar orbit.

A more immediate application of ERTS images lies in map editing or revising of previous small-scale maps. In regions of the world where rock exposures are sharply defined (mainly in deserts or other areas of low vegetation), the correspondence of ERTS-viewed surface geology patterns with those in the maps is almost self-evident (Figure 6a and b). But, close comparison of image to map frequently points to serious discrepancies in the map version. Reality resides with the ERTS image.

(2) <u>Lithologic Identification</u>: Identification of rock types from aerial or space platforms has long been a goal that consistently remains elusive. The high hopes that at least the major rock groups could be recognized with presently used remote sensor data have met with varied success. Depending on the experience of the interpreter and his awareness of the rock types known to occur in the imaged area, the photogeologist frequently has been able to correctly identify basalts, granitic rocks, some metamorphic types, limestones, shales, and sandstones. This ability will, of course, decrease considerably as resolution becomes too poor to single out individual lithologic units. However, in some geologic terrains, generally homogeneous rock units are exposed over wide enough surfaces to produce distinguishing tones and patterns in ERTS imagery, as illustrated in Figure 7.

It is not likely that remote sensors operating from space will ever achieve a high degree of reliability in rock type identification. Unlike laboratory methods, such as x-ray diffraction in which unique solutions to component mineral identity result from fundamentally different combinations of atomic structure, there is little that most remote sensing devices can measure that is exclusive to any given rock type. In the spectral range scanned by the ERTS MSS, the only rock properties directly measured are color and brightness; indirectly, derivative properties such as relative erodability (expressed topographically), surface stains, soil associations, structural response, vegetation preferences, etc. are taken into account in making identifications. However, it is not possible to set up a meaningful working classification of rocks based primarily on typical colors and relative brightness (most classifications are built from mineral assemblages, textural aspects, and field relationships). Thus, granites and schists, sandstones and limestones, shales and slates, and other lithologically or genetically dissimilar rock pairings may have roughly the

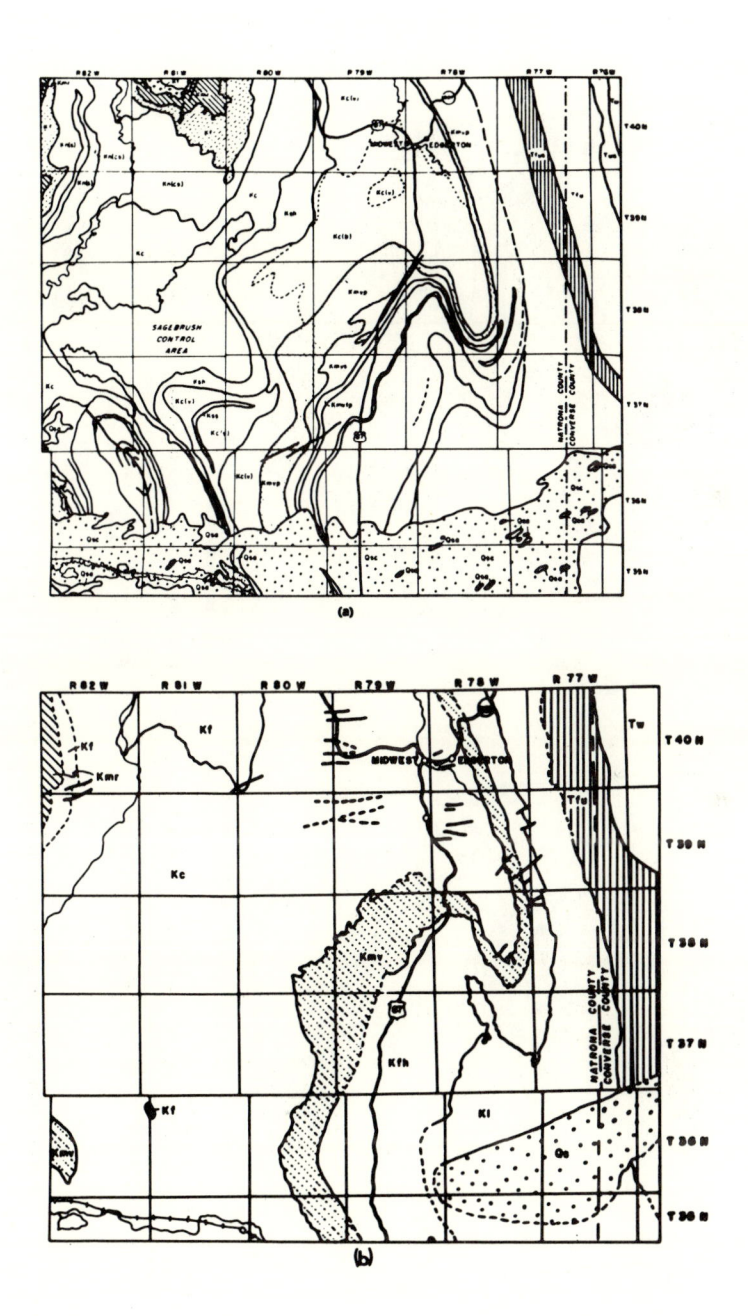

Figure 4. Details of the southeast 1/4th of the map of the Arminto area in central Wyoming as prepared from ERTS imagery (top) compared with the equivalent area taken from the state geologic map (bottom) published in 1955 (courtesy R. S. Houston, University of Wyoming).

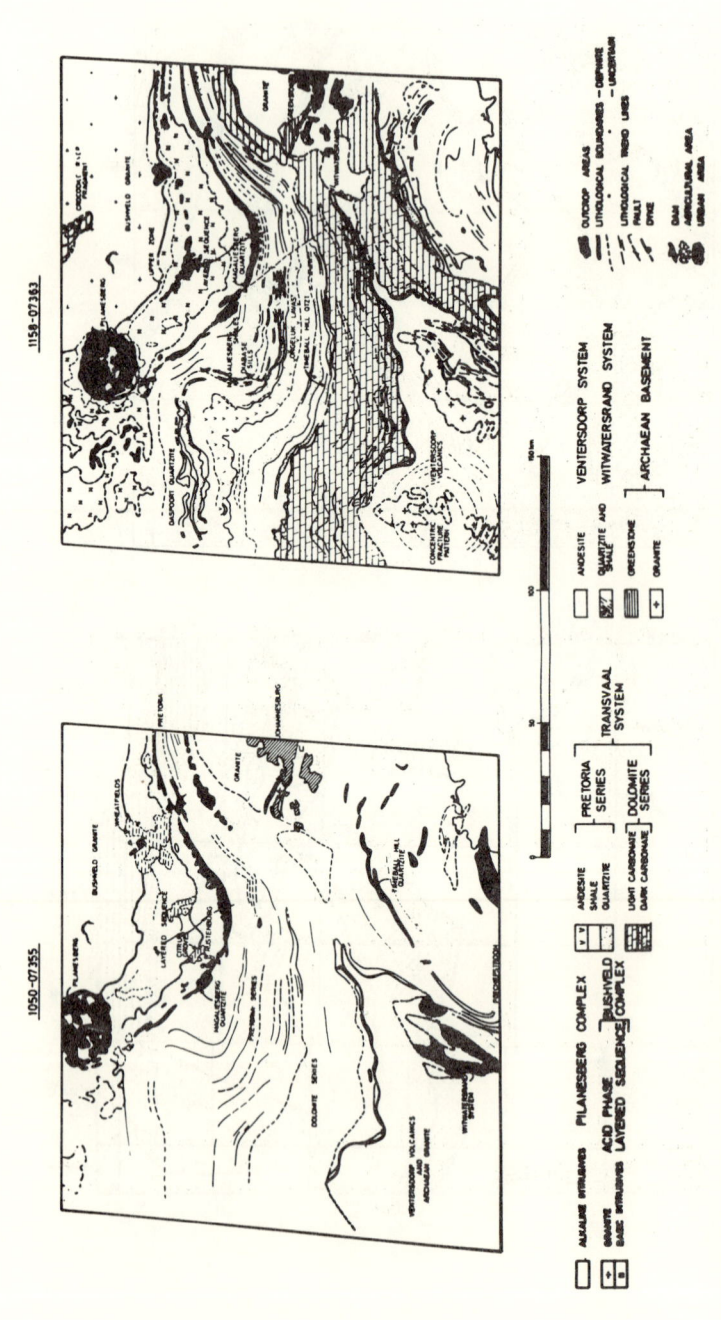

Figure 5. Sketch maps of geologic interpretations made from ERTS images of part of South Africa west of Johannesburg acquired in early September, 1972 during the dry season (left) and late December, 1972 during the wet season (right).

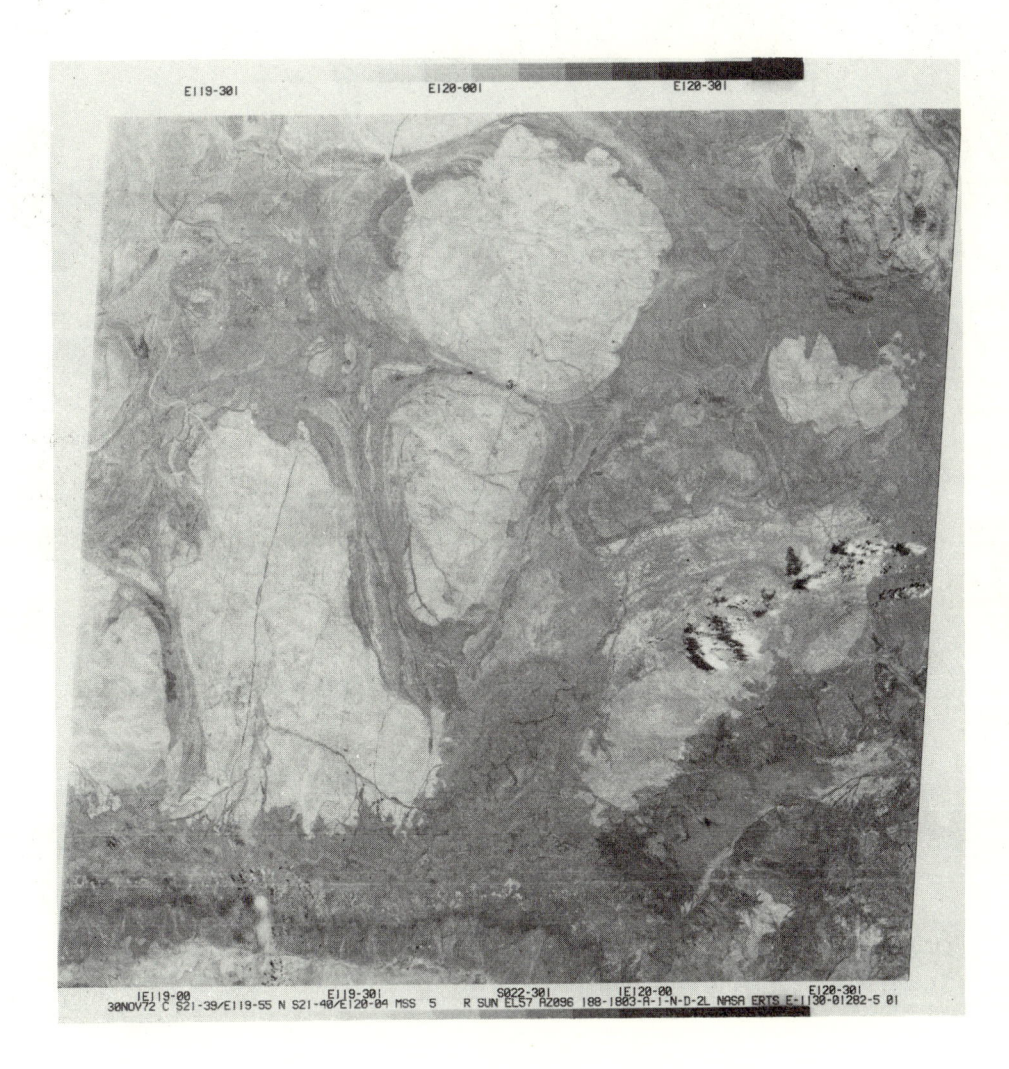

Figure 6a. ERTS red band image of a region in western Australia just south of the northwest coast near Port Hedlund showing several large igneous plutons cutting into metamorphic rocks (mantling bands).

Figure 6b. Part of the Tectonic Map of Australia that includes almost all of the area imaged in Figure 6a. Ag, As, and A refer to Archean granitic and metamorphic rocks; P denotes Proterozoic metasediments; Cz relates to Cenozoic sediments. Close comparison between mapped unit boundaries here and those in the ERTS image reveals that improvements in the precise locations of these boundaries can be made from the ERTS view.

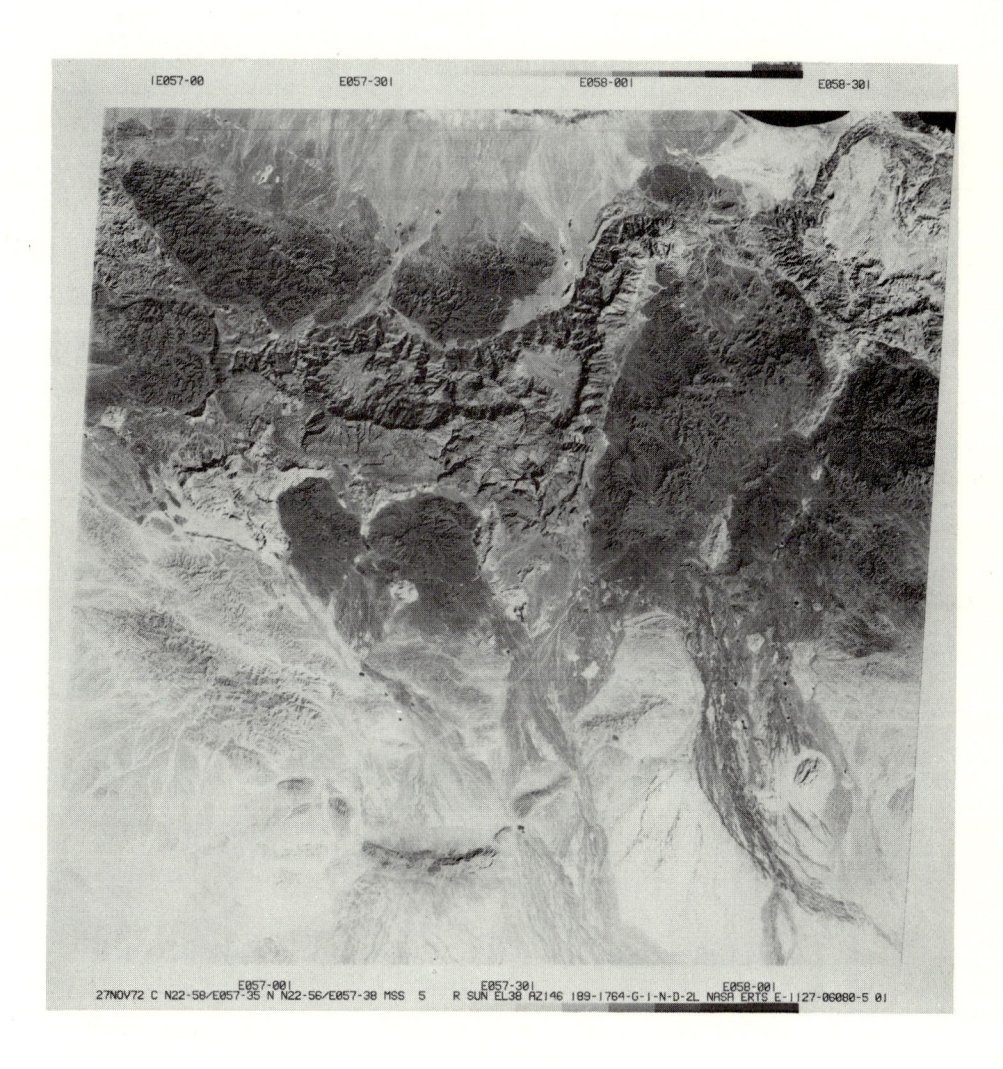

EXPLORATION FOR FOSSIL AND NUCLEAR FUELS FROM ORBITAL ALTITUDES 203

Figure 7. The Oman Mountains in Oman along the Gulf of Oman in the northeast sector of the Arabian Peninsula. The darker areas are mostly ophiolites and serpentines. Several folded belts of Paleozoic and Mesozoic carbonate-shale sequences lie between the basic igneous intrusives; one near top center has been breached to exposed Paleozoic dolomites. Tightly folded radiolarities and other sedimentary rocks are evident in center left.

same colors and brightnesses. Conversely, one given rock type may have many color variants as, for example, green, red, buff, gray, and black shales, or white, dark-gray, buff, and red limestones.

(3) <u>Structural Geology</u>: Experience with earlier space imagery had disclosed the exceptional value of synoptic imagery for displaying extended structural elements such as closed anticlines, domes, and intrusive bodies, folded mountain belts, fault zones, regional joint patterns, and other fracture systems in their regional context. In arid regions, especially, the surface expression of structurally disturbed parts of the crust was often better revealed in the images than in maps of the same areas. The interplay among underlying structure, topography, vegetational distributions, and solar illumination commonly enhanced the appearance of structural elements, so that subtle relationships not apparent on the maps were made to stand out. New lineaments of considerable magnitude and extent were picked out in the images because their breadth and continuity were commonly overlooked on the ground where only small, localized effects of a segment exposed discontinuously from one outcrop or topographic expression to the next were insufficient to manifest the "whole from the parts". In some areas of the world (e.g., Gulf of Oman; the Afar; Afghanistan) space imagery has brought about improved understanding or even fundamental resynthesis of the tectonic framework. ERTS has broadened these observations to sections of the globe never before imaged in detail from space. Four outstanding ERTS views of complexly folded and faulted parts of the crust are documented in Figure 8 a-d.

As expected, the principal output so far from examination of ERTS images for new structural information is the recognition of numerous linear features, ranging in length from 1-2 miles up to several hundred miles. Almost every ERTS image having notable geologic content is marked also by occasional to frequent linears. In the first rush to report significant results, investigators usually equated these linears with structural features such as faults, joints, or inclined strata. Many of these interpretations have stood the test of field-checking. But others have been abandoned when individual linears were found to be lighting artifacts, spurious alignments of diverse ground features, or man-made objects; the degree to which a region is vegetated represents another factor that influences the apparent occurrence of linears.

The geologic studies of the state of New York by Y. W. Isachsen provide a case in point. At the March 1973 ERTS Symposium, Isachsen displayed a mosaic of the state which has been analyzed for structural features. His efforts concentrated on the Adirondack and Catskill Mountains (Figure 9) where large crustal fractures were well-known and mapped, in part because of their control on regional topography. He has since presented an updated ERTS-based map (Figure 10) of linears of all kinds observed in the eastern half of New York. His first appraisal of these linears had indicated 1) many of those already recorded from geologic studies were recognizable, 2) while other known ones which should be visible did not show up, and 3) still others prominent in ERTS images were completely new features not recorded on any maps. However, after careful field-checking and examination of maps and photos, he concluded that less than 1/3 of the new linears are likely to be strictly structural in nature. Many undiscerned faults and lineaments fail to be expressed in ERTS images because of unfavorable illuminations and/or degradation effects in the 3rd- and 4th-generation images used.

Another study reported by N. H. Fisher and his colleagues in the Australian ERTS investigations is particularly instructive. Several test areas that have been thoroughly studied and mapped over the years on the ground and from aerial photographs were chosen for comparison with the information content extractable from ERTS. In the case of linears, it was found that only 30% and 10% of the previously known faults and major fractures were recognized in the 1:1,000,000 and 1:250,000 scale ERTS images respectively covering the same area. Furthermore, many of the ERTS-identified linears were new and did not coincide with the field-mapped lineaments. Also, radar-detected lineament patterns generally were not compatible with the ERTS linear sets. This discrepancy between ERTS and ground truth, while it seemingly casts doubt on the validity and reliability of the space imagery as a discriminator of structural features tends to be reduced when it was realized that the mapped lineations 1) were often defined by criteria discernible only on the ground (e.g., fault gouge, slickensides), 2) included many short lineaments below the maximum length (1-2 miles) detectable from ERTS imagery, 3) often consisted of close-spaced sets counted as single linears in ERTS scenes, and 4) did not suffer from the bias of one-time-a-day coverage at mid-morning. A related Australian study disclosed an additional effect of operator bias: the same interpreter picked out (or missed) different linears when ex-

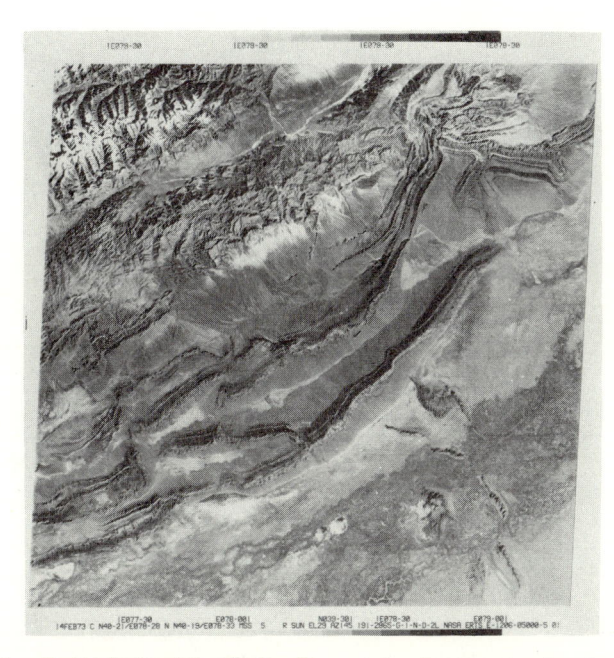

Figure 8a. Broad folds, offset by faults, of the K'op'ing Shan, a series of ranges set against the edge of the Takla Makan desert of western Sinkiang Province in westernmost China. The higher mountains to the north are part of the Tien Shan which passes along the border with the Kirgiz Republic in the USSR.

Figure 8b. Part of the folded Appalachian Mountains, the Blue Ridge, the Piedmont, and the dissected Appalachian Plateau in western Virginia (Roanoke appears in the right center) and West Virgina.

Figure 8c. The Kuruk Tagh mountain range, a complex of Precambrian igneous and metamorphic rocks and infolded Ordovician rocks in the Sinkiang Province of China. The great east-west wrench or tear fault can be traced for more than 300 miles. Lake Baghrash lies to the north.

Figure 8d. A band 7 (IR) ERTS image of the eastern end of the Great Slave Lake in Canada's Northwest Territories. Glacially-scoured fractures in the Precambrian crystalline rocks and glacial lakes in moraines are filled with water in this scene; the water appears very dark in this infrared and thus tends to delineate these structural and geomorphic features.

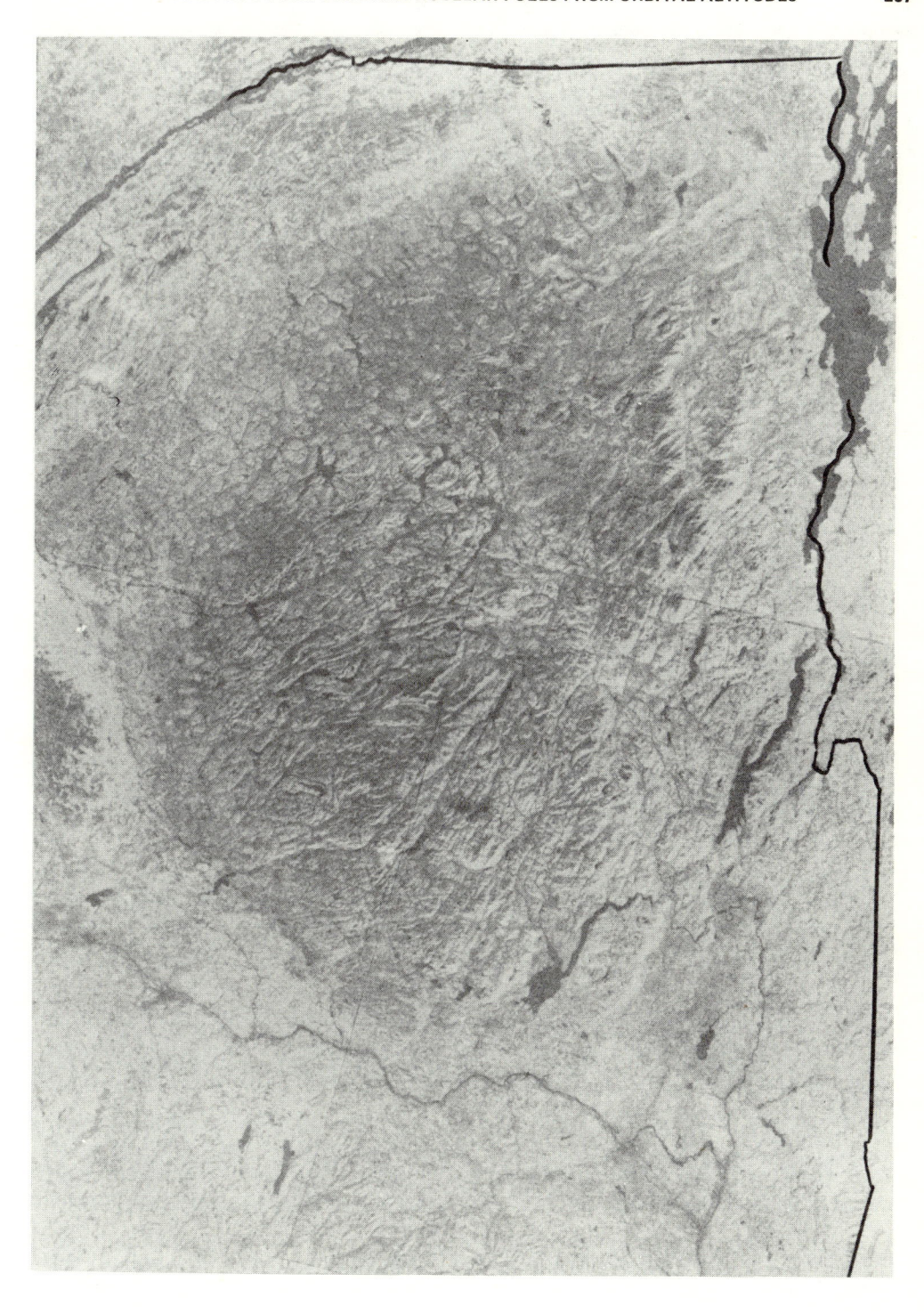

Figure 9. ERTS mosaic showing the entire Adirondack Mountains of Eastern New York. The St. Lawrence River appears at the upper left, Lake Champlain at the upper right, and the Mohawk River near the bottom (courtesy Y. W. Isachsen).

Figure 10. Sketch map of previously mapped lineaments (solid lines) and new "linears" (dotted lines) not all of which have proved to be rock fractures, observed in ERTS images of eastern New York (courtesy Y. W. Isachsen).

aming the same image at intervals several months apart and several interpreters tended to produce different, subjective linears maps from the same image.

The Wind River Mountains of Wyoming provide a dramatic indication of the rapidity with which a mapping effort can be accomplished using ERTS imagery. Dr. R. Parker of the University of Wyoming has been mapping in the high country of this range for five years—a task carried out on pack mule and "shanks-mare" in the grand tradition. His labors led to the map produced in Figure 11 left. After receipt of ERTS imagery covering this range, he completed the map shown in Figure 11 right in just 3 hours. Although this map should be rated as "preliminary" because most of the lineaments have not been verified, some confidence in the correctness of identification is afforded by field checks at several localities where evidence of fracturing was then obtained. Still, a note of caution has been added to this work following receipt of some Skylab imagery of the same area. A linears map made from the Skylab scene is compared with the ERTS version in Figure 12. The difference in numbers of linears detected results from the higher resolution of the Skylab metric camera. The differences in orientation of prevailing linears is almost certainly due to the times of day when the data were acquired—the morning sun direction from ERTS favors enhancement of northeast sets while the afternoon sun-illuminated images from Skylab emphasize north to northwest sets.

ERTS has shown a special — almost unique — facility for calling attention to circular as well as linear features. Most of these are volcanic or intrusive in nature and many are newly recognized. The circular or arcuate markings traced in Figure 13 are thought to be fractures in the country rocks overlying a series of intrusives in central Colorado. Similar curved linear features, consisting of segments that commonly encompass a full 360° are in places associated with porphyry copper deposits in many parts of the world. Again, these may have developed over stocks and diapirs but many are believed to be fractures within the initial subsurface or "roots" zone of volcanoes long since eroded below their cones or craters.

ERTS mosaics are ideal for getting a perspective on the tectonic framework of large regions of the crust (Figure 14). Thus, through-going lineaments that continue for hundreds of miles can be integrated into a unified network which reflects the influence of fractures in an ancient basement or results from stress systems developed from more recent plate tectonic movements. Seen in a broad context, where diversities of topography and surface geology tend to be filtered out, a new synthesis of structural data can emerge. Various investigators are even now building revised structural models from ERTS for their regions of interest. Some first results of one such effort applied to all of the conterminous United States are depicted in Figure 15, although no follow-up verification of the existence of heretofore unrecognized lineaments has been carried out as yet.

(4) <u>Mineral Exploration</u>: Work to date has pinpointed two potentially useful ways in which ERTS data could help to locate conditions favorable to the concentration of metals and other mineral materials. First, the recognition of new crustal fractures and, especially, intersections in lineaments systems improves the probability of finding ore <u>if</u> one believes in the commonly held view that such fractures control localization of mineralizing solutions. Each new fracture or intersection provides new targets for exploration. Point intersections, particularly, represent a significant narrowing in on promising zones of concentration so that exploration of vast areas can be greatly compressed. An example of this approach has already been presented in Figure 12.

Second, many shallow mineral deposits give rise to distinctive surface stains (gossans and blooms) caused by alteration or secondary enrichment. If broad enough, some of these stains should be detectable as color-brightness anomalies—subject to the caveats raised in the section on lithologic identification. A simple test of this capability would be to look at ERTS imagery for any evident visual (tonal) differences around known mineral deposits that single them out from their surroundings. Caution must be maintained in examining active mining areas to avoid confusion between surface conditions at man-made workings (excavations; mine dumps; dried-up lakes, etc.) and natural stains present before exploitation.

A. F. H. Goetz (Jet Propulsion Lab) and L. C. Rowan (U.S. Geological Survey) have de-

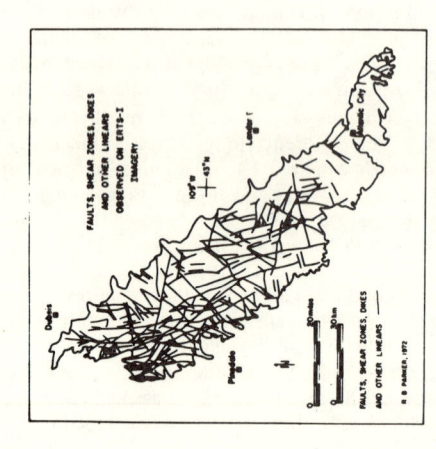

Figure 11. Interpretation of major fracture systems in the Wind River Mountains of western Wyoming made by R. B. Parker of the University of Wyoming from field studies prior to 1972 (left) and then updated by analysis of a single ERTS image (right).

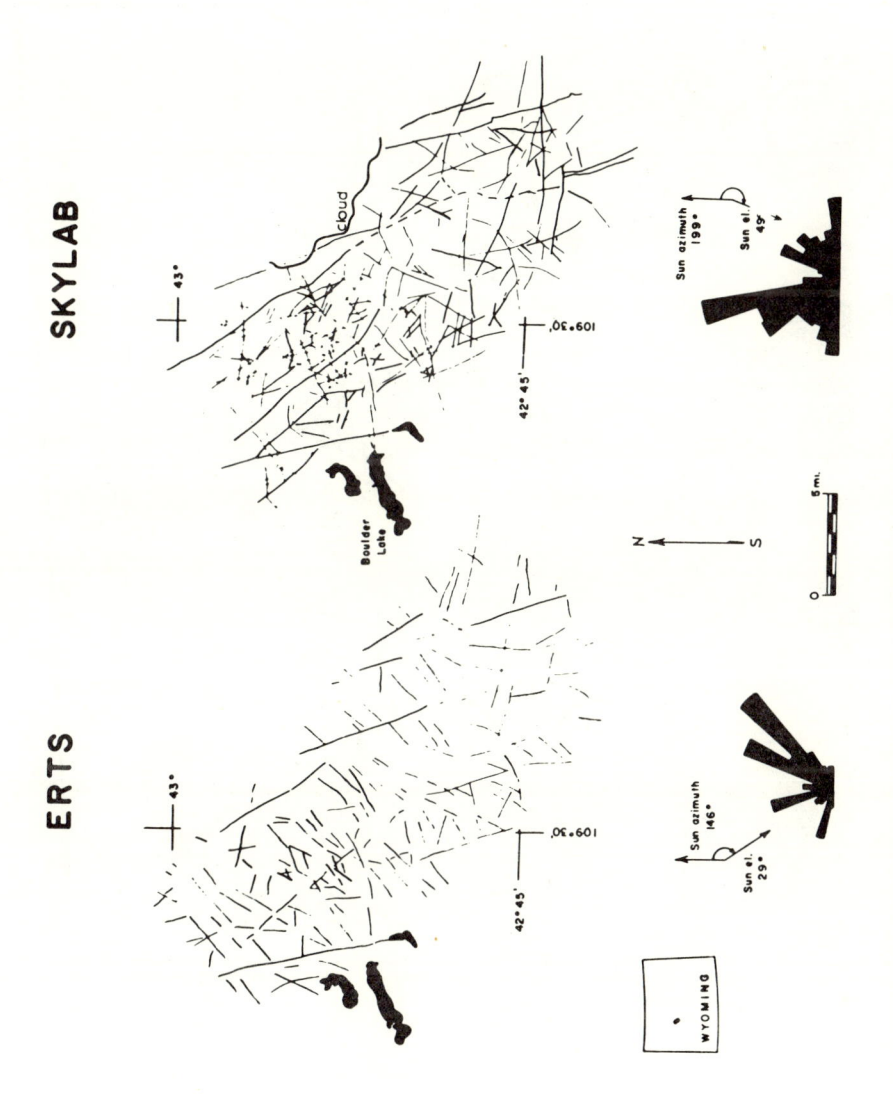

Figure 12. Comparison of linear features mapped using ERTS images and Skylab S-190B photographs of a part of the Wind River Mountains. The rose diagrams indicate the sun-azimuth bias introduced by acquiring the images in the morning and the photographs in the afternoon (courtesy R. S. Houston, University of Wyoming).

Figure 13. Tracing of straight and curved or arcuate linear features recognized in ERTS imagery on a winter ERTS scene over the Central Mineralized district of Colorado. The area is south of Denver; Colorado Springs appears in the right center. Higher frequencies (or densities) of linears have been circumscribed (from S. Nicolais, 1973).

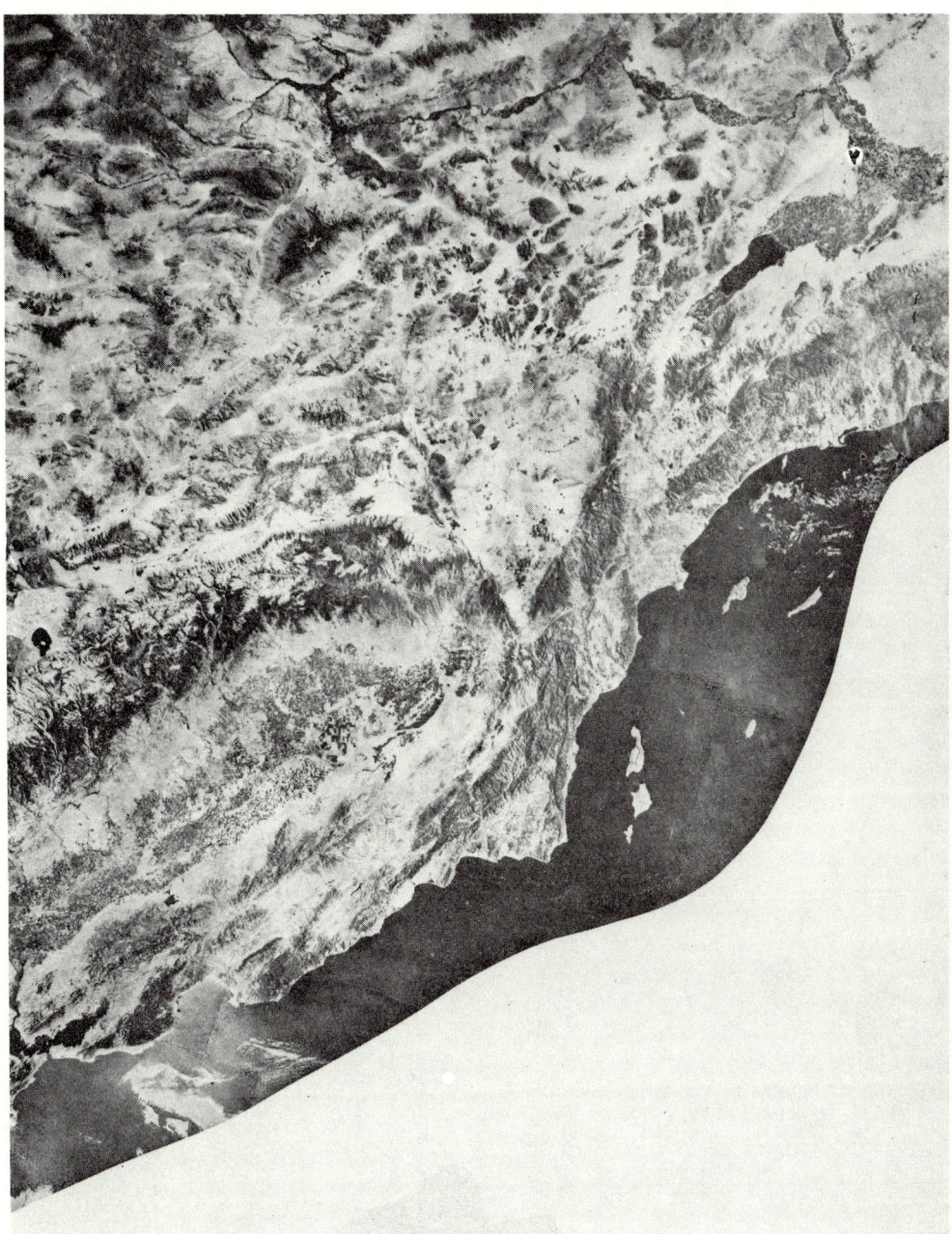

Figure 14. ERTS Band 5 mosaic of much of the southwestern United States, including central and southern California, southern Nevada, and small parts of Arizona; northern Mexico around the mouth of the Colorado River is also shown. This is part of the mosiac of the United States prepared by the Soil Conservation Service.

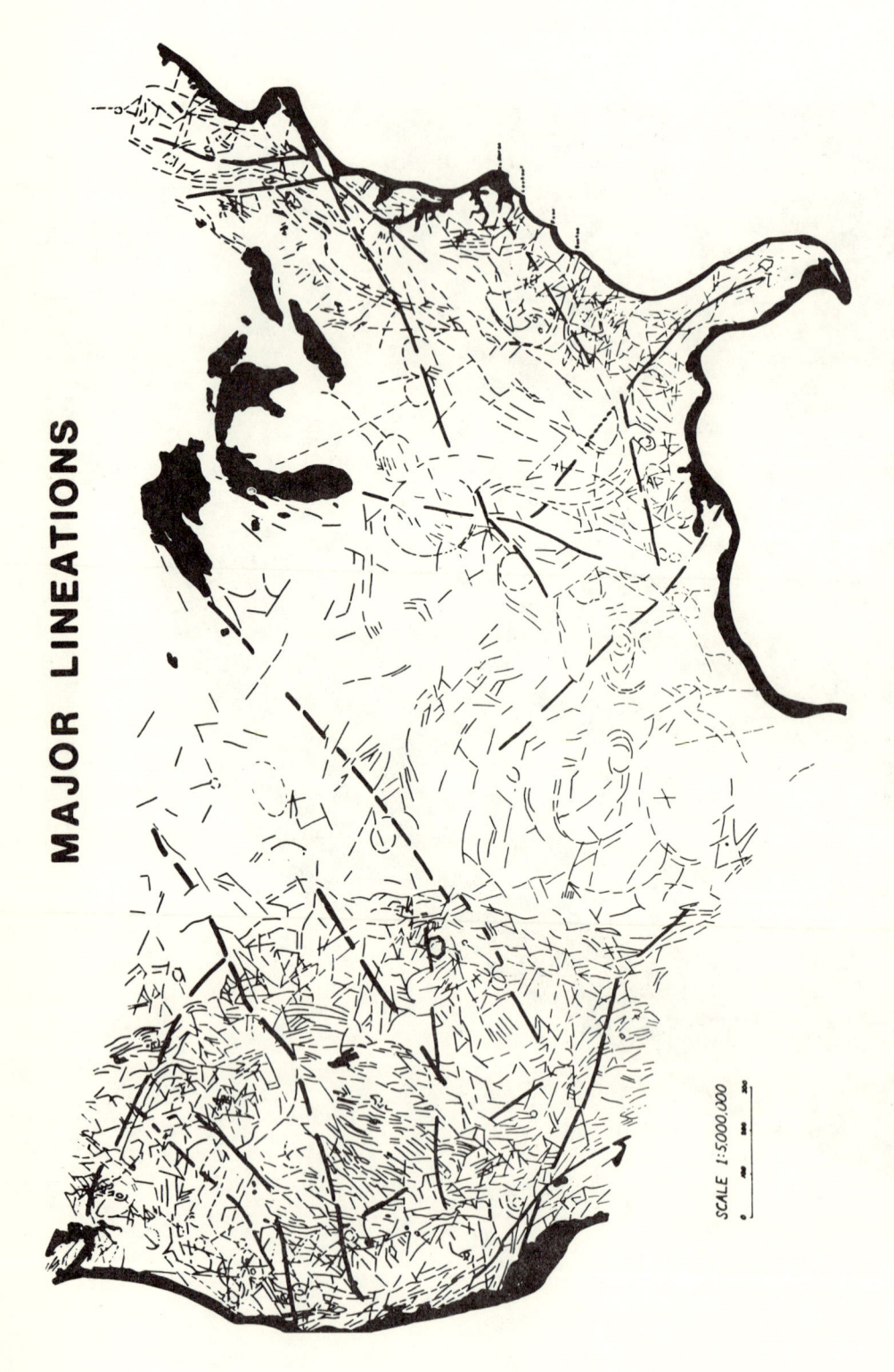

Figure 15. Sketch map outlining the United States on which W. D. Carter of the United States Geological Survey, Reston Virginia, has plotted the major or regional straight and curved linear features that he could recognize in a preliminary examination of the ERTS mosaic of the entire United States.

veloped a computer-based method for enhancing ERTS imagery to bring out the effects of surface alteration; usually subtle accumulations of hydrated iron oxide (iron rust or gossan) associated with sulphide deposits are emphasized. From field and laboratory measurements with a reflectance spectrometer, these investigators have confirmed that limonite has a spectral response quite unlike that of most other common minerals. This response can be made more sensitive to small difference when several ratios of different ERTS MSS band pairs are calculated. Each resulting ratio represents a variable signal which, like individual MSS band analog signals, can be used to construct photo images. The individual ratio images are passed through an optical processor using color filters to produce a color composite. Filter combinations have been found that cause the tones or grey levels representing limonite in three different ratio images to appear a yellowish-brown much like that of iron stain. Different rock types, vegetation, etc. also take on distinctive hues; in particular, clay alteration products also indicative of certain ore deposits can be made to take on a characteristic set of colors. Goetz and Rowan have tried out their method on an ERTS scene of central Nevada that includes the Goldfield mining district (gold and silver accompanied by iron sulphide). Prominent yellow-brown color patterns are observed around Goldfield (principally in a ring or aureole that roughly outlines that alteration zone surrounding the underlying intrusive) and other areas where surface iron stains were known before. The color composite has now been field-checked from the air and ground during which many of these color anomalies were verified. Insofar as gossan often indicates mineralization (including uranium ores), this enhancement technique, if it bears up under further testing, may well prove to be a major breakthrough in mineral prospecting.

PART 2: SPECIFIC RESULTS IN FUELS EXPLORATION

Both direct and indirect information bearing on the search for gas and oil and for uranium can be gleaned from ERTS data. Some types of direct information are obvious to anyone familiar with the use of aerial photographs in geologic exploration. Most have been alluded to in Part 1, along with certain reservations in their accuracy and utility. Among those guides to presence of underground fuel sources recognizable at the surface are: 1) lithologic units (reservoir rocks or host beds) whose subsurface extensions elsewhere can be inferred, 2) folded structures that persist at depth, 3) relatively short lineations and fracture or joint sets that may localize mineralization or trap petroleum below, and 4) generally longer regional linear systems that afford clues to basement trends responsible for controlling structural and/or stratigraphic traps. Previous illustrations provide examples of the ability of ERTS to detect and define such features. Another relevant example appears in Figure 16 which shows how readily ERTS can pick out some of the classic salt domes in the Gulf Coast from which petroleum products have been recovered.

Still another example is that of the actual presence of oil at the surface. This is a rare condition on the land (tar sands and pits being the exception) but natural seeps on the seafloor give rise to surface slicks analogous to the better publicized oil spills. Geologists at the Conoco Oil Company Research Laboratory in Ponca City, Oklahoma, working with the NASA aircraft facility at Johnson Spacecraft Center in Houston, have demonstrated the detectability of marine natural seeps as these are manifested on the ocean surface. Images obtained through narrow band blue filters on photographic cameras during aircraft flights over known seeps in the Gulf of Mexico readily display the outlines of the surface collection of oil (Figure 17) but film exposed to longer wavelengths fails to define the same spots.

As examples of diverse indirect information types, one can cite 1) possible effects of petroleum on vegetation—a contemporaneous phenomenon, and 2) recognition of modern-day sediment distribution patterns (Figure 18) in coastal, estaurine, or lacustrine waters, from which comes new insight into the accumulation of source beds or reservoir units in depositional basins of the geologic past.

Of the 72 ERTS-1 NASA-approved investigations in geology, perhaps more than half provided useful information of immediate interest to a petroleum geologist looking for principles or demonstrations of the exploration capabilities of space imagery. Some of the same information is applicable to the geologist searching for uranium. A few of the investigations dealt specifically with suggested applications to fuels exploration as a secondary concern or by-product of their

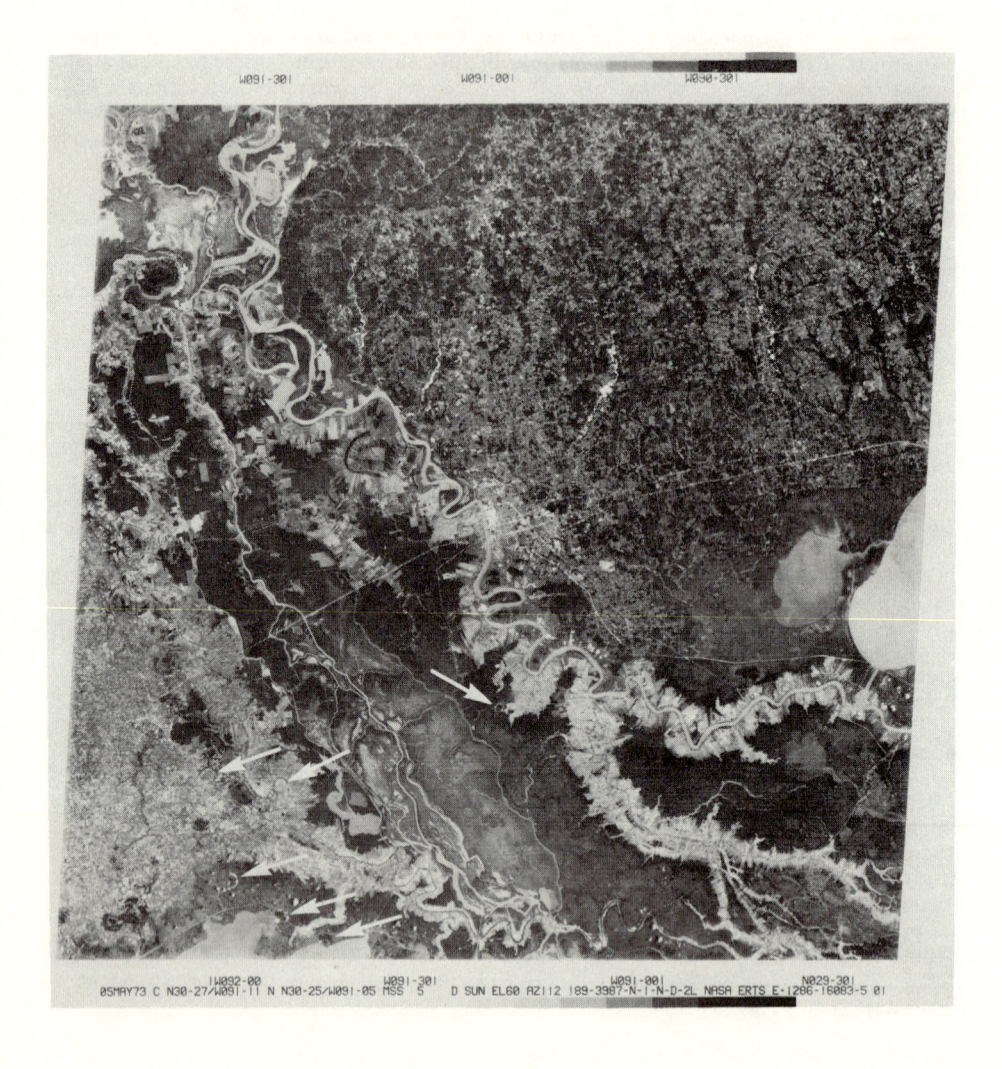

Figure 16. ERTS image of southern Louisiana (just west of New Orleans) and southwest Mississippi (top) through which flows the Mississippi River past Baton Rouge and the Atchafalaya River to the west. Numerous salt domes occur in this region; several which are easily identified in this band 5 image are indicated by arrows.

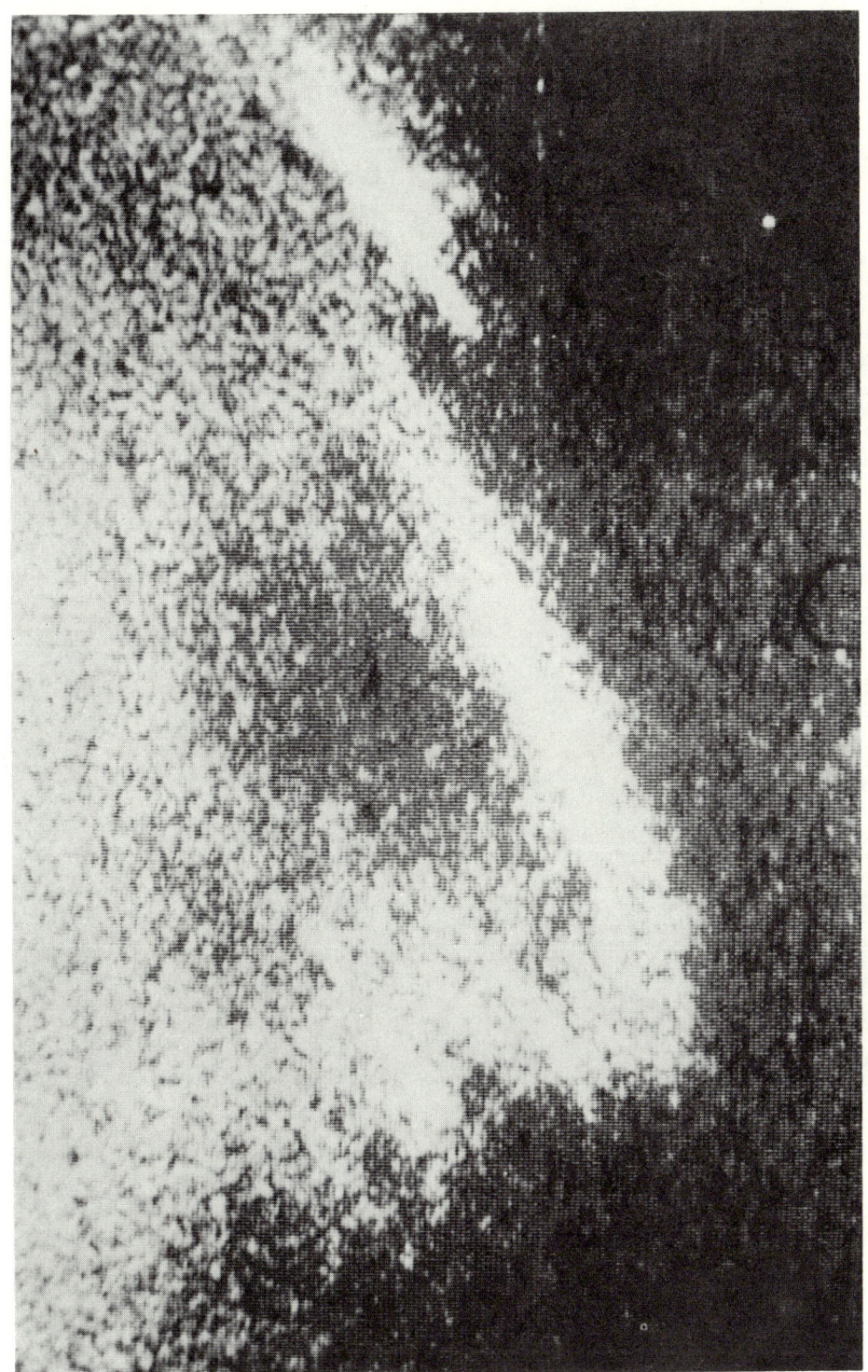

Figure 17. Black and white version of a color-enhanced photograph of the surface "slick" formed from a natural oil seep in the Gulf of Mexico. The photo was made during a NASA aircraft overflight over known seeps to test methods for detection of oil on the sea surface by remote sensing. As shown here, best detection was achieved with a No. 35 filter (near UV-blue) over the KA62 camera loaded with panchromatic film (courtesy J. A. Eyer, Continental Oil Company).

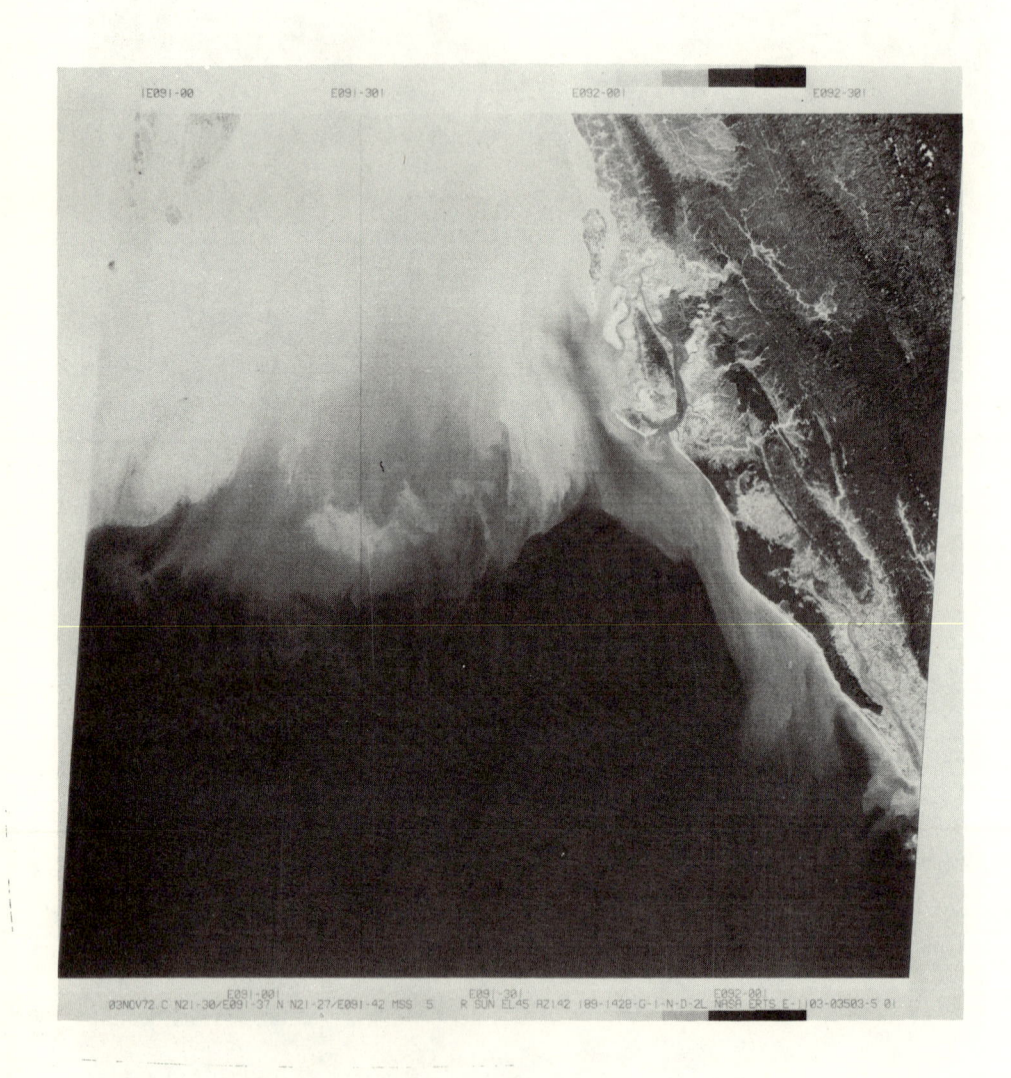

Figure 18. Sedimentation pattern of discharge zone of the Ganges River into the Bay of Bengal south of Bangladesh, ERTS band 5.

main objectives. Only one investigation was completely dedicated to the use of ERTS data for petroleum exploration while no investigation considered the question of prospecting for uranium from space.

The sole ERTS investigation in petroleum exploration was conducted by R. J. Collins, Jr., president of the Eason Oil Company of Oklahoma City, and his colleagues. Under his direction, the field and interpretive work was carried out by F. P. McCown, L. P. Stonis, and G. J. Petzel of the company and J. R. Everett of the Earth Satellite Corporation, Washington, D.C.

Their philosophy underlying the objectives of the study followed a sound and proper approach: Concentrate on a major oil basin already in production as though it were an unknown or virgin territory earmarked for the first phases of exploration. This outlook tends to reduce the bias of familiarity but still retains the eventual opportunity to compare their findings with established ground truth, namely, the location of productive fields and the data on the geologic factors responsible for the petroleum accumulations.

The choice of test area was dictated in part by company interests and proximity but was also based on a set of conditions that offered an optimum appraisal of the potentialities of exploration from space. Thus, the Anadarko Basin of southwest Oklahoma and the Texas Panhandle was selected for detailed analysis. This is one of the older oil-producing regions in the United States—having been developed in the 1920s. More than 25 major producing fields occur in the structural traps within the basin and many more are found in stratigraphic traps.

The basin consists of a west-northwest-trending subsidence trough that was filled with more than 45,000 feet of epicontinental sedimentary rocks from the early Cambrian through the Pennsylvanian. Maximum sedimentation took place in the Pennsylvanian when about 11,000 feet of clastic rocks were deposited. Subsequent to this the basin underwent strong deformation, more or less contemporaneous with that in the Ouachita Basin further east, leading to a steepening of the southern basin flank accompanied by extensive basement faulting that carried crystalline rocks upward at the Wichita Mountains. Following rapid erosion, about 3000 feet of Permian red beds, saline deposits, and carbonates were laid unconformably over the now asymmetric basin—these units are only slightly deformed. Thin deposits of Tertiary sediments cover the western part of the basin.

The modern-day surface, therefore, consists of flat-lying Permian and younger sedimentary rocks which show little or no direct evidence of the subsurface conditions around the structural traps (except for several fields along the south limb). Most traps lie below 2000 feet but many occur at depths of 8000-12000 feet, and, more recently, production has reached 17000 feet and deeper. The present surface is used both for farming (wheat, etc.) and ranching. Scattered woods mixed with shortgrass are interspersed with farmlands on the eastern side. This gradually gives way westward to country used primarily for grazing in open rolling long grass prairie and sagebrush. The test area, then, is characterized as a low to high plains on which several different vegetation covers have developed on soils derived from near-surface rocks that have no involvement with the more deeply buried and "masked" producing zones. This type of petroleum environment provides an exacting test of the capability of ERTS to "sense" clues to hidden petroleum reserves.

The Eason Oil investigators have compiled one of the most comprehensive and analytical reports in the geology phase of the ERTS program. They have thoroughly documented the variety of techniques used in their study. They have demonstrated the degree of reliability to which small-scale mapping can be done from space imagery and have shown the value of multi-seasonal coverage in separating and identifying surface units. They have also evaluated the presumptive cost-effectiveness of the ERTS approach to petroleum exploration—major savings in the initial (or reconnaissance) stages of exploration are indicated and further savings are suggested for later stages, as for example a potential reduction in the number of seismic lines that may otherwise have been planned which could then be eliminated over areas rated as unfavorable for accumulation. However, in the remainder of this survey of their results, we shall confine our attention to the two most promising observations that could lead directly to the discovery of petroleum.

First, as did most other investigations, the Eason Oil group gathered a great deal of new information on surface linears. This is almost self-evident in comparing Figures 19a and b. The increase in known linears as defined in ERTS images is impressive, so much so that it raises a question of doubt in any petroleum geologist (by nature a skeptic made so through the experience of many dry holes) as to their validity. The immediate reaction is to wonder why only a few of the linears have been found in the field or through subsurface drilling. One obvious answer is that they have—but the policy of propietary use practiced by the oil companies has prevented the linears from being recorded on published maps. However, the investigators discount this as a prime explanation. Another answer resides in the definition of linears—many may be non-geological and only a fraction might be fracture zones, faults, or surface expressions of basement lineations. The investigators again counter this by claiming that almost everywhere they field-checked roadcuts along the linears path they found some verification of a recorded linear even though that same linear was usually "invisible" in aerial photos or by visual inspection from an aircraft. The question remains open but, suffice to say, the likelihood that many of the linears actually exist as structural features seems reasonably high in the Anadarko Basin. The importance of this in oil-finding is that 1) more oil traps can now be looked for, 2) the linears themselves—if they are fracture zones—can be loci of increased porosity, especially at intersections, and 3) the patterns revealed by the linears should better define regional stress fields and trends of structural adjustment.

The second important result concerns several types of surface anomalies defined by the investigators. A few individual examples can be seen in the ERTS image reproduced in Figure 20, although most are revealed only by inspection of transparencies on a light table. Most of these anomaly types have been categorized as closed anomalies, in that they can be circumscribed by a boundary line. One type clearly is due to drainage patterns, some of which represent adjustments to subsurface structures, such as an anticline over which younger strata have unevenly settled. Another type is topographic, caused in some instances by differential erosion. Still another comes under the heading of geomorphic and is based on subjective interpretations by the investigators. Less definitive, but nevertheless recognizable in the imagery are the two classes termed tonal and textural anomalies. Tonal anomalies are expressed as unaccounted-for differences in grey level whereas textural anomalies appear as streaked, mottled, or rough patterns in the imagery. These may have a variety of causes, from cultural activities, or as peculiarities in soil, rock, or vegetation cover, through geological effects of uncertain nature, to unusual reflectances of unexplained origin. Figure 21 presents a map of the large or more conspicuous closed anomalies over the Anadarko Basin. Many of these, however, show no positive correlation with, or superposition on, underlying oil and gas fields.

Two classes of anomalies yield a high correlation with the location of producing areas. One is the geomorphic-topographic type. The other is the so-called hazy anomalies (see Figure 20)—described by the investigators as resembling a "blurring" or "smudging" of the photo-image. A map of their distribution in the basin appears in Figure 22. The extent to which both classes correlate with known fields is outlined in Table 2 along with data on the other anomaly types. Some of the hazy anomalies occur over fields producing from structural traps as much as 8000-14000 feet down; anomalies over stratigraphic traps are rare. The correlation is so strong that one must accept some kind of relationship between a surface effect and the presence of petroleum beneath, even though the nature of that relationship is not yet settled and is still open to suspicion.

The investigators note that the hazy anomalies are best seen in dry weather fall images and are also well displayed in west spring imagery. They are best developed in grazing lands on the west but are less easily found in farmlands. Almost invariably, when visited on the ground, they coincide with local areas of sandy soils that, in places, even have formed into dune-like deposits. They also tend to associate with the Ogallala formation or in Plio-Pleistocene terrace deposits.

Already these hazy features have engendered considerable debate as to their nature, cause, and value as a guide to petroleum. Two diverse views will be discussed here—the opinion of the present author is intertwined with those of other supporters or critics.

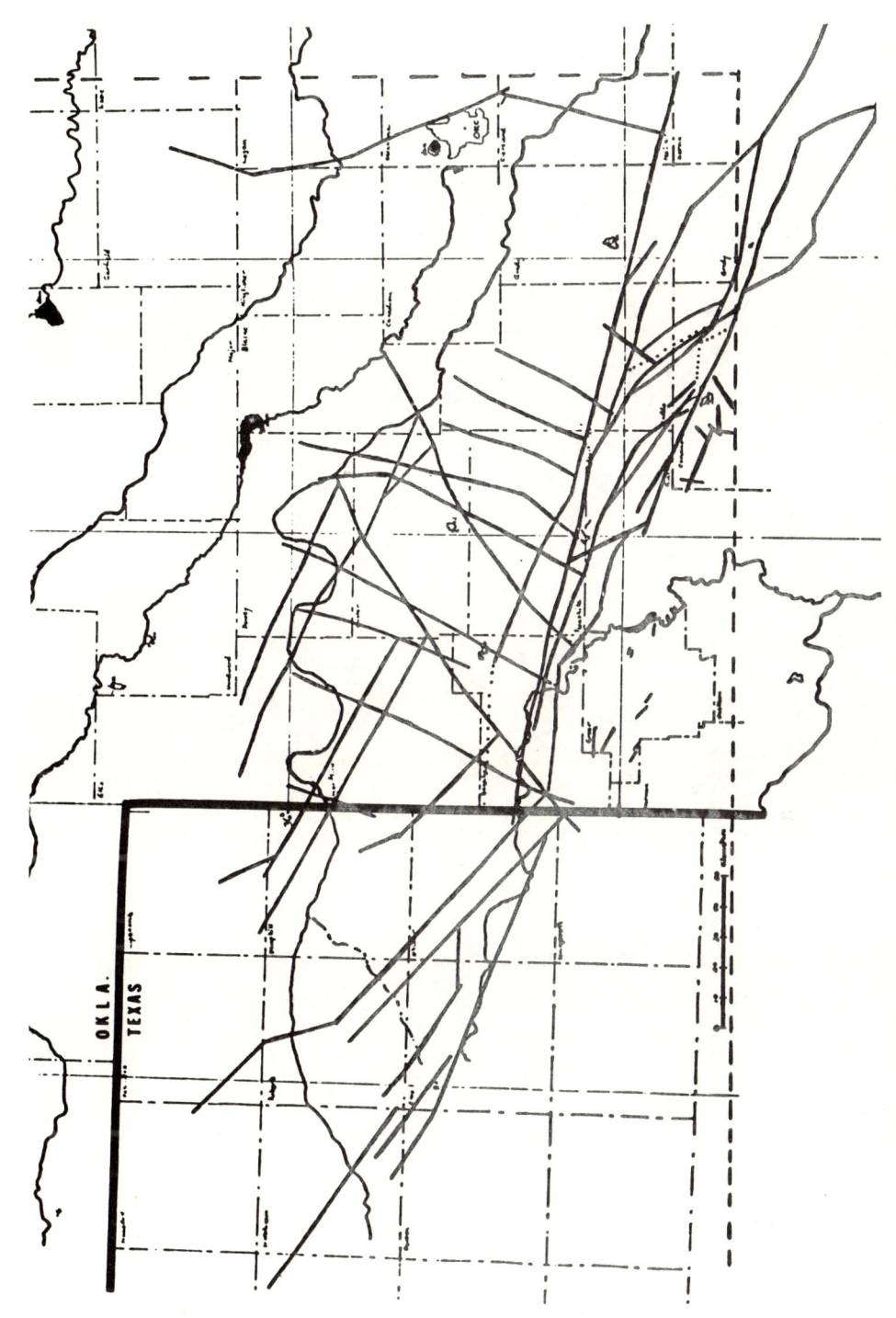

Figure 19a. Tracing on a geographic base map of the major known and hypothesized faults and linears expressed at the surface over the Anadarko Basin (from Eason Oil Company final ERTS report).

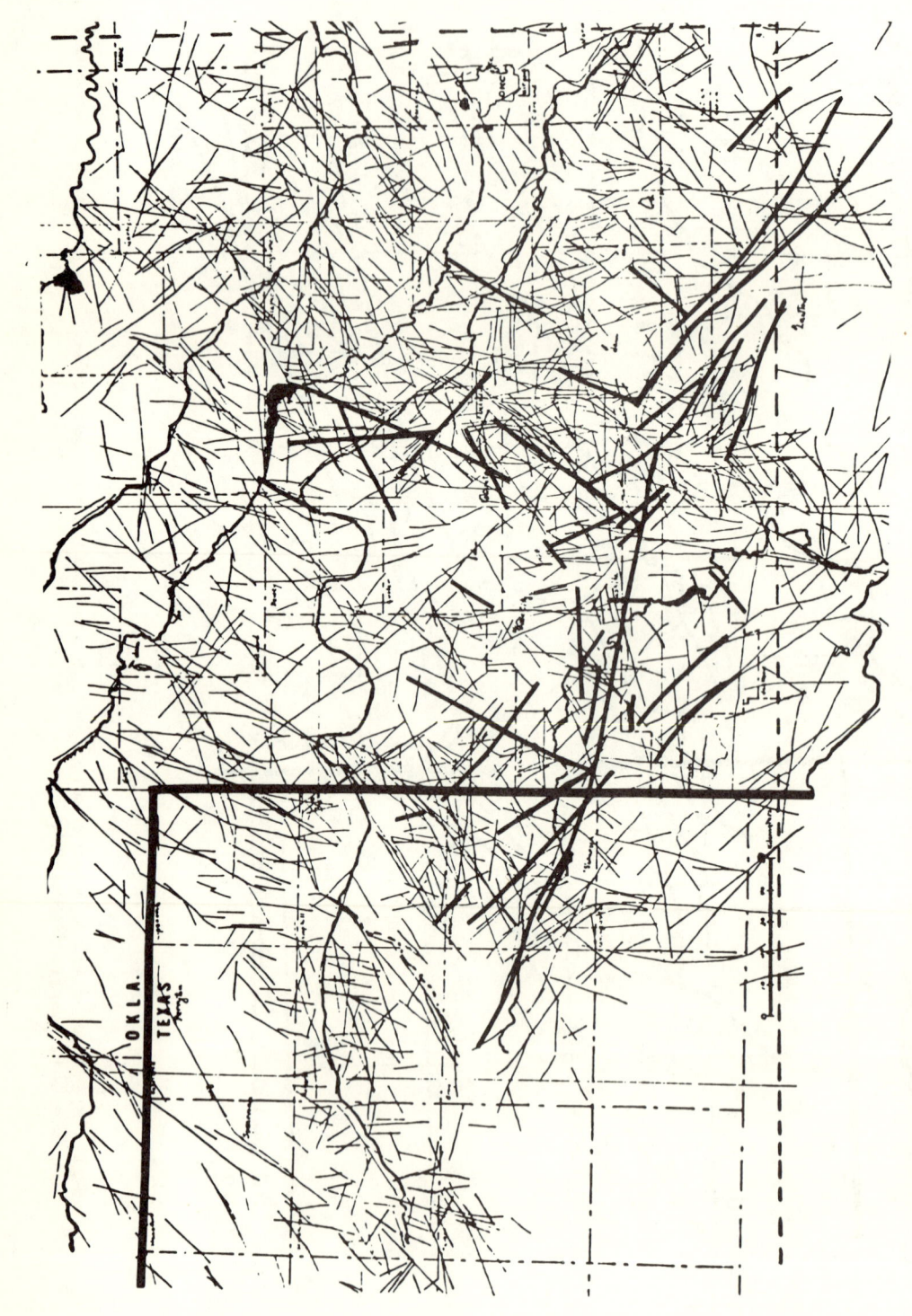

Figure 19b. Linears recognized in ERTS imagery covering the Anadarko Basin and surrounding areas in Oklahoma and Texas. Heavy lines emphasize ERTS linears coincident with or extending from linear features plotted in Figure 19a.

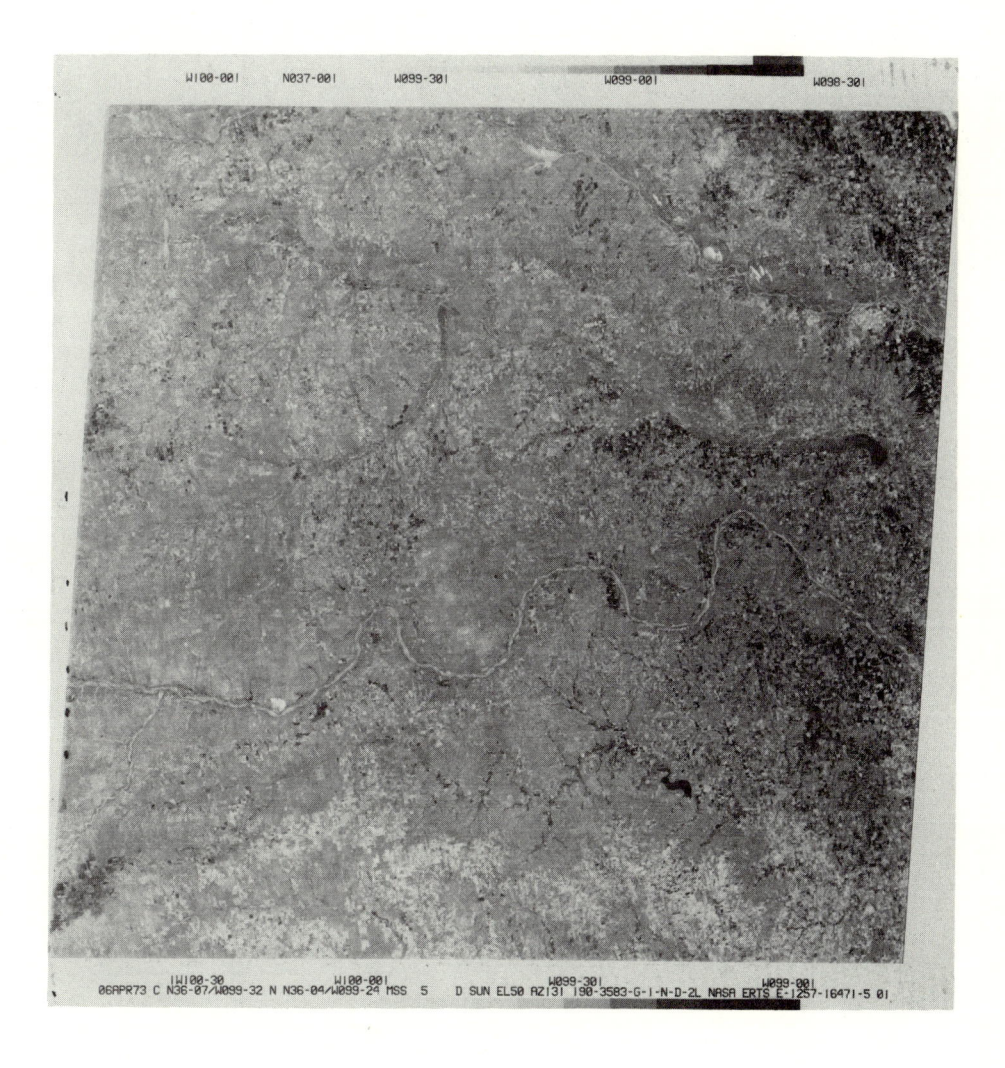

Figure 20. ERTS band 5 image covering part of western Oklahoma and some of the Texas panhandle. The meandering Canadian River and the North Canadian River are the principle drainage features. "Hazy" anomalies are particularly evident in the bend of the Canadian River near the center of the image.

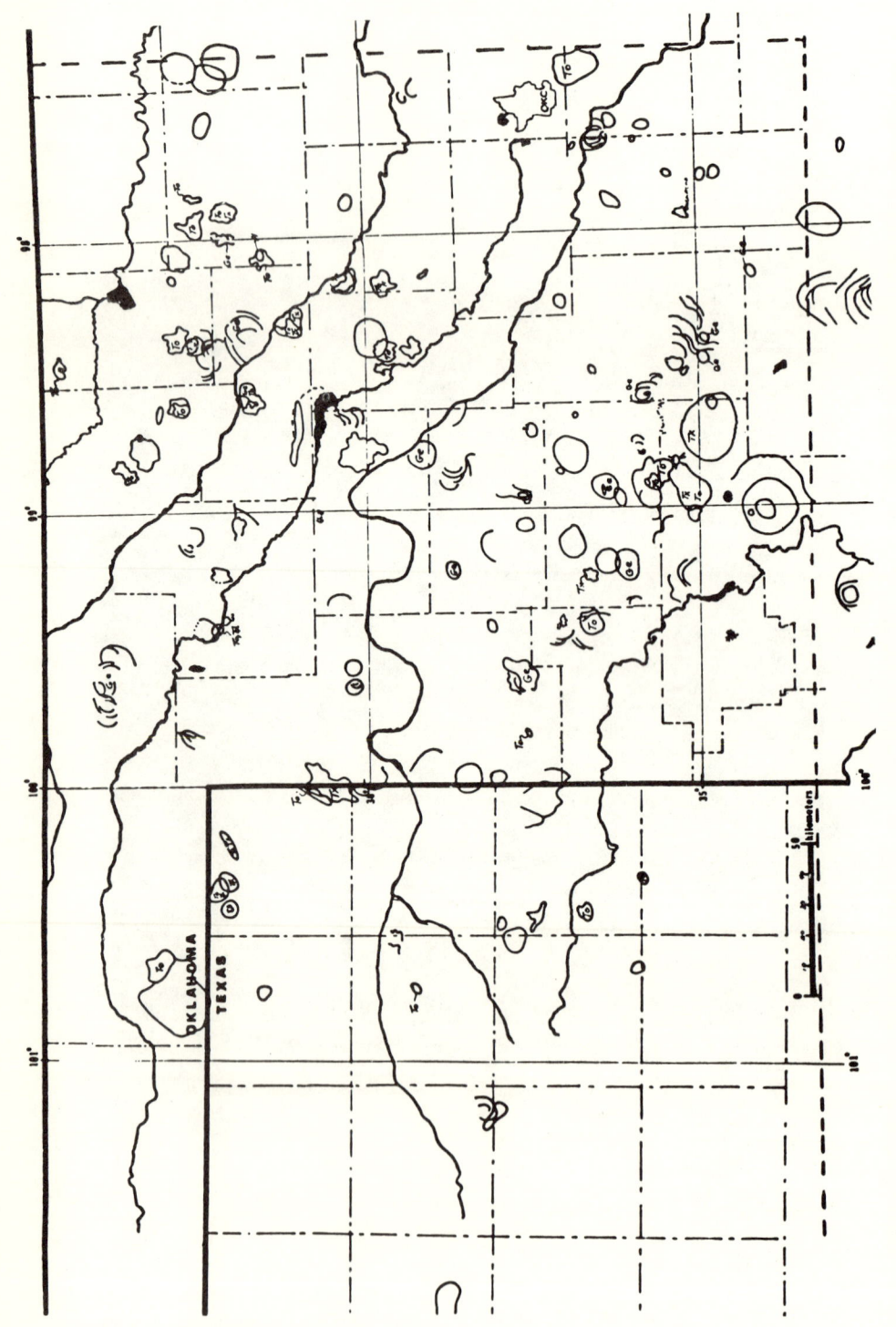

Figure 21. Sketch map showing the traced outlines of many of the closed anomalies recognized in the Anadarko Basin during the Eason Oil Company ERTS investigation. To refers to tonal anomalies, Tx to textural anomalies, and Ge to geomorphic anomalies (from Eason Oil company final ERTS report).

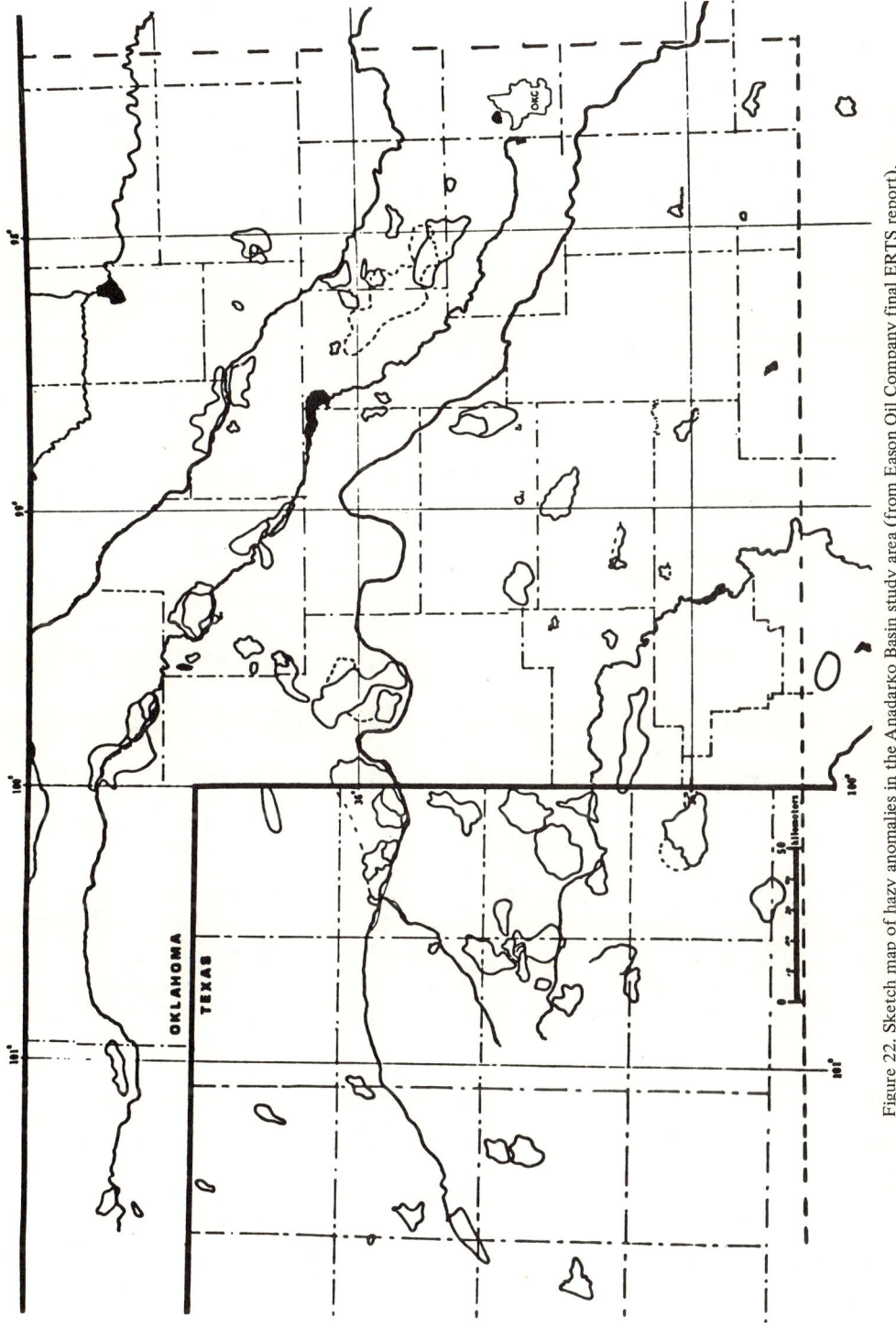

Figure 22. Sketch map of hazy anomalies in the Anadarko Basin study area (from Eason Oil Company final ERTS report).

Table 2

Closed Anomalies

Survey 1 (Fall)	Geomorphic, Tonal & "Hazy" Anomalies
76	Total Anomalies
59	Producing Fields
11	Nonproductive Structures
6	No Coincidence
33 of 37	Geomorphic ⎫
33 of 35	"Hazy" ⎬ Anomalies coincide with field or Structure
0 of 4	Tonal ⎭
Survey 2 (Fall and Spring)	"Hazy" Anomalies
57	Total Anomalies
42	Producing Fields
6	Nonproductive Structures
9	No Coincidence

One piece of evidence is given as a computer-processed enlargement (Figure 23) of the ERTS view of the most prominent hazy anomaly shown in Figure 20, that in the bend of the Canadian River near Webb, Oklahoma. This rendition is remarkably similar in details to the view presented in an RB-57 high altitude aerial photo. Both show extensive drilling and pumping sites along a network of roads. Ground inspection indicates the presence of stabilized sand dunes and very sandy soils. An hypothesis to explain this—and other hazy anomalies elsewhere—simply holds them to be consequence of man's activities in extracting the oil, such as damage to the terrain by countless vehicles and bulldozers plus possible effects from escaping hydrocarbons, that may have occurred prior to current land use and restoration regulations. However, the explanation does not adequately account for the extensive sand deposits unless it is assumed that so much vegetation has been stripped off that the area(s) converted to a sand waste during the dust bowl days (or at a later time).

A competing explanation is more exciting and desirable to those seeking to show the utility of orbital exploration. This postulates that hydrocarbons have leaked or migrated up and out from the trap and upon reaching the surficial layers introduce chemical and/or botanical changes. This idea is old—the notion of telltale alterations by escaping gases or fluids has long been advocated despite meager proof. One variant holds to geochemical changes in bedrock or soil resulting in bleaching or staining to produce color anomalies. Another envisions the escaping substances as reacting with the country rock to induce distinctive metasomatic products. A third considers the hydrocarbons as capable of moderating surface vegetation—either by killing off plant life and therefore accelerating erosion or by "fertilizing" the soil and thereby stabilizing growth. This last view might seem to apply to the sand-rich hazy anomalies, if toxic hydrocarbons have damaged the grasses and sage, allowing an increase in loss of the fine fractions from the soils or if beneficial hydrocarbons have fostered thicker vegetation that causes the sand to pile up.

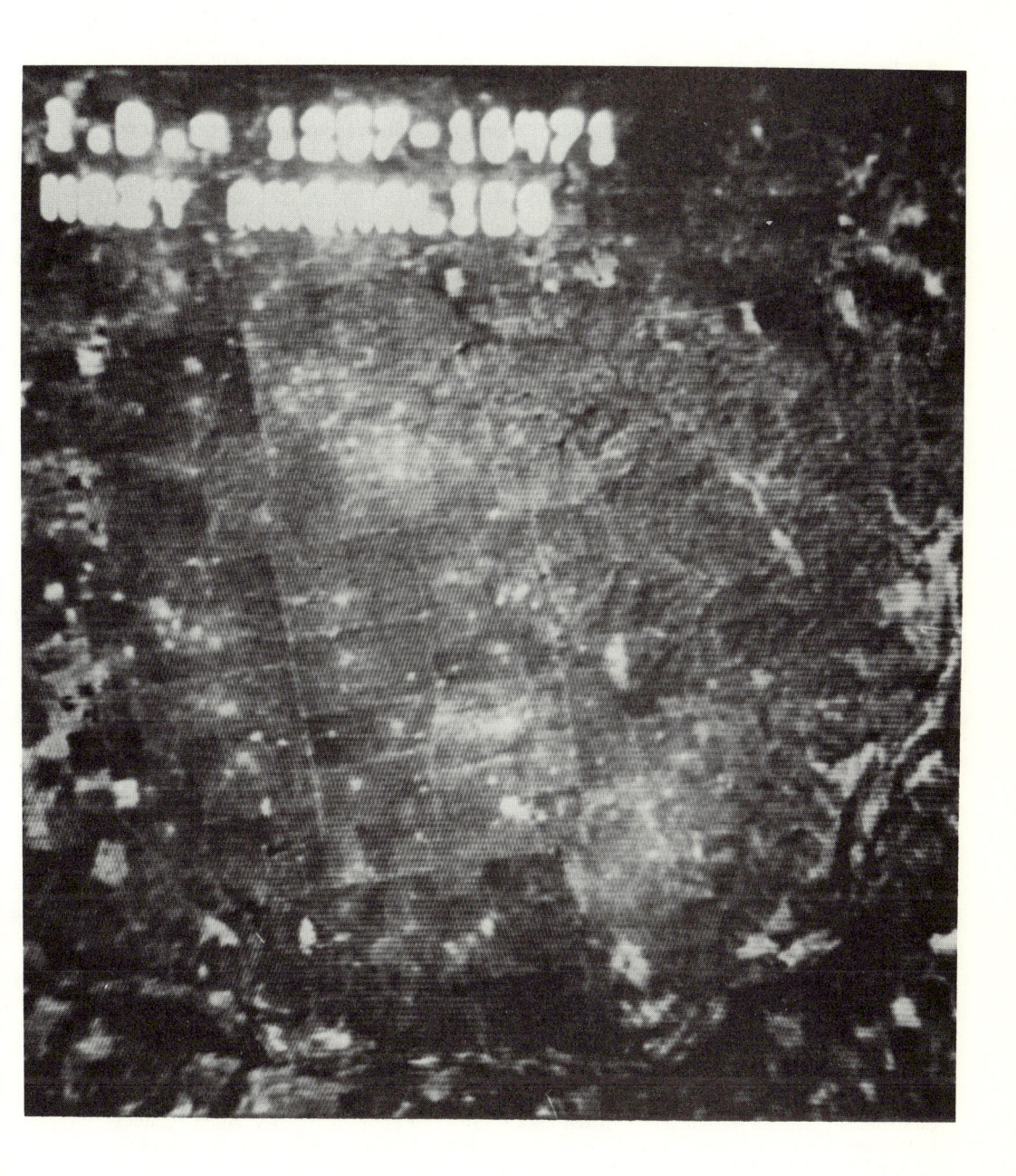

Figure 23. Black and white version of color image photographed from the TV monitor screen of the Image 100 computer-interactive processing system at the General Electric Space Sciences Laboratory in Beltsville, Maryland. This scene shows a bend of the Canadian River in western Oklahoma enlarged to a scale of approximately 1:30,000 from the original scale of the same ERTS image shown in Figure 20; however, this rendition is made entirely from the ERTS computer-compatible tape of the April 6, 1973 pass over Oklahoma.

So far, no decisive data have been gathered to explain the specific nature of these anomalies. However, T. J. Donovan of the U.S. Geological Survey reports in the March, 1974 issue of the American Association of Petroleum Geology Bulletin (v. 58, pp. 429-446) the results of some highly relevant analyses at the Cement field near the axis of the Anadarko Basin. This field occurs within a long doubly plunging anticline which, unlike most in this basin, has prominent surface expression. Exposed bedrock consists of Permian red sandstones and gypsum beds. In the field in many places, the sandstone has been bleached to a yellowish-brown to white color (Figure 24). This is ascribed to the reducing action of hydrocarbons carried in expelled reservoir pore waters. The gypsum beds are converted to low porosity carbonate rocks. Sandstones are also re-cemented by migrating carbonate solutions. Both the metasomatized gypsum and these sandstones tend to form more resistant rocks that stand above the terrain as butte-like outliers.

Such effects have been observed before but their relationship to hydrocarbons has never been firmly established. However, Donovan has determined the carbon and oxygen isotope compositions (Figure 25a and b) of the Permian sandstone and has been able to explain the extremely anomalous isotopic compositions—a deficiency of C^{13} and exceptionally high O^{18}—by a plausible geochemical model dependent on the reducing action of CH_4 and other hydrocarbons. He has also found anomalies at six other localities overlying producing fields.

Strangely, the Cement field itself is detected only with difficulty in the ERTS imagery. This is unexpected in view of the prominent color and rock type changes noted at the surface. Special enhancement techniques (e.g., those developed by Goetz and Rowan) may be needed to discriminate these surface alterations in some kinds of terrain or surface materials. However, Donovan's results look highly promising as a general guide to petroleum accumulation if it is ultimately proved that 1) reservoirs leak, 2) the escaping products reach the surface, 3) these products interact with surface materials to bring about discernible changes, and 4) the changes are detectable from space.

In closing Part 2, I shall state briefly that some of the observations made at the Anadarko Basin, together with the results of Goetz and Rowan discussed in Part 1, have a direct bearing on prospecting for certain kinds of ore deposits. Like most mineral deposits, uranium ores tend to be localized by fractures—hence any new information on linears will be helpful in exploration. One type of uranium deposit—the sedimentary roll deposits of Wyoming and Colorado are excellent examples—in most instances is accompanied by secondary hydrated iron oxides released from the original source rocks or the eventual host rocks in the migrating solutions that concentrate the ore. This gossan is commonly located at the surface over the roll front. Present exploration methods include a search for iron staining of soils or surficial bedrock and in some instances the limonite alteration products are widespread enough to be detectable from aircraft (and by presumption from space). The band ratio method that worked in the Goldfield, Nevada district has not yet been tried on an area of known uranium roll deposits but it is reasonable to expect some success in its use.

PART 3: EVALUATION OF THE POTENTIAL FOR PETROLEUM EXPLORATION FROM SPACE

Too little work has been done to date to allow a definitive prediction about the outlook or prospects for success of petroleum prospecting from satellites or other orbiting platforms. However, enough has been learned already from the ERTS investigations in geology, and particularly those carried out by Eason Oil Company and by Goetz and Rowan to permit a summary to be made of the positive factors which offer real promise for the approach. These are:

(1) Improved perspective of exposed surface structural expressions.

(2) Subtle indications of subsurface structures through drainage control (circular or offset stream patterns) and other geomorphic anomalies.

(3) Direct indications of linears or circular features that can be related to local to regional fractures and lineations which also influence the localization of petroleum.

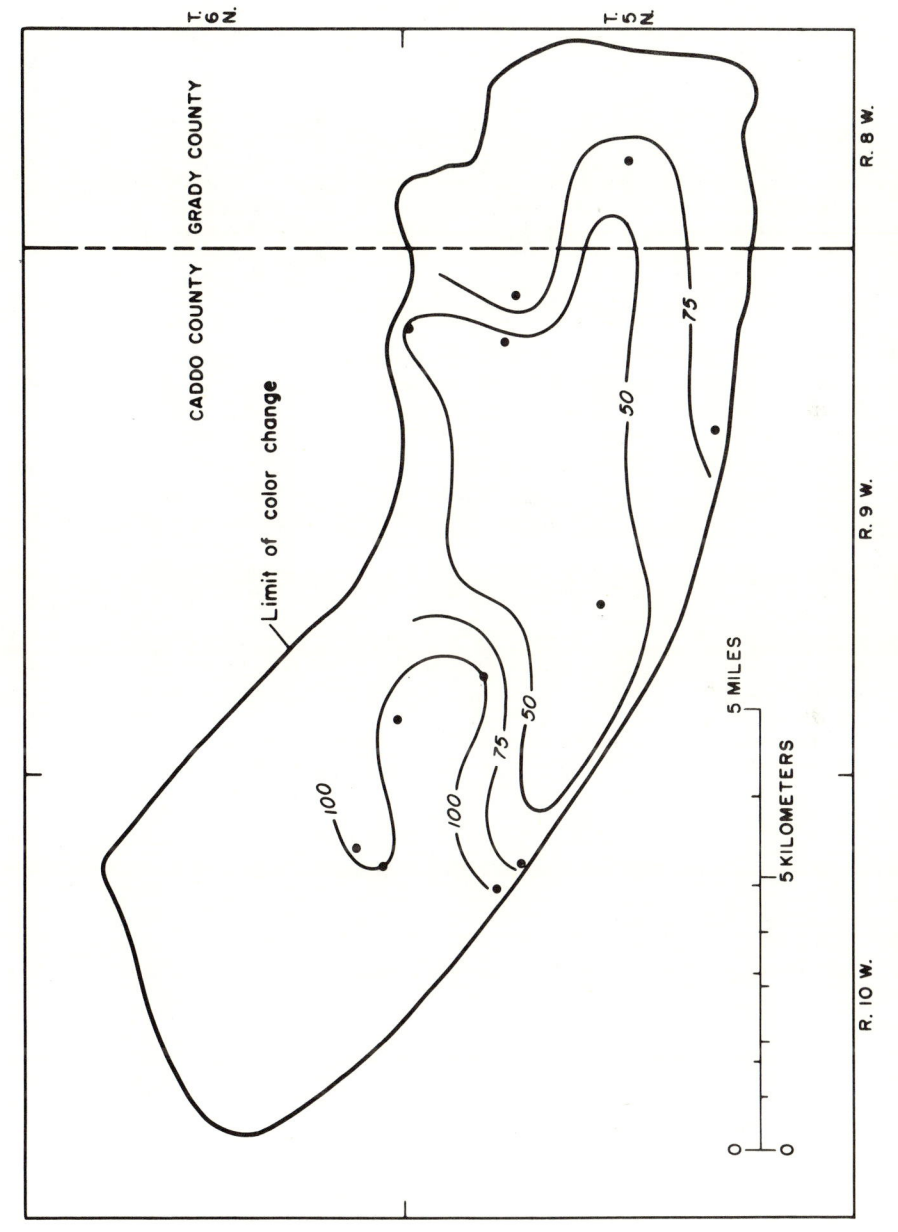

Figure 24. Contoured values of total iron content (in parts per million; each number should be multiplied by 100) within the "bleached" areas of the Rush Springs sandstone (Permian) exposed at the surface over the Cement field, Andarko Basin, Oklahoma. Solid dots represent sample locations (courtesy T. J. Donovan, U.S. Geological Survey).

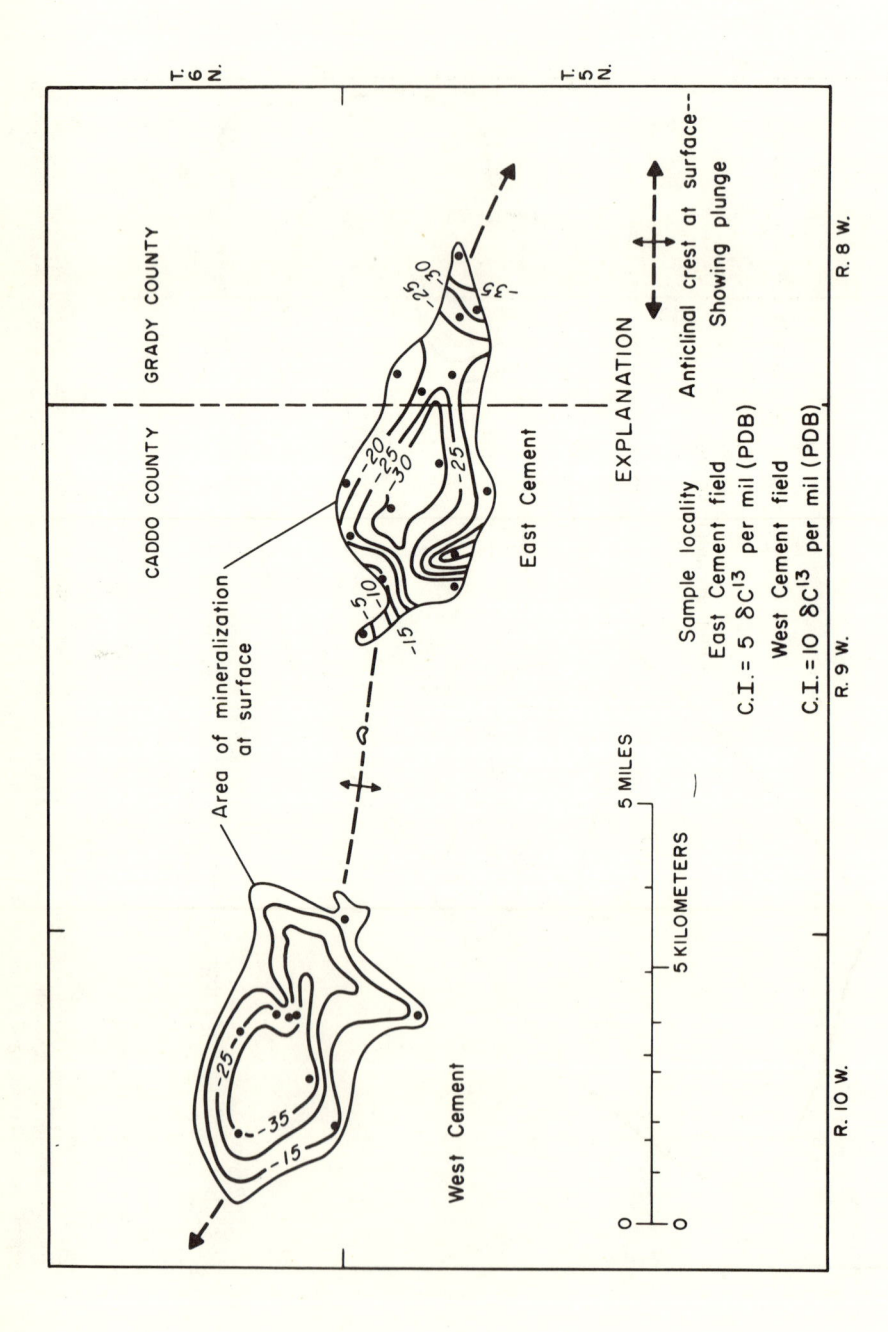

Figure 25a. Variations in carbon-isotope composition within the area of strong carbonate mineralization of the Rush Springs sandstone at the Cement field (courtesy T. J. Donovan, U.S. Geological Survey).

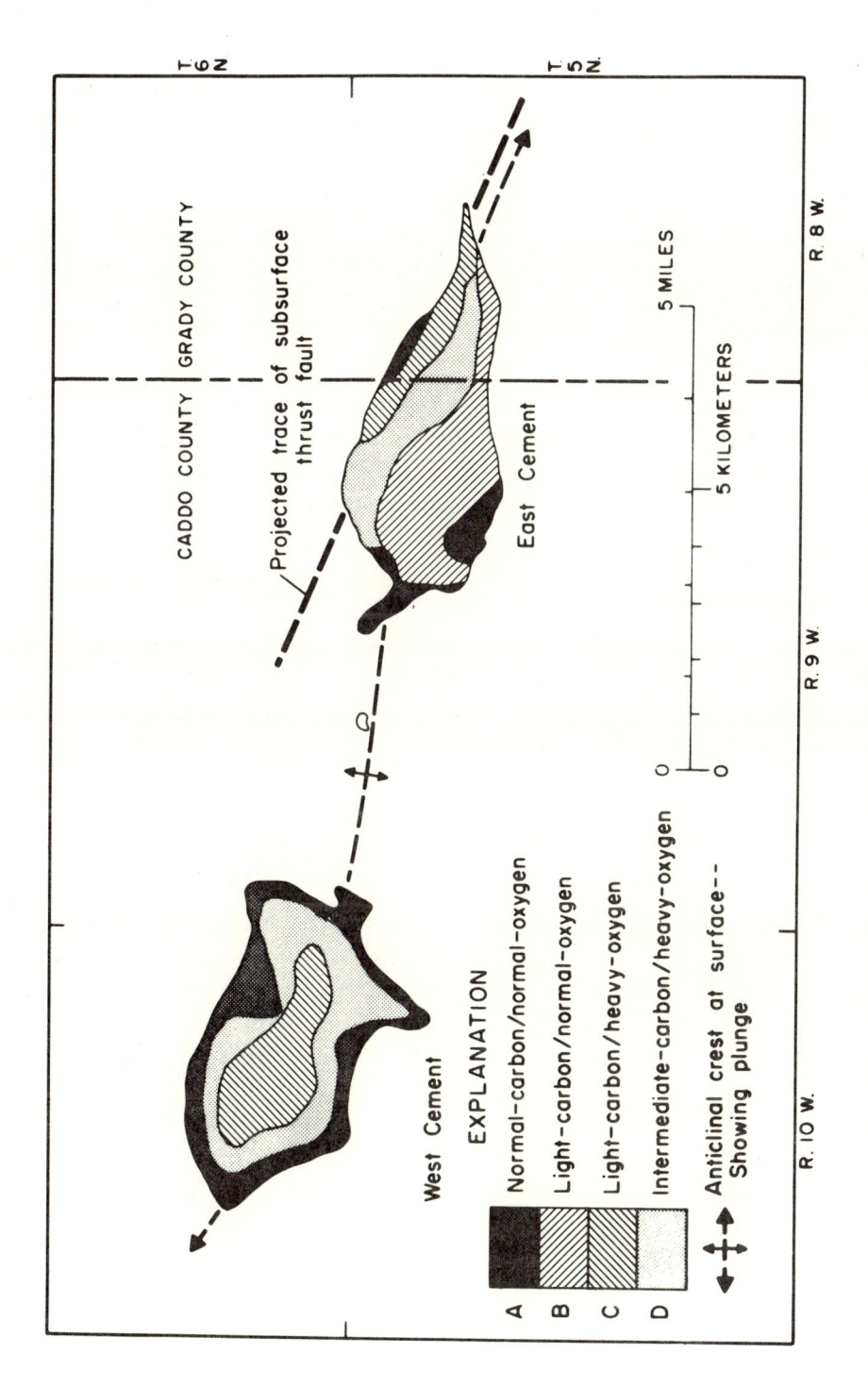

Figure 25b. Distribution of different types of carbonate cement in the Rush Springs sandstone based on the relative proportions of different carbon and oxygen isotopes (courtesy T. J. Donovan, U. S. Geological Survey).

(4) Recognition of distinctive tonal or hazy anomalies, especially through enhancement techniques, that are eventually proved to be caused by or related to natural petroleum occurrences.

(5) Association of geobotanical anomalies with the subsurface presence of petroleum; this has not yet been established.

Enough experience is at hand to suggest a suitable working procedure for the analysis of ERTS imagery in search of guides or clues to petroleum. The procedure should begin with a standard photogeologic interpretation of ERTS transparencies and paper prints leading to maps of surface materials units, linears, and various types of closed anomalies. This should be accompanied by preparation of appropriate color composites. Special processing should follow: this should include edge enhancements (by optical, electronic, or computer techniques) to bring out more information on linears and band ratioing (with color output) and other computer-based data reformatting methods.

At this point it is wise to reiterate emphatically that remote sensing from aircraft and/or spacecraft is not a "magic black box" method that will supercede the many tried and true field and instrumental methods used over the years in petroleum exploration. Remote sensing is still just another "tool" in the workbox of skilled, intuitive exploration geologists. It will be put to its best use when it is considered as another data source to be integrated with field mapping, subsurface data, and methods that can detect geophysical, geochemical, and geobotanical anomalies.

A word of caution about the limitations already set forth by doubting critics (which include from time to time even some of us who are actively trying to apply remote sensing to petroleum exploration). Among these are:

(1) Surface indications of oil and gas are relatively uncommon and most have probably already been detected in the more accessible parts of the world.

(2) Today, most new oil discoveries are being made primarily through use of geophysical methods and by drilling inasmuch as the most decisive data now needed relates to deeply buried subsurface conditions that generally have poor surface expression.

(3) The precise role of linears in localizing oil has not been fully established; depends on time of formation, depth to producing zone(s), caprocks characteristics, etc.

(4) Geological complexities (unconformities, glacial cover, etc.) unrelated to petroleum accumulation frequently mask oil and gas traps.

(5) The hypothesis that hydrocarbons can escape to the surface and cause recognizable alteration effects is unproved; the rate of escape may be less than the rate of surface erosion or other removal factors.

(6) It is difficult to assess man's role or that of vegetation in producing (or obscuring) tonal and hazy anomalies.

SUMMARY STATEMENT

Evaluation of investigation results from ERTS-1 leads to the conclusion that this satellite can do many of the same prime tasks in geological applications done in the past four decades with aerial photography. A broad challenge has been thrust upon geologists to more fully exploit and extend these accomplishments. The ever-rising demands for more energy sources and related raw materials places petroleum exploration near the top of the list of priority applications of space-acquired data that must still be demonstrated and implemented.

SURVEILLANCE OF THE MISSOURI RIVER BASIN USING REMOTE SENSING

JOSEPH H. SENNE

Chairman, Department of Civil Engineering
University of Missouri-Rolla
Rolla, Missouri, U.S.A.

ABSTRACT

Proper management of the Missouri River and the adjacent flood plain area is an important aspect of energy conservation. The maintenance of the navigational channel has provided a means of economical transportation for large quantities of material, thus providing a savings in fuel costs. The reservoir and flood control program has also furnished hydroelectric power and a continuing supply of water for atomic and fossil fuel plants built near the River.

Remote sensing, which is essentially a data-gathering tool, is being used to monitor certain aspects of the River and to identify environmental changes. This includes identification of vegetation adjacent to it, sediment concentrations carried by the River and its tributaries, sources of pollution, recreation sites, urban and industrial expansion, and assistance in flood plain management.

ERTS and SKYLAB imagery, along with U-2, RB-57, and low-altitude flights, have all produced useful data for this monitoring process.

INTRODUCTION

This remote sensing study was undertaken as part of an extensive environmental inventory investigation of the Missouri River from Rulo, Nebraska to the mouth at St. Louis, Missouri. Included in the overall study were the areas of aquatic biology, water quality, fisheries, terrestrial biology (birds, mammals, reptiles and amphibians, botany), recreation, and remote sensing. The project involved investigators from five university campuses and covered a period from August 1972 to June 1974. The work was done under a grant from The Department of the Army, Kansas City District, Corps of Engineers, under the direction of Paul R. Munger, professor of civil engineering (1), University of Missouri-Rolla.

The Missouri River plays an important part in our energy production and conservation programs. A stabilized channel depth of 8 feet and a 220-width permits economical barge transportation as far as Rulo, Nebraska. The reservoir program on the River and its tributaries has aided in flood control and currently produces some 13 billion kilowatt hours of hydroelectric power yearly.

The bank stabilization and navigation project officially started in 1912 and, with several modifications, is continuing at the present time with completion scheduled within the next few years. System reservoir regulation and flood control levees were combined to form a comprehensive plan for water management in 1944. The activities of the Corps of Engineers relative to their efforts on this plan as authorized by the United States Congress all fell within the broad category of water resources planning and development.

Traditionally, from a strictly engineering standpoint, water resources projects have been evaluated on the basis of which alternative was not only sound technologically but which also satisfied the benefit-cost criteria. While such considerations are very important, of equal significance are the environmental impacts associated with each alternate choice of action.

Typical water resource projects associated with the Missouri River such as flood control, navigation control, irrigation and low-flow augmentation, are known to have effects on the environment. Not only do these projects have effects on the River proper, but they also are reflected in aquatic life; hence fisheries, water quality, terrestrial biota, and even recreation are involved.

The purpose of the remote sensing study was to investigate the application of these techniques as a method of monitoring environmental changes in the Missouri River Basin. In general, remote sensing can be considered a data-gathering tool, one not intended to replace all on-site data collection but having the advantage of being able to cover large areas, thus filling gaps between data stations. Eventually it is hoped that it will reduce the number of data-collecting sites.

More specifically the study involved the identification of vegetation adjacent to the Missouri River, sediment concentrations carried by the River and its tributaries, sources of pollution, recreation sites, urban and industrial expansion, and ways to assist in floodplain management. Use was made of all forms of available imagery, including ERTS, SKYLAB, U-2, and other NASA flights, plus special low-altitude flights.

SATELLITE SURVEILLANCE OF THE MISSOURI RIVER BASIN

The first ERTS satellite was launched on July 23, 1972 from Vandenberg Air Force Base in California in an orbit which is nearly circular at 566 statute miles (911 km) above the earth. The period is 103.2 minutes and the inclination is $99.1°$. This particular orbit has the property of crossing the same latitude parallel 105 miles further west each day and permits the taking of a swath of successive day photographs 115 miles wide with a 10 percent overlap. There is also repetitive coverage of the same area every 18 days at the same time of day. Therefore, sun angle changes are due only to the seasonal variations of the sun's declination. While the resolution is low compared to airborne imagery, it does provide an inexpensive way to monitor environmental changes over a large area for an extended period of time. During the length of the project, good cloud-free imagery over the entire river reach was obtained about seven times. No space will be devoted here to the ERTS imagery systems since they are by now well-known and are described in detail in other publications (2,3). Figure 1 is included to show the ERTS spectral coverage as related to a typical leaf reflectance.

In general, ground resolution for ERTS varies from 400 to 500 feet for low contrast, 300 feet for medium, and 250 for high contrast. The large area covered by these photographs permitted a macro-examination of the entire river and flood plain from St. Louis to Rulo, Nebraska, located at River Mile 498. At the same time three sites each 10 miles long were examined in more detail, using aircraft imagery and collected ground-truth data.

Urban areas along the river were also examined and compared with high altitude aerial photography. At present it seems that only limits of urban growth and not the types of growth can be detected at this time. However, the determination of urban limits is very clearly defined, which makes this a valuable tool in observing urban expansion over a period of repetitive coverage. Again, the direction and speed of urban movement provides indirect data on types of urban use. For example, it is a simple matter to observe in the ERTS photos the gravel roads leading to the Missouri River's edge. If a road in an area should appear more pronounced in a subsequent photograph, this would be

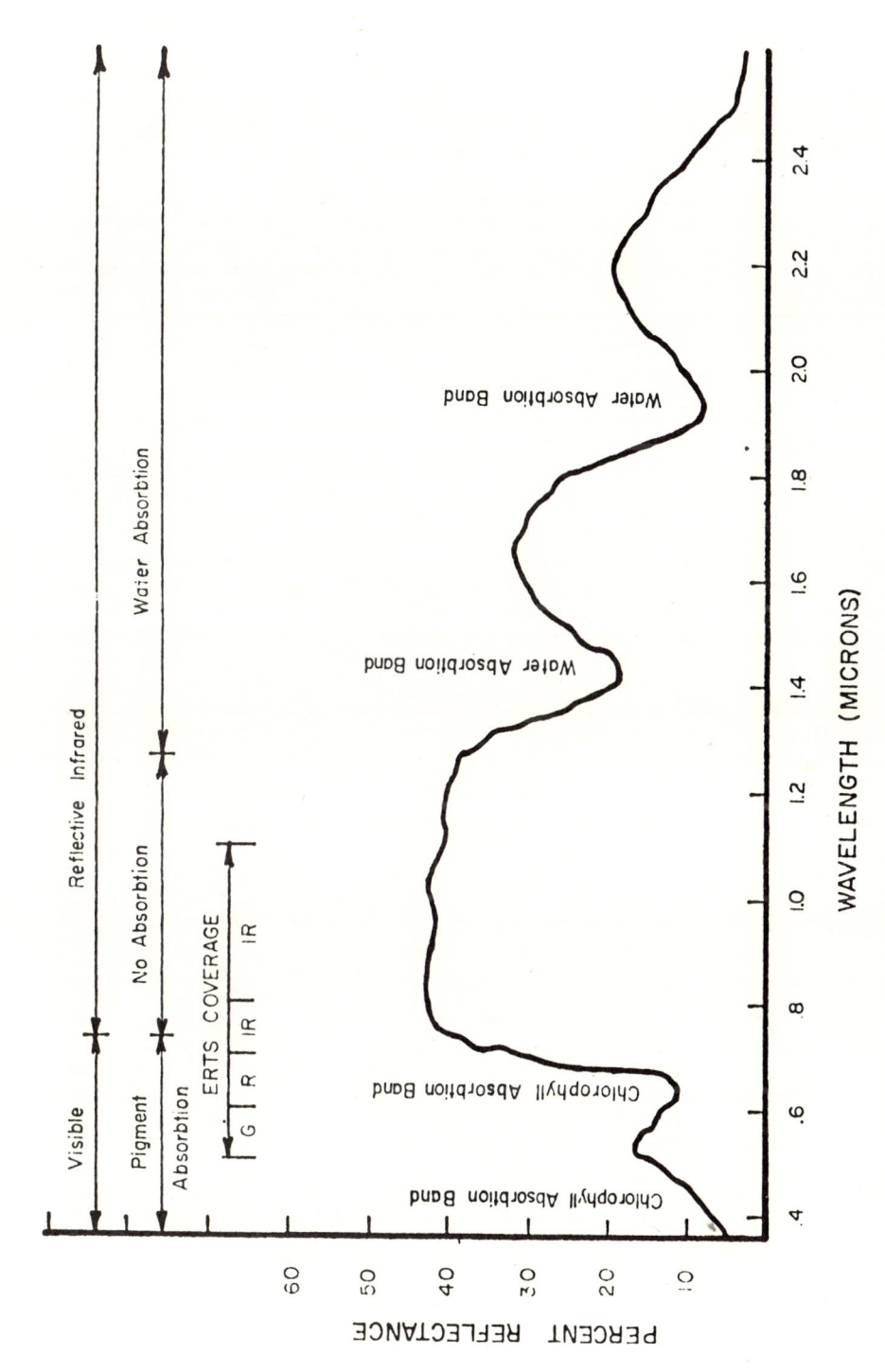

Figure 1. Three Primary Regions of Response in Relation to Leaf Reflectance

evidence that the road is probably receiving higher use for one of the following reasons: (1) increased recreational use of the site, (2) building of cottages close to the river, (3) moving in of some specialized industry, or (4) illegal dumping in the river. At any rate the area could be pinpointed for additional investigation. An excellent example of this was noted on an ERTS image set taken on September 18, 1972 of the Kansas City-Parkville site. Just 2,000 feet northeast of Mile 376, an irregular white shape was observed that did not correspond to anything on the aerial photos taken of the area on January 7, 1972. A check on the NASA flight coverage made in October 1972 revealed that apparently a new housing development area was under way.

Although stereoscopic relief is practically nonexistent in ERTS images, the scale of subsequent photos is so exact that they can be viewed stereoscopically for blink comparator purposes. The blink comparator method, which has been used by astronomers for many years, permits the observer to see alternately right and left images, which then reveal any changes that may have occurred over a period of several months. Since infrared shows all lakes and large ponds in high contract, the same technique was used to observe the water level in oxbows and other bodies of water that lie adjacent to the river.

To date the general consensus is that the ERTS, with the exception of some geologic formations, is not uncovering any new phenomena that cannot be located by high or low altitude aircraft remote sensing. <u>Its great advantage appears to be in the repetitive coverages to detect environmental changes</u> and in its ability to cover large areas at one time. Much of this information would be too costly if obtained by conventional aircraft.

In general, Band 5 (red) seemed to provide the most information and detail. While Band 4 (green) works well in the western United States and is useful for sediment detection in water, it appears to be of limited use in the humid atmosphere of Missouri and Iowa. IR seems useful mainly in the detection of streams and lakes, since these show up in very strong contrasts due to the absorbtion of IR by water. A sample of the four-band ERTS imagery is shown in Figure 2. The pictures show the Missouri River from Jefferson City to St. Louis, Missouri, which is at near right-center of the photograph. Other major rivers included in the area covered are the Mississippi, Illinois, Osage, Gasconade, Meramec, and Salt. The dark areas in the red band represent forest land, the lighter shades are generally farm land, and the lightest shades are urban areas. All oxbows show plainly along the Missouri River in the IR bands. This is particularly true of the reach between Rulo and Kansas City.

The extensive flooding of the Missouri River from February through May 1973, while hampering the gathering of ground-truth data, provided an excellent opportunity to monitor the river during this period, using ERTS imagery. The progressive changes are shown in a series of ERTS color composites, Figures 3 through 6. These composites were made at UMR by combining Bands 4, 5, and 7 of ERTS black and white transparencies. The false color overlays were made using the Diazachrome process which permits any normal color positive to be made from a black and white positive. In this case, yellow was used for the green band, red for the red band, and blue for the infrared. When the three color transparencies are overlaid, a false color composite results. Identification is done by noting the various colors and their intensities.

Figure 3 shows the Missouri reach from Jefferson City to St. Louis as it looked on August 28, 1972. At this time both the Missouri and Mississippi Rivers were at normal flow. In the color coding process, dark red represents forest areas while pink in the flood plains represents cropland (mainly corn). The dark color for rivers and lakes indicates clear water while blue indicates turbid conditions. Notice that the Mississippi River above St. Louis is clearer than the Missouri, while the Illinois River appears about the same as the Missouri. Tributaries such as the Osage and Gasconade appear dark and are therefore much clearer than the Missouri. Urban areas look blue while highways, particularly the interstates, show as white lines. Figure 4 pictures the

Green 0.5 - 0.6 μm 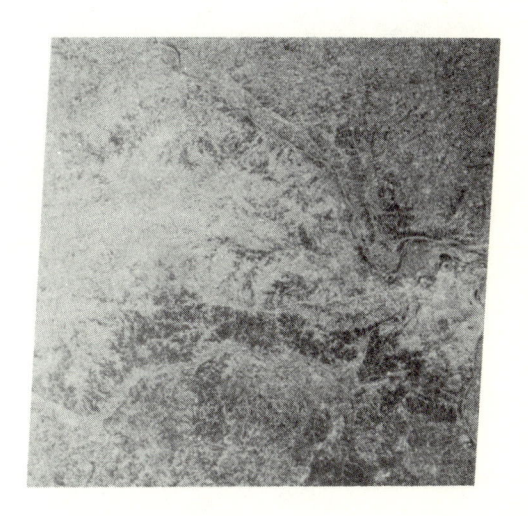   Red 0.6 - 0.7 μm

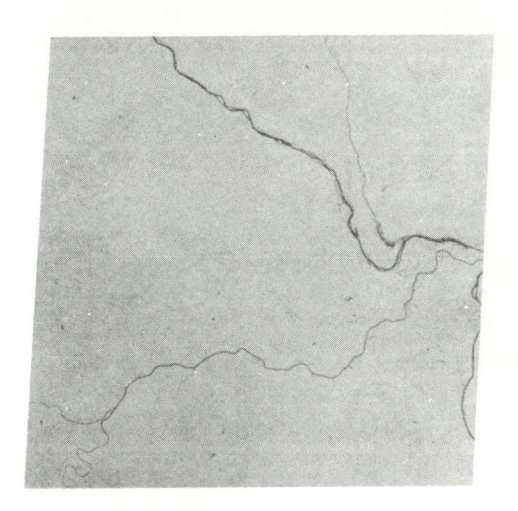

Infrared 0.7-0.8 μm 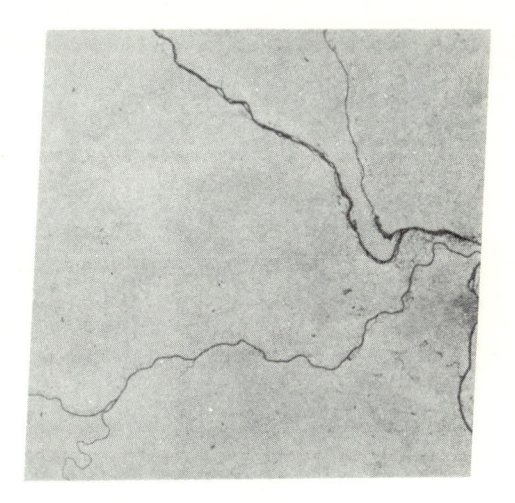   Infrared 0.8-1.1 μm

Figure 2.   ERTS Four Band MSS Imagery for the Missouri River from Jefferson City to St. Louis, Missouri, taken 28 August 10:16 AM CST.

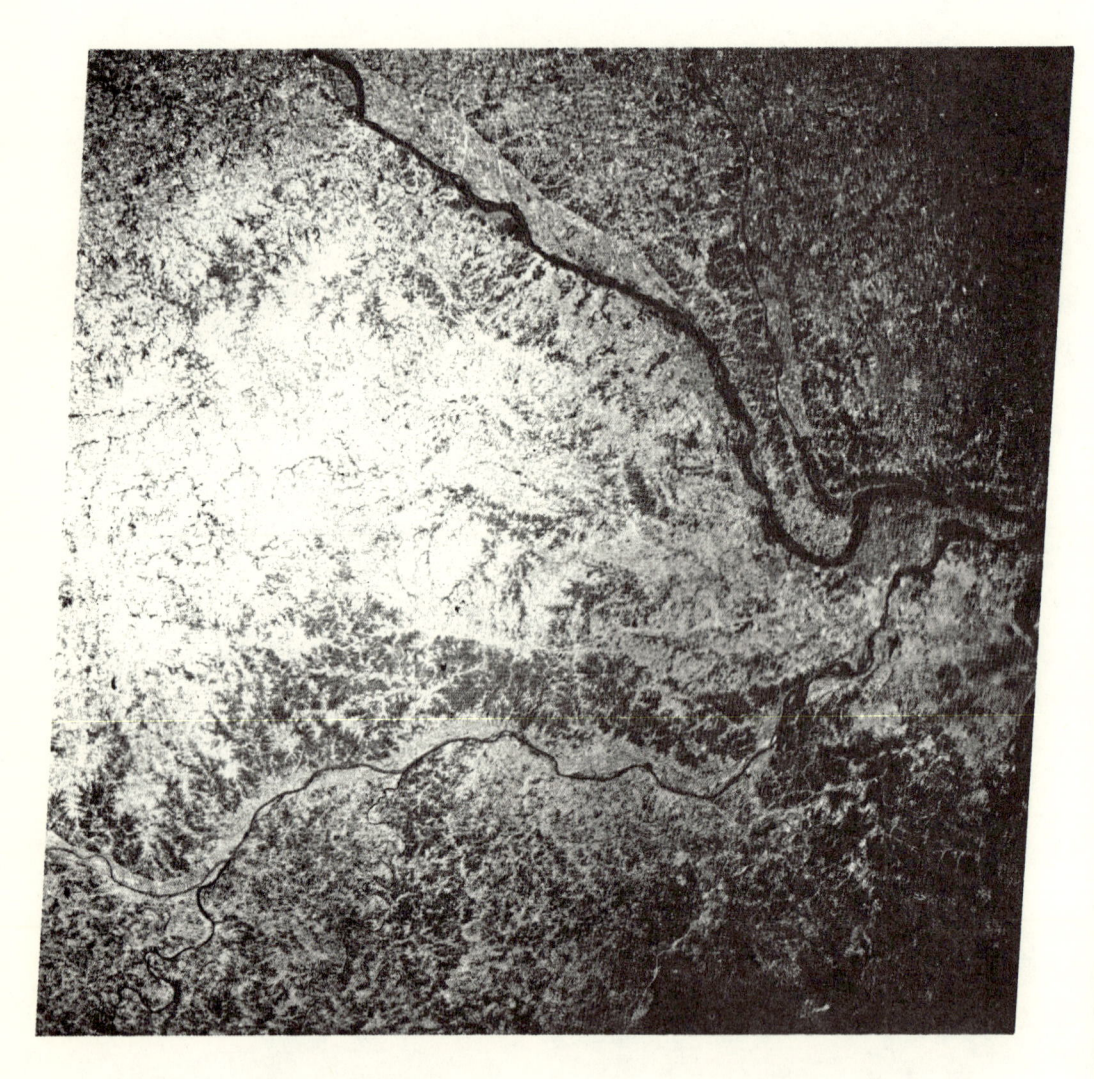

Figure 3. ERTS color composite formed by superimposing the images of Bands 4, 5, and 7 (green, red, and IR) shown in Figure 2. Clear water is shown as black, turbid water as light blue, forest as dark red, cropland as light red, and other blue areas generally denote urban limits. There is some color and definition loss in the prints from the original transparencies. The photo covers about 10,000 square miles and extends from Jefferson City at the left to St. Louis on the right; taken 28 August 1972.
(from original color print)

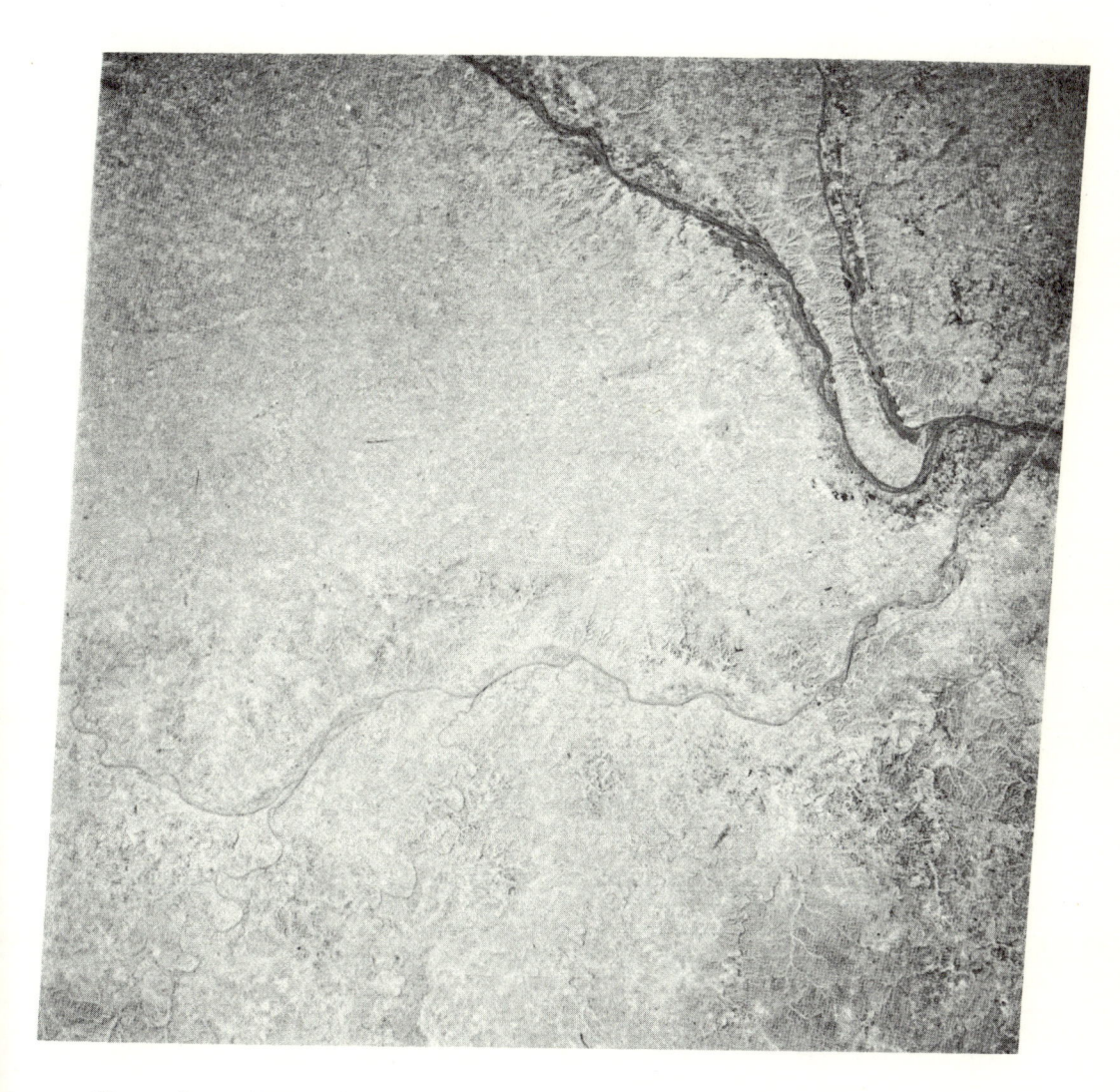

Figure 4. ERTS color composite taken 24 February 1973 showing the early stages of flooding. Note lack of vegetation cover, typical of the winter months.
(from original color print)

same area as it appeared on February 24, 1973. At this time some flooding is evident by the darkened appearance of the flood plain area. Note the absence of green vegetation and the increased detail of drainage patterns and other geological features. Unfortunately there is always some loss in resolution and color when making prints from the original transparencies. Figure 5 shows this area on March 14 when heavy flooding is quite apparent over most of the flood plain. The Gasconade is particularly noticeable and appears to be carrying a heavier silt concentration than the Missouri. The contrast between the Osage and the Gasconade is quite striking. Note the lack of vegetation in March with the corresponding increase in drainage pattern detail. Figure 6 shows the area on May 25, 1973. The flood has passed its peak and is beginning to subside. Most tributaries are back to normal and the flood plain on the upper Mississippi and Illinois Rivers is beginning to drain but still shows up quite wet, while most of the Missouri River flood plain is still under water. The wet season has created lush forest vegetation which is apparent by its bright red appearance. Also, since most of the grasses and brush in the flood plain are still recovering from inundation, the flood plain forests of mostly willow and cottonwood trees are easily distinguished as red clusters. This is true in both the Mississippi and Missouri reaches.

In addition to using the color additive process on ERTS imagery to determine features in the Missouri River Basin, a computerized enlargement of the Jefferson City area was made. This enlargement is shown in Figure 7 and is a small segment of the lower left portion of the August 28 photograph shown in Figure 3. The picture was produced by taking digital data directly from the four ERTS bands on tape and translating all of it into a television image with color identification similar to those used in this report. In this image the flood plain bordering the River is the habitat of willows and cottonwoods. The River itself is blue, indicating turbid water, while dark areas in the flood plain indicate plowed or bare soil. The light blue in the flood plain represents some grain (wheat), grasses, and pastures that have turned yellow. Jefferson City itself is plainly visible as a mottled blue with the highway system clearly visible, making it very easy to determine the urban limits. Cedar City and the municipal airport are also easily visible as light blue to white, just north of Jefferson City. Both the Moreau and Osage Rivers show that their water is much clearer than that of the Missouri. To the north of the flood plain, the hardwood forests can be distinguished as light red, and the drainage patterns are clearly visible. The picture was made using equipment at the Purdue University Remote Sensing Laboratory.

SKYLAB began with the launch of the laboratory on May 14, 1973 and ended with the return of the last team of astronauts on February 8, 1974. One of the prime experiments conducted during the flights was on the earth resources program, which utilized remote sensing as a tool to collect data. The fact that SKYLAB's altitude (270 miles) was about half that of ERTS, coupled with its high-quality sensors, provided a resolution of one order of magnitude over ERTS. Resolution in some cases has been to less than 30 feet.

On June 9, 1973, the first astronaut crew made a remote sensing run on a line from Omaha, Nebraska through St. Louis, Missouri. This provided partial Missouri River coverage from the Chariton River to the mouth. Imagery from this flight from both the multi-spectral and the earth terrain cameras was examined and found to be remarkable in its resolution. All individual streets in St. Louis could be distinguished, as well as considerable detail in the flood plain areas. Smoke plumes from power plants along the River could be traced for miles.

Flights over Missouri were made by succeeding crews of SKYLAB, but they were not available in time to be included in this project. The SKYLAB program has indicated excellent possibilities for high-resolution space photography; however, the main disadvantage is that there are no plans by NASA to continue such work in the immediate future.

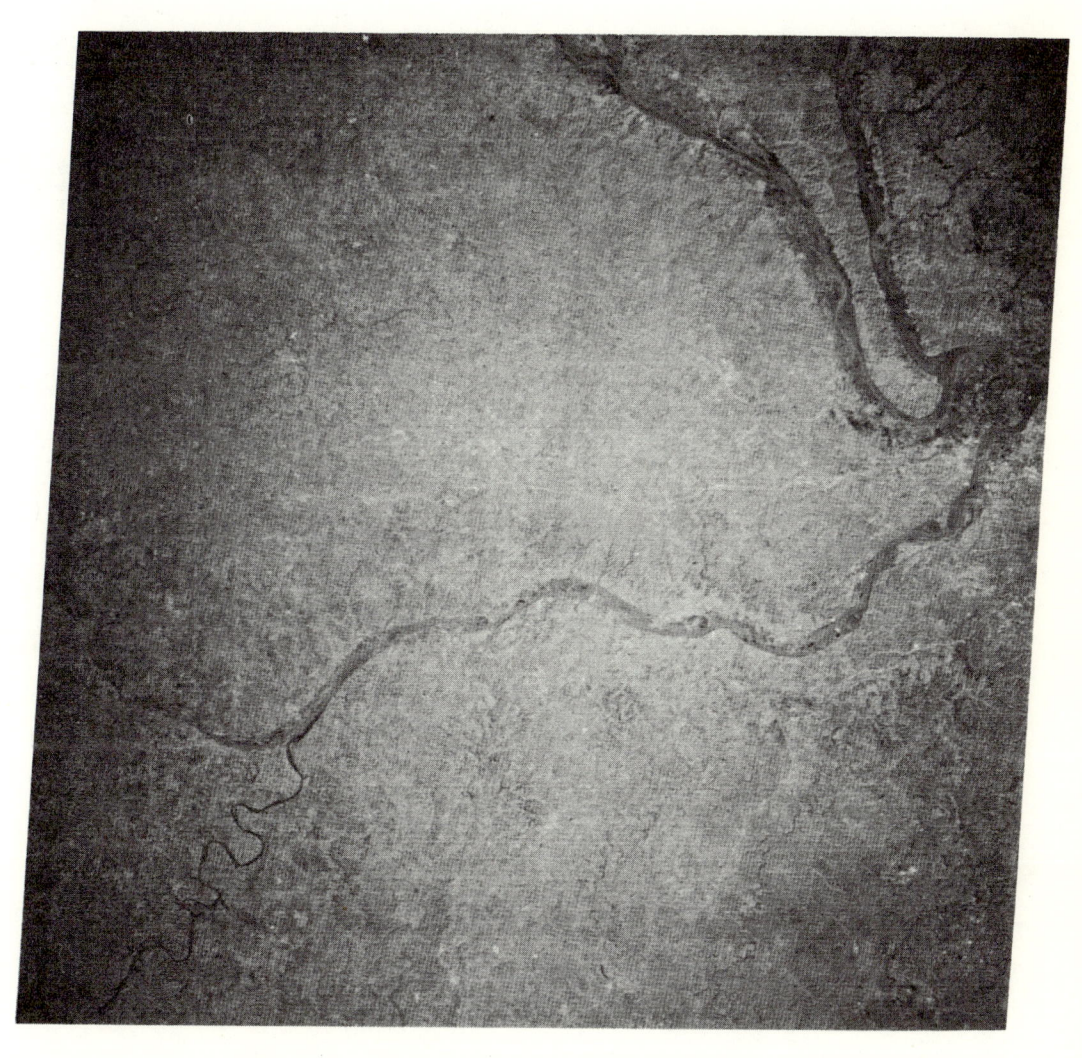

Figure 5. ERTS color composite taken on 14 March 1973. Flooding has increased, particularly in the tributaries. The Gasconade River is at a high stage and is carrying considerable sediment. Note the drainage patterns visible in the absence of vegetation.

(from original color print)

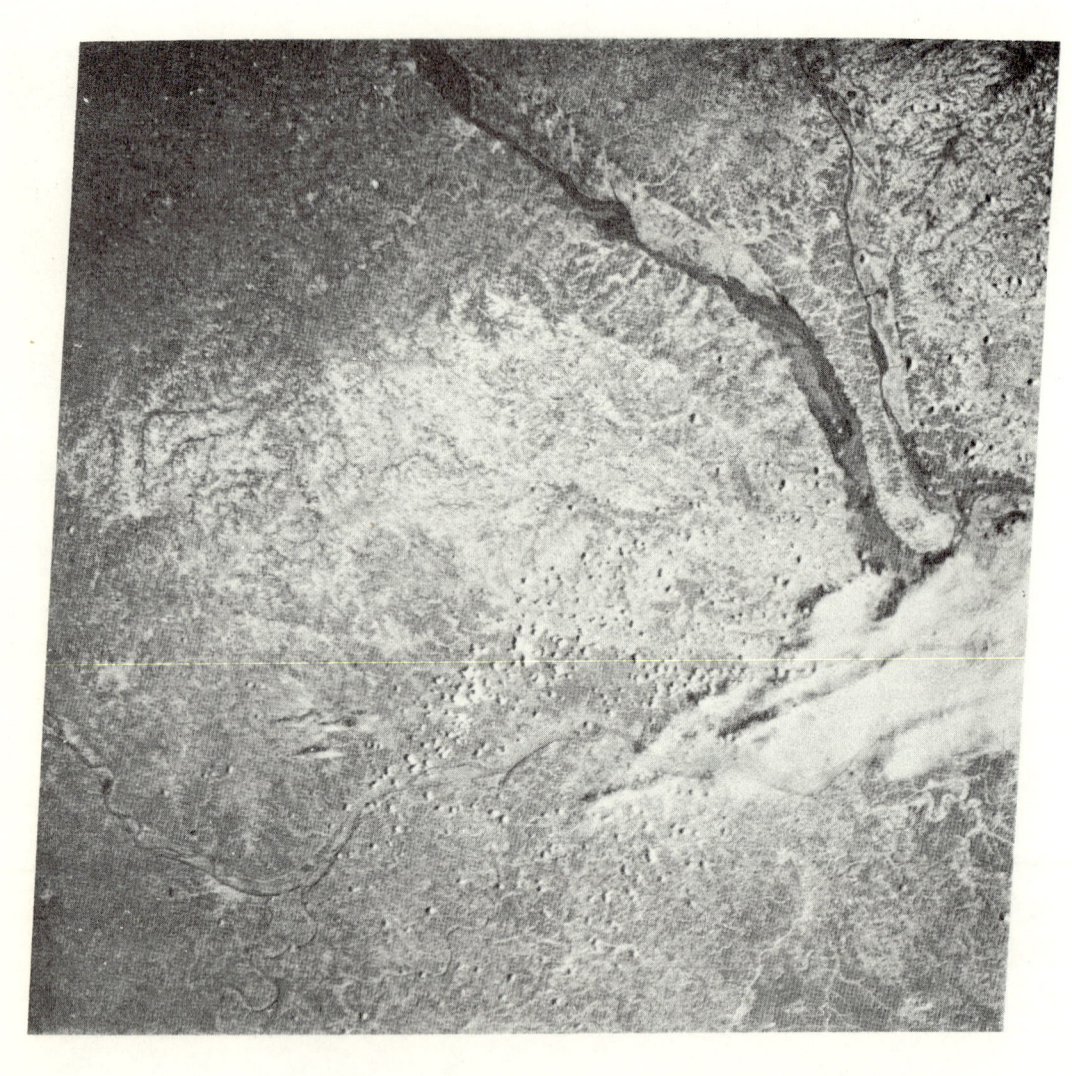

Figure 6. ERTS color composite taken on May 25, 1973 following the flood peak. Flooding in the upper Mississippi is decreasing; the tributaries are down but the Missouri is still carrying considerable flood water. The rich vegetation is due to the unusually wet spring.
(from original color print)

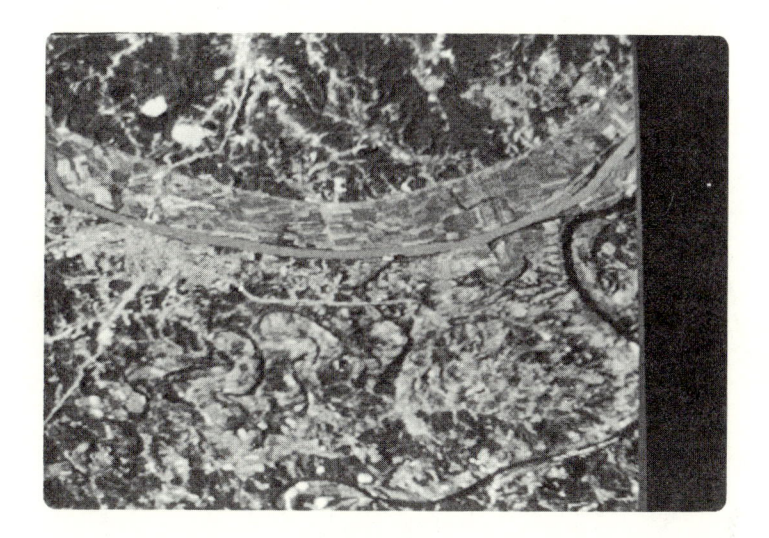

Figure 7. A computerized enlargement of the lower left corner of Figure 3, showing details of the Jefferson City area, crop lands, and forests in the flood plain from a height of 560 miles.
(from original color print)

REMOTE SENSING FROM AIRCRAFT

To study the uses of low-altitude photography, three sites were selected. They were (1) the area just north of Leavenworth, Kansas (Mile 394-407), (2) La Benite, east of Kansas City (Mile 352-360), and (3) the McBaine area east of Boonville (Mile 166-180). The Leavenworth and McBain sites were chosen to overlap the intensive study sites selected by the terrestrial biology team in order to make use of the ground truth data thus obtained. The LaBenite site was chosen for the purpose of examining a portion of the River close to a large urban and industrial area.

This investigation was divided into vegetation identification, recreation areas, pollution sources, and flood plain flood-flow management.

A low-altitude flight (6000 feet) was made over the selected sites using a cluster of four Hasselblad cameras equipped with color IR and B & W film with green, red, and IR filters. In addition, film from a NASA RB-57 flight was studied, as well as some U-2 IR photography taken in April 1973 for the Missouri State Geological Survey.

VEGETATION IDENTIFICATION

Initially the sites were classified into broad areas of forest, cultivated land, fallow ground, and pasture. This was not difficult by direct interpretation, and compared favorably with the ground truth data obtained by the terrestrial biology team. More difficult was the identification of specific trees, forest types, and crops, since the films available were not well-suited for this type of recognition. For some identification work, film density profiles were measured in different bands for known vegetation, and were extended to other areas. In this way it was possible to distinguish between some willows and hardwoods, and to identify crops, particularly wheat and corn. Figure 8 shows some of the vegetation classifications for one of the selected sites.

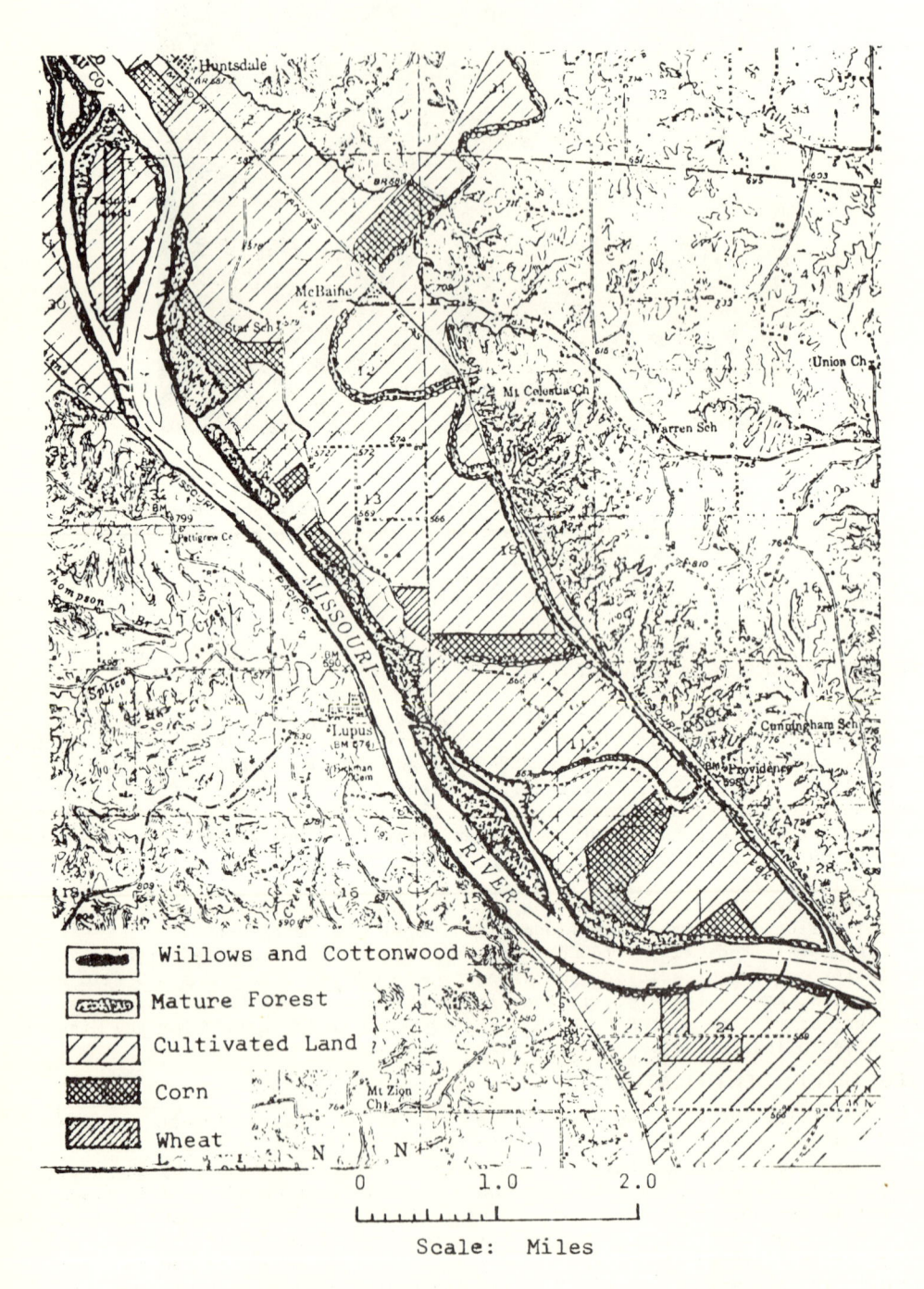

Figure 8. Vegetation Identification for McBaine Area east of Boonville, Missouri.

RECREATION AREA IDENTIFICATION

The identification of major recreational areas was relatively simple. All facilities listed by the recreation study team lying within the remote sensing site area were located. In addition, many small private boat ramps were noted along with numerous small roads and trails leading to the riverbank. The imagery taken in February was particularly useful in locating small trails since much of the forest canopy was absent at that time. Color IR (film type 2443) appeared to give the best contrast in identifying details of this type.

POLLUTION SURVEY

Because of the high turbidity of the Missouri River, it was not possible to determine quantitatively the sediment or pollution content in the river itself. It was of course possible to examine the quality of streams feeding into the main river and to note the appearance of the effluent as it entered. It was noticed that all of the streams such as the Grand, Chariton, Osage, and Gasconade fed relatively clear water into the Missouri; and there was no discernible clear discharge past the mouth. The only exception noted was the Kansas River, which appeared more turbid than the Missouri, with a noticeable separation existing for several thousand feet downstream. Searches for sources of pollution were made along the river in the area of the selected sites, photographs of two different dates being used. However, only one small source was noted west of the La Benite recreation area.

In the U-2 IR photography, taken for the Missouri State Geological Survey during the flood, it was possible to identify seep water behind dikes and roadways by the change in color. This was particularly true in the McBaine area between Providence and the river, where there was no discernible levee break. Colors also varied somewhat, depending on velocities in different parts of the flood plain.

Because there was considerable cloud cover during the flood season, several flights over the lower 200 miles of river were made, using radar scanning equipment. The imagery thus obtained was useful in determining the extent of flooding and the condition of dikes in the areas.

FLOOD PLAIN FLOOD-FLOW MANAGEMENT STUDIES

Since vegetation and man-made structures may change flood plain characteristics, it is important that surveillance of the flood plain be maintained in order to keep the flood-flow channel at optimum conditions. The purpose of this study was not to set the flood plain flood-flow criteria but to determine the value of remote sensing and photo-interpretation as a tool in laying out and monitoring the system, once the criteria have been set. With the flow area delineated and the tolerance of forest lines and levees established, it would be a relatively simple matter to note any variance that might affect the flow of water. For example, in the area around the McBaine site an extensive dike system was observed which narrowed the channel considerably; also, the forest strips around Tadpole Island could collect brush and hinder flood-flow. This is not to say that such arrangements are detrimental, but only that any change such as unusual tree or brush growth or illegal dikes could be observed and, if necessary, removed.

Although selected areas could be computerized to note selected changes in the flood plain, it would probably, at the present state of the art, not be economical to computerize the entire river. Until such a system is developed, a well-trained photo-interpreter could undoubtedly maintain a close scrutiny of questionable areas.

SUMMARY

The ERTS satellite has proven extremely useful in monitoring macro-changes in the Missouri flood plain area. The spring 1973 flood is an excellent example of this surveillance in which the progressive flooding was analyzed on successive multi-spectral photographs. In addition, it was possible to monitor vegetation changes and the flooding of tributaries during this period. The ERTS and Skylab-type photography will, if continued, serve as a long-term record of changes in the Missouri River Basin.

The ERTS photography also has provided a means of examining large areas and pinpointing changes. Although it may not be possible to identify the exact type of change taking place, it would be possible to examine the area by direct visitation or by low-altitude photography.

The use of multispectral imagery and conventional aerial photography at various altitudes has provided a means of identifying and monitoring in detail the plant life existing on the flood plains. During the flood period, it was possible to identify relative flood water conditions by noting the turbidity of the water. Thus it was possible to distinguish seep water behind dikes and other standing water from sediment-carrying water moving at higher velocities. Because of the high turbidity of the Missouri River it was not possible to measure quantitatively this condition. However, areas of pollution could be detected indirectly by observing the condition of tributaries feeding into the Missouri River (for example, the Osage River, being clearer, would tend to reduce turbidity). Other factors such as cities, industrial sites, and areas where damage to plant life is observed were also found to be indicators of possible pollution conditions.

In the investigation of flood plain flood-flow characteristics, it was determined that remote sensing could aid in flood plain management, both in delineating that area set aside to carry flood water, and monitoring and maintaining it in such a condition that it will be able to carry floods of designated levels. While it was not possible to determine roughness coefficients for the type of growth and other obstacles in flood-flow, it was possible to identify conditions that affect flood-flow.

It has been recommended that a file on space photography be set up for periodic checking of changing conditions. This, coupled with lower altitude photography, would serve as a continuing record on flora and flood plain changes. Selected areas could be computerized to note these changes.

REFERENCES

1. Munger, P. R., et al, "A Base Line Study of the Missouri River: Rulo, Nebraska to Mouth near St. Louis, Missouri," University of Missouri-Rolla (1974).

2. Finch, W. A., Jr., "Earth Resources Technology Satellite-1", Symposium Proceedings, Goddard Space Flight Center (1972).

3. Lyon, R. J. P., "The Multiband Approach to Geological Mapping from Orbiting Satellites: Is It Redundant or Vital?", Journal, Remote Sensing of Environment, $\underline{1}$:237-244 (1970).

DETERMINING POTENTIAL SOLAR POWER SITES IN WESTERN HEMISPHERE OCEAN AND LAND AREAS BASED UPON SATELLITE OBSERVATIONS OF CLOUD COVER

H. W. HISER and H. V. SENN

Remote Sensing Laboratory
School of Engineering and Enviromental Design
University of Miami, Coral Gables, Florida, U.S.A.

ABSTRACT

A system is presented for using the two Geostationary Operational Environmental Satellites (GOES-1 and -2) to determine potential geographic locations for solar power stations in ocean and land areas of the western hemisphere. These satellites can measure the numbers of daytime cloud-free hours throughout the area of study with a resolution of approximately 0.5 nautical mile. Cloud density, time of day of minimum cloud cover, and seasonal distribution are considered through the use of digital data processing techniques. Low density cirrus clouds are less detrimental than other types; and clouds in the morning and evening are less detrimental than those during midday hours of maximum insolation.

Techniques are proposed for correlating satellite cloud observations with solar radiation in Langleys arriving at the earth's surface at any geographic point within the area investigated. The sparse nature of solar radiation measurement stations makes this an important item in the system. The geostationary satellite measurements are far superior to ground observer reports of clouds for these correlations. Ground observations are too sparse and the mesoscale differences that are not presently observed can be of major significance in solar power station locations.

INTRODUCTION

It is obvious from all surveys and reports that we are using our fossil fuels at a tremendous and ever increasing rate so that in the not too distant future those supplies of energy, so vital to our present growth of civilization will be depleted. In 1972, the Federal Power Commission [1] reported that in the U.S.A., essentially no more gas will be available in 1990 than there is today and that its price will then be doubled. Oil and coal will increase in cost 50 percent and electrical rates will increase 19 percent by present dollar values. If annual inflation rates are considered, the increases are likely to be much greater than these figures. Also, pollution may be more than quadrupled since by 1990 the higher quality coal and oil will have been exhausted.

In 1972, Faltermayer [2] anticipated that by 1985 the U.S. will import 50 percent of its oil. If it does not import fuels, a Rand Corporation Study [3] for the state of California stated that the nation's gas resources could be exhausted between 1990 and 2000, and petroleum supplies could be exhausted between 1990 and 2010. Once synthetic fuels are produced from coal to replace natural gas and petroleum, the nation's supposedly abundant supplies of coal will undergo rapid depletion and domestic coal could be exhausted in less than 70 years.

Unfortunately, in less than two years, many of these once unbelievable predictions have become almost a reality to us today. For this reason, it is of utmost importance that we look for other permanent sources of energy and learn to use them before the dire need arises. Solar energy is a logical choice as it is readily available, well distributed and essentially inexhaustible. Furthermore, it can be collected and converted to useful energy with a minimum impact on the environment and in many cases with no disposable wastes.

Solar energy arrives on the surface of the United States at an average rate of 4100 Kgcal/m^2-day or about 41×10^8 Kgcal/km^2-day [4]. Over a period of a year, a square kilometer would receive an average of 1.5×10^{12} Kg calories. In 1970, the total energy consumed by the United States for all purposes was about 17×10^{15} Kg calories [5]. Thus, 11,300 square kilometers of land receives on the average in one year the equivalent of all U.S. energy needs. At a 10% conversion efficiency, 113,000 square kilometers or about 3% of the land areas of the 48 contiguous states could provide the equivalent of the 1970 U.S. energy consumption.

By proper choice of location at a place receiving a higher than average daily solar radiation, the area required to receive a given amount of energy can be considerably reduced. Also, by locating energy collection in southern United States or at a lower latitude, a higher sun angle can be obtained with less atmospheric attenuation and less seasonal variation of solar radiation. For example, there are places in southern and southwestern United States that receive an average of 5000 Kgcal/m^2-day or more per day. A proper choice of location in one of these areas could reduce the land area required to receive a given amount of solar radiation by at least twenty-five percent.

In response to the President's energy message to the U.S. Congress, the National Science Foundation (NSF) and the National Aeronautics and Space Administration (NASA) established a Solar Energy Panel in January of 1972 to assess the potential of Solar Energy as a national energy resource. The panel has concluded: that with adequate R & D support over the next thirty years, solar energy could provide at least 34% of the heating and cooling of future buildings or greater than 30% of the methane and hydrogen needed in the U.S. for gaseous fuels, and greater than 25% of the electrical power needs of the U.S. [6]. All of this could be done with a minimal effect on the environment and a substantial savings of non-renewable fuels.

In the 1973 report of the Working Party to Advise the U.N. Educational, Scientific and Cultural Organization on a Programme in Solar Energy [7], it was concluded that recent events in many countries make it apparent that a take-off point for large scale solar energy development may be at hand. The report states that there is a widespread appreciation of the limited lifetime of fossil fuel supplies and in many countries there is also concern about the pollution engendered by the use of fossil and nuclear fuels. Therefore, there is a growing recognition that solar energy as a renewable and non-polluting source may play a major role along with other new forms of energy in meeting future world wide power needs.

Solar energy development is now receiving increased scientific attention and financial support. As a typical example, the National Science Foundation (NSF) has experienced a growth in its budget in this area from about $300,000 in FY 1973, to $12.5 million in FY 1974, to about $30 million in FY 1975. A major component of the President's new Energy Research and Development Agency (ERDA) has been identified as solar energy. Some international cooperation has been initiated on an informal basis and a number of agreements for bilateral cooperation have shown that it should be possible to efficiently concentrate and collect the solar energy on a large scale. For example, Hildebrandt [8] has investigated the use of collected solar heat without storage as a fuel displacement in a conventional oil burning electrical plant during sunlight hours. Oman and Bishop [9] have conducted a feasibility study on converting solar power efficiently to electric power with heat engines. A Solar Sea

Power Plant Conference and Workshop was held at the Carnegie-Mellon University in June 1973 to investigate the large scale sea-based solar power plant. Escher [10] has even proposed a model energy facility.

The major concern in the large scale solar energy development is the high initial installation cost and large size of area involved. The United States Solar Energy Panel [4] has estimated that solar thermal plants will require 25 square kilometers of land per 1000-megawatt installation capable of operating on the average of 70% capacity. Installed costs are estimated to range from $900 to $2000 per kilowatt. Photovoltaic electric power generating plants will also be very expensive and require 50 square kilometers or more of land per 1000-megawatt installation, but would not require cooling water or towers. Photovoltaic terrestrial systems, using today's space technology but not incorporating the high reliability specifications for flight hardware, could be built for about $70,000 per installed kilowatt.

Because of this tremendous amount of capital investment involved, it will be necessary to conduct a careful study on site selection to assure a high return of investment before such a solar power plant is planned. The ideal area to be selected should be most cloud free, with the longest hours of sunshine and receive the most intense radiation. Such information is presently available for only a fraction of 1% of the possible solar plant sites. However, it can be readily obtained by remote sensing. Remote sensing is superior to a finite number of ground stations in that it can give a continuous survey over a large area, obtaining more detailed and widely distributed information. Furthermore, remote sensing can be used to monitor the vast sea surface while it is very difficult and expensive, if not impossible to be achieved by in-situ measurements. A mapping of these solar-intensity distributions will be invaluable to the future planning of solar energy development.

A very important aspect of this project is to develop transform equations for relating cloud cover or hours of various intensities of sunshine as observed by satellite to daily solar radiation received at the earth's surface. Surface observations of clouds, although more numerous than solar radiation measurements, are too sparse and the mesoscale differences that are not presently observed can be highly significant in selecting solar power station locations. Also, the available surface observations of clouds do not indicate the position of clouds with respect to the station. Therefore, they do not indicate whether or not the sun is shining at the station at a given time. Much better calculations of solar radiation reaching a given geographic location should be attainable from gridded, digitalized cloud observations by satellite than from the meager and poorer quality surface observations.

DATA SOURCES

Daytime satellite imagery from the Geostationary Operational Environmental Satellites (GOES-1 and -2) can be used to provide the cloud data. GOES-1 is positioned at 75°W longitude and GOES-2 will be at 135°W longitude over the equator. Figure 1 shows the approximate positions and areas of coverage of these two satellites which define the area of study for this project.

Data from five solar radiation measurement stations in Florida plus some from stations in other parts of the United States can be used for developing the relations between satellite cloud observations and solar radiation reaching the earth's surface. Data for stations at different latitudes and seasons of the year must be considered.

DATA ANALYSES

The analog gray scale satellite data can be converted to digital intensities of cloud cover for processing. Three to five levels of cloud intensity or reflectivity in the 0.55 to 0.75-micron visible range should be used. One will indicate thin clouds with weak reflectivity, two more dense clouds, etc. A space resolution of the order of 2 to 4 miles should be used since this is adequate for the survey and is compatible with the resolution of the GOES system in the visible range of the spectrum for places considerably removed from the center of the camera's field of view.

Figure 2 shows the GOES Centralized Data Distribution System (CDDS) with a readout in Miami at the National Environmental Satellite Service (NESS) Satellite Field Service Station (SFSS) located in the National Hurricane Center on the University of Miami campus. The VISSR data is from the Visible/Infrared Spin-Scan Radiometer which will be used for this project.

As the GOES satellite spins at 100 revolutions per minute, the VISSR's scanning mirror will face the earth for about one-twentieth of each complete 360-degree rotation, scanning from west to east in eight identical visible channels and two redundant IR channels. Immediately, the scan data will be transmitted in digital form to the Wallops Island, Va., Command and Data Acquisition station (CDA). While the spacecraft is completing its revolution, the mirror will move to the next southward step and scan again when it is looking at the earth once more.

Within 18.2 minutes, the radiometer will accomplish the 1821 scan steps required to provide an image of the coverage area, Figure 1. The resulting visible images, made only in daylight, will contain 14,568 lines and have a resolution of nearly 1/2-mile at the Nadir point. IR images, acquired in darkness as well as in daylight, will have a total of 1821 lines, with 5-mile resolution. Allowing time for the scanning mirror to return to its starting point, and for correction of any "wobble" which may be caused by this retracting action, the Satellite Service plans to schedule GOES picture coverage at 30-minute intervals. This picture rate will provide far more data than will be required for our studies.

At the CDA station, the eight lines of visible data acquired while the spacecraft looks earthward will be gridded automatically and the rate of data transmission reduced, "stretched". As the satellite is completing its revolution and the VISSR is looking toward space, the stretched visible data signals will be retransmitted from the CDA station through the satellite to the NESS at Federal Office Building number 4 in Suitland, Maryland and then relayed by microwave to the NESS Central Facility a few miles away.

Figure 3 is a VISSR daylight image of clouds over southeastern United States taken by GOES-1 on 30 January 1975 at 1830 GMT. This image in the visible portion of the spectrum was made with the 0.5-mile-resolution sensor. The computer-produced geographic grid is not accurately matched to this image but can be adjusted for our purposes. The "sea breeze front" can be observed in the cloud pattern inland along the east coast of Florida. Also, the reflective and heat-sink effects of Lake Okeechobee and Tampa Bay can be seen. The result is an area over and immediately downwind from each of these water bodies that is free of low clouds. There are also some smaller water bodies in central and northern Florida that show the same effects. These illustrate the significant mesoscale differences in cloud cover and sunshine that can be monitored from the GOES imagery.

The GOES stretched VISSR data with geographic grids can be digitized for this cloud cover study. From these data, maps can be prepared showing hours of sunshine for both land and ocean areas. The results will give more detailed time and space distributions than Figure 4, and will include extensive ocean areas, South and Central America, and parts of Canada as indicated in Figure 1. Figure 4 is based upon measurements of hours of sunshine at a number of National Weather Service observation stations in the United States [11]. Transform equations discussed in the next section, can be used to convert the satellite-derived sunshine data to maps of solar energy received at the earth's surface.

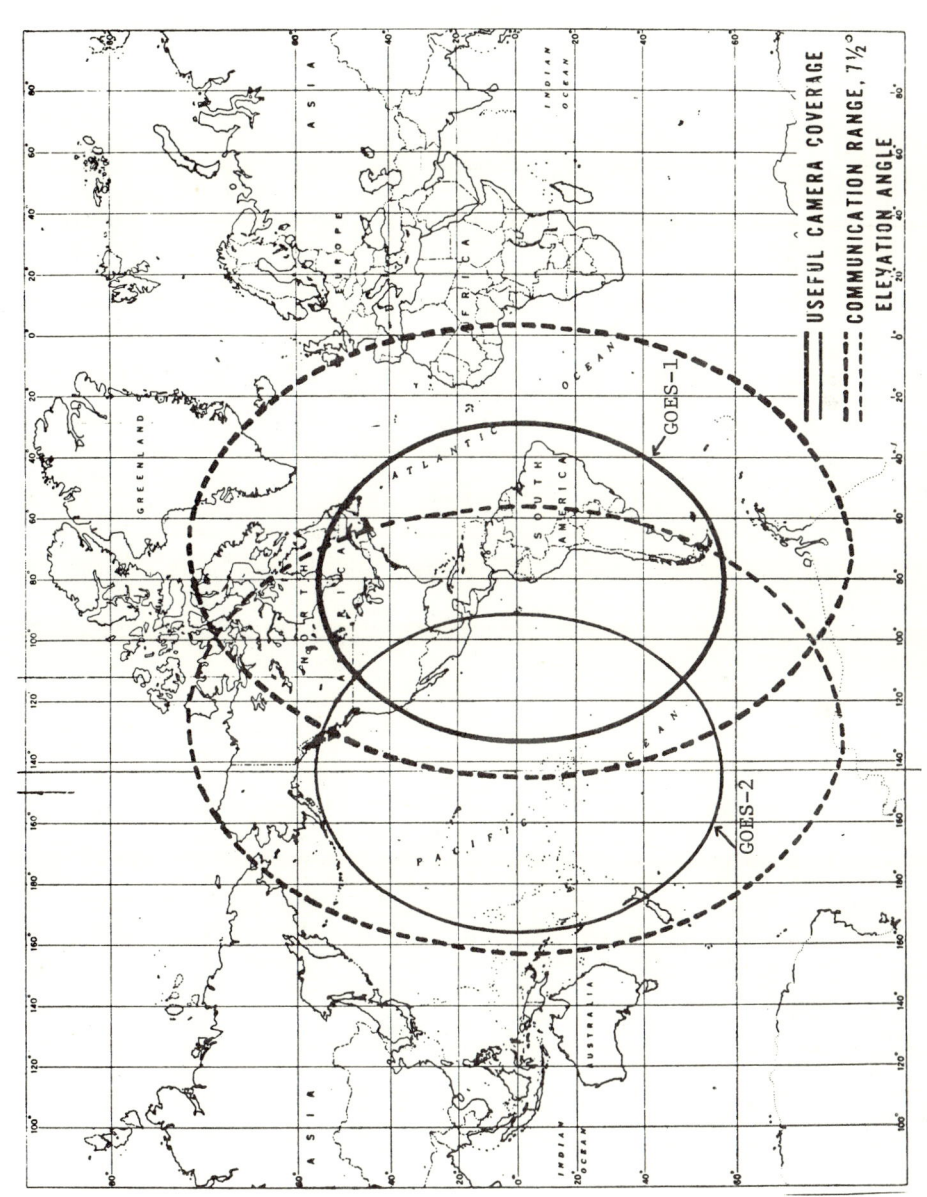

SOLAR POWER SITES BASED UPON SATELLITE OBSERVATIONS OF CLOUD COVER

Fig. 1. — Typical coverage from a 2-GOES system showing area of useful camera coverage and communications range at 7.5-degree antenna elevation for data collection and relay.

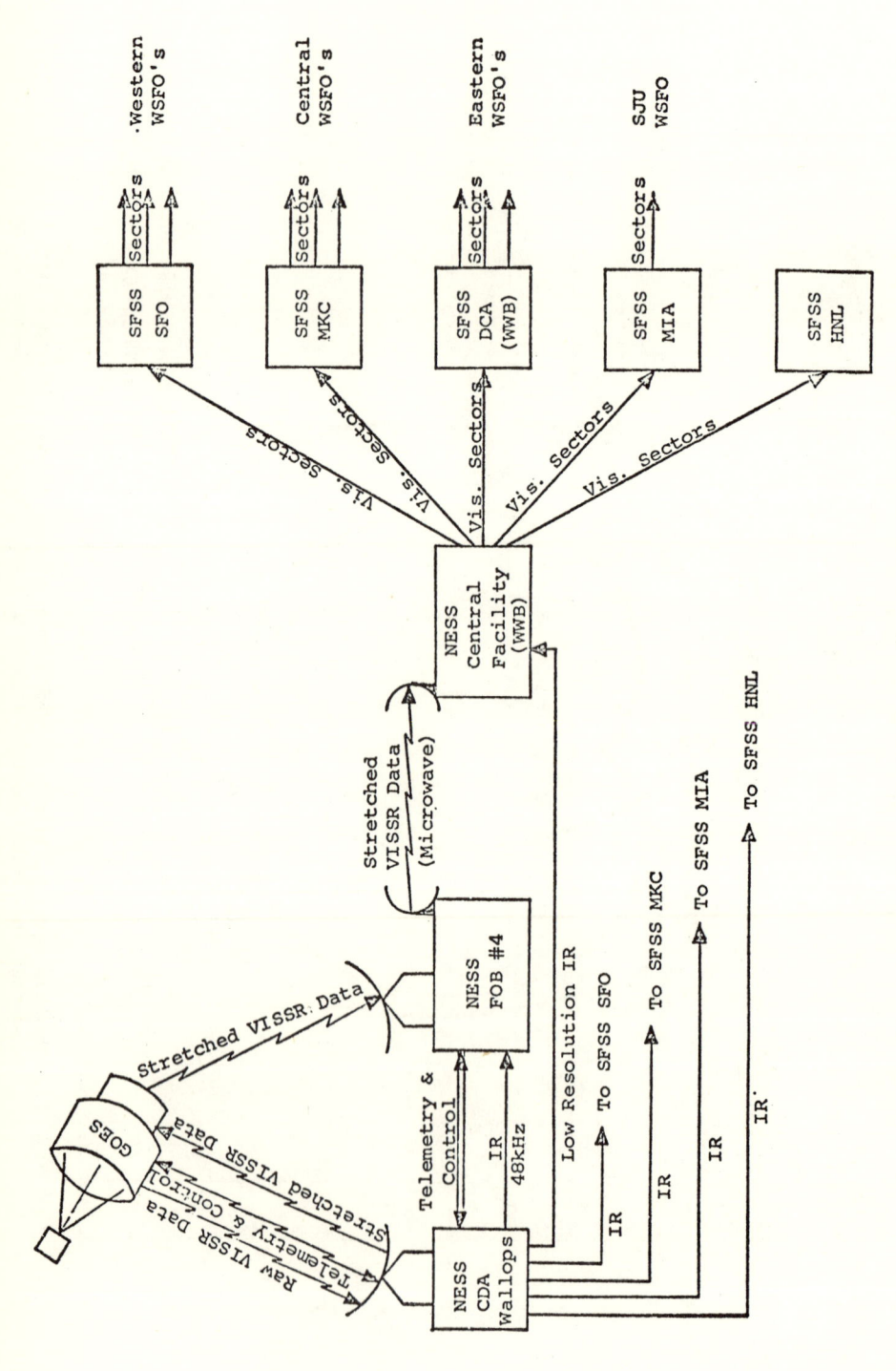

Fig. 2. – GOES Centralized Data Distribution System (CDDS)

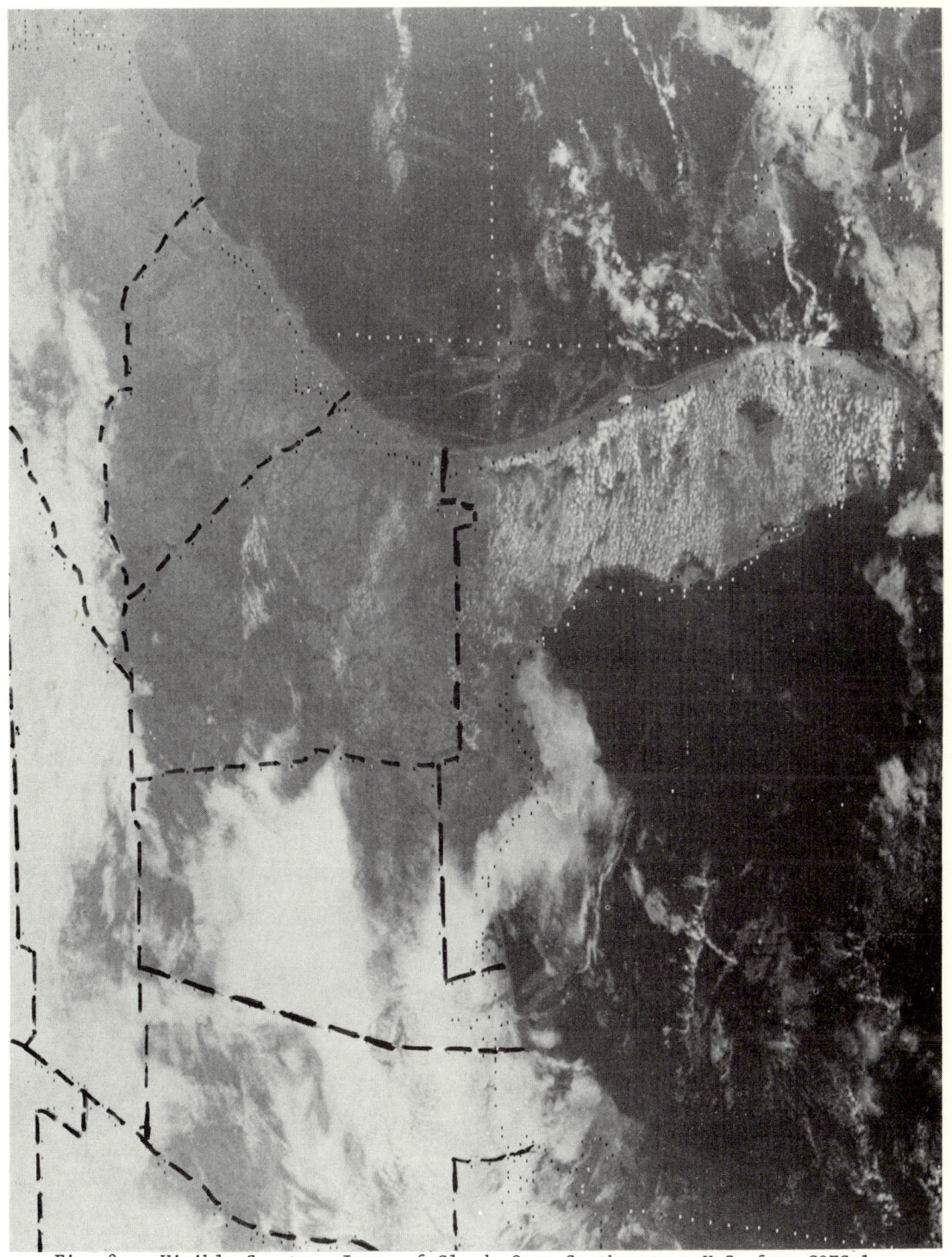

Fig. 3. - Visible Spectrum Image of Clouds Over Southeastern U.S. from GOES-1, 30 January 1975, 1830 GMT. Sensor Resolution 0.5-n.mi.

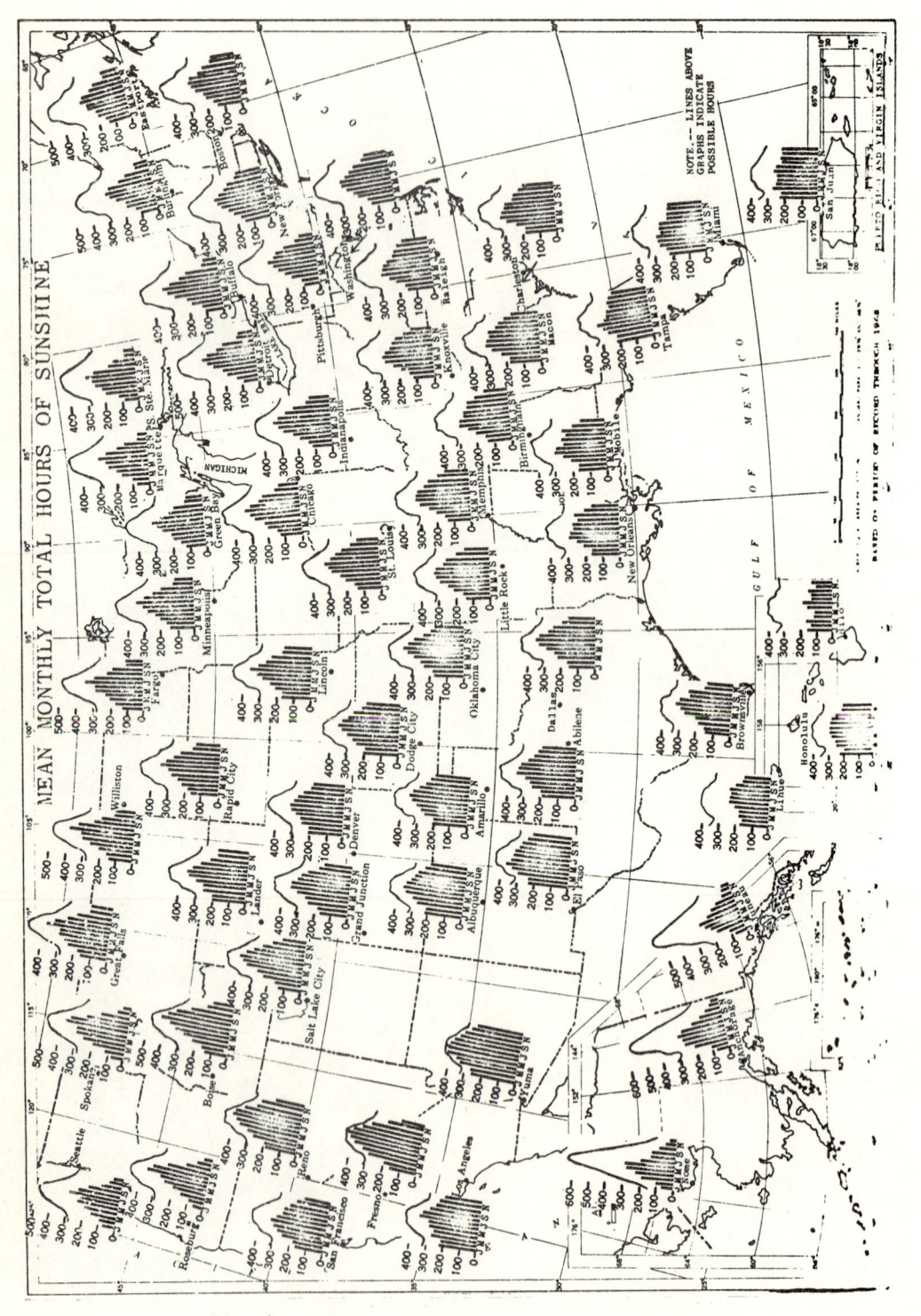

Fig. 4. - Mean Monthly Total Hours of Sunshine.

RELATIONS BETWEEN HOURS OF SUNSHINE AND SOLAR ENERGY

Relations can be developed between the hours of sunshine per day and the daily solar radiation in gram calories per square centimeter (Langleys). Figure 5 gives the annual mean daily solar radiation in Langleys based upon solar radiation measurements at a number of National Weather Service observation stations in the United States [11]. Considering the fact that only five stations routinely record pertinent data in Florida, it can be seen that very much interpolation and extrapolation of the meager data was necessary to produce Figure 5. Conversely, with good sunshine and solar radiation data from only a few stations, the vastly more comprehensive satellite data on hours of sunshine of various intensities can be converted to available solar energy arriving at the earth's surface in Langleys.

In addition to the 2-mile resolution of the derived radiation data, it will be possible to determine time variations in intensity occurring within each day which would be pertinent to the design of a plant at a given location.

Several things will have to be considered in developing equations for relating cloud cover to insolation received at the earth's surface. One important consideration is the time of day of maximum sunshine, whether morning and evening or midday. Two places with the same number of hours of sunshine per day can receive quite different amounts of solar energy depending upon this factor. Also, time of the year, station altitude, latitude, air pollution, and atmospheric moisture content will need to be considered. Probably, an equation can be developed with different constants for different climatic and geographic locations [12].

Lund has studied relationships between insolation and other surface weather observations for a 22-year period at Blue Hill, Massachusetts [13]. He found the highest monthly positive correlation, in the mid ninety percentile range, was with total hours of sunshine as one might expect. Also, the highest negative correlation, in the middle and lower eighty percentile range, was usually with average daytime sky cover. In April, May, and July, there was a slightly better negative correlation between daytime hours of precipitation and insolation than between sky cover and insolation.

Bennett found that the percentage of opaque sky cover, which does not include cirrus and thin altostratus clouds, gave a better correlation with insolation measured at the ground than was obtained by use of total sky cover data [14]. The numerical values of his negative correlations between opaque sky cover and insolation were nearly equal to his positive correlations between hours of sunshine and insolation.

Lumb [15] has studied the influence of cloud on hourly amounts of total solar radiation at the sea surface and has proposed the following relation:

$$Q = 135 \text{ fs} \quad (\text{miliwatt hr/cm}^2) \quad (1)$$

where $f = a + bs$ is the fraction of solar radiation transmitted through the atmosphere, s is the mean of the sines of the solar altitude at the beginning and end of the hour, and a, b are constants. The constants a and b are determined by least squares best fit of Q plotted against hourly mean sine of solar altitude for a given category of sky condition. Lumb used nine categories from clear to thick layers of low cloud usually accompanied by rain. Thus he obtained nine pairs of constants. His standard error of estimate of Q for all nine sky categories combined, ranged from 1.7 miliwatt hour/cm^2 at low sun angles to 7.6 at high sun angles.

Norris [16] has tested Lumb's and other methods for correlation of solar radiation with hourly cloud observations at Melbourne, Australia. All methods tested performed poorly because of difficulties in fitting complex cloud observations to a sky category. Since all of the approaches failed to give the desired accuracy of prediction of solar radiation and since it has been found that on a basis of monthly mean values a good correlation exists between total

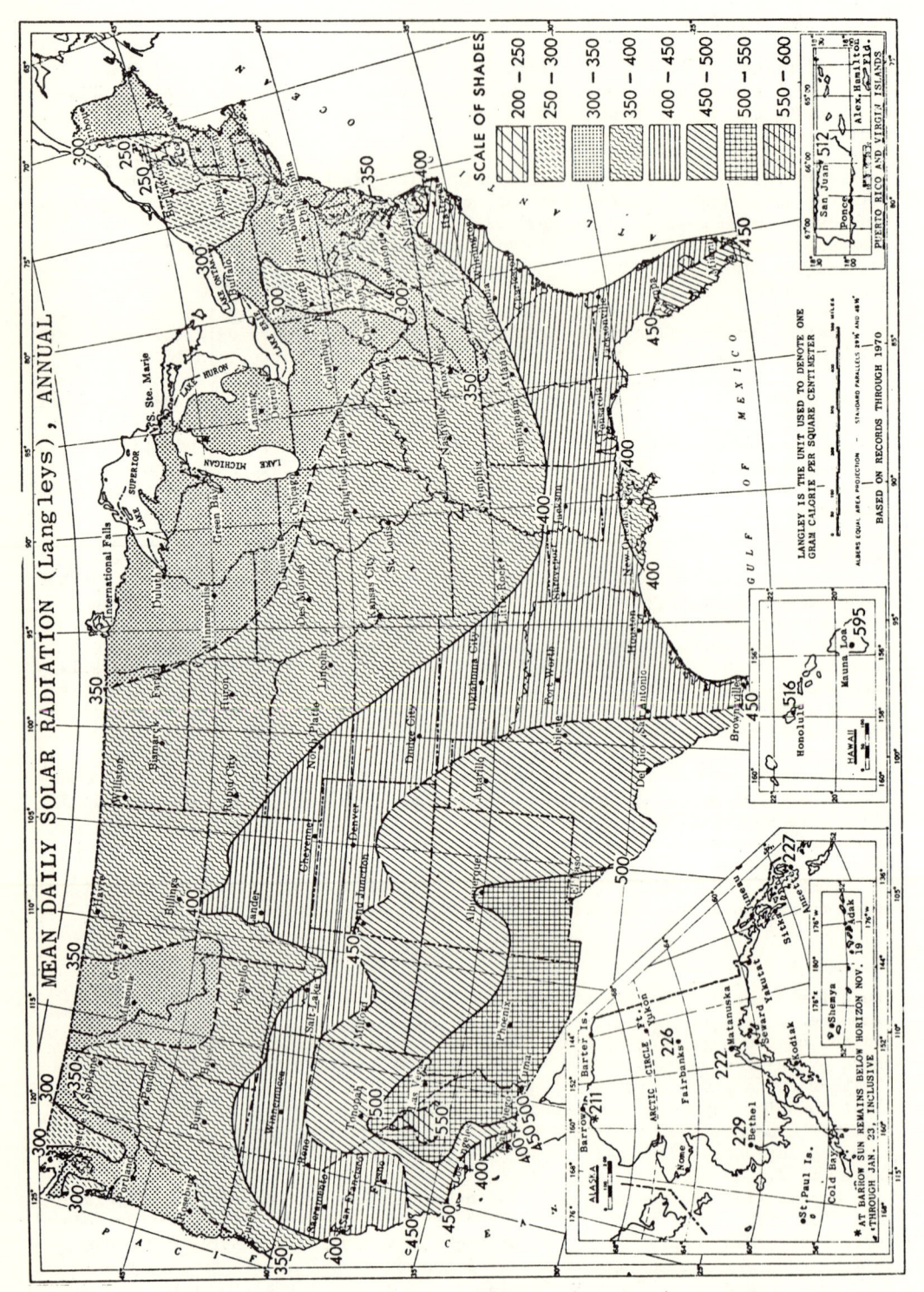

Fig. 5. - Mean Daily Solar Radiation (Langleys), Annual.

A-14

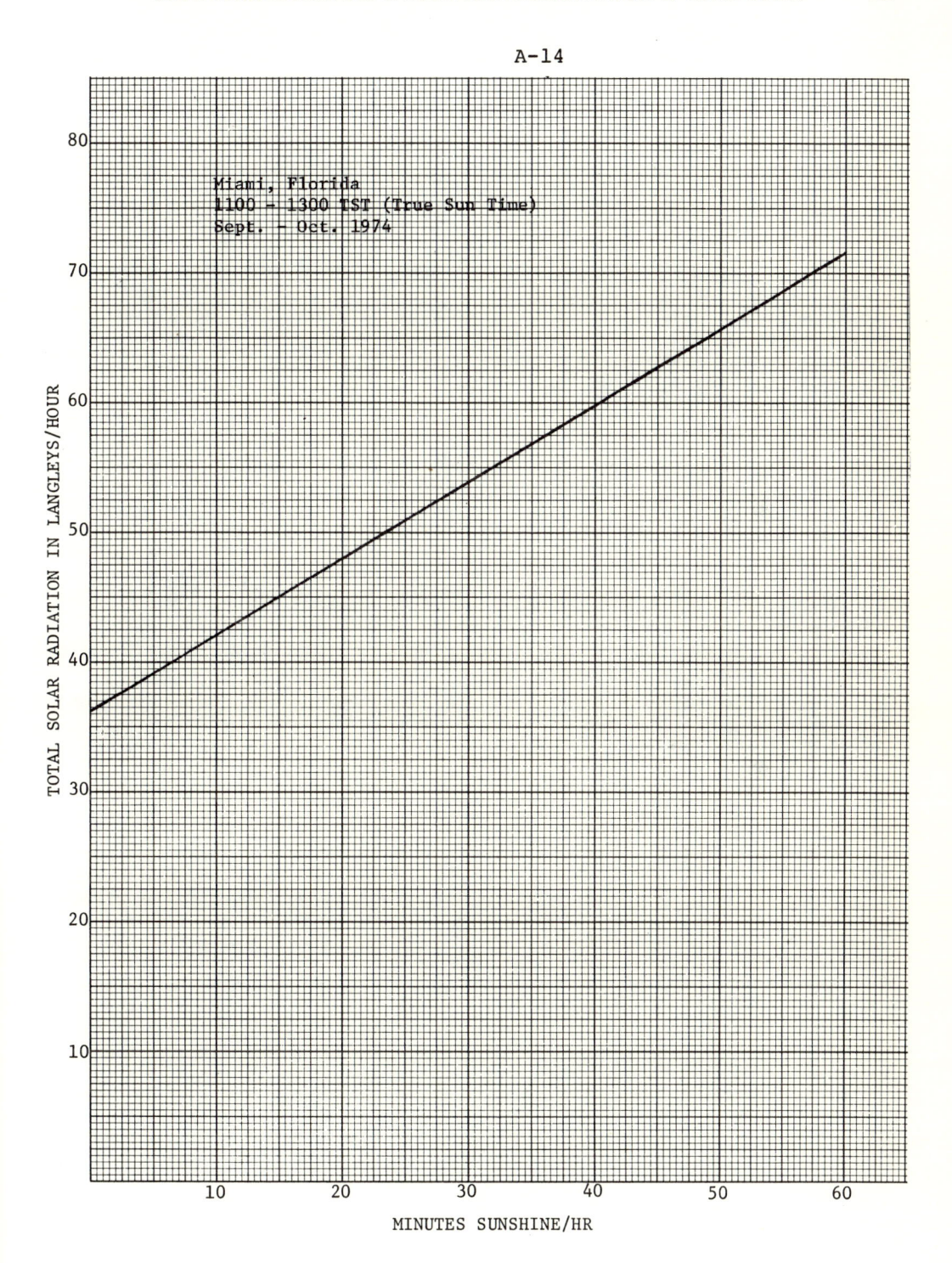

Miami, Florida
1100 - 1300 TST (True Sun Time)
Sept. - Oct. 1974

Fig. 6. - Approximate Relationship Between Solar Radiation and Sunshine for Mid-day Period at Miami, Florida.

cloud amounts and solar radiation, Norris proceeded to analyze daily total cloud amounts. From the 0900, 1200, and 1500 total cloud observations, a mean was taken and a number assigned to each day. This was compared with daily total radiation. A computer program was devised to determine the number of days that must enter the mean to obtain a correlation coefficient greater than 0.85 between cloud amounts and solar radiation. No number of days less than thirty gave this degree of correlation at Melbourne.

Norris recognized some of the problems in trying to use standard meteorological observations of clouds to estimate solar radiation received at the earth. One major problem is not knowing where the clouds are located with respect to the station. Local conditions may cause a partial sky cover to persist over the station or sunshine may predominate at the station with a high percentage of clouds persistent near the station. A geostationary satellite can do much better by monitoring the number of cloud-free hours (or hours of sunshine) in detail throughout an area. Thus it can map local geographic differences in insolation that may be highly significant.

The National Weather Service measures total solar radiation in Langleys at the Miami Airport using an Eppley Pyranometer. They also measure minutes of sunshine per hour at the National Hurricane Center on the University of Miami campus. Unfortunately, these stations are located approximately five miles apart so that accurate correlations of the two sets of data cannot be made.

An approximate relationship between these two parameters has been developed from the Miami data by plotting Langleys per hour versus minutes of sunshine per hour, Figure 6. Forty days of data for September and October 1974 were used in this study. Two mid-day hours of data 1100-1300 true sun time (TST) were used so that the sun angle was approximately the same for both hours. The regression line in Figure 6 gives a relation:

$$\text{Langleys/hr.} = 0.59 \text{ (minutes sunshine/hr.)} + 36 \qquad (2)$$

Similar analyses can be made for times two hours before and after noon, three hours before and after noon, etc., so that relationships can be obtained for different sun angles throughout the day. Also, correlations can be made for June and December, the maximum and minimum insolation months respectively in the northern hemisphere, so that interpolations can be made for other months. In order to obtain more reliable results, data from stations such as Lakeland and Tampa, Florida, where the sunshine and insolation measurements are co-located, will be used for these additional studies.

Sunshine is registered by the recorder when there is sufficient sunlight to cast a shadow during midday hours. Therefore, the total minutes or hours of sunshine recorded per day includes times when thin clouds such as cirrus or cirrostratus are present. Although thin clouds of these types usually will transmit sufficient light to register as periods of sunshine, they may slightly reduce the insolation received at the earth's surface.

The SMS-1/GOES geostationary satellites will monitor sunshine (cloud free areas) at 30-minute intervals with approximately 0.5-mile resolution in the visible wavelengths during daylight hours. The solid ellipses in Figure 1 show the extensive areas of coverage of the first two of these satellites. These satellite cloud observations can be applied to Figure 6 as follows. If all three observations at the beginning, middle, and end of the hour show no clouds in the grid square of interest, use 60 minutes sunshine. If one observation shows clouds and the other two do not, use 40 minutes sunshine. If two observations register clouds, use 20 minutes sunshine. If all three register clouds, use zero sunshine. A computer can scan the digitized satellite cloud reflectivity data and tabulate the results for any chosen geographic grid system. Very thin clouds which give low reflectivities to the satellite sensor can be programmed in the computer to register as sunshine.

Probably it is adequate to break the day into two-hour time blocks with one hour equally spaced before and the other after solar noon as in Figure 6. This method will provide two data bits per day in each chosen geographic grid for each of the 2-hour relationships between sunshine and insolation. Hourly solar radiation and sunshine recordings for June and December from several stations at different latitudes can be analyzed to obtain the necessary latitudinal, seasonal, and hourly sun angle relationships needed to process the satellite data. More than one and preferably several years of satellite data should be used to acquire optimum results from this study.

CONCLUSIONS

1. A high positive correlation exists between recorded sunshine and solar energy received at a given place.
2. Geostationary satellites can monitor sunshine (cloud free areas) over large expanses of continents and oceans with great detail, on the order of 0.5-mile resolution.
3. Relationships can be derived between solar radiation and sunshine for different latitudes, seasons, and times of day by use of some existing records at a few stations. Computer techniques can apply these relationships to the satellite observations in order to obtain detailed maps of land and water areas showing the best possible locations for solar power stations.

ACKNOWLEDGEMENTS

The authors wish to acknowledge the excellent assistance of Donald C. Gaby, Manager of the Satellite Field Services Station, NESS/NOAA, Miami and his staff in providing SMS/GOES satellite photographs. Eileen Kavlock assisted in preparing the final manuscript and illustrations.

REFERENCES

[1] U.S. Federal Power Commission, The 1970 National Power Survey, Part I, Government Printing Office, Washington, D.C., 1972.

[2] Faltermayer, E., "The Energy 'Joyride' is Over," Fortune, Vol. 86, No. 3, pp. 99, 1972.

[3] Hay, H.R., "Solar Energy, Solar Power, and Pollution," Solar Energy Conference, Paris, France, August 1973.

[4] Cherry, W.R. and Morse, F.H., "Conclusions and Recommendations of the United States Solar Energy Panel," Solar Energy Conference, Paris, France, August 1973.

[5] Associated Universities, Inc., Reference Energy Systems and Resource Data, AET-8, April 1972.

[6] NSF/NASA Solar Energy Panel Report, An Assessment of Solar Energy as a National Energy Resource, Mechanical Engineering Department, University of Maryland, December 1972.

[7] United Nations Educational, Scientific and Cultural Organization, Report of Working Party to Advise UNESCO in a Programme in Solar Energy, Unesco, Paris, June 1973.

[8] Hildebrandt, A.F., and Vant-Hul, L.L., "A Tower-Top Point Focus Solar Energy Collector," Proceedings of THEME Conference, University of Miami, March 1974.

[9] Oman, H. and Bishop, C.J., "Feasibility of Solar Power for Seattle, Washington," Solar Energy Conference, Paris, France, August 1973.

[10] Escher, W.J., "Model for an Ocean Based Solar Hydrogen-Energy Facility," Proceedings of Carnegie-Mellon University, pp. 96-125, June 1973.

[11] Climates of the United States, Environmental Data Service, NOAA, 1973.

[12] Reddy, S.J., "An Empirical Method for the Estimation of Total Solar Radiation," Solar Energy, Vol. 13, p. 289, 1971.

[13] Lund, I.A., "Relationships Between Insolation and Other Surface Weather Observations at Blue Hill, Massachusetts," Solar Energy, Vol. 12, pp. 95-106, 1968.

[14] Bennett, I., "Correlation of Daily Insolation with Daily Total Sky Cover, Opaque Sky Cover and Percentage of Possible Sunshine," Solar Energy, Vol. 12, pp. 391-393, 1969.

[15] Lumb, F.E., "The Influence of Cloud on Hourly Amounts of Total Solar Radiation at the Sea Surface," Royal Meteorological Society Quarterly Journal, Vol. 90, p. 383, 1964.

[16] Norris, D.J., "Correlation of Solar Radiation with Clouds," Solar Energy, Vol. 12, pp. 107-112, 1968.

PART IV

ENVIRONMENTAL QUALITY MONITORING

SATELLITE DETECTION OF AIR POLLUTANTS

WALTER A. LYONS

University of Wisconsin-Milwaukee
Milwaukee, Wisconsin 53201, U.S.A.

ABSTRACT

NASA's ERTS-1* satellite, with its high resolution and multispectral capabilities, has been found useful in the detection and analysis of smoke from large point sources (power plants, steel mills, etc.), and widespread atmospheric stagnations. Smoke plumes from the Chicago-Northern Indiana industrial complex have been tracked over Lake Michigan for more than 100 km. This has important implications with regard to air pollution control strategems. If indeed there is significant interregional pollution transport, our concepts of Air Quality Control Regions may have to be revised.

Experience has shown that smoke plumes are relatively easy to detect in ERTS imagery over water (in the 0.6-0.7 micrometer band) but much more difficult over land surfaces. Pattern recognition techniques (cluster analysis) were applied to the digital ERTS data and it was found that the smoke plumes indeed had a unique spectral signature. Studies are currently underway examining the use of measured plume geometries to obtain quantitative estimates of diffusion over water surfaces.

ERTS data have also shown great utility in the detection of inadvertant weather modification, in particular, the formation of convective clouds within steel mill plumes. Contrails are also easily identifiable.

INTRODUCTION

Since the launch of the first meteorological satellite, TIROS I, in 1960, the meteorological community has been increasingly involved in the field loosely referred to as "Remote Sensing." The earliest vidicon images of the earth's cloud systems yielded signatures of the structures and patterns of atmospheric motions. As image resolution gradually increased to its present 0.3 mile maximum for operational meteorological satellites, more and more detail about the all-important mesoscale atmospheric processes could be discerned.

Even with the increased image resolution, meteorologists concerned with air quality control and monitoring of atmospheric pollutants took relatively little note of the potential of satellite monitoring. Various sensors and platforms have been employed for sensing pollution (see Colvocoresses [1]). These have included free floating balloons, tethered systems, kites and aircraft. It took the successful launching of NASA's first Earth Resources Technology Satellite (ERTS-1) with its vastly increased resolution and multispectral sensors before the meteorological community developed its current interest in the use of spacecraft as reliable pollution monitors.

---
*Now designated Landsat-1

The above is not to say that spacecraft imagery has not been applied to air quality problems, however. Numerous 70 mm photographs of smoke plumes were returned from the manned Gemini and Apollo earth orbital missions (Fig. 1). These clearly demonstrated that smoke plume detection from these altitudes was feasible given adequate sensor resolution. The Nimbus IV satellite made a number of relevant observations. A massive turbidity "cloud" associated with Sahara dust storms was observed drifting westward over the tropical Atlantic for a five-day period in April, 1970. An ash plume over 300 km long was found emanating from the erupting Beerenberg Volcano on Jan Mayen Island (September, 1970). Massive clouds of smoke associated with southern California brush fires and agricultural burning on Yucatan Peninsula have been noted by several satellites.

McClellan [2] was among the first researchers to attempt the use of digital data from the Applications Technology Satellite (ATS-III) to study urban scale atmospheric pollution. Analysis of radiance values over the Los Angeles basin, after comparison with other target areas, suggested that some correlation did exist between radiance values and atmospheric aerosol loading (as indicated by visibility reports). The comparatively low resolution (2.8 km) and difficulties in geometric navigation have frustrated attempts at using these data for anything but the most qualitative use.

While imagery obtained in the visible portion of the spectrum (roughly 0.4 to 0.7 micrometers) from meteorological satellites has occasionally detected particulate air pollution, there have been only meager returns in terms of the detection of gaseous pollutants. The Nimbus IRIS (Infrared Interferometer Spectrometer Subsystem) is typical of a developing family of sensors that have the capacity of detecting either total gas content or profiles. Using the IRIS, maps of ozone content have been made on a global basis. These have proven valuable in studying synoptic scale distribution of total ozone (the vast percentage of which resides within the stratosphere). Such systems have not yet been routinely applied to the mapping of urban scale boundary layer pollution concentration, but a growing family of sensors suggests that this may be possible in the future. Limb scanning techniques are being developed which can detect such gases as H_2O, CO, CO_2, N_2O, NO_2, etc., especially in the stratosphere [3]. Lawrence and Ward [4] detail as many as nine sensing systems that are in various stages of development and testing for possible use in spacecraft as monitors or urban scale pollution levels.

At this writing, the only reliable system for making even semiroutine observations of air quality parameters is the ERTS-1 satellite. This imagery has yielded valuable information on atmospheric aerosol patterns. These will serve as the basis of this paper.

THE EARTH RESOURCES TECHNOLOGY SATELLITE (ERTS-1)

ERTS was designed specifically for environmental monitoring. It was placed in a nearly circular, 99.11 degree orbit, nominally about 915 km with a period of 103.267 minutes. The sun-synchronous orbit has a descending node time of 0942 LST. Images are 185 by 185 km on a side. It takes 251 revolutions (18 days) to make one complete global coverage. Thus, every portion of earth (between 81° north and south latitude) is viewed at least once every 18 days. At the latitude of Chicago, there is approximately 35% horizontal image sidelap, so some locations can be seen on successive days. A one-year system lifetime was contemplated, but as of November 1974, more than 60,000 images had been collected. All images are characterized by a zero or near-zero zenith angle with illumination depending on the solar elevation angle (a function of date and latitude of observation). A second system, Landsat-2, was launched in January, 1975, with the original Landsat-2 now being used on a limited basis.

Fig. 1. - 70mm photograph taken during Gemini manned orbital mission showing smoke plumes along a shoreline.

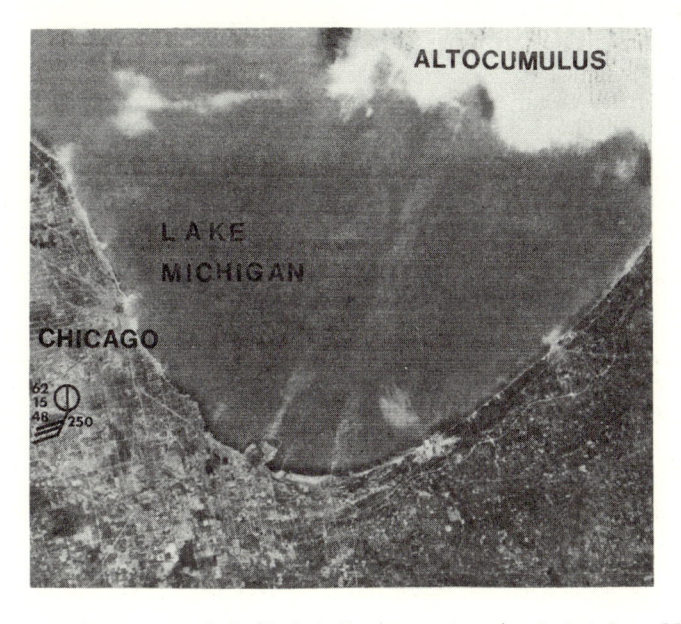

Fig. 2. - ERTS-1 image, band 5 (0.6-0.7 micrometers), 1 October 1972, showing smoke plumes from steel mills drifting over Lake Michigan.

The two great advantages of ERTS are its extremely high resolution and a multispectral imaging capability. The original specifications for the multispectral scanner (MSS) called for 100-200 m resolution. Initial results have shown some high contrast targets as small as 50 m. Highways, airport runways, small ponds, jet contrails, harbor breakwaters, etc., are routinely visible. The four spectral bands are (1) MSS-4, 0.5-0.6 μm ("green" band); (2) MSS-5, 0.6-0.7 μm ("red" band); (3) MSS-6, 0.7-0.8 μm; and (4) MSS-7, 0.8-1.1 μm ("near infrared" band). ERTS products are available in several formats including digital tapes, 9½ by 9½ inch black and white and color prints and transparencies, and most routinely, 70mm negative and positive transparencies [5].

Because the four spectral bands are viewed simultaneously in space and time, it is possible to use color-additive viewing techniques to produce color-coded results. Combining MSS bands 4, 5, and 7 results in a false-color infrared image. The red color associated with foliated vegetation makes this color analysis a valuable diagnostic tool in agriculture, forestry, and land-use studies, but for the meteorologist the ERTS multispectral imaging techniques are most useful in penetrating thick haze, revealing cloud shadows, delineating snow cover from vegetation, and demarcating land/water boundaries.

The author has followed the lead of Pease and Bowden [6] in using Kodak 35mm Ektachrome Color Infrared [CIR] film for environmental monitoring. The greatest success has been found using a 160 ASA, and 80 B, Wratten 12, and polarizing filters. Numerous CIR pictures of ground targets and smoke plumes have been made from the University of Wisconsin-Milwaukee (UWM) instrumented aircraft coincident with ERTS overflights.

The various design characteristics of ERTS prompted the UWM Air Pollution Analysis Laboratory to submit a proposal to study ERTS-1 data. It was hoped that a satellite with these characteristics would be capable of detecting major plumes of suspended particulates, making possible synoptic studies of interregional pollution transport over the southern Lake Michigan basin.

DETECTION OF SMOKE PLUMES BY ERTS

The heart of the Chicago-Gary industrial complex stretches from the southeastern part of the city of Chicago eastwards along the shoreline of Lake and Porter Counties, Indiana, to east of Gary. Figure 4 is an aerial view of a part of this region taken from an NCAR Queen Air at 6000 ft AGL on the morning of 15 July 1968 when brisk southwesterly flow was advecting numerous smoke plumes over the lake.

A total of 16 major particulate sources have been located in the study area. The size of some of these sources is truly remarkable. Source 3, a cement plant, discharges over 140,000 tons/year of particulates. In comparison, all of Milwaukee County, Wisconsin, a relatively industrialized area, had a 1970 suspended particulate emission of only approximately 45,000 tons/year.

During the morning of 1 October 1972, brisk southwest surface flow covered the southern Lake Michigan basin area (Fig. 3). A bank of altocumulus clouds was present to the north and east of the Chicago region and was moving rapidly northeastwards. A strong nocturnal radiation inversion had been present at 0600 CST according to the Peoria, Illinois (PIA) sounding (inset, Fig. 3). Figure 2 is a portion of the ERTS images taken at approximately 1003 LST. A number of particulate plumes can be seen streaming northeastward, disappearing beneath the altocumulus cloud deck at a distance of about 60 km.

It should be noted that due to the various degradations involved both in making the photographic print and in publication, the plumes do not appear as clearly as they do in the original 70 mm negatives. The prints were also photographically dodged to maximize details over both land and water.

Fig. 3. - (inset). Synoptic conditions, 1000 LST, 1 October 1972. One wind barb equals 5 kts. Peoria, Illinois, radiosonde taken 0600 LST shown.

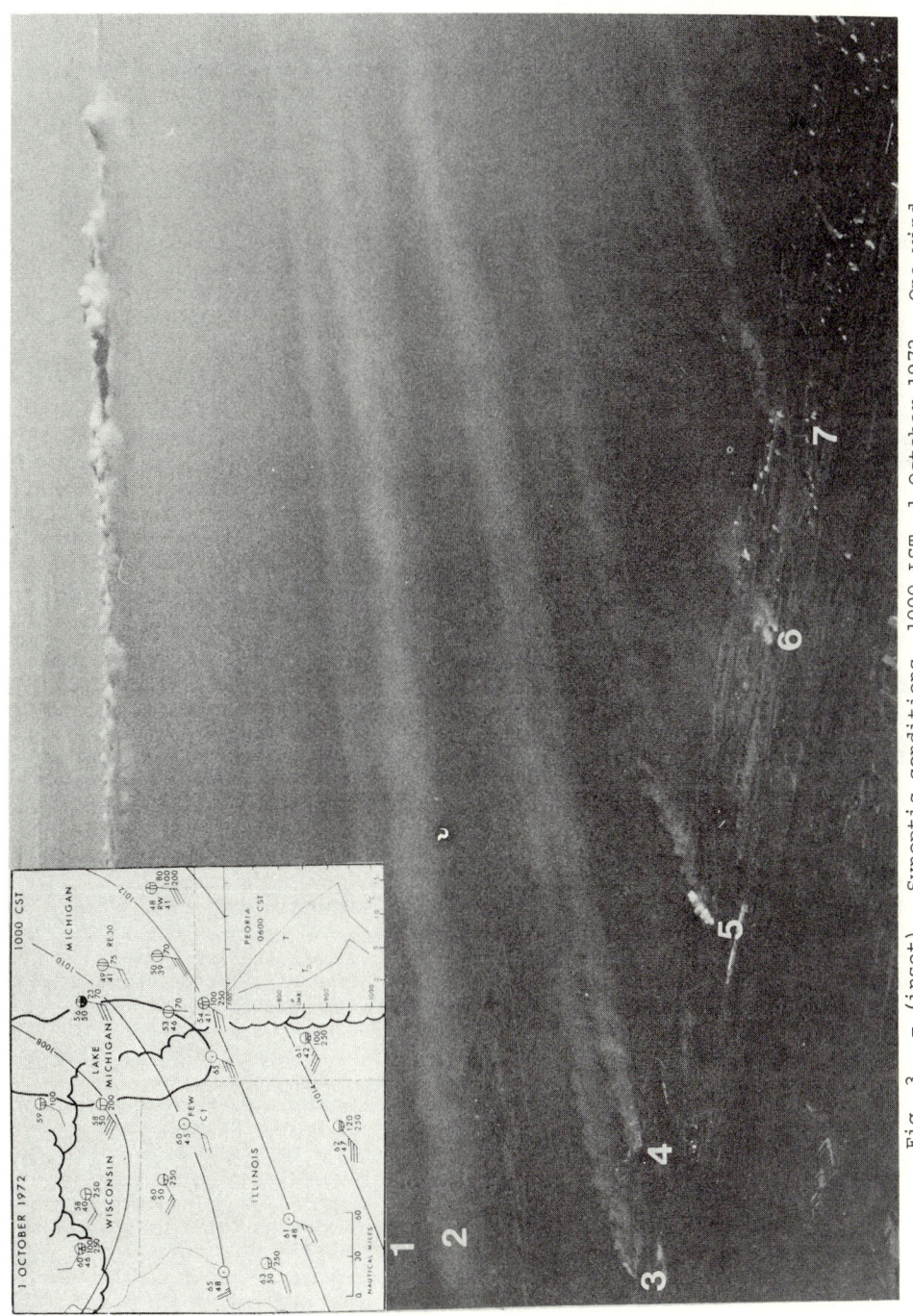

Fig. 4. - Photograph, looking north, of a portion of the Chicago-Gary industrial complex, 0700 LST, 15 June 1968, from NCAR Queen Air, at 6000 ft MSL.

The plumes visible in this image, for the most part, are not from single stacks, but from entire steel mill complexes. The total spread of the visible plume at 60 km downwind is about 4.5 km. Measurements of plume spread can be converted into useful information regarding overwater mesoscale diffusion of pollutants. The visible plume may be correlated to a parameter such as the point where concentration drop off to 10% of the centerline value, or 2.15 σ_y, in the parlance of Pasquill and Gifford. If that were the case, then this plume would appear to have a diffusion rate characteristic of Class E (rather stable atmosphere) in the empirical classification of atmospheric stabilities used in Gaussian plume diffusion calculations. Work is currently underway to subject transparencies of these and other plume images to microdensitometric analysis. This would yield profiles of optical density (presumably related to particulate loading) as a function of distance, from which estimates of diffusion rates could be made. These are valuable data sets inasmuch as little data for the overwater spread of plumes from large point sources (such as power plants) is available. Brookhaven National Laboratory has for the past several years attempted to make actual measurements over the waters around Long Island. The formidable operational difficulties have made data collection expensive and tedious. Use of ERTS data could be a valuable adjunct to these programs.

An important consideration is the choice of the appropriate ERTS spectral band to provide optimum discrimination of a particulate plume against the underlying surface, in this case water. The plume will be most visible on the photograph when the difference between the optical density of the plume image and the optical density of the lake-surface image is greatest. Plume visibility over a water surface will thus be enhanced by: (1) decrease in the spectral albedo of the lake surface, (2) increase in overall image contrast, and (3) increase in the amount of radiation scattered and/or reflected vertically upwards from the solar beam by the plume. In the case of a plume advecting over a surface of very high spectral albedo, a fourth factor would have to be considered: the extent to which a plume attenuates solar radiation reflected vertically upwards from the surface, a factor dependent on the geometry of scattering and absorption of radiation by the plume.

The first of these major considerations, the spectral albedo of the lake surface, shows a marked variation by wavelength. Direct reflection of solar radiation from the lake surface makes a relatively minor contribution to the variation of lake spectral albedo in the four ERTS bands. Reflection vertically upwards from the lake surface is small (approximately 2%) and for a solar elevation angle of 40° shows only slight wavelength dependence with very slightly higher values in the lower wavelengths of the visual spectrum.

Far more important in influencing the spectral albedo of the lake is the wavelength dependence of solar radiation absorption within the lake itself. The lower the spectral absorption coefficient within the lake for any wavelength, the greater is the penetration of incident solar radiation into the lake and the greater is the likelihood of backscattering upward by water molecules and hydrosols, i.e., Rayleigh and Mie scattering, respectively. For distilled water, maximum transmissivity occurs at 0.46 μm, and absorption increases rapidly with increasing wavelength, resulting in a darker lake image. As an example, the mean absorption coefficient for pure water in ERTS band 4 is approximately six times less than the coefficient for band 5, and the lake thus should appear brighter. Increased amounts of suspended and dissolved matter shift the wavelength of minimum absorption and maximum lake albedo to longer wavelengths. Sea water (and lake water?) typically has minimum absorption near 0.55 μm, which is the center of ERTS band 4.

An additional consequence of the greater transparency of water in the blue and green portion of the spectrum occurs in the case of shallow water, where reflection from the lake bottom may further increase values of surface lake albedo and limit the ability to detect pollution plumes near shore.

The net result is that the lake generally appears brightest in band 4 and nearly jet black in band 7. An exposition of the spectral dependence of backscattering from suspended particulates becomes even more complex. A complete discussion of the topic is presented by Lyons and Pease [7]. Various theoretical arguments suggest that the highest inherent brightness of smoke plumes would probably be in the "red" portion of the visible spectrum.

An inspection of the four spectral bands for the image shown in Fig. 2 reveal that in band 4 the smoke plumes were virtually invisible due to the extreme "brightness" of the lake which eliminated almost all inherent contrast. While the plumes were seen over water in bands 6 and 7, the best plume contrast over moderately clear and deep lake water was found in band 5.

Figure 5a shows a large coal-burning power plant located on the southern shore of Lake Erie near Loraine, Ohio. Most of the particulate matter emanates from a single stack. In this view, the plume is drifting northeast over the lake (with patches of fog forming over the cold lake surface). This picture was taken from about 32,000 ft using 35mm CIR film. An ERTS-1 band 5 image of the same plant taken on 4 September 1973 shows the plume remarkably well. This ERTS image was specially processed by projecting the 70mm positive transparency onto a Kodalith negative from which the print was made. This greatly increases the contrast, and though there is a severe loss of ground target resolution, the smoke plume is greatly enhanced. A pall of smoke from sources in Cleveland is also visible, as is the considerable turbidity in the lake water. The power plant plume is visible in the original for over 15 km downwind.

The examples discussed so far have all been for plumes from major point sources traveling over water. It has been found that it is far more difficult to detect plumes over land. Copeland, et al. [8] have had some success, but these were for unusually dense smoke plumes. The generally higher albedo of the land surface generally reduces inherent contrast between the smoke and the underlying surface. The mottled and uneven nature of the terrain further complicates the matter. Smoke from brush and forest fires has been detected occasionally, sometimes made easier by the large areas of blackened earth left behind according to Wightman [9].

To illustrate the contrast between over water and over land plume detection, refer to Fig. 6. In Fig. 6(a), we see the same Gary, Indiana, industrial complex, this time taken from 33,000 ft using 35mm CIR film at 0950 LST, 7 March 1974. Extremely stable northeasterly flow off cold Lake Michigan had trapped these elevated plumes and they were traveling on inland without undergoing significant mixing. Fig. 6(b) is an enhanced ERTS-1 band 5 image taken at virtually the same time. Several plumes are visible. The short, very white plumes labeled 1 are clouds of condensate from large cooling towers which are evaporating only a short distance from their sources. The smoke plume labeled 2 is from the steel mill complex in downtown Gary. It can only be tracked for about 10 km over land before becoming invisible. Thus, while smoke and condensate plumes are visible over land, they must be unusually large and dense. In any case, their detectability is far less than an equivalent plume over a water surface.

MACHINE PROCESSING OF ERTS DATA

The mere visual inspection of ERTS images can be most helpful in gaining insight into the behavior of air pollutants, but even more advanced methods are being developed to treat such data. The Robotics and Artificial Intelligence Laboratory (RAIL) of the University of Wisconsin-Milwaukee has established both hardware and software capability for automatic image interpretation using ERTS digital tapes. Pattern recognition is a powerful statistical approach to classification of image features, especially those obtained by multispectral

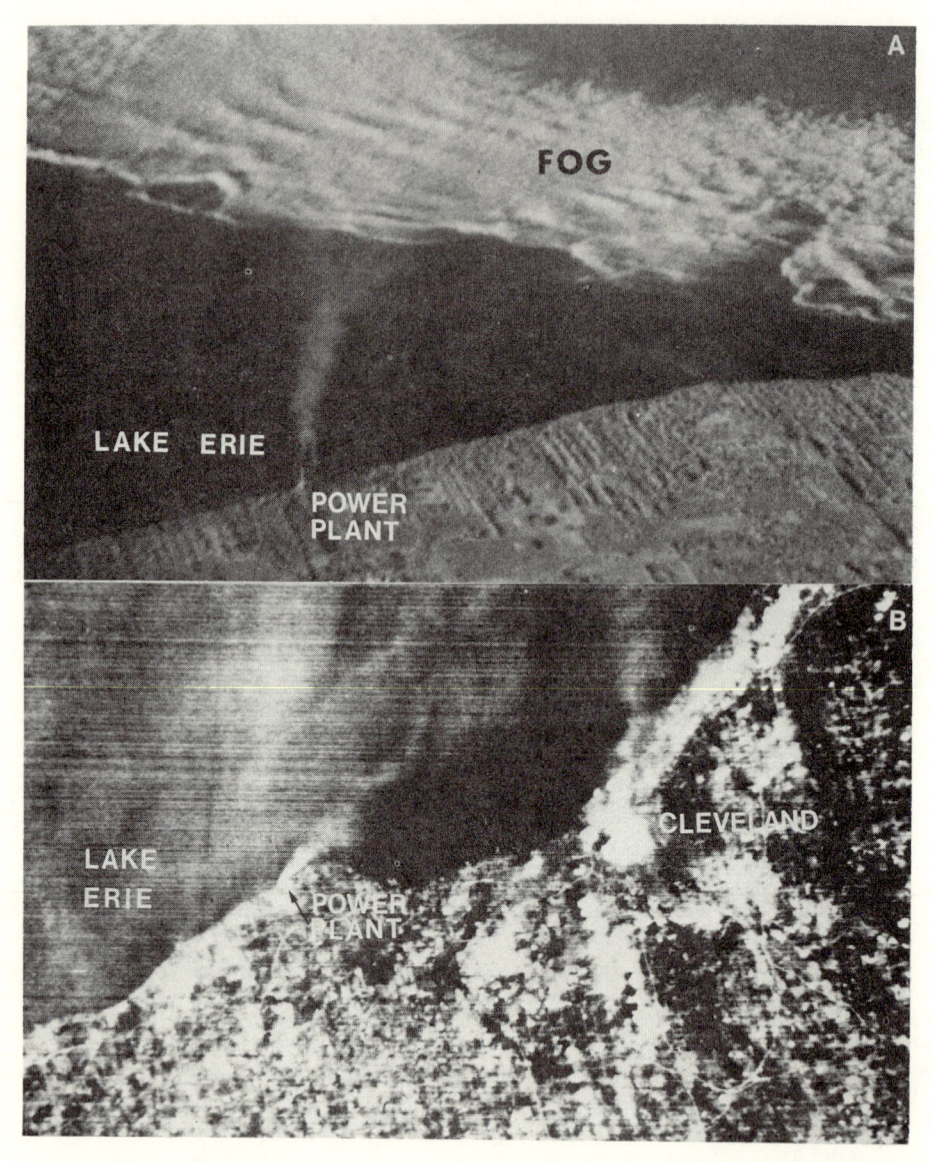

Fig. 5(a)- Large fossil fuel power plant near Loraine, Ohio, on southern shore of Lake Erie with smoke plume drifting northeastward over fog shrouded lake.
  (b)- ERTS-1 image band 5, showing plume from same plant advecting over the lake on 4 September 1973.

Fig. 6(a) - View looking east, at 33,000 ft of Gary, Indiana plumes, using CIR
film, 0950 LST, 7 March 1974. Plume 1 is condensation from large
cooling tower; plume 2 is smoke plume from steel mill.
  (b) - ERTS-1 image, band 5, taken at 1003 LST, 7 March 1974 of same area.

sensors. Identification of crop types using ERTS digital data processed entirely by computer often exceeds 80% accuracy.

As a test of the RAIL system, a small portion of the image shown in Fig. 2 was processed using cluster analysis techniques developed by Eigen and Northouse [10]. A section of the image which showed an industrial landfill peninsula (with several steel mills) and two major smoke plumes was chosen. All four ERTS bands were available in digital tape format. Processing was accomplished on a MODCOMP II/25 computer with line printer output. After several runs experimenting with the best clusters, a final version was selected (Fig. 7). Here all land features (roads, corn fields, etc.) were suppressed, accounting for the blank spaces. Water with its distinct spectral characteristics was easily isolated from other features, and was printed everywhere as ----. The smoke from the sources (steel mills and a cement plant to the east) also were found to have rather homogeneous spectral signatures. These are indicated in Fig. 7 by a ⊕ and are clearly delineated. This technique has also been used to automatically distinguish clean from "polluted" water in and around Milwaukee harbor.

Computer image processing has other capabilities besides simply making an automatic classification of the nature of the scenes in each pixel of an ERTS image (60x80m). For instance, since radiance values are the raw material of the tape data, digital density slicing can be accomplished easily, and with color CRT output, graphic displays can be obtained. The next step is using the system as a digital microdensitometer combined with pattern recognition algorithms to automatically detect "smoke" and make estimates of total particulate loading in the atmosphere below the satellite. For this, "ground truth" is necessary, and a Cessna 182 aircraft has been instrumented for this purpose. It has both total aerosol mass monitors and 2 channel particle counters (recorded in digital format on magnetic tape). It is hoped to obtain a complete profile of the Gary plumes simultaneously with an ERTS overflight so that image radiance values could be calibrated and yield some measure of total aerosol mass loading.

INTERREGIONAL POLLUTION TRANSPORT

The Federal Clean Air Act of 1967 (amended 1970) divided the United States into Air Quality Control Regions (AQCR). The boundaries of these regions ideally were chosen to delineate independent, self-contained "air sheds" where air quality levels were determined by emissions from sources within that region. The actual boundaries selected were often compromises between meteorological and political realities. An AQCR's final compliance with national Air Quality Standards is ultimately dependent upon the successful formulation and implementation by regional authorities of the necessary emission control strategy for sources within that region. In reality, however, two adjacent AQCR's might occasionally find themselves breathing each other's effluents due to long-range interregional transport of pollutants. A case in point is the Southeast Wisconsin AQCR (the seven counties of southeastern Wisconsin, including metropolitan Milwaukee) and the Chicago Interstate AQCR (northeast Illinois and the two northwesternmost counties of Indiana). In the latter, there exist the numerous extremely large point sources mentioned above. The evidence has gradually been accumulating that these sources are indeed more than just local problems.

In August 1967, the author participated in the flight of an aircraft from the National Center for Atmospheric Research instrumented to monitor ice nuclei. The object of the study was the dispersion of anthropogenic ice nuclei emanating from the Chicago-Gary steel mill complex. Under conditions of brisk southwesterly flow over southern Lake Michigan, with a rather strong synoptic-scale subsidence inversion around 1300 m, a clearly defined plume of ice nuclei was easily tracked to the vicinity of Battle Creek, Michigan (some

Fig. 7. - Computer processed ERTS-1 digital data, small segment of frame shown in Fig. 2. Water of lake is indicated by (-), all land signatures are suppressed (and therefore blank), and smoke is denoted by (⊛). Note man-made peninsula and breakwater protruding into lake.

180 km downwind). On several summer days with south-southeasterly flow over the lake, elevated layers of red iron-oxide smoke have been seen drifting past Milwaukee [7]. Again the most likely origin of this smoke was the mills at the southern end of the lake -- almost 165 km away.

The effect of the much larger Chicago Interstate AQCR upon Milwaukee is probably significant, but most difficult to ascertain qualitatively. It has been estimated that approximately 10% of the suspended particulates measured in the Milwaukee area originate in and around Chicago. Frequently in Milwaukee, several hours after the surface winds shift to the south or southeast, there is a rapid increase in haze and smoke, presumably the influence of interregional transport from the Chicago area. Certainly in any megalopolis, such as found along the East and West coasts, adjacent AQCR's could indeed be exchanging pollutants on a regular basis.

In the emerging Great Lakes megalopolis, which shows signs of extending from Green Bay, Wisconsin, to Buffalo, New York, in the not-too-distant future, the peculiar meteorological effects of the Great Lakes often exacerbate this interregional transport. When continental air masses advect across the relatively warm lakes in winter, any plume moving over a Great Lake will be rapidly dispersed. Turbulence generated by the convection rising from the surface can be extreme, sometimes to the point of generating a myriad of miniature waterspouts or "steam devils." Thus, if plumes from northern Indiana are to pass over Lake Michigan into southwestern Michigan or Wisconsin, they probably arrive diluted to a great degree. From early spring through late summer quite the opposite situation prevails; air temperatures frequently exceed those of the lake by 10, 20, and sometimes 30°C. Extremely intense, though shallow (100 to 200 m), surface conduction inversions form over the lake (Lyons [11]). Air streams advecting over cold lakes not only are rapidly cooled in their lowest layers but, due to the absence of upward convective heat transport, do not warm and destabilize in the overlying layers as occurs over land during the day. The almost total lack of cumulus clouds over the Great Lakes on summer afternoons is one manifestation of this. A plume from a large elevated point source such as a steel mill or power plant may travel for long distances over water with relatively little dilution and arrive above a downwind shoreline in still very high concentrations. If such a plume arrives on a downwind shore during mid-day, and solar radiation is sufficiently intense, the plume is then rapidly mixed to the surface by the turbulence forming over the heated ground in a process called fumigation. This causes high pollution levels several kilometers inland, the origin of which could be quite baffling to local control officials (Lyons, Keen, and Northouse [12]).

Figure 2 clearly suggests that plumes can travel for great distances over water during the "warm season" with relatively little dilution (in this case some 60 km). Figure 7 is an even more convincing example, an ERTS-1 band 5 image acquired at 1003 LST, 14 October 1973. On this day, skies were clear, the overlying air mass was warmer than the lake water, and brisk southwesterly flow advected the Gary plumes into southwestern lower Michigan, almost 100 km away. The almost total lack of penetrative convection over the cold, smooth lake surface allowed the plumes to arrive near Benton Harbor, Michigan in still high concentrations. It is almost a certainty that these plumes were fumigated to the surface beginning about five kilometers inland. This effect, which could have continued for most of the day, would undoubtedly have caused considerable consternation to any local control officials who might have been monitoring in that region.

ERTS has also been useful in studying the interregional transport of photochemical pollutants from the Chicago-Gary complex northwards into southeastern Wisconsin. These invisible gases cannot be detected by the ERTS sensors. The extremely high resolution images have shown, however, the details of cloud structures associated with lake breeze fronts. This information in turn assists in the formulation of models of the complex local wind fields in the area which apparently results in the advection of massive amounts of ozone and other materials along the western Lake Michigan shoreline (Lyons and Cole [13]).

The problem that is raised by these discoveries is a legal and political, as well as meteorological one. Simply stated, if one AQCR fails to meet the prescribed Federal Air Quality Standards, will it be penalized due to the "sins of emission" of an adjacent one? Should one AQCR be forced to have stricter clean-up regulations simply because another up-wind has numerous sources?

It should be reiterated that sensors of the type employed by the ERTS system cannot be used to detect gaseous pollutants. However, some indirect information can be gathered. By using color infrared's capability to detect changes in vegetation characteristics, severe forest damage in Ontario has been used by Murth [14] as a bioindicator of sulfur dioxide dispersion from major point sources (smelters).

SYNOPTIC SCALE POLLUTION EPISODES

Up until this point we have been discussing the use of ERTS data in the detection and measurement of pollution on the mesoscale (less than 100 km); however, it is possible for entire synoptic scale air masses to become polluted. This generally occurs when a high pressure system stagnates over a several state region for a period of several days to over a week. The combination of light winds and reduced mixing depths due to the overlying subsidence inversion allows suspended particulates (and other pollutants) from thousands of sources to commingle and steadily increases the pollution levels over a wide area. Such a large scale weather phenomena usually results in the issuance of an Air Stagnation Advisory by the National Weather Service. Conditions such as these resulted in the 1966 New York City Thanksgiving Day episode when over 400 excess deaths were recorded due to the build-up of SO_2 and particulates in the atmosphere over a wide region.

During early September 1973, a late season heat wave associated with stagnant anticyclonic conditions over the eastern U. S. prompted the issuances of numerous Air Stagnation Advisories. ERTS-1 provided a graphic illustration of the extent to which entire air masses may become polluted. Figure 10(a) shows a typical view of eastern Lake Ontario during a time when relatively unpolluted cP air covered the region. Figure 10(b) is the identical region, on a day with only a few clouds scattered through the area. On 1 September 1973, the air mass was so filled with particulate matter that image contrast was reduced to the point where the lakeshore was barely visible. Both views were in band 4 (0.5 to 0.6 μm) which accentuates the scattering effects of both molecules and larger aerosols. If problems associated with varying turbidity levels of water can be solved, comparison of contrast and total radiance values for selected water (and perhaps ground) targets from image to image shows promise in making satellite estimates of total suspended aerosol burden on a large scale (Griggs [15]).

One of the sources for the aerosols seen by ERTS on 1 September 1974 was most likely the complex of coal-fired power plants which have proliferated in Ohio and West Virginia near the strip mines which provide their fuel (Fig. 9). The generally southwest winds on these days caused these plumes (along with others from Pittsburgh, etc.) to merge into one massive smoke dome which gradually covered New York State, Lake Ontario, and southern Ontario. This last fact illustrates that significant pollution exchange also occurs across international boundaries.

Fig. 8. — ERTS-1, band 5, image of Chicago, Gary and southern Lake Michigan, 1003 LST, 14 October 1973. Smoke plumes advect across relatively cold lake showing little diffusion, only to fumigate on downwind shoreline near Benton Harbor, Michigan.

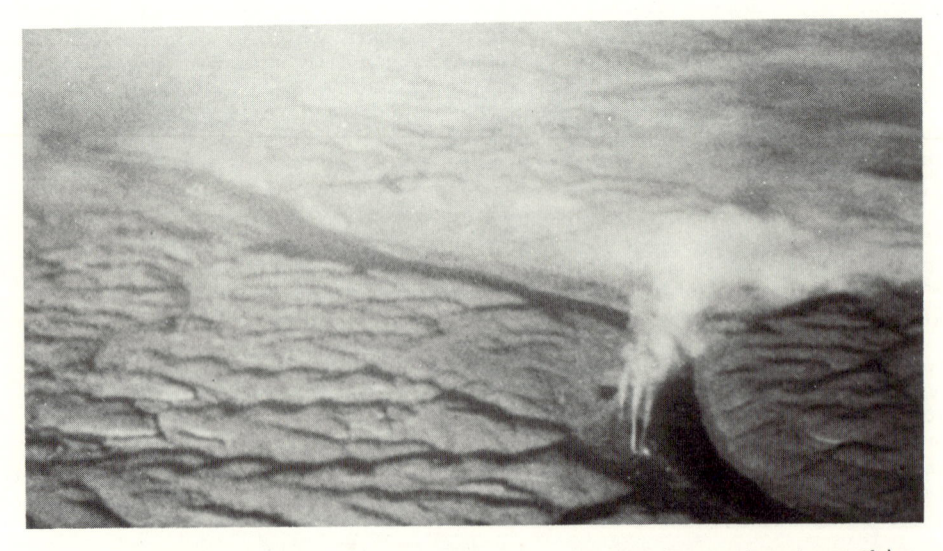

Fig. 9. — Just one of many coal-burning power plants in southeastern Ohio and West Virginia which contribute to the general turbidity of the atmosphere during frequent air stagnations in the Appalachian region.

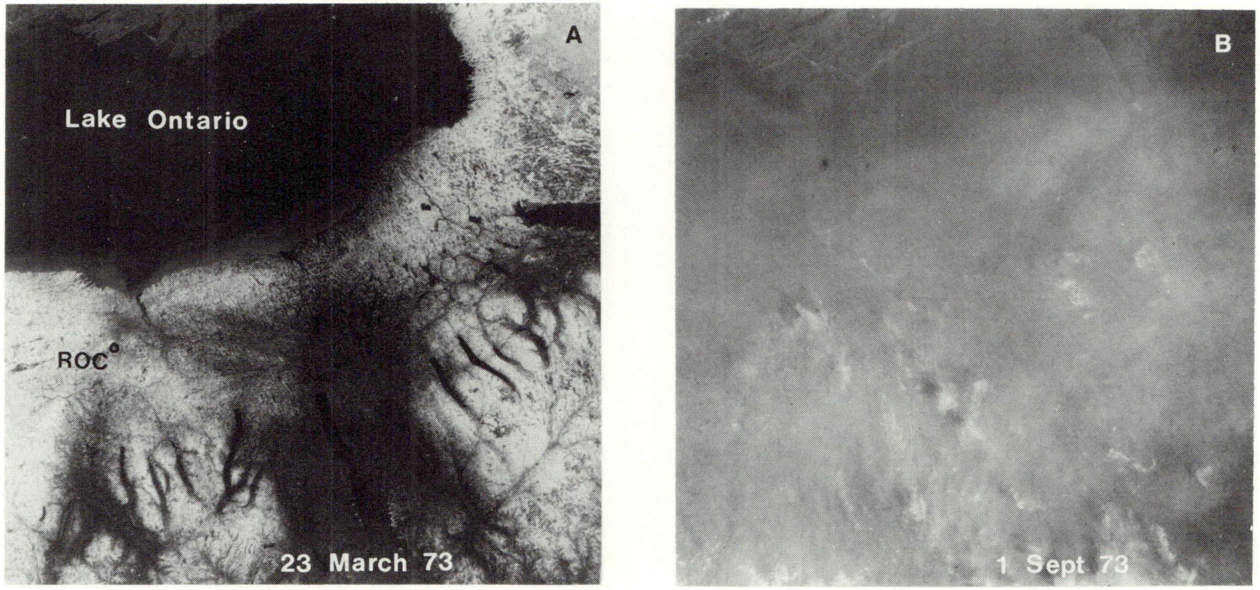

Fig.10(a) – ERTS-1 image, band 4, 0945 LST, 23 March 1973, of eastern Lake Ontario and the Finger Lakes, on a cloud free day with low atmospheric turbidity. Snow cover is spotty through the region. Rochester, N.Y. indicated by ROC.
(b) – Identical geographical area, band 4, 0945 LST, 1 September 1974, on a cloud free day but with a highly polluted atmosphere associated with a synoptic scale air stagnation episode.

OBSERVATIONS OF INADVERTENT WEATHER MODIFICATION

Man's inadvertent modification on the earth's climate and weather has recently become a much studied topic in the meteorological community. As summarized by the 1970 Presidential Council on Environmental Quality, there exists a wide spectrum of potential mechanisms. Proper assessment of the many theories, however, must await the collection of a considerable body of appropriate climatological data and the development of reliable numerical models of global climate. More amenable to immediate study are local climatic variations and the specific weather alterations from which they arise. In the area of man-made influences on cloud and precipitation processes, however, speculation seems still to outstrip proven fact. The La Porte, Indiana, precipitation anomaly is a case in point. The existence of the La Porte anomaly as a physical reality, rather than an example of spurious data, has not at all been universally accepted. Some doubt even the existence of the La Porte anomaly. In any case it is still an effect searching for a cause. The same can be said for apparent precipitation increases downwind of numerous large urban areas, estimated as high as 27% of the summer season rainfall. The list of possible causes include: (1) anthropogenic sources of ice and condensation nuclei from industrial and perhaps automotive sources, (2) inputs of sensible heat from urban sources, (3) inputs of moisture from certain industrial activities, and (4) changes in boundary layer wind flows due to variations in surface thermal and roughness characteristics. Project METROMEX[1] is currently trying to assess the magnitude and isolate the cause(s) of urban-induced weather changes in the St. Louis, Missouri, area.

The La Porte and St. Louis studies have one feature in common. Proof of any effect will be obtained as a result of sophisticated statistical tests on vast amounts of climatic data correlated with numerous case studies involving extensive physical measurements. It is a relatively rare event when the atmosphere is actually "caught in the act" of responding to man's activities. Some exceptions can be noted, such as cirrus contrails spreading over a sky, cooling ponds and towers producing localized fogs, clouds, and light rain or snow, and valleys filled with fog from industrial sources of moisture and nuclei. Small areas of light snow resulting from power plant and factory effluents entering supercooled fogs are sometimes noted. However, any observations of direct cause-effect relationships between man's activities and atmospheric processes are usually fortuitous. Without data gathering systems designed to detect specific incidences of inadvertent weather modifications, this "catch-as-catch-can" situation will continue.

Some photography obtained by manned orbital missions occasionally saw evidence of inadvertent weather modification. Figure 11 shows the east coast of the U. S. during a period of cold westerly flow over the warmer ocean. Cumulus formed, as expected, over the water. What is unusual is the two cumulus lines which begin considerably closer to the shore, and downwind of the Baltimore and eastern Virginia industrial complexes. ERTS, while actually developed for studies of land usage, water quality, agriculture and the like, thus has the capability of making observations of sufficiently high resolution to detect many instances of inadvertent weather modification.

ERTS-1 took a remarkable series of images over Lake Michigan at 1003 CST, 24 November 1972, which apparently show modification of 1.5 km wide cumulus cloud streets by the effluents of the industries at the southern end of the lake (Fig. 13). On this day, the Lake Michigan region was under the influence of southwest flow about a cold high pressure cell centered in Kentucky (Fig. 12). Minimum temperatures west and south of the lake during the prior night

---

[1] See "Project METROMEX: A Review of Results," Bull. Amer. Meteor. Soc., 55, February 1974, pp. 86-121.

SATELLITE DETECTION OF AIR POLLUTANTS

Fig.11. - Apollo manned orbital flight photograph of east coast, from Virginia to New York, during period of cold westerly flow over relatively warm ocean. The expected cumulus clouds form, but the two bands of clouds forming closer to land downwind of major industrial centers.

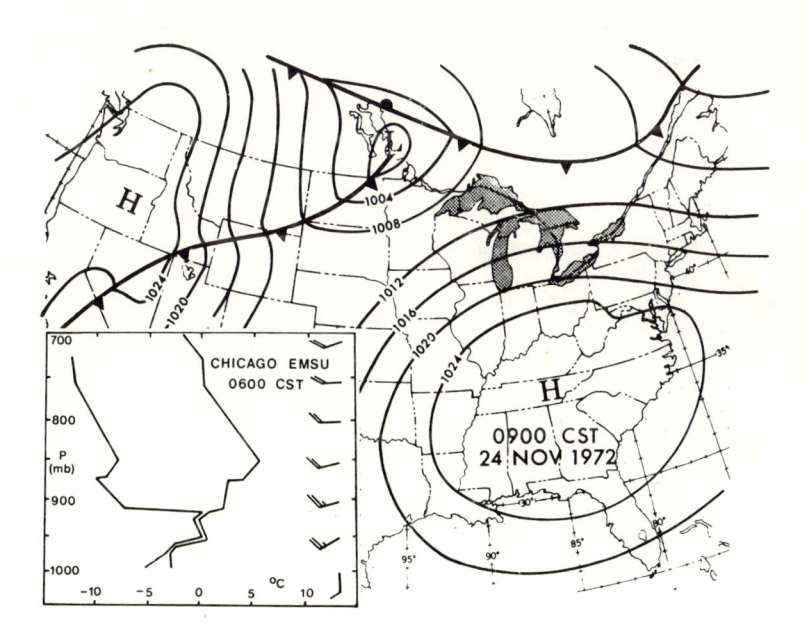

Fig.12. - Synoptic situation, 0900 LST, 24 November 1974. Inset shows Chicago radiosonde ascent at 0600 LST, with strong inversion present in lowest 4000 ft.

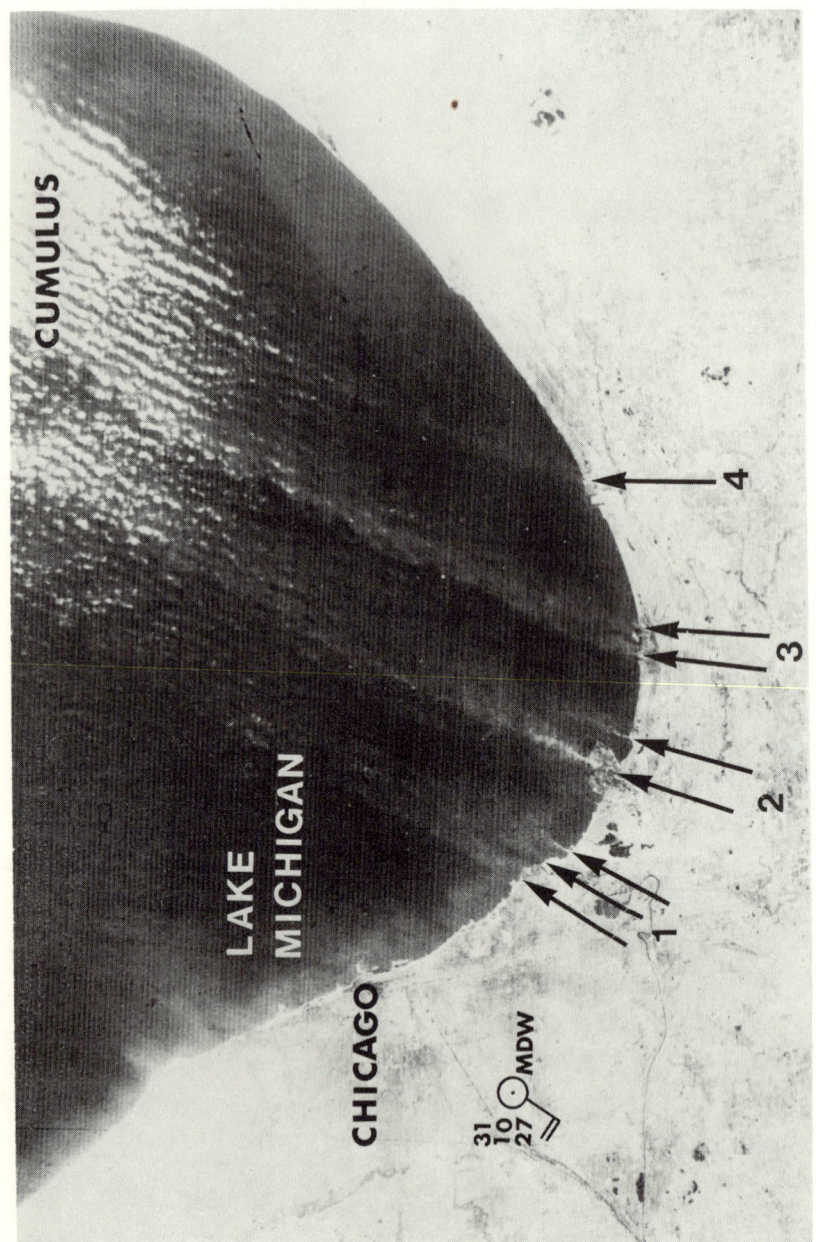

Fig.13. - ERTS-1 image, band 6 (0.7-0.8 micrometers), over southern Lake Michigan, 1003 LST, 24 November 1972. Image photographically enhanced. Only a very light snow cover actually was present. Midway Airport (MDW) shows clear skies, 31 degrees, and southwest winds. Four major particulate plumes emanate from shoreline sources. Cumulus clouds form over relatively warm lake, and those arising out of smoke plumes begin forming closer to shore and become larger and brighter.

ranged from 20F to 29F. A few scattered ship water temperature reports indicated that the lake surface had cooled to about 39-45F. Thus, the presence of cumulus over the relatively warm lake is as expected.

A pilot report over the western Michigan shoreline at the time of the ERTS image noted cloud tops of 2700 ft AGL, with clear skies above. The vertical extent of the clouds was apparently limited by the strong synoptic-scale inversion (Fig. 12).

A closer inspection of Fig. 13 reveals numerous interesting phenomena, the most important of which is the apparent alteration of the cumulus clouds along the axis of the major pollution plumes. The general cloud cover was largely restricted to the southern portion of the lake, perhaps a reflection of somewhat warmer water temperatures there. Experience and theoretical indications suggest that the optimum contrast between smoke and underlying water should occur around 0.65 µm, that is, within band 5 (Lyons and Pease [7]). In this case, however, the turbidity of the water was so considerable that MSS-band 6 (0.7-0.8 µm) achieved slightly better plume/water contrast.

In Fig. 13 it is quite clear (especially in the original 70mm transparencies) that four clusters of industries form large combined plumes that advect northeastwards over the lake while diffusing. The following behavior was noted along these plume axes: (1) the clouds began forming closer to the shore, the first fragments appearing within 15 km of the plume sources, and (2) the individual cloud elements forming the lines were larger and had higher reflectivities (and thus presumably were thicker). No particular enhancement of the cloud streets is noted within plume 4. This plume, however, is from an isolated steel mill complex which has very well-controlled particulate emissions and is smaller than those to the west. See Lyons [16] for additional details.

The ERTS observation showed that the cumulus cloud lines forming within the pollution plumes began somewhat closer to the shore and were better developed than those presumably not affected by significant industrial effluent. Exactly what caused this behavior, however, is not discernable from an ERTS picture. Heat, moisture, ice nuclei, and cloud condensation nuclei are all candidates. Isolating the specific cause(s) awaits further study. The simple fact is that a mere image is not adequate to determine what has caused the effect. In situ measurements of nuclei concentrations and cloud microphysical parameters in conjunction with such satellite imagery would provide a powerful research tool.

What is exciting about the ERTS image is the fact that something indeed has been clearly related to industrial activity. A continued inspection of ERTS images in the vicinity of Chicago (and also Cleveland, Buffalo, etc.) could very well prove extremely enlightening.

If downwind cloud modification from industrial activity does indeed occur frequently, it would appear that any future attempts at management of Great Lakes water levels by operational cloud seeding might run into some very "interesting," or rather challenging, problems of verifying seeding effectiveness.

An extensive radar-echo climatology for the southern basin of Lake Michigan during three summer seasons revealed slightly greater frequencies of thunderstorm-related precipitation over the lake northeast of Gary than in other quadrants. What is rather startling about this finding is that the project set about to prove that lower echo frequencies should exist due to the well-documented suppressive effects of the lake on certain warm season convective rain systems. This tantalizing shred of evidence suggests that perhaps these same plumes act also as effective enhancers of warm season precipitation systems to the extent that the lake effects are more than cancelled out. The "La Porte Anomaly" could conceivably be part of a much larger area of urban precipitation enhancement northeast of Chicago-Gary which is unrecognized

Fig.14. – ERTS-1 image, band 6, Chicago and southern Lake Michigan, 1003 LST, 16 July 1963. There was a general northeasterly flow throughout area, although a weak lake breeze was just pushing onshore along eastern shore. This caused the abrupt edge of cumulus there. Cumulus gradually reformed along the southern and western shore, with noticeable lines of clouds forming out of the plumes of major steel mills and industrial complexes (arrows).

at this time due to lack of precipitation data over the lake itself.

Lyons and Olsson [17] reported occasionally seeing what appear to be lines of cumulus clouds growing out of the steel mill and power plant plumes around Chicago during periods of onshore flow in summertime. Figure 14 confirms these observations. The ERTS-1 band 6 image taken 1003 LST, 16 July 1973, shows cumulus around the southern basin of the lake. There was a light northeasterly flow present during the day along the southern and western shores. Cumulus appeared in a random manner after about 20 km of overland fetch - except downwind of several major steel mills and industries in Gary and Chicago where there were definite lines (arrows). What relationship these cloud lines might have, if any, to the urban enhancement of precipitation cannot yet be ascertained, but another effect upon convective clouds by industrial processes is clearly shown.

An additional effect of man's use of energy upon the environment has been reported by Lyons and Pease [18]. Man-made cirrus (ice crystal clouds) produced by the condensation trails(contrails) of high-flying jet aircraft are being considered as a potential modifying factor on both local and global climates. Under certain conditions of moisture, these artificial clouds may diffuse over much of the sky in thin films at heights above 20,000 feet. Denver has reported a significant increase in high cloudiness since 1958, when commercial jet traffic was inaugurated. Cirrus clouds, even though thin enough to see through, are highly reflective and can cause a significant amount of the sun's radiation to be returned to space. It is conceivable that man could be unwittingly altering the global energy budget in such a way as to induce disasterous changes in the earth's mean temperature. Assuming that the many computerized models of climate can someday be perfected, they will still be of little use if proper input parameters are not available. Much of the furor over the SST resulted from the lack of solid data as to how much increased cloud cover might result. ERTS has been shown capable of detecting jet contrails, and thus may serve as a prototype of a monitoring system able to make these arrangements.

Figure 15 shows pictures taken with regular 35mm Kodachrome II film just after take off from Milwaukee on a commercial flight. This was coincidental with the ERTS-1 passage of 6 November 1972. Clearly visible in Fig. 14(a) are several very thin jet contrails amid a broken deck of natural cirrus clouds. Figure 14(b) shows the much denser natural cloud cover in the Chicago vicinity. In Figure 15(a), the band 4 ERTS-1 image, a close inspection of the original clearly reveals the same jet contrails visible. Figure 15(b) is the same scene but for the near-infrared band 7. The cirrus clouds, both natural and man-made have a definite spectral dependency in their transmissivity, with thin ice crystal clouds being virtually transparent in the 0.8-1.1 μm range. This spectral signature of ice clouds, combined with various edge finding algorithms used in pattern recognition, might allow the possibility of an automated jet contrail detection system. A reliable census of increased cirriform cloudiness due to man's air traffic activity would indeed be a valuable contribution.

Figure 16 shows another scene of the Milwaukee area on 20 October 1972. On this day considerable natural cirrus was present. An inspection of the band 4 image shows numerous contrails which, due to the high ambient humidity, were able to diffuse to great widths. Some were at least 11 km wide. In band 7 distinct contrail shadows could be seen on the ground.

NEEDS FOR THE FUTURE

A comprehensive evaluation of remote sensing for resource and environmental surveys (especially with respect to the ERTS System) has been prepared by the National Academy of Science [19]. It concurs that such systems as ERTS

Fig.15(a) - View looking east, at 2000 ft MSL, just east of Milwaukee airport
(MKE), 1000 LST, 6 November 1972. Note several small contrails.
(b) - View looking south, from 5000 ft MSL, about 15 km northwest of
Chicago O'Hare airport (ORD) about 1020 LST, 6 November 1972. Note
dense cirrus cloud cover above.

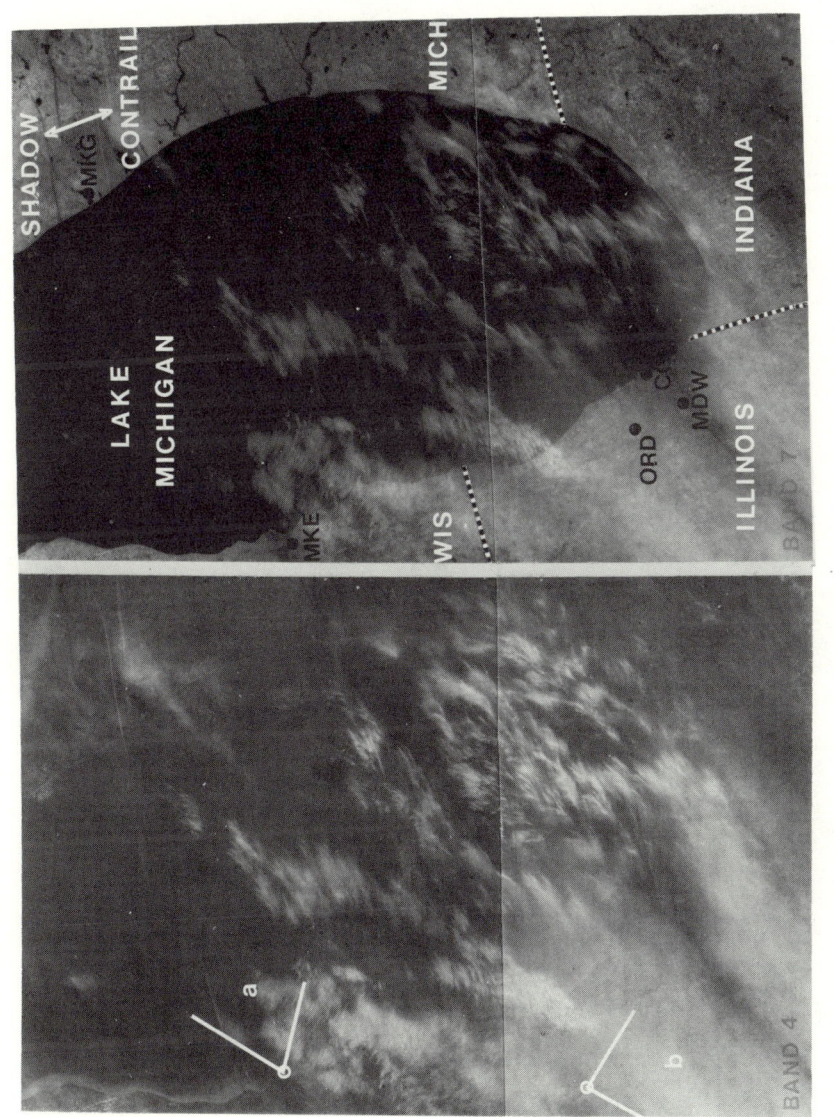

Fig.16(a) - ERTS-1, band 4 image, 1003 LST, 6 November 1972, showing thin cirrus cloud cover over Lake Michigan. Location of pictures shown in Fig. 15 indicated. Close inspection east of (a) shows contrails visible.
(b) - Same, but for band 7, in which the surface is visible through all but the densest cirrus. Contrail and shadow visible near Muskegan, Michigan (MKG).

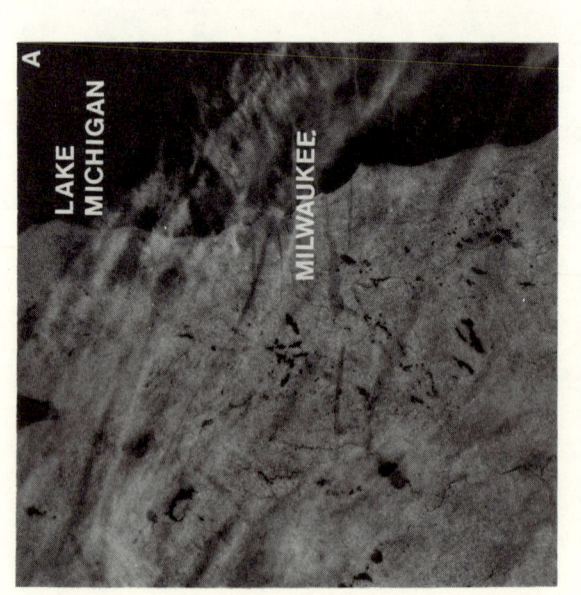

Fig.17(a) – ERTS-1 image, band 7, 1003 LST, 20 October 1972, over southeastern Wisconsin, revealing surprising degree of transparency of cirriform (ice) clouds in the near infrared. Also visible are distinct shadows from jet contrails.
(b) – Same, but for band 4, showing natural cloud cover commingled with jet contrails. In the moist environment at about 25,000 ft, some contrails have diffused to widths of at least 11 km.

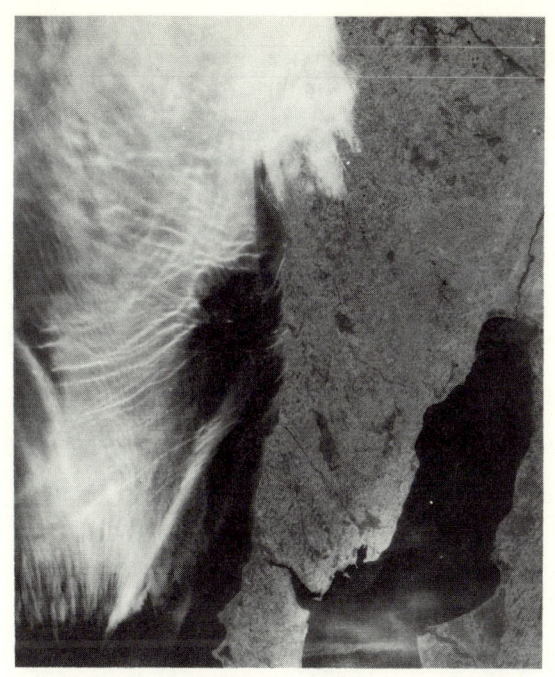

Fig.18. - ERTS-1 image, band 5, 18 April 1973, over Lake Michigan near Door County Peninsula, Wisconsin. Shallow fog within lake conduction inversion manifests complex wave patterns. Fog also extends inland several miles further down the coast.

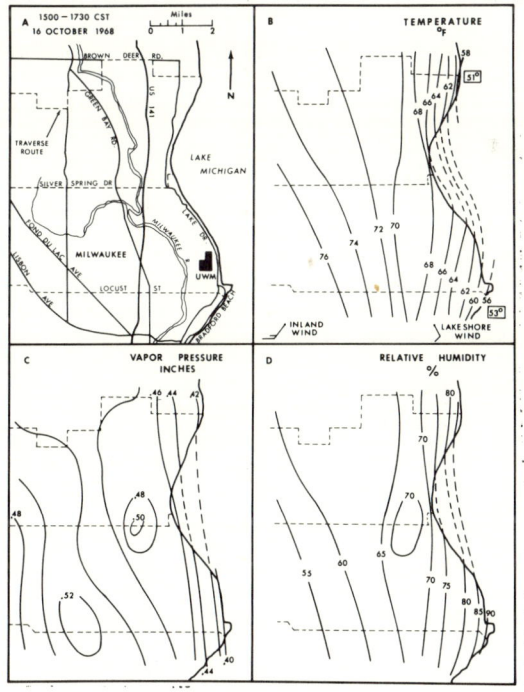

Fig.19 -- Summary of observations of temperature and humidity made in Milwaukee on 16 October 1968 on a day with southeasterly onshore flow over a relatively colder lake. Notice strong gradient in relative humidity from 55% in western part of city to 90% at lakeshore. Data courtesy of Mr. John Chandik.

have shown considerable promise in assisting in the solution of many of the environmental problems being faced today. The multispectral scanner approach has been proven effective in its first full scale test from orbital altitude. The Academy recommended that work proceeded on several fronts including: (1) improving the sensor resolution by a factor of two to four, which should greatly help in the detection of smaller particulate plumes, (2) moving the ERTS system from an experimental to an operational basis, thus allowing for the receipt of data on a more frequent basis than the current 18 day cycle, and (3) the development of a thermal channel to expand the multispectral approach.

This last point is very important for many reasons. A sensor working in the 10-12 µm region with a resolution approaching that now achieved in the visible range would immensely add to the overall detection and identification powers of subsequent systems. It is unlikely that it would be of significant assistance in the detection of pollutants per se, but could be of enormous help in understanding the environment in which plumes disperse. In particular, if high resolution maps of lake water surface temperature could be obtained, it would be possible to better estimate atmospheric thermal stability over water. The extreme differences in atmospheric states existing between land observing stations, even just a few miles from a lake shore, and those over the water proper are just now being fully appreciated. For instance, Figure 18 shows a section of an ERTS-1 image of Lake Michigan, near Door County Peninsula, Wisconsin, on 18 April 1973. Southeasterly winds were advecting humid 65F land air over 35-45F lake water. An intense conduction inversion formed in the lowest 100m [11]. Conditions became saturated within this layer and a thin fog deck formed. Unusual wave patterns can be seen in the fog, indicative of the complex motions that can occur in stable air. More importantly, one sees that the fog bank extended inland only a few miles along the Wisconsin shoreline. Figure 19 is a summary of observations made on the Milwaukee shoreline on a similar type day. In the space of about 10 km, relative humidities vary from 55% to over 90%. Such intense local patterns, which occur frequently throughout the "warm season," are of considerable importance when one must design such energy related facilities as fossil and nuclear power plants, cooling towers, cooling ponds, etc. In almost all shoreline regions of the Great Lakes, official National Weather Service stations are too far inland to adequately measure shoreline meteorological conditions. The addition of a thermal channel to an ERTS successor system would be a big step towards obtaining the necessary climatological data base for this region. We could then make a more intelligent assessment of the environmental impact of numerous transportation and power production projects.

SUMMARY AND CONCLUSION

It has been shown that the monitoring of major suspended particulate plumes is feasible from satellite altitudes at least under ideal conditions. Increases in resolution and frequency of the data will be necessary in order for such to reach even a remotely operational status. At this time ERTS should not be considered for applications such as routine monitoring of such local events as cooling tower stratus and fog. But for studies of such phenomena as large (or even urban) scale smoke and turbidity patterns, jet contrails, air mass turbidity, and some aspects of inadvertent weather modification, numerous "targets of opportunity" do present themselves. It is not yet feasible to detect gaseous pollutants on the scale discussed in this paper, but there are rapid developments in sensor technology which may make such detection possible in the not too distance future.

It will be many years, perhaps decades, before satellite systems can be expected to become the prime air quality monitoring tool. The requirement for large numbers of in situ sensors will remain. But as sensor quality and data processing techniques increase in sophistication, satellite observations of at least certain air quality parameters will become both routine and vital additions to our information pool.

ACKNOWLEDGEMENTS

This work was partially supported by the National Aeronautics and Space Administration (Contract NAS5-21736), the Environmental Protection Agency (Grant R-800873), the National Science Foundation (Grant GA-32208), and the University of Wisconsin-Milwaukee Center for Great Lakes Studies.

Special thanks are tended to Professor Richard A. Northouse and Roger Gersonde of RAIL for digital data processing. Also photographer Peter Tolsma III, and technical typists Ms. Jean Kenney and Nancy Hirt are credited with excellent work. Mr. Steven R. Pease has also contributed to these studies.

REFERENCES

1. Colvocoresses, A. P., 1974: "Remote Sensing Platforms," U.S. Geological Survey Circular 693, U.S. Government Printing Office, 75 pp.

2. McClellan, A., 1971: "Satellite Remote Sensing of Large Scale Local Atmospheric Pollution," paper CP-38C, Procs. of the Second International Clean Air Congress, Academic Press, New York, 1354 pp.

3. Derr, V. E., editor, 1972: "Remote Sensing of the Troposphere," Wave Propogation Laboratory, NOAA Environmental Research Lab, Boulder, Col., U.S. Gov. Printing Office.

4. Lawrence, G. F., and E. Ward, 1974: "Remote Sensing of Urban Ambient Air Pollution," paper #74-23, 67th Annual Meeting of the Air Pollution Control Association, Denver, Colorado.

5. MacCallum, D. H., 1973: "Availability of ERTS-1 Data," Bulletin of the American Meteorological Society, 54, 112-114.

6. Pease, R. W., and L. W. Bowden, 1969: "Making Color Infrared Film a More Effective High-Altitude Sensor," Remote Sensing of the Environment, 1, March, 23-30.

7. Lyons, W. A., and S. R. Pease, 1973: "Detection of Particulate Air Pollution Plumes from Major Point Sources Using ERTS-1 Imagery," Bulletin of the American Meteorological Society, 54, November, 1163-1170.

8. Copeland, G. E., R. N. Blais, G. M. Hilton, and E. C. Kindle, 1974: "Detection and Measurement of Smoke Plumes in Aerial and Satellite (ERTS-1) Imagery," paper #74-240, 67th Annual Meeting of the Air Pollution Control Association, Denver, Colorado.

9. Wightman, J. M., 1973: "Detection, Mapping and Estimation of Rate of Spread of Grass Fires from Southern African ERTS-1 Imagery," Symposium on Significant Results Obtained from the Earth Resources Technology Satellite, NASA Goddard Space Flight Center, 1730 pp.

10. Eigen, D. J., and R. A. Northouse, 1973: "Unsupervised Discrete Cluster Analysis," TR-AI-73-2, Robotics and Artificial Intelligence Lab Report, University of Wisconsin-Milwaukee.

11. Lyons, W. A., 1970: "Numerical Simulation of Great Lakes Summertime Conduction Inversions," Proc. 13th Conference on Great Lakes Research, International Association for Great Lakes Research, 369-387.

12. Lyons, W. A., C. S. Keen, and R. A. Northouse, 1974: "ERTS-1 Satellite Observations of Mesoscale Air Pollution Dispersion Around the Great Lakes," preprint volume, Symposium on Atmospheric Diffusion and Air Pollution, American Meteorological Society, Santa Barbara, California, 273-280.

13. Lyons, W. A., and H. S. Cole, 1974: "The Use of Monitoring Network and ERTS-1 Data to Study Interregional Pollution Transport of Ozone in the Gary-Chicago-Milwaukee Corridor," paper #74-241, 67th Annual Meeting of the Air Pollution Control Association, Denver, Colorado.

14. Murth, P. A., 1973: "SO_2 Damage to Forests Recorded by ERTS-1," NASA Third ERTS Symposium, Washington, D. C., (Abstracts).

15. Griggs, M., 1973: "Determination of Aerosol Content of the Atmosphere," Symposium on Significant Results Obtained from the Earth Resources Technology Satellite, NASA Goddard Space Flight Center, 1730 pp.

16. Lyons, W. A., 1974: "Inadvertant Weather Modification by Chicago-Northern Indiana Pollution Sources Observed by ERTS-1," Monthly Weather Review, 102, 503-508.

17. Lyons, W. A., and L. E. Olsson, 1973: "Detailed Mesometeorological Studies of Air Pollution Dispersion in the Chicago Lake Breeze," Monthly Weather Review, 101, 387-403.

18. Lyons, W. A., and S. R. Pease, 1973: "ERTS-1 Views the Great Lakes," GLUMP Report No. 15, University of Wisconsin-Milwaukee, 7 pp.

19. CORSPERS, 1974: "Remote Sensing for Resources and Environmental Surveys: A Progress Report," National Academy of Sciences, Committee on Remote Sensing Programs for Earth Resource Surveys [CORSPERS], Washington, D.C., 101 pp.

REMOTE SENSING BY ERTS SATELLITE OF VEGETATIONAL RESOURCES BELIEVED TO BE UNDER POSSIBLE THREAT OF ENVIRONMENTAL STRESS

PREMSUKH POONI and WALTER J. FLOYD

Bethune-Cookman College, Daytona Beach, Florida, U.S.A.

ROYCE HALL

Federal Electric Corporation, Cocoa Beach, Florida, U.S.A.

ABSTRACT

The distribution of natural vegetation, a primary storage system for solar energy, which is possibly under some threat of environmental stress due to intensive utilization of other forms of energy, on North Merritt Island, Florida, was studied by analysis of ERTS satellite Multispectral Scanner data on the IMAGE-100 Computer System. The boundaries of six distinct plant associations were located on photos made on the Image Analyser, with a non-significant mean error of -24.38 meters. The six plant associations are described as I. Aquatic Estuarine Association, II. Mangrove, III. Spartina swamp, IV. Wooded swamp, V. Sabal Hammock, VI. Oak-palmetto, each having a characteristic spectral signature. The difference in average reflectance 'grey level' between the lowest of the four spectral scanning bands (Channel 1 of IMAGE-100) and the highest spectral scanning band (Channel 4) for the six vegetation types were I. + 8.30, II. + 4.50, III. + 0.42, IV. - 1.97, V. - 6.22, VI. - 12.53. The decreasing trend of the differences is strongly negatively correlated with height of land, the coefficient of correlation being -.9696.

INTRODUCTION

Under an Educational Grant made by the National Aeronautics and Space Administration, Bethune-Cookman College is conducting a program for the study of the environment in the neighborhood of the John F. Kennedy Space Center which includes remote sensing of plant communities with a view to observing possible deterioration which may result from extensive mechanical ground operations and in due course, exhaust fumes from space vehicles. The present paper deals only with the aspect of remote sensing of the plant associations in the North Merritt Island area of the Kennedy Space Center with the purpose of developing a technique for the measurement of vegetational resources in general, resources which constitute a continuous primary storage system for solar energy [6] and which are at the same time subject to deterioration under conditions of intensive utilization of other forms of energy in their immediate neighborhood.

MATERIALS AND METHODS

The present study was based upon the vegetation types of North Merritt Island which is shown as the diagonally shaded area on Figure 1. Because of the present intensive construction program which is in progress in the area and because of the large volumes of exhaust gases of space vehicles which are expected to influence the area in the future, the plant associations will be

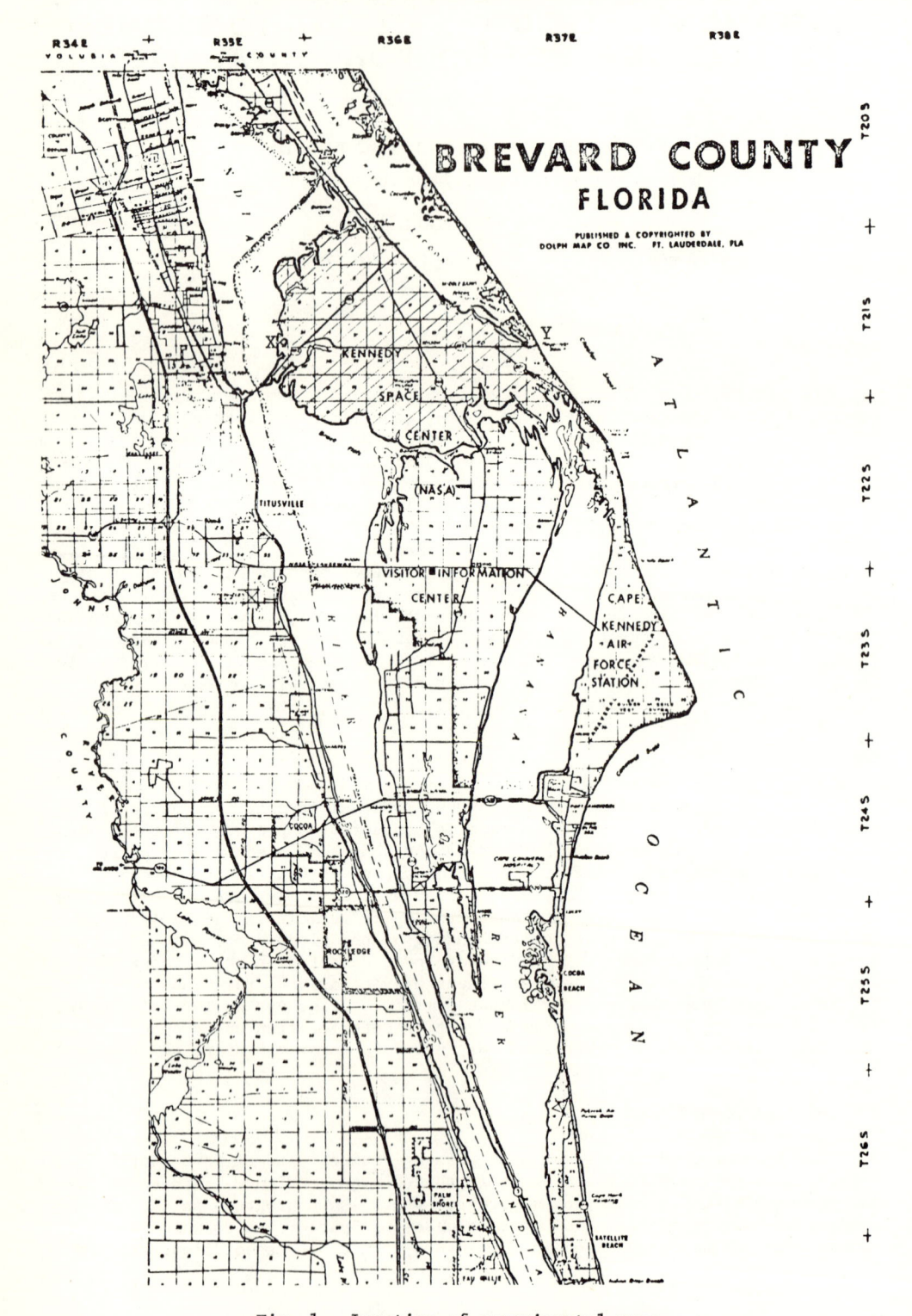

Fig. 1. Location of experimental area

studied over an adequate period of time to observe possible changes. One of the more important methods of study will be remote sensing. In the first stage of such a study, data acquired by the four-channel Multispectral Scanner of the ERTS Satellite [1] was analyzed on the IMAGE-100 Interactive Multispectral Image Analysis System [3].

Band 1   0.5 - 0.6 micrometers

Band 2   0.6 - 0.7 micrometers

Band 3   0.7 - 0.8 micrometers

Band 4   0.8 - 1.1 micrometers

The Multispectral Scanner of the ERTS Satellite is a four-band scanner which operates in the solar-reflected spectral band region between 0.5 and 1.1 micrometer wavelengths. The earth's surface is scanned in the four spectral bands simultaneously at an orbital velocity which permits a global coverage every 18 days. The four bands are: -

The data received by the ERTS scanner are reformatted and written on magnetic tape by the Ground Data Handling System at Greenbelt, Maryland. It is finally transmitted to the NASA Data Processing Facility for cataloging and dissemination to users.

The IMAGE-100 is an Interactive Multispectral Image Analysis System which is capable of extracting information from multispectral imagery data loaded from magnetic tape. Thematic images extending over parts of the ERTS scene under study, may be developed and stored temporarily to be used for the creation of composite thematic imagery of contrasting colors representing the different types of vegetation which occur in the area. The imagery is observable on the primary display device, a color cathode ray tube and it may be photographed for further study. Next to the Image Analyser Console is a Graphics Display Terminal which shows the spectral signature of each theme in the form of four frequency histograms which refer to that part of the scene enclosed within the training area selected by the operator with the use of a moveable cursor. The four histograms give for each of the four multispectral channels, characteristic reflectance 'grey levels' on the horizontal axis and number of picture elements associated with the 'grey levels', on the vertical axis. The term 'grey levels' refers to operator selected density levels of reflectance which quantize the multispectral data into discrete ranges. A printed output of the Graphic Display is also available giving among other data, the spectral boundaries, means and variances of each of the four channels. Thus it is possible to obtain by use of the IMAGE-100 System, color themes representing vegetation types and spectral characteristics for each theme.

The ERTS data on which image analysis was carried out for the purposes of this paper were acquired on 18th March, 1974. In the process of image analysis on the IMAGE-100 System it was found in this project that the themes and spectral signatures produced by single pixel training and 36 pixel training on the ERTS scene, were not appreciably different from each other. A pixel is a picture element 79 meters square. The cursor was used for systematic exploration of the ERTS scene until a set of more or less mutually exclusive themes were developed which occupied the total surface. Each theme was photographed with High Speed Ektachrome film having a sensitivity of ASA 160. A printed output of the spectral characteristics of each theme was also obtained.

Legend:

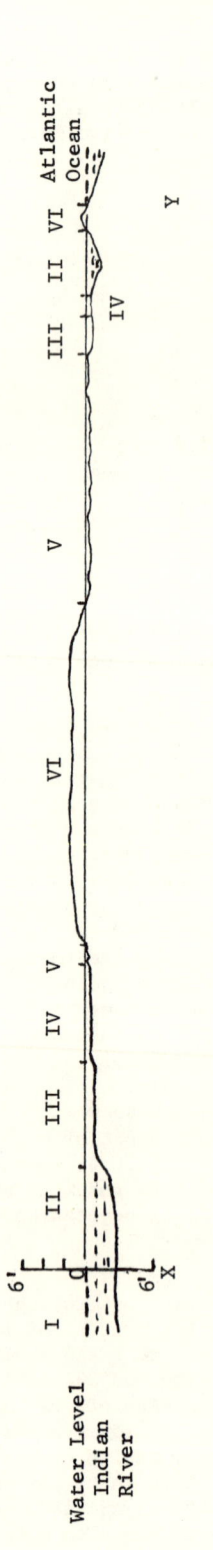

I. Aquatic Estuarine Association.
II. Mangrove Association.
III. Spartina Typha Swamp.
IV. Swampy woodland.
V. Sabal-Oak-Red maple-Magnolia hammock.
VI. Oak-Palmetto Association.

Fig. 2. Relative widths of vegetation types which occur along a line near and parallel to State Road 402, Brevard County, Florida.

Fig. 3 - Color themes produced by IMAGE-100 representing 6 plant associations

I. Blue = Aquatic Estuarine Association
II. Red = Mangrove
III. Brown = Spartina swamp
IV. White = Wooded swamp
V. Yellow = Sabal - Acer Hammock
VI. Purple = Oak-palmetto

Although there was a certain amount of prior knowledge of the distribution of vegetation types on North Merritt Island, it was nevertheless necessary to carry out extensive field studies in order to determine the degree of correlation which may exist between color themes and distribution of plant associations in the field. The principal plant associations are illustrated in Photos 1 to 6.

In order to compare theme distribution with the distribution of vegetation, measurements were made from the point X to the point Y shown on Figure 1, both across the themes shown on Figure 3 and across the corresponding strips of vegetation in the field. The widths of theme strips were converted to field scale and the differences between the derived values and corresponding vegetation strips were used for computing a mean difference and its standard error [4].

Based upon the observation that the height of the land above water level or the water table increases from the point X on Figure 1 towards the center of the Island, that vegetation types change as one moves Eastward from X and that the spectral characteristics of the themes appear to alter systematically with changes in theme from X toward Y, a correlation study was carried out between mean height of land, x, for each vegetation type and difference between average 'grey levels' of channel 1 and channel 4, y, for each vegetation type.

RESULTS

Multispectral data received from the NASA Data Processing Facility for North Merritt Island on ERTS tape for 18th March, 1974, were analysed on the IMAGE-100 Interactive Multispectral Image Analysis System for comparison of imagery themes with plant associations which they may represent.

In examining the correspondence between imagery produced on the IMAGE-100 from ERTS multispectral data and vegetation types, three related approaches were adopted. In the first instance the general distribution of the individual imagery themes on the ERTS scene were compared with the distribution of corresponding vegetation types in the field. In a more precise study, widths of theme strips measured along the line X Y shown in Figure 1, coverted to field scale, were compared with actual field widths of corresponding vegetation types. The third approach in comparing image and field features consisted of correlating height of land of the vegetation types with their spectral characteristics. The results of these three aspects of the study are discussed separately below.

COMPARISON OF COLOR THEMES WITH VEGETATION TYPES

In developing the color themes on the ERTS scene representing North Merritt Island and shown as the diagonally shaded area in Figure 1, one pixel, 79 meters square, and 36 pixel size cursors were used systematically across the scene until a number of more or less mutually exclusive themes were produced. A composite of the six themes thus produced is shown in Figure 3. The composites produced with one pixel and 36 pixels are not appreciably different. An examination of the colors on Figure 3 along the line X Y shown in Figure 1, starting at X and proceeding East, shows that the following colors occur in order:-
I. Light blue, II. Red, III. Brown (theme not developed), IV. White, V. Yellow, VI. Purple. The colors I to VI may be written in that order to represent the vegetation types I to VI in Table 1. Ground truthing expeditions indicated that the vegetation type corresponding to each color theme is consistently confined to its own theme.

Photo 1. Aquatic Estuarine Association

Photo 2. Mangrove Association

Photo 3. Spartina-Typha Swamp

Photo 4. Wooded Swamp

Photo 5. Sabal - Acer Hammock

Photo 6. Oak-Palmetto Association

Table 1. Principal plant associations of North Merritt Island and their more dominant species.

I. Aquatic Estuarine Association

    Thalassia testudinum. Turtle grass.
    Spartina alternifolia. Cord grass.
    Zostera marina.
    Ulva lactuca.
    Gracilaria confervoides.
    Acetabularia crenulata.

II. Mangrove Association

    Laguncularia racemosa. White mangrove.
    Avicennia nitida. Black mangrove.
    Baccharis halimifolia. Groundsel.
    Schinus terebinthifolius. Brazilian pepper.
    Borrichia frutescens. Seaside ox-eye.

III. Spartina-Typha Swamp

    Spartina bakerii. Bunch grass.
    Juncus roemerianus. Black rush.
    Typha angustifolia. Cat tail.
    Acrostichum danaeaefolium. Leather fern.

IV. Swampy woodland

    Salix caroliniana. Willow.
    Baccharis halimifolia. Groundsel.
    Typha angustifolia. Cat tail.
    Ilex glabra. Gallberry.
    Sabal palmetto. Sabal.
    Myrica cerifera. Wax myrtle.

V. Sabal-Quercus-Acer-Magnolia hammock

    Sabal palmetto. Sabal palm.
    Quercus virginiana. Seaside live oak.
    Acer rubrum. Red maple.
    Salix caroliniana. Willow.
    Magnolia virginiana. Swamp magnolia.

VI. Oak-Palmetto Association

    Quercus sp. Scrub oak.
    Serenoa repens. Palmetto.
    Smilax auriculata. Green briar.
    Vitis rotundifolia. Wild grapes.

CORRESPONDENCE BETWEEN THEME BORDERS AND ACTUAL PLANT ASSOCIATION BORDERS

A more precise correlation of imagery color themes and distribution of vegetation in the field was carried out by comparing widths of theme strips on Figure 3 measured along the line X Y of Figure 1, with the widths of corresponding vegetation strips in the field. The differences between the widths of theme strips converted to field scale and the widths of vegetation strips, were utilized for calculating the mean difference and its standard error. Nine such strips were studied along the line X Y of Figure 1 and the nine corresponding differences are shown in Table 2. The mean difference between actual widths of vegetation strips and widths as estimated from ERTS imagery is shown at the bottom of Table 2 to be - 24.38 meters. This value is not significantly different from zero, on a probability of P = .05 [4] , [7].

Table 2. Estimate of standard error of widths of vegetation strips as measured on IMAGE-100 picture produced from ERTS tapes.

Differences between actual width
and width obtained from the
picture (meters)

```
+     4.39
+     8.78
-    57.06
-   118.51
-     8.78
-    21.95
-    26.33
-    21.95
+    21.95
-   219.46
```

Mean difference = -24.38 meters
Std. Error of mean diff. = ± 14.07
Significant difference (P = .05, n = 8) must exceed ± 32.45.

CORRELATION BETWEEN HEIGHT OF LAND AND SPECTRAL FEATURES OF VEGETATION TYPES.

A third method of studying the correspondence between ERTS multispectral data and plant associations consisted of examination of the degree of correlation between height of land of each association and spectral features of the association. Figure 2 gives relative widths of color theme strips along the line X Y shown in Figure 1. The numbers I to VI shown on Figure 2 refer to the vegetation types of Table 1 and to the colors light blue, red, brown, white, yellow and purple respectively, of Figure 3, one color corresponding with one vegetation type. Figure 2 shows that the six vegetation types correspond with six characteristic land levels, the height of the land increasing from the point X towards the East. Because an examination of the multispectral data of the six vegetation types showed that there was a systematic change in the spectral features from low lying land to higher land, the degree of correlation between spectral signatures and height of land was determined. The feature of the spectral characteristics which appeared to be the best correlated with height of land was the difference between the average 'grey levels' of Channels 1 and 4 of the IMAGE-100 output. The pairs of values recorded for height of land, x, and differences in 'grey levels' of Channels 1 and 4, y, are given in Table 3. The correlation coefficient between the factors x and y is shown at the bottom of the table to be -.9696 which is a large negative and significant value with 95% confidence limits of -.7612 to +.7612 [2] .

Table 3. Correlation between height of land above water level or water table in meters (x) and difference in average 'grey levels' of Channels 1 and 4.

Plant Association	Height of land in meters (x)	Mean 'grey levels'* Channel 1	Mean 'grey levels'* Channel 4	Channel 1 - Channel 4 (y)
I. Aquatic flora	-1.22	9.33	1.03	+8.30
II. Mangrove	- .61	12.00	7.50	+4.50
III. Spartina swamp	- .30	13.03	12.61	+0.42
IV. Swampy woodland	- .23	10.47	12.44	-1.97
V. Sabal hammock	- .08	12.06	18.28	-6.22
VI. Oak-palmetto	+ .91	10.50	23.03	-12.53

*'Grey levels' refer to operator selected density levels of reflectance which quantize the multispectral data into discrete ranges.

Correlation coefficient (r_{xy}) = -.9696

For significance (d.f. = 6, P = .05) r_{xy} must transcend $\pm$ .7612 [2]

CONCLUSIONS

1. Data acquired by ERTS Satellite Multispectral Scanner were processed on the IMAGE-100 Interactive Multispectral Image Analysis System to identify and define the distribution of six principal plant associations on North Merritt Island, Florida.

2. The borders of the six plant associations were located with a mean error of -24.38 meters, a value which is not significantly different from zero on a probability of P = .05.

3. The spectral signatures of the six plant associations are systematically related to each other as reflected in the very high negative and significant correlation coefficient of -.9696 between the height of land of the association on the one hand and the difference in reflectance 'grey level' between Channels 1 and 4 of the IMAGE-100 output, on the other hand.

ACKNOWLEDGEMENTS

The work discussed in this paper was carried out under a financial grant made by the National Aeronautics and Space Administration to Bethune-Cookman College. Data processing was conducted at the Data Analysis Facility at the Kennedy Space Center. Mr. Cliff Dillon, Chief of the DAF scheduled the use of the Facility. Mr. Bob Butterfield and Mr. Joe Bartoszek, members of Mr. Dillon's staff operated the IMAGE-100. Thanks are also due to Mr. Tom Hammon, an Earth Resources Project Manager for providing color IR tapes which were used in conjunction with ground-truthing. The students of Bethune-Cookman College who participated in the operations were Mr. William Smiley, Mr. Richard Paige, Mr. Mark Stephens, Mr. Dealmus Dixon, Mr. Vishwmitr Poonai and Mr. Michael Adegbite.

REFERENCES

1. General Electric Space Division. 1972. Earth Resources Technology Reference Manual.

2. Hoel, P. G. 1966. Elementary Statistics. John Wiley and Sons, Inc., New York.

3. Ground Systems Department, General Electric Space Division. 1974. IMAGE-100 Users Manual.

4. Snedecor, G. W., and Cochran, W. G. 1969. Statistical Methods. The Iowa State University Press.

5. Steel, R. G. D., and Torrey, J. H. 1960. Principles and Procedures of Statistics. McGraw-Hill Book Company, Inc., New York.

6. Szego, G. C., and Kemp, C. C. 1973. Energy forest and fuel plantations. Chemtech 3(5): 275-284.

7. Yule, G. W., and Kendall, M. G. 1950. An Introduction to the Theory of Statistics. Charles Griffin and Company Ltd. London.

REMOTE SENSING APPLIED TO THERMAL POLLUTION

S. S. LEE, T. N. VEZIROGLU, S. SENGUPTA and N. L. WEINBERG

Clean Energy Research Institute, School of Engineering and Enviromental Design, University of Miami, Coral Gables, Florida

I. INTRODUCTION

Many industrial plants, especially electric power generating plants, use large quantities of water for cooling purposes (1).* When heated waters from such industrial plants are discharged into rivers, lakes and estuaries, they can disturb the ecological balance and destroy the natural habitat of aquatic life by changing the temperature levels, dissolved oxygen levels and the biochemical oxygen demand (2). In rivers, the hot discharge waters can cause miles of hot sections which act as thermal barriers and prevent fish from going upstream to their spawning grounds. In regions of high ambient temperatures, such as along the East Coast of Florida, the hot discharges can cause excessive estuarine temperatures. This, in turn, increases the evaporation and results in an overall increase in the level of salinity. Turbidity of mineral origin can be generated by the dislodging and suspension of sediments by currents such as those caused by hot discharges. Turbidity of biological origin can be generated by phytoplankton growth in thermally suitable environments (3). The above outlined processes can ruin the marine environment as far as ecological, fishing and recreational interests are concerned.

A major source of the thermal pollution of rivers, lakes and estuaries is the hot water discharge from the condensers of fossil and nuclear fueled power plants. About 5,500 BTU of thermal energy is produced for every kilowatt hour of electricity generated by a conventional power plant (4). Nuclear powered plants, being less efficient, produce about 10,000 BTU per kilowatt hour. Large volumes of water are used as a transfer medium to remove the heat from the condenser and dissipate it into the environment. The cooling water is drawn from a large source, such as a river or an estuary, passed through the condenser cooling system of the plant and returned to the receiving water with its thermal load. Most units in use today raise the temperature of the cooling water between 10 and 25°F. The pumping rate varies widely according to the plant design but it is estimated that electric power plants are now discharging 50 trillion gallons of heated water a year into rivers, lakes and estuaries (1). The demand

---

*Figures in brackets refer to references at the end of the text.

for electric power in the United States has doubled about once every ten years since 1900 and indications are that the rate of increase might even be greater in the coming decades (5,6). The cooling water requirements of the power industry will be on the order of 100 trillion gallons per year by 1980, a volume of water equal to one-fifth of the total fresh-water runoff of the continental United States. To meet these water needs, more plants will be built near estuaries and coastal areas where almost limitless supplies of water are available. Picton (7) projects that by 1980, thirty-two percent of all steam electric stations will be located adjacent to estuaries or on open sea coasts. The problems associated with the release of these large volumes of heated water are compounded by recent trends within the power industry. Large plants, including more nuclear powered generators, are being built, and groups of these units are being located at a single site. The United States and the world thus face a potential problem in environmental alteration of enormous proportions, particularly in estuarine and coastal marine waters. This knowledge, together with information that thermal pollution can adversely affect ecological systems (1,8,9) has focused the attention of many scientists and engineers on thermal pollution. In order to deal with these thermal discharge problems, one important step is to be able to detect thermal pollution and to alleviate it. Hindley and Miner (10) compared different techniques of thermal pollution detection and concluded that the remote sensing by infrared scanning would provide more accurate and more complete information of water surface temperature fields than the in situ measurements. Using infrared scanners, it is possible to obtain a continuous and complete temperature distribution of a given region. On the other hand, if the same data is to be obtained in situ, a large number of field measurements must be taken to obtain even a less complete representation of the temperature field. Infrared sensors can operate under both day and night conditions. Their resolutions and accuracies are continuously being increased.

II. WATER SURFACE TEMPERATURE MEASUREMENTS BY INFRARED IMAGING

The infrared portion of the electromagnetic spectrum lies between the visible and microwave regions (0.78 micron to 1,000 micron). The existence of this "invisible" radiation was discovered in 1800 by Sir Frederick William Herschel, who used a thermometer to detect the energy beyond the red portion of visible "rainbow" produced by a prism. In 1861, Richard Bunsen and Gustav Kirchoff established the principles underlying infrared spectroscopy. After a century of further progress, the use of infrared technology has been extended to a host of applications. Among them is the use of infrared sensors for producing images of remote scenes. These devices (remote-sensing systems) have been used on the ground, in airplanes, and in space vehicles. These devices are relatively small, lightweight, and require little electric power. Infrared systems can operate under both daytime and nighttime conditions. However, they are hampered by clouds and rain. The two important features of infrared surveillance systems are the high spatial resolution achievable with relatively simple designs and the fact that they operate passively: e.g., it is not necessary to illuminate the ground scene artificially in order to record an image.

1. BASIC PRINCIPLES

Passive methods of heat measurements are based upon the fact that all bodies above absolute zero radiate electromagnetic energy to their surroundings: the higher their temperature, the more they will radiate. The most efficient radiator at a given temperature is the so-called "black body" which is also a perfect absorber. The radiation of a black body is described by the famous Planck Radiation Laws:

$$I_{\lambda T} = \frac{2hc^2}{\lambda^5} \cdot \frac{\Delta \lambda}{\exp(\frac{hc}{k\lambda T}) - 1} \tag{1}$$

Here, h is the Planck's constant, c is the velocity of light in vacuum, λ is the wavelength, $\Delta\lambda$ is the wavelength range, k is the Boltzman's constant, T is the absolute surface temperature and $I_{\lambda T}$ is the intensity of radiation, i.e., the energy emitted per unit time from a unit area of surface at absolute temperature T, at wavelength λ in a wavelength band of width $\Delta\lambda$ into unit solid angle. Substituting in values for a 1 micron band we obtain,

$$I = \frac{1.192 \times 10^{-16}}{\lambda^5 (e^{1.44/\lambda T} - 1)} \text{ watts/cm}^2 \text{ ster. } \mu \tag{2}$$

For example, at the sun's surface if we take the black body temperature to be 5500°K at λ = 6000 Å (0.6μ) we obtain I = 6000 watts/cm² · ster. μ.

The emission from a black body versus wavelength is plotted in Figure 1 with temperature as a parameter. One can see from this figure that the emission maxima move toward longer wavelengths with decreasing temperature. The emission maxima from the sun with surface temperature about 5500°K are near 0.5 micrometer (in the green-yellow region of the visible spectrum). At the earth's ambient temperature (approximately 300°K), the maximum is at about 10 microns. Therefore, the region of most interest in the passive measurement of surface temperatures is in the region 8-14 microns, the so-called "thermal infrared region."

Looking at Figure 1, one might conclude that the sun's own light would tend to drown out the emission from the earth's surface, but this is not so. This is because the sun is at a great distance, and the curves in Figure 1 are calculated for the hot surfaces themselves. At the distance of the earth the sun subtends a solid angle about 7.6 x 10⁻⁵ steradian. The maximum irradiance arriving at the top of the atmosphere from the sun is therefore 6000 x 7.6 x 10⁻⁵ or about 1/2 watt/cm² per micron. The sun's radiation is compared to the emission at 50°F (300°K) in Figure 2. We notice that at 10 microns, the arriving sunlight is more than 100 times less intense than the emission from the earth. In the near infrared, around 1 micron, the situation is reversed and the sunlight is more than a million times brighter than the earth's emission. At much longer wavelengths than 10 microns, the amount of emission from the earth is very small, so the advantage of the 8-14 micron region for passive temperature measurements is clear.

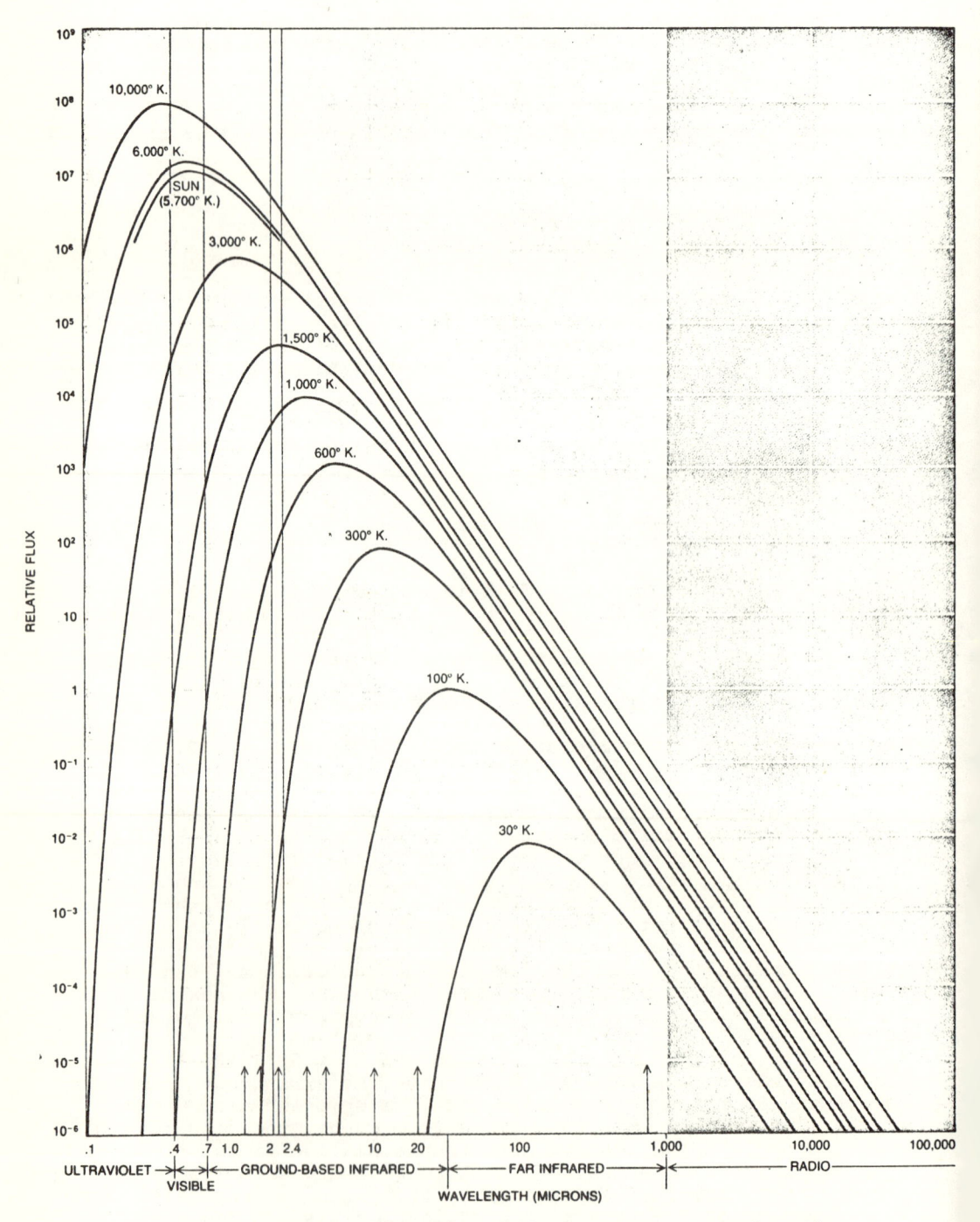

Fig. 1. Emissions from a blackbody versus wave-lengths.

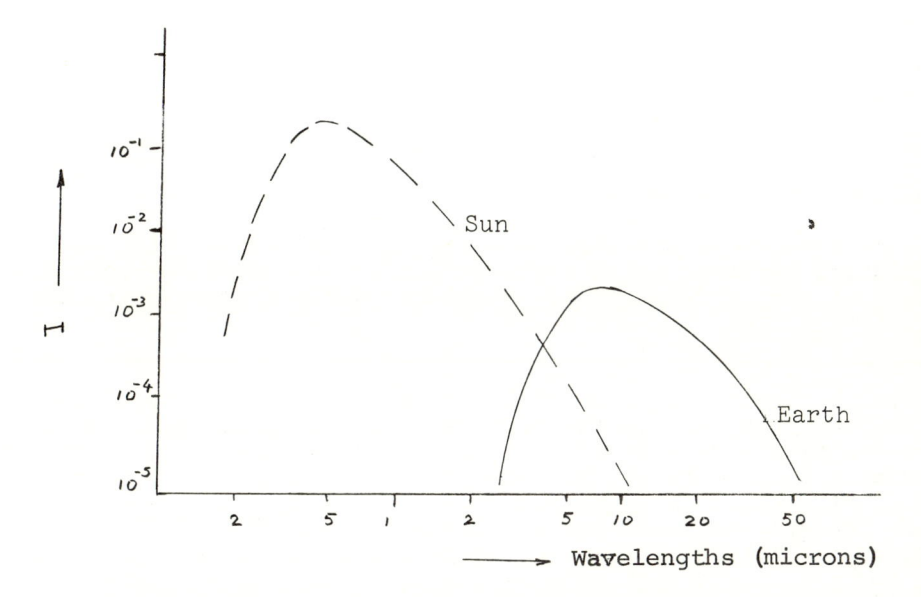

Fig. 2. Comparison of radiation at earth surface from the sun and the emission of blackbody at 50°F (300°K).

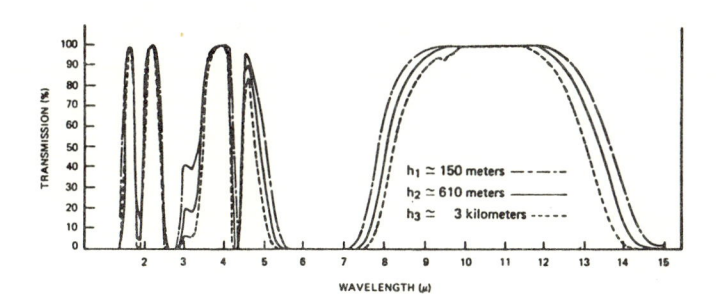

Fig. 3. Plot of atmospheric transmission spectra as a function of altitude, H.

Corrections must be applied to equation (1) in order to measure the real surface temperatures. These corrections are principally as follows: 1) Departure of the sea from a true black body, and 2) Effects of the atmosphere through which the electromagnetic waves pass.

2. CORRECTION FOR NONBLACKNESS

A radiometer measures radiance, R, that is the flux of radiation per unit solid angle per unit wavelength per unit area. The radiance is composed of an emitted and reflected part, namely

$$R = EI + rN \qquad (3)$$

Where R is the radiance of the surface, I is the radiance of a blackbody at surface temperature T, given by eq. (1), E and r are the emittance and reflectance of the surface, and N is the radiance falling on the surface (the sky radiance). The emittance of a plane water surface viewed at normal incidence has been measured as 0.986 and its reflectance as 0.014 (11). The effect of reflection on water temperature measurements was studied by Lorenz in 1968 with an 8-14 µ radiometer (Barnes portable radiation thermometer PRT-4) for clear sky conditions. In Lorenz's experiments (12) the air temperature near the radiometer is used to represent the much lower radiation temperature of the clear sky.

This simplification may be justified since the influence of the long-wave sky radiation is not very strong. It was determined that corrections for measured values of radiation temperature may be derived from radiation temperature and air temperature only. The corrections for surface temperature measurements of smooth water due to reflectivity (specular reflections) for a cloudless sky are shown in Figure 4. For a rough surface (diffuse reflection) the corrections are 0.05 ∿ 0.1°C lower than that prescribed above. In general, the corrections are below 1°C. They become smaller when, for a given surface temperature, the air temperature increases. They increase when, for a given air temperature, the surface temperature increases.

Using a Model PRT 5 Airborne Radiation Thermometer (ART) manufactured by the Barnes Engineering Company, Saunders (11) calculated the nonblackness correction values for various conditions. Some typical computed values are shown in Table 1.

Table 1. Art nonblackness correction for PRT 5

Cloud Type	Cloud Height km	Range of Correction °C
Clear		0.5-0.7
Dense cirrostratus overcast	8	0.4-0.55
Altocu or altostratus overcast	6	0.25-0.4
Stratus or stratucumulus overcast	3	0.2
Stratus or stratucumulus overcast	2	0.1
Stratus or stratocumulus overcast	1	0.1

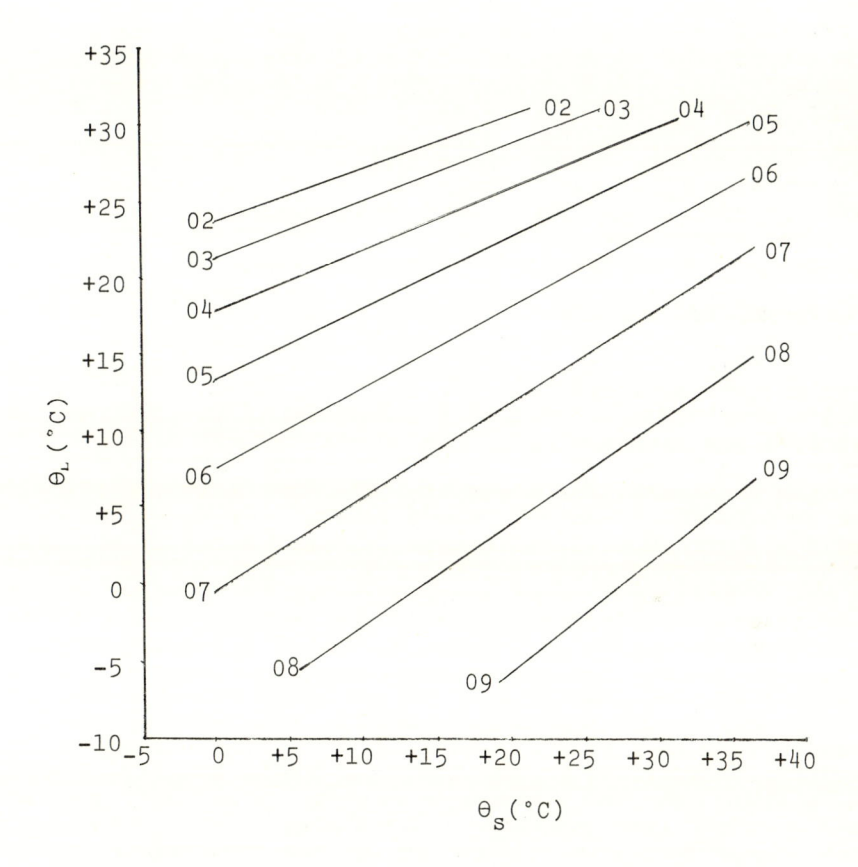

Fig. 4. Corrections for surface temperature measurements of smooth water due to reflectivity vs radiation temperature Θ_s (abscissa) and air temperature Θ_L (ordinate) for clear sky conditions, reproduced from Lorenz (12).

It can be seen from this table that the necessary corrections are relatively small, usually less than half a degree centigrade.

The above corrections are calculated for measurements vertically from above. When the radiometer is inclined at an angle, the influence of reflectivity grows, since the reflectivity of water increases with increasing inclination. Radiometer measurements of water under an angle of 45° from the normal require corrections nearly one and a half times those for vertical measurements. Under 60°, the correction becomes two times or a little more. The quality of water surface temperature measurements may also be improved by a decrease in the spectral bandwidth of the radiometer. Bell (14) has found that the spectral reflectivity of water has a minimum of 0.8% at 11μ wavelength. It increases rapidly with increasing wavelength up to 3.5% at 14μ. If the radiometer's spectral response is limited to the vicinity of 10μ, e.g., from 9μ to 11μ, the difference between the true surface temperature and radiometer reading becomes nearly one-third of the corrections needed for the 8μ to 14μ band. However, this would greatly reduce the signal to noise ratio.

3. ATMOSPHERIC TRANSMISSION

The infrared region of the spectrum is characterized by regions of intense atmospheric absorption. At these wavelengths, molecular absorption becomes very significant; in addition, atmospheric gases play a role. Each of the various gases that comprise the atmosphere has its own characteristic infrared-absorption spectrum. The gases that have the most significant effect include CO_2, N_2O, H_2O, and O_3. For those portions of the infrared spectrum where one or more of these gases is strongly absorbing, the atmosphere is essentially opaque. The spectral intervals at which the atmosphere is reasonably transparent are (approximately, in microns), 0.7-1.35, 1.35-1.8, 2.0-2.4, 3.5-4.1, 4.5-5.5, 8-14, 16-21, and 780-1,000. Those transparent spectral regions are called windows as shown in Figure 3. It is interesting to see that the 8-14 micron band, which is useful for surface water temperature measurements as described above, is also provided with a window. However, the atmosphere is still not totally transparent in the spectral region between 8μ and 14μ. There is absorption and emission mostly by water vapor and carbon dioxide. Corrections for these atmospheric effects still have to be made if accurate remote measurements are required.

The expression for the radiance measured through a column of air is

$$N_s - N = - \int_0^u (N_s - B_o) \frac{d\tau}{du} \, du \qquad (4)$$

where $N_s - N$ is the correction to the observed radiance due to the air column, B_o is the radiance of a black body at the air temperature, τ is the transmittance of the air, du is the incremental intervening mass of air, and u the total mass of the air between the sensor and the sea surface. Some of the absorption of the air in the infrared is due to water vapor, so that to calculate τ, the absolute humidity in the air column should be measured. Also, since B_o is dependent on temperature (equation 1) the temperature lapse-rate must also be known. Fortunately, the corrections are small if the height from which the measurements are made is not too great, and approximate values can often be used. Further, in the area where horizontal mixing is uniform, one correction will suffice for the whole region as long as the height of observation is constant.

If the distance to the target is less than 154 m (500 ft), the atmospheric effects on radiometric surface temperature measurements is negligible. For longer distances, corrections can be determined empirically and by computation. Correction diagrams for flight levels of 305 m (1,000 ft) and 944 m (3,000 ft) was derived by Lorenz (12) from a radiation diagram by Müller and Raschke. He showed that, for a given flight level, the most important factors influencing the radiometer measurements are surface and air temperatures, whereas variations of the humidity and the temperature gradient of the air layer between target and radiometer are less important. The corrections are small when the differences between surface temperature and air temperature are small.

In 1967, Saunders (13) developed a method to make corrections for both the nonblackness of the ocean surface and the absorption and emission of radiation in the air path between water surface and sensor. Based on his findings that a radiometric measurement of a water surface inclined at an angle near 60° from the normal approximately doubles both the influence of the air layer and the influence of reflectivity, he suggested that the difference between a normal and a 60° measurement is the total correction required. In his method successive radiometric measurements both normal and inclined are made as well as oblique to the water surface to determine the correction. Saunders estimates the accuracy of his method to be ± 0.2°C. Later, Saunders (11) revised his procedure to view the surface only normally with an improved Airborne Radiation Thermometer (Model PRTS by Barnes Engineering Company). Saunders also suggested a simple empirical approach to obtain a correction for atmospheric absorption and emission. From the measured radiometric temperature (T) and atmospheric temperature at flight level (T_a), the correction is given by

$$C = \alpha (T - T_a) \qquad (5)$$

where α is expected to have a value in the range 0.06 to 0.15 at an altitude of 300 meters. By making a series of passes over the same piece of water, the value of C can be measured. It should be noted that in this expression, the air temperature measured at flight level is used to represent the temperature of the entire column of atmosphere below. Therefore, the greater the altitude the poorer this approximation will be.

Another approach for atmospheric correction is suggested by Shaw and Irbe (15). They have developed an atmospheric model and a computer program to evaluate theoretically the relative importance of the environmental effects under given meteorological conditions and obtain an adjusted water surface temperature. Based on the theoretical study, a relatively simple graphical method for estimating the water surface temperature was developed. Using airborne radiation thermometers (ART manufactured by Barnes Engineering Company) water surface temperature, it was found that their model of environmental adjustment reduced the root-mean-square difference between Art readings and a ship bucket thermometer from 1.7°C to 0.6°C. Such an improvement can be of great importance when monitoring water surface temperature near a critical value, such as the freezing point.

On the basis of the two models discussed above, Tien (16) has established simple analytical forms for environmental corrections as given below:

$$R_{bo} - R_h = \frac{A}{\varepsilon(1-A)} (R_h - R_{ba}) + (\frac{1-\varepsilon}{\varepsilon}) (R_h - R_s), \qquad (6)$$

where R is the radiance and A is the absorption on emission of the entire air column between surface and sensor. Subscripts o and s refer to sea surface and sky; h the altitude of sensor; bo and ba the blackbody at surface and the air temperatures respectively, and ε the water surface emittance. In the above expression, the first term on the right-hand side characterizes the effect of the intervening atmosphere, while the second term accounts for the sky radiation effect due to non-blackness of the water surface. Based on the measured values of R_h, R_{ber} (from air temperature), and R_s, and the tabulated information of ε and A (a function of air temperature, humidity and pathlength), R_{bo} can be obtained and the surface temperature can be calculated. By taking two-term truncated Taylor series expansions of R(T)'s around $R_h(T_h)$, direct temperature corrections may be obtained by the following approximation:

$$T_{bo} - T_h = \frac{A}{\varepsilon(1-A)} (T_h - T_{ba}) + (\frac{1-\varepsilon}{\varepsilon}) (T_h - T_s) \qquad (7)$$

The equivalent sky temperature ranges from -50°C in a clear sky condition to 0°C in a low cloud condition, resulting in a sky radiation correction term of about 0.1 to 0.7°C obtained from the above equation. For the correction for atmospheric absorption and emission, the effective absorption, A, for an air path of 300 meters ranges from 0.05 to 0.15, thus indicating a numerical range of the correction coefficient A/ε(1-A) from 0.06 to 0.15.

III. AIRBORNE INFRARED REMOTE SENSING SYSTEMS

There are a variety of airborne infrared systems commercially available at the present time. Among them, the Bendix Thermal Mapper (BTM), Barnes Airborne Radiation Thermometer (ART) and the Daedalus Scanner Systems, have been used by many research groups for water surface temperature measurements. A brief discussion of these systems are given below.

1. BENDIX THERMAL MAPPER AND MULTIBAND SCANNER

A. BENDIX THERMAL MAPPER

The Bendix Aerospace System Division first introduced its thermal infrared Line Scanner, the Bendix Thermal Mapper (BTM), in 1966. It consists of three basic modules: the Scan head, the control-console power-supply module, and the roll-compensation unit. The standard BTM has a temperature difference sensitivity of 0.5°C with indium centimonide detector filtered to 3 to 5.5 microns. It is small, light and simple enough to go on a light, twin-engine aircraft. It includes a self-contained film cassette so that the system could be handled much as a standard aerial photographic system, with a roll of 70mm film as the immediate output, ready for processing upon landing. The standard model does not have the capability of magnetic tape recording and multiple channels.

This thermal mapper was used in February 1969 to obtain Thermal Infrared Imagery and Thermal Contours at Turkey Point Power Plant, south of Miami, Florida. Isotherms of surface water temperature were derived from thermal mapper imagery (3.7 to 5.5 micron region) by means of measured water temperatures and densitometer values. The actual water temperatures were provided by the University of Miami and six surface measurements were used as "ground truth" for the densitometric analysis. Twenty densitometer profiles were made parallel to the scan lines on the imagery and the resulting percent transmittance values were converted to temperature values. Fluctuations in density from one scan line to the next required that the gray area at the edge of the image be used as a reference assumed to be 25% transmittive. The densitometer profiles were converted to temperature at intervals of 1°C and plotted on a base map and used to manually construct the thermal contours. The resulting isotherms were within ± 0.5°C of the actual surface temperatures. This proved to be a good method for acquiring quantitative temperature measurements over a large area. Isotherms derived from this mapping is shown in the Figures.

The Bendix Company later introduced two modified models, Bendix Thermal Mapper Model LN-2 and Model LN-3. They are equipped with the more popular Mercury-Cadmium-Telluride (Ng:Cd:Te) detector (peaked at 10 microns) with a filtered spectral bandwidth of 8 to 12.5 microns. The data from these thermal mappers can be tape recorded.

B. BENDIX MODULAR MULTIBAND SCANNER

Bendix recently introduced a new scanner system called Modular Multiband Scanner (M^2S). It is a true (fully calibrated) imagery radiometer. M^2S not only has blackbody references, it also carries its own UV visible reference, a quartz-halide lamp which is transfer calibrated from the National Bureau of Standards. There is also a skylight reference in this same spectral region.

The M^2S is a fully digitized machine. The data is recorded on high density digital tape. This system has a dynamic range over eight times that of the thermal mapper analog film recorder.

One drawback of the M^2S system is its weight. With the tape recorder and an inevitable camera or two, the payload in a light aircraft, equipped with M^2S, will be about 300 lbs and up, depending on the cameras. This is apt to exceed safe operating limits for a single engine aircraft. However, a light twin should be quite satisfactory.

The Bendix Company has been flying its system in a Beach Queenaire aircraft and the Swiss Air Photo Agency is using it in the same type of aircraft. NASA Lewis Center plans to install its scanner in a DC-3, and NASA Houston is using a P3A. Researchers from Japan and Argentina have also adopted the M^2S for their remote sensing usage.

The aforementioned scanners can be either purchased or leased from Bendix. They have equipped aircraft and can also fly missions for their customers.

2. THE AIRBORNE RADIATION THERMOMETER (ART) BY THE BARNES ENGINEERING COMPANY

The Barnes Engineering Company first introduced Model IT-2S infrared thermometers around 1966. It was later modified and designated IT-3 and PRT-5. The airborne radiation thermometer (ART) is considered to have the capacity of reliably measuring the apparent blackbody surface temperature of a large body of water in a relatively short period of time. It has a band-pass filter system which limits the wavelengths detected by the radiometers to the general region from 7.5 to 16μ. This filter window is centered approximately on the "water vapor window" at 10μ and is close to the maximums in the Planck emission curve for 300°F.

The ART has been used by many research groups and the experiences of some of these groups in using it for water surface temperature measurements are discussed briefly below.

Richards and Massey (17) gave a complete description of the installation of the ART in the aircraft, the problems encountered in obtaining reasonably smooth and accurate temperature readings, and the conduct of ART test.

Shaw and Irbe (15) of the Canada Atmospheric Environment Service have used ART in their investigation since 1966. In their operation, the ART was installed initially in a Lockheed 14 aircraft. Later it was flown in a Beechcraft 18 and Piper Aztec C aircraft. A strip chart recorder conducting a continuous temperature trace was connected to the ART. Comparisons between ART temperature readings and surface temperature measurements were obtained in several ways. During the early evaluation of the instrument in 1965-1966, the aircraft was flown over the Great Lakes research vessel CCGS Porte Dauphine at several heights above the water. Simultaneous ART temperature readings and ship bucket temperatures were recorded. On later routine ART survey flights in 1967-1971, comparisons were made between ART readings and bucket temperature on the Ponte Dauphine and on the limnological ships of the Canada Center for Internal Waters (CCIW). Readings were taken from floating thermistors on CCIW buoys in Western Lake Ontario and from thermistors monitored on dams across the St. Lawrence River at Cornwall, Ontario and Lachine, Quebec. Upon adjustment of ART readings for environmental effects, the airborne remote water surface measurements had an accuracy within 0.5°C of the observed water temperature.

Saunders (11) of Woods Hole Oceanographic Institution has also used ART for remote measurements of ocean surface temperatures. After corrections for nonblackness of the ocean and for absorption and emission in the intervening atmosphere were made, they could achieve an absolute accuracy of ± 0.2°C when the measurements were made carefully from low altitudes.

3. THE DAEDALUS SCANNER SYSTEM

A Daedalus infrared thermal scanner system was used by a University of Newcastle research team (18) to study Thermal

Plumes. This system is equipped with a 8-14 micrometer detector. It projects through the floor of a survey aircraft, and has a scan angle of 120° centered about the vertical. The scanner contains a horizontally mounted telescope with its axis aligned along the direction of flight of the aircraft. A mirror rotating at 3,600 RPM is mounted at 45 degrees to the telescope and directs heat radiation from the earth into the system. A one-third rotation of the mirror covers a complete step perpendicular to the scanner axis. Optical resolution is approximately 1.7 milliradians so the ground area from which the detected signal is averaged is a function of height. Considerable overlap occurs between successive sweeps. The video signal from the infrared detector is amplified and recorded on magnetic tape in the aircraft. This is later played into a printer-viewer which has been modified to produce either a positive or negative transparency, or heat picture, of the scene overflown. A microdensitometer is used to extract grey scale information from the transparency and to present this information in the quantitative form needed for contouring. If accurate quantitative information is required from this thermal scanner system, two problems will have to be considered, namely: 1) the geometrical distortions of the image and 2) the calibration of the grey scale to give a temperature scale. Non-linear distribution of position across the flight path is inherent in the standard system but this error is insignificant over the central 80° or so of the total 120° scan. A further source of image distortion can arise from the aircraft drift angle. Variation in the aircraft flight track from a straight line will also introduce distortion into the image. An aircraft mounted drift sight and a Doppler navigation system can monitor actual drift angle and speed, and consequently enable the operator to eliminate drift angle distortions and along track versus across track scale errors. The grey scale to temperature scale calibration in water is achieved relatively easily by recording several temperatures chosen to be as markedly dissimilar as possible.

A Daedalus Scanner System is also being used by a University of Miami research team to study the thermal discharges in South Florida. This is a Daedalus DS-1250 Multispectral Line Scanner System owned and operated by NASA Kennedy Space Center. It includes a DS-1220 Basic Dual Channel Line Scanner System, a DS-1050 Basic Spectrometer System and a Roll Correction System. In the Dual Channel Line Scanner System, a split-beam optical system is equipped to provide simultaneous operation of two detectors within a single scan head. Broad-band coverage from 0.38 μ to 14 μ is attainable with this system by selecting the appropriate detectors. In this study, a Hg:Cd:Te detector is used for sensing 8-14 μ radiation.

IV. SURFACE TEMPERATURE SENSING FROM SATELLITE RADIOMETER

The space resolution of thermal IR temperature measuring system used in the present generation of satellites does not provide as much detail as is desirable. However, the resolution of these thermal sensors is improving with each new generation of satellites. Clouds may blank satellite-borne sensors from seeing the ocean surface on some occasions but the frequency of observations is still better than obtainable by other means for similar costs. It is believed that, ultimately, satellites can be used not only for global ocean surface temperature mapping, but should be able to provide the accurate, high-resolution water temperature data needed for monitoring thermal pollution from power plant hot discharges and for model studies. Some Spacecraft programs which provide I.R. systems useful for water surface temperature measurements are discussed here below.

1. THE EARTH RESOURCES SURVEY FLIGHT PROGRAMS

In the Earth Resources Survey Program of the National Aeronautics and Space Administration (NASA), three Space Flight experiments are included: The Earth Resources Technology Satellite A (ERTS-A), launched in March, 1972; the Earth Resources Experiment Package (EREP) of the manned Skylab Orbital Facility, launched in April, 1973; and ERTS-B which has not yet been launched.

The ERTS-A and B sensors payload consists of (1) a multi-spectral TV system using return-beam vidicon (RBV) cameras, (2) a multi-spectral scanner (MSS) system, and (3) a data collection system for collecting data from sensors at known locations in the earth. The MSS is sensitive to the following wavelengths: band 1, 05 $\sim$ 0.6 μ m; band 2, 0.6 $\sim$ 0.7 μ m; band 3, 0.7 $\sim$ 0.8 μ m, and band 4, 0.8 $\sim$ 1.1 μ m. Comparable ground resolutions for the Scanner are approximately 80 meters. In addition to this, the ERTS-B was originally designed to have a band 5 with 10.4 to 12.6 μ m spectrum and 220 meter ground resolutions. In view of this, the ERTS-B should be useful in remote sensing sea water surface temperatures. However, it is learned that the ERTS-B would not provide the 10.4 $\sim$ 12.6 μ m thermal I.R. band as expected. Until a thermal I.R. band is provided in its future flights, the ERTS series of satellites will not be suitable for thermal pollution studies.

2. THE AIR FORCE DEFENSE METEOROLOGICAL SATELLITE PROGRAM (DMSP)

This program was formerly known as the Data Acquisition and Processing Program (DAPP). This unique and valuable data system is employed by the Department of Defense and the United States Air Force's Air Weather Service (AWS), and has been made available to the public recently. These sun-synchronous satellites are in polar orbit, approximately 450 nautical miles above the earth and have a period of 101 minutes. In the operational mode, two satellites provide imagery data every six hours over any spot on the globe (sunrise, sunset, local noon-midnight). This imagery includes the visual, near infrared and infrared spectral intervals over a 1,600 to 1,800 nm swath below the satellite. Real time data within the acquisition range, approximately a radius of 1,500 miles of the receiving station, is provided to several tactical sites, while the Air Force Global Weather Central (AFGWC) at Omaha, Nebraska, receives stored data of global coverage.

The present capability of the DMSP includes the following:

DATA TYPE	SPECTRAL INTERVAL	RESOLUTION
VHR Data	.4 ∿ 1.1 μ	0.3 nautical mile
WHR Data	8 ∿ 13 μ	0.3 nautical mile
HR Data	.4 ∿ 1.1 μ	2.0 nautical mile
IR Data	8 ∿ 13 μ	2.0 nautical mile

The Very High Resolution data (VHR) "sees" in the visible and near infrared. The product output is a positive transparency that can be received at 1:7,500,000 or 1:15,000,000 scale. Sensor altitude and attitude variations are compensated for and foreshortening at the edges is removed. Flexibility is further provided in that the data can be enhanced to bring out cloud/land and cloud/cloud significant contrasting features. The infrared products (WHR and ZR) receive emitted thermal energy from 310°K to 210°K. Flexibility is provided to receive 1°K increments in grey shade steps from 2 to 16 or a 25°K span of temperatures can be divided into 16 grey shade steps. Electronic circuitry in the sensors converts the sensed thermal IR energy directly into equivalent blackbody temperature, making temperature the directly displayed parameter within the limits indicated (19,20).

The present generation of these satellites uses the 0.3 nm and 2.0 nm resolution lenses interchangeably. Normally, the 0.3 nm lens is used to acquire WHR data on the 8 ∿ 12 μ thermal IR channel only for the midnight pass. During daylight passes, the 0.3 nm lens is usually switched to the 0.4 ∿ 1.1 μ visual wavelength channel and the 2.0 nm lens is used on the IR channel. In 1975, a 0.3 nm lens is to be available at all times on the IR channel of a new series of these satellites. Presently, the 0.3 nm lens can be used for IR data during the daylight passes by special request.

A University of Miami research team has been using I.R. dara from this program in its thermal pollution study.

3. THE NATIONAL OCEANIC AND ATMOSPHERIC ADMINISTRATION NOAA SERIES SATELLITE DATA

The National Oceanic and Atmospheric Administration NOAA-2 and NOAA-3, ITOS series of satellites provide thermal IR data in the 10.5 to 12.5 micron region. This modified version of the improved TIROS operational satellite (ITOS D-G) became operational in 1972. A Very High Resolution Radiometer (VHRR) with 0.5 nm resolution provides the thermal IR data. These satellites are in sun synchronous, 790 nm polar orbit with a period of 115.14 minutes (21). Thermal IR data from these satellites are also being used by the University of Miami research team in its thermal pollution program.

These satellites also have a Vertical Temperature Profile Radiometer (VIPR) for obtaining atmospheric temperature and water vapor profiles by measuring spectral radiance in eight intervals in the infrared vapor. Energy is measured in the 11 micron atmospheric window to deduce radiance free from effects of clouds in the other seven channels. Then six cloud-free, discreet, narrow intervals in the 15 micron CO_2 region are measured to infer the atmospheric temperature profile. Finally, radiance in a single spectral interval near 18.7 microns in the rotational water vapor absorption band is measured and used to estimate the water vapor concentrations in the troposhere (21,22,23,24). This VTPR System is presently used by National Environmental Satellite Service (NESS) primarily to obtain temperature profiles and moisture data over the oceans or other places where no radiosonde balloon measurements are available. However, if needed, spectral computations can be carried out to obtain water vapor data for applying absorption corrections to thermal IR measurements of sea surface temperatures.

V. THE UNIVERSITY OF MIAMI THERMAL POLLUTION PROGRAM

An interdisciplinary team of researchers at the University of Miami has been carrying out a program in Thermal Pollution Studies. Among their activities is a research project sponsored by the NASA Kennedy Space Center aiming at the development of a generalized, three-dimensional, predictive, numerical model for determining water temperature, velocity and salinity distribution in coastal regions receiving hot discharges from power plants. Remote Sensings from aircraft as well as satellites are used in parallel with in-situ measurements to provide information needed for the development and verification of the mathematical model. In addition, experiments are conducted in order to improve the accuracy of the thermal remote sensing systems by directly relating them to the thermal radiation from the sea surface, and by better accounting for the absorption in the atmosphere. Additionally, there is a study program underway for the development of active remote sensing systems to measure the vertical distribution of the water turbidity and temperature below the water surface. Three sites are used for the model development and experimental studies. These three power plant sites are the Florida Power and Light Company, Turkey Point and Cutler Ridge facilities and Port St. Lucie Nuclear Power Plant units which are under construction on Hutchinson Island. The Turkey Point and Cutler Ridge facilities are located at a shallow lagoon-type estuary, while the Hutchinson Plant will discharge into the off-shore continental shelf. The contrast of these sites should provide a good test for the generalized nature of the mathematical model. While the mathematical model development may be beyond the interest of the present topic, the experimental part of the study will be discussed briefly here.

Two experiments were conducted to measure surface temperature in Biscayne Bay using aircraft and satellite I-R pictures together with in-situ measurements from a boat. The in-situ measurements were synchronized with aircraft overflights.

1) July 29, 1974: The area of experimentation covered the thermal plume at Cutler Ridge and a corridor normal to the shoreline leading to the Atlantic Ocean.

2) October 24, 1974: The ground truth measurements were made on a transect across the Cutler Ridge thermal plume and at selected locations in the Bay. The aircraft IR data was gathered over a large area covering the perimeter of the Bay, as well as the Cutler Ridge thermal plume.

These experiments are discussed in the next two sections from the point of view of aircraft and satellite remote sensing.

1. AIRCRAFT REMOTE SENSING EXPERIMENTS

The Airborne Scanner system used in this study is a Daedalus DS-1250 Multispectral Line Scanner System owned and operated by the NASA Kennedy Space Center. It includes a DS-1220 Basic Dual Channel Line Scanner System, a DS-1050 Basic Spectrometer System and a Roll Correction System. In the Dual Channel Line Scanner System, a split-beam optical system is equipped to provide simultaneous operation of two detectors within a single scan head. Broad band coverage from 0.38 µ to 14 µ is attainable with this system by selecting the appropriate detectors. In this study, a Hg:Cd:Te detector is used for sensing 8-14 µ radiation. This system is installed in a C-45H TRI-BEECHCRAFT (NASA-6) together with the photographic sensors. It is also equipped with an environmental measurement system for meteorological measurements such as air temperatures and dew point. Its data output can be magnetic tape imaged on film or computer processed. It can also provide real time CRT image.

The data analysis facility of the Kennedy Space Center is being utilized for this study. It includes a digicol, a microdensitometer, an additive color viewer, a zoom transfer scope, a photointerpretation station, a stereo viewer, a versatile plotting table, a variscan, a thermal scanner, and analyzing tape drivers. A G.E. Image 100 Interactive Multispectral Image Analysis System has recently been added to this facility.

The first research flight was carried out on July 29, 1974, over Biscayne Bay at Miami, Florida. The aircraft flew at 1,500 ft altitude and made twelve west to east data passes across Biscayne Bay between 0800 EDT(1200 Z) and 1300 EDT(1700 Z) while NOAA and Air Force satellites with thermal IR radiometers were passing overhead and a boat was gathering ground truth data in the Bay.

The 8-14 micron IR Thermal Scanner was used to remotely sense the sea surface temperature from the aircraft. Its field of view is 2.5 miliradians, and it scans through 90 degrees of arc normal to the flight path of the aircraft. It has an 18°F usable dynamic range and for this study it was calibrated to measure temperatures from 76°F to 94°F. The analyzing readout was recorded on magnetic tape aboard the aircraft. These data were later transferred from the tape to 70mm film at the Kennedy Space Center. This film was analyzed by a densitometer with a calibration of the film grey scale density in terms of temperature. A 9" x 9" aerial camera was also used to take color photographs along the flight path on some data passes. The purpose of this was to record the positions of the boat and some key landmarks in addition to possibly showing turbidity patterns in the Bay.

In the in-situ measurements, besides current, salinity, turbidity and other related data, water surface temperature and vertical temperature profiles were gathered by means of a thermistor string from the boat. The thermistors are standard 4,000 ohm semiconductors whose resistance is a function of the temperature. In addition, a portable thermal IR radiometer was used to sense water surface temperatures from the boat for atmospheric correction study.

The second experiment was carried out on October 24, 1974, over Biscayne Bay. Experimental data gathered in this experiment are being analyzed.

2. SATELLITE I.R. REMOTE SENSING EXPERIMENTS

Up to the present, thermal IR sea surface temperature measurements from satellites have been used for macroscale studies over the broad oceans where one reading per degree of latitude and longitude was usually adequate. The study being carried out at the University of Miami is to identify problems and limitations of these satellite data and to develop methods for enhancing and enlarging thermal IR displays for mesoscale sea surface temperature measurements. This is a pioneering effort. Even if it may turn out to be impractical with the presently available resolution, the study should set the foundation for the future when spacecraft thermal IR will be improved to provide more accurate information with better resolution.

Two kinds of satellite data are used in the University of Miami study. The thermal IR data from the NOAA-2 and NOAA-3 satellites are in the 10.5 to 12.5 micron region with 0.5 nautical mile resolution. It is received at Wallops Island, Virginia, and processed at the National Environmental Satellite Service (NESS) facility in Suitland, Maryland. Data for the studied area are available on southbound passes at approximately 1945 EDT(1345 Z). For minimum distortion and maximum resolution, experiments are scheduled whenever possible, to have the satellite pass directly over the research area. The NOAA-2 and NOAA-3 orbits are not identical and both have a precession of about one degree of longitude per day in order to remain sun-synchronous. Therefore, the orbital track of one of these satellites is usually better than the other for the experiment on a given day. In the first preliminary experiment on the morning of July 29, 1974, the NOAA-3 satellite pass was too far west of the research area for optimum use. Thermal IR photographs were obtained from the NOAA-2 satellite which passed east of Florida as shown in Figure 5. South Florida is in the upper left-hand corner of this photograph with Cuba s-own near the center of the photograph. Unfortunately, a trough of low pressure was along the east coast of Florida and it produced extensive clouds and rain. The area of interest at approximately 25.5°N and 80.3°W is marked by a NE-SW band of clouds over the coastal area. The National Environmental Satellite Service in Washington is preparing an enlarged and enhanced thermal IR photograph of the Miami area in place of Figure 5. However, there is little hope of using it to measure sea surface temperatures in Biscayne Bay because of the cloud cover. Nonetheless, this preliminary experiment has served its objective to develop satellite data processing techniques to be applied to future experiments. Two more experiments have been conducted since then. Data gathered is presently being analyzed.

Also used in the study are data from the Air Force DMSP satellite thermal IR data. These data are in the 8 to 13 micron spectrum with 0.3 nautical mile resolution. At the present time, DMSP thermal IR data are obtained through the STAFFMET Section Detachment 11, 6 Weather Wing, Air Weather Service, Patrick AFB, Florida. A DMSP local readout and data processing van is scheduled to be located at Kennedy Air Force Station, Florida, in 1975. The Air Force DMSP satellite thermal IR data for the July 29 experiment are presently being processed at Patrick Air Force Base, Florida. A DMSP satellite thermal IR photograph with one-degree Kelvin gray scale steps taken on March 24, 1972, is presented in Figure 6, for informational purposes. This illustrates the capability of the DMSP and is typical of the quality of sea surface temperature data

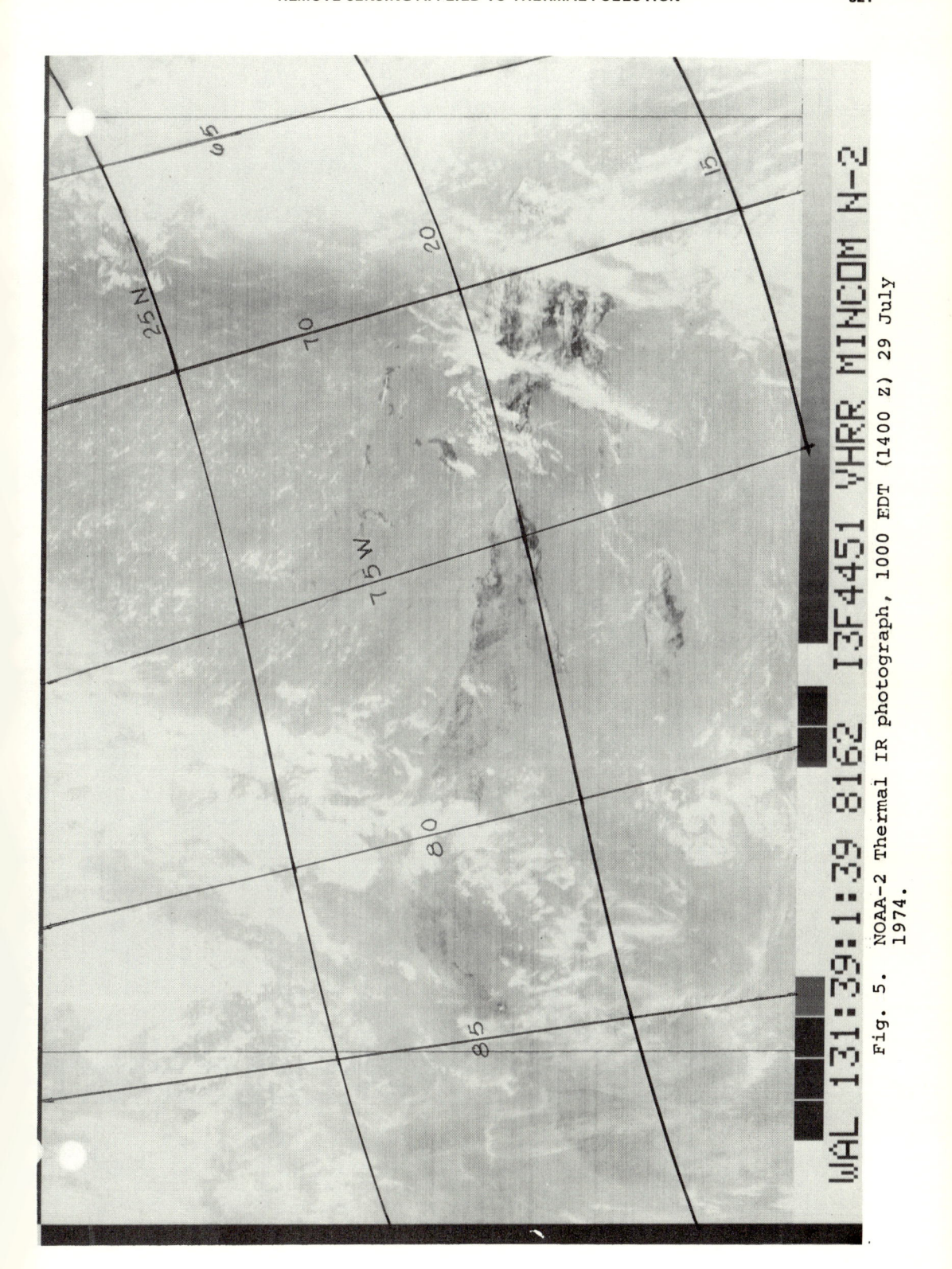

Fig. 5. NOAA-2 Thermal IR photograph, 1000 EDT (1400 Z) 29 July 1974.

Fig. 6. Thermal IR (8-13µ) data from the air force DMSP satellite early on March 1972. (White colder than 294°K, light gray 294-295°K, dark gray 295-296°K, black warmer than 296°K). 0.3 nm resolution.

that can be obtained in the absence of clouds.

3. RADIOSONDE DATA FOR WATER VAPOR CORRECTIONS OF IR MEASUREMENTS

Radiosonde balloons are launched daily at Miami International Airport and Kennedy Air Force Station at 0800 EDT (1200 Z) and 2000 EDT (0000 Z). They obtain pressure, temperature, humidity, wind speed, and wind direction data up to altitudes in excess of 50,000 ft. Most of the atmospheric moisture is contained in the troposphere below 30,000 ft. Therefore, these soundings can be used to calculate the precipitable water that will attenuate the thermal IR measurements of sea surface temperature from aircraft or spacecraft on a given day.

Figure 7 is the Miami sounding for 0800 EDT on July 29, 1974, when the NASA-6 aircraft research flight was made over Biscayne Bay. The precipitable water content of the atmosphere was obtained by graphically determining the mean vapor pressure (e) in milibars of the atmospheric layers between plotted points on this sounding. The water content for each of these layers was then calculated using the equation:

$$W = \frac{0.622}{g} e \ln (P_2/P_1) \qquad (8)$$

where g is the gravitational acceleration, e is the vapor pressure of water, P_1 is the pressure at the top of the layer, and P_2 is the pressure at the bottom. W, the total amount of water precipitable in centimeter, was then obtained by summing these values for all layers of the sounding.

Table 2 presents the precipitable water computed by layers for the 1200 Z sounding, July 29, 1974. These data can be applied to the nomograms in Figure 8 to obtain the temperature corrections for the thermal IR sea surface temperatures measured from satellites. Since the atmosphere was extremely moist on this occasion, it is necessary to extrapolate the Kelvin temperature curves in Figure 4 up to the 5 cm precipitable water line. Figure 8 also provides data for lower altitudes that can be used to apply corrections to IR temperature measurements made from aircraft.

4. DIGITIZED NOAA-3 SATELLITE DATA

Figures 9 and 10 are digitized and contoured temperature maps prepared by computer from the NOAA-3 VHRR, 10.5-12.5 micron, IR data in the vicinity of Miami, 24 October 1974. Figure 9 is a 1-km grid based upon the minimum resolution of the satellite and Figure 10 is approximately a 2-km N-S by 3-km E-W grid that includes space-scale data averaging. The coldest areas are radiation from clouds and can be ignored for our purposes. The isotherms are at 4°C intervals and all areas where the initial satellite measured temperatures are 16°C or colder have been stippled to indicate clouds. A north arrow is shown on each of these illustrations because the NOAA-3 satellite track is inclined 16° from the meridians.

The data in both of these figures are uncorrected for water vapor absorption. The Miami radiosonde at 12 GMT (7 AM Eastern Standard Time) on 24 October 1975, which was less than two hours

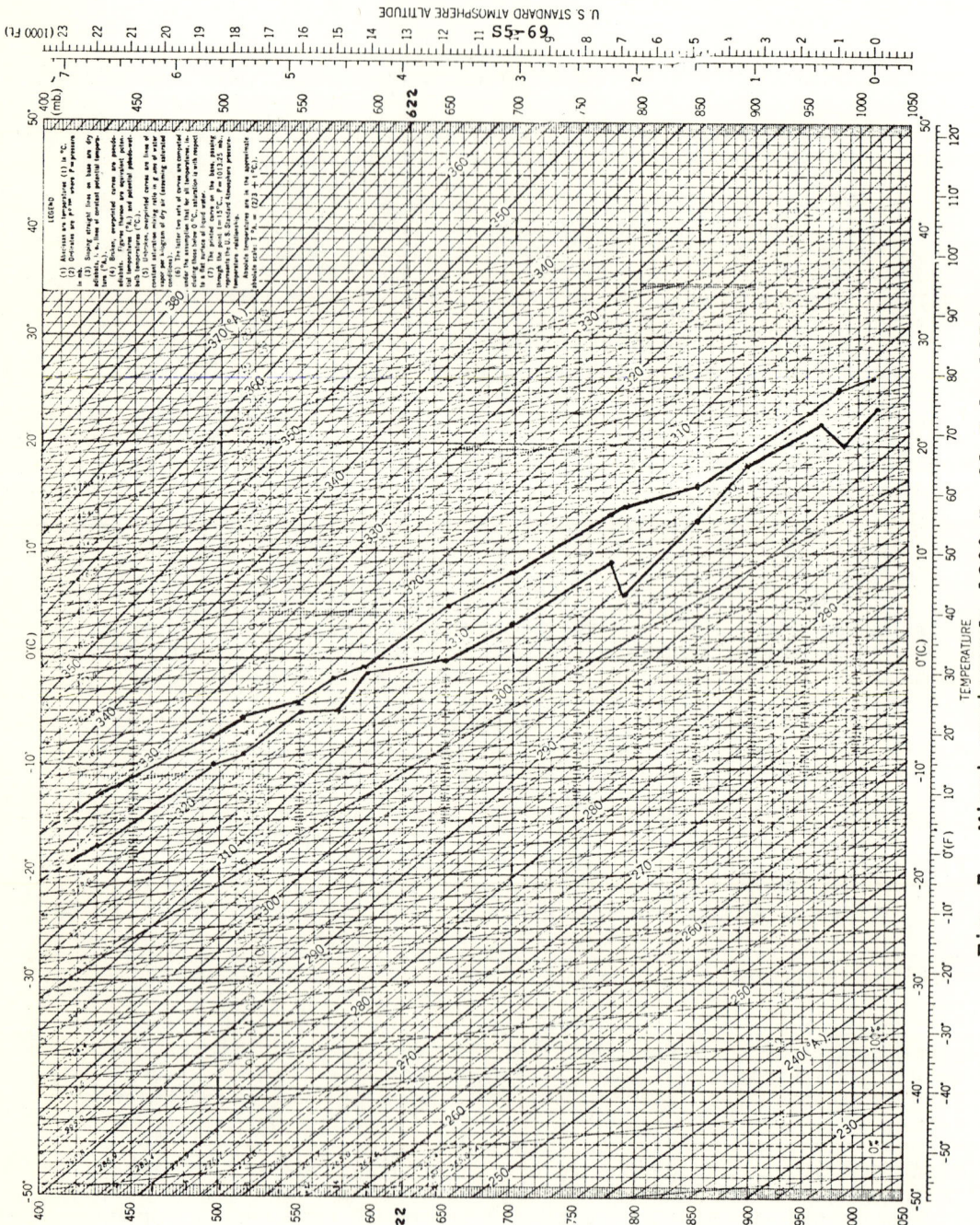

Fig. 7. Miami rawinsonde 1200 Z, 29 July 1974.

Table 2. Precipitable water at Miami, 1200 Z, 29 July 1974

Approximate Altitude Feet	Layers P_1/P_2	Mean e for Layer	Accumulative W cm for Layers
23,600	400	1.38	5.376
21,400	438		
	438	2.25	5.299
18,300	500		
	500	3.1	5.113
17,400	518		
	518	3.7	5.045
16,000	550		
	550	4.4	4.907
14,600	580		
	580	5.1	4.762
13,800	598		
	598	6.2	4.665
11,400	658		
	658	7.0	4.296
9,900	700		
	700	9.8	4.027
7,100	780		
	780	10.6	3.367
6,700	790		
	790	12.4	3.283
4,800	850		
	850	17.8	2.718
3,300	895		
	895	22.8	2.1479
2,000	940		
	940	29.0	1.4516
700	985		
	985	28.0	.6081
Surface	1020		

before the satellite pass, gave a total liquid water content of 4.09 cm. Since the satellite measured surface temperatures that we are interested in are near 290°K and the Nadir Angle is near 0°, the correction to be added is 9°C, see Figure 8. In Figures 9 and 10 the solid line is the 20°C indicated, 29°C water-vapor corrected, isotherm, and the dashed isotherm in Figure 10 is 24°C indicated or 33°C corrected. The temperature resolution of the VHRR sensor is approximately ± 0.5°C for a 300°K scene. This can cause the computer to produce some small contoured areas that may not be significant. The area of these is usually one resolution cell of the sensor and they are indicated to be only one degree colder or warmer than their surroundings. Most of these small contoured areas are not visible in the reproduction of Figure 9 and the computer averaging in Figure 10 eliminates them.

These computer printouts were prepared by the National Environmental Satellite Service, NOAA, Washington, D.C. Because of the great enlargement of a small area for mesoscale analysis, a latitude and longitude grid coordinate system was not available in the program for these printouts. They must be located geographically by use of photographs made from the visual spectrum band of the multi-spectral scanner aboard the satellite. Such photographs were provided along with the digital printouts for this purpose. Nevertheless, precise geographic location of the grids remains a problem. Future mesoscale systems will need geographic coordinates programmed into the computer. The large-scale oceanographic grid system presently used for these satellites does include geographic coordinates in the program.

VI. DISCUSSION AND CONCLUSIONS

Remote sensing of thermal pollution is an effective technique in monitoring large areas of ecological interest. Satellite IR imagery brings into perspective large-scale thermal anomalies. The airborne IR systems are, however, more useful in near field and mesoscale analyses. Especially in the case of thermal plumes, remote-sensing by airborne radiometers is indispensable. The resolution is sufficient to define thermal-plume boundaries and therefore map the regions of ecological disturbance.

The two most important areas of suggested development in remote sensing technology are:

1) <u>Refinement of resolution</u>: This is imperative for satellite imagery since thermal plumes are often of maximum extent of a few miles. The resolution of present-day satellite IR imagery cannot identify the structure of thermal plumes effectively. The airborne IR imagery though of a finer resolution can give a general picture of the plume structure. However, for applications involving mathematical modeling, detail thermal mapping in the plume region is necessary.

2) <u>Correction for atmospheric path-lengths</u>: The effect of atmospheric factors on the IR imagery depend on the path-length and meteorological parameters. For overflights at altitudes of 500 ft or less the effects are negligible. However, since flights are usually at higher altitudes correction techniques need to be improved. The need for accurate correction procedures is critical for satellite IR data because of the very large path-lengths.

Extensive research is at present underway at the University of Miami and at other institutions to improve the state-of-the-art of remote sensing, both in terms of higher resolutions and

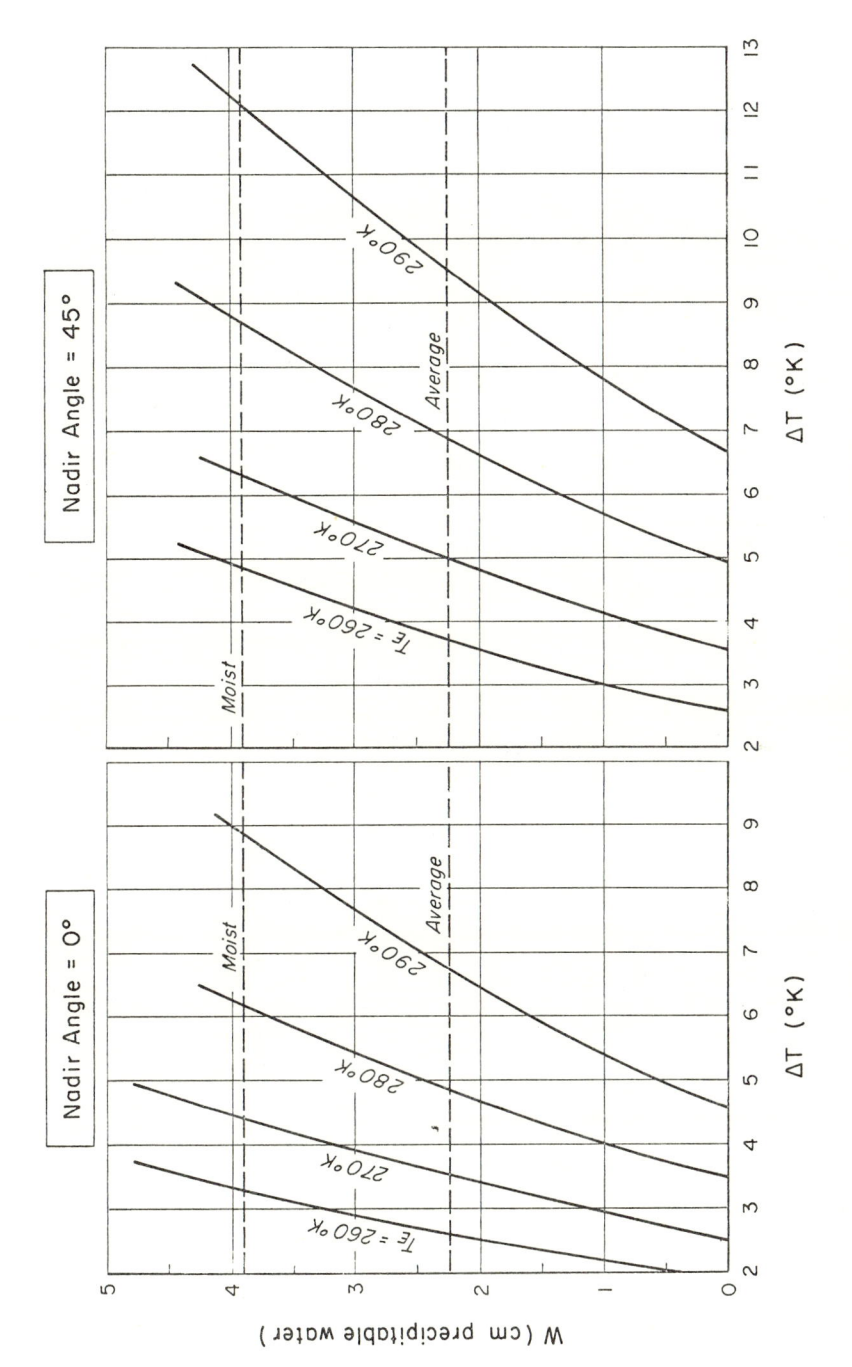

Fig. 8. Clear-sky temperature corrections.

REMOTE SENSING THERMAL POLLUTION

Fig. 9. Digitized IR temperatures from NOAA-3 satellite, 24 October 1974, 1342 Z (0842 EST). 1-km Grid squares, temperatures not water-vapor corrected.

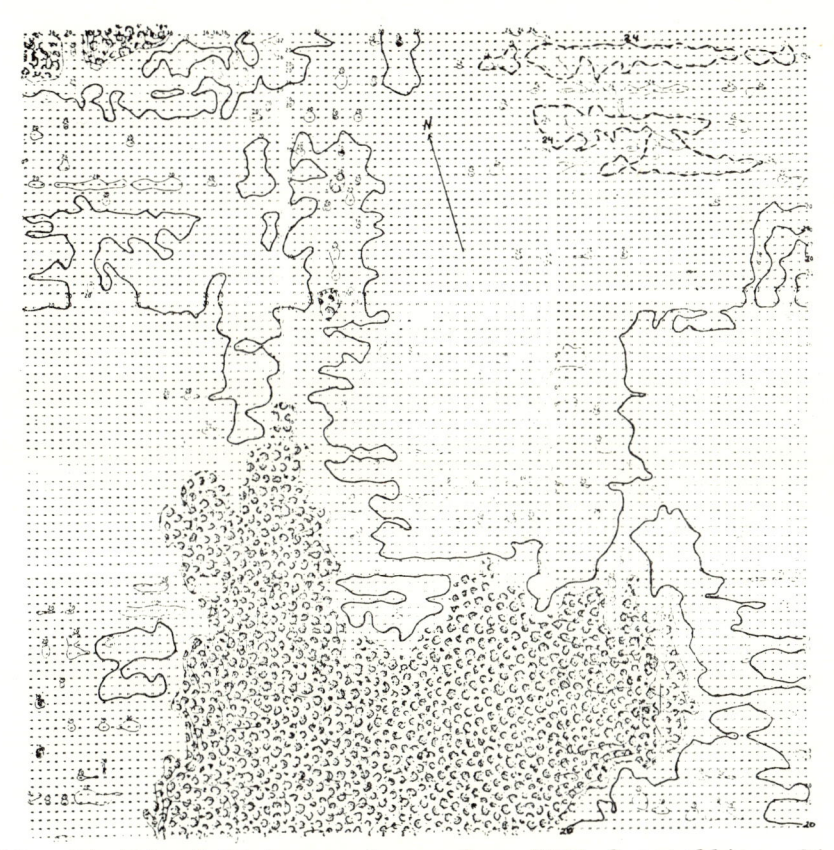

Fig. 10. Digitized IR temperatures from NOAA-3 satellite, 24 October 1974, 1342 Z (0842 EST). 2 x 3-km grid, temperatures not water-vapor corrected.

better atmospheric correction procedures.

Table 3. Precipitable water at Miami, 00 A, 25 October 1974
(7 PM, 24 October 1974)

Approximate Altitude Feet	Layers P_1/P_2	Mean e for Layer	Accumulative W cm for Layers
23,600	400	0.46	3.94
21,400	430		
	430	0.65	3.91
18,300	500		
	500	0.99	3.81
16,000	550		
	550	0.75	3.72
9,900	700		
	700	1.38	3.54
8,300	760		
	760	5.5	3.42
6,400	800		
	800	7.6	3.14
5,100	840		
	840	11.0	2.77
4,800	850		
	850	12.5	2.64
3,000	910		
	910	15.4	1.79
200	1000		
	1000	17.0	.34
Surface	1020		

REFERENCES

(1) Krenkel, R. A. and Parker, F. L., 1969. Biological Aspects of Thermal Pollution, Vanderbilt University Press.

(2) Baltzer, R. A., Leendertse, J. J., Lockett, J. B., Wilson, B. W. and March, F. D., 1970. "Research Needs on Thermal and Sedimentary Pollution in Tidal Waters," Journal of Hydraulics Division, Proc. Am. Soc. Civil Engineers, pp. 1539-1548 (July 1970).

(3) Dietrich, G., 1963. General Oceanography, Interscience Publishers.

(4) Parker, F. L. and Krenkel, P. A., 1969. Engineering Aspects of Thermal Pollution, Vanderbilt University Press.

(5) U.S. Senate Committee on Public Works, 1963. "A Study of Pollution: Water," U.S. Government Printing Office, 100 pp.

(6) Federal Water Pollution Control Administration, 1968. "Industrial Waste Guide on Thermal Pollution," U.S. Department of Interior, September 1968 (Revised).

(7) Picton, W. L., 1960. "Water Use in the United States: 1900-1980," U.S. Department of Commerce Water and Sewage Division.

(8) Federal Water Pollution Control Administration, 1970. "Report on Thermal Pollution of Interstate Waters, Biscayne Bay, Florida," U.S. Department of Interior (February 1970).

(9) de Sylva, D. P., 1969. "Theoretical Considerations of the Effects of Heated Effluents on Marine Fishes," Institute of Marine Sciences Report, University of Miami.

(10) Hindley, P. D. and Miner, R. M., 1972. "A Comprehensive Thermal Model Study," ASME Winter Annual Meeting, Paper 72-WA/NE-11 (November 197-).

(11) Saunders, P. M., 1970. "Corrections for Airborne Thermometry," J. Geophys. Res. $\underline{75}$, 7596-7601.

(12) Lorenz, D., 1968. "Temperature Measurements of Natural Surfaces Using Infrared Radiometers," Applied Optics, Vol. 7, No. 9, 1705-1710 (September 1968).

(13) Saunders, P. M., 1967. J. Geophys. Res. $\underline{72}$, 4109.

(14) Hell, E. E., 1957. "An Atlas of Reflectivities of Some Common Types of Materials," Interim Engineering, Rep. Contract No. AF33(616)-3312, Ohio State University, Columbus, Ohio, 15 pp.

(15) Shaw, R. W. and Irbe, J. G., 1972. "Environmental Adjustments for the Airborne Radiation Thermometer Measuring Water Surface Temperature," Water Resources Research, Vol. 8, No. 5, 1214-1225 (October 1972).

(16) Tien, C. L., 1974. "Atmospheric Corrections for Airborne Measurements of Water Surface Temperature," Applied Optics, $\underline{13}$, 1745 (August 1974).

(17) Richards, T. L. and Massey, D. G., 1966. "An Evaluation of the Infrared Thermometer as an Airborne Indicator of Surface Water Temperatures," Department of Transport, Canada, CIR 4354, TEC, 592 (January 1966).

(18) Summers, W. R., et al., 1972. "Infra-Red Scanning Techniques Applied to Quantitative Analysis of Thermal Plumes," Thermo-Fluids Conference.

(19) STAFFMET Section, Detachment 11, 6 Weather Wing, Air Weather Service, Defense Meteorological Satellite Program (DMSP), Patrick Air Force Base, Florida, 1973.

(20) Meyer, Walter D., 1973. "Data Acquisition and Processing Program: A Meteorological Data Source," Bulletin of the American Meteorological Society, Volume 54, Number 12 (December 1973).

(21) Schwalb, A., 1972. "Modified Version of the Improved TIROS Operational Satellite (ITOS D-G)," NOAA Technical Memorandum NESS 35, U.S. Department of Commerce, Washington, D.C. (April 1972).

(22) McMillin, L. M., et al., 1973. "Satellite Infrared Soundings from NOAA Technical Report NESS 65, U.S. Department of Commerce, Washington, D.C. (September 1973).

(23) Weinreb, M. P. and Crosby, D. S., 1973. "Estimation of Atmospheric Moisture Profiles from Satellite Measurements by a Combination of Linear and Non-Linear Methods," National Environmental Satellite Service, NOAA, Washington, D.C. (June 1973).

(24) Fritz, S., et al., 1972. "Temperature Sounding from Satellites," NOAA Technical Report NESS 59 (July 1972).

(25) Veziroglu, T. N. and Lee, Samuel S., 1973. "Feasibility of Remote Sensing for Detecting Thermal Pollution," NASA CR-134453 (December 1973).

(26) Veziroglu, T. N. and Lee, Samuel S., 1974. "Application of Remote Sensing for Prediction and Detection of Thermal Pollution," NASA CR-139182 (October 1974).

BIBLIOGRAPHY

Blythe, Richard and Kurath, Ellen. "Infrared Images of Natural Subjects," Applied Optics, Vol. 7, No. 9, 1769-1777, September, 1968.

Chow, Ming-Dah. "An Iterative Scheme for Determining Sea Surface Temperatures, Temperature Profiles, and Humidity Profiles from Satellite-Measured Infrared Data," J. Geophysical Research, Vol. 79, No. 3, 430-434, January, 1974.

Fritz, S., et al. "Temperature Sounding from Satellites," NOAA Technical Report NESS 59, July 1972.

Greaves, James R., et al. "The Feasibility of Sea Surface Temperature Determination Using Satellite Infrared Data," NASA CR-474.

Lorenz, Dieter. "Temperature Measurements of Natural Surfaces Using Infrared Radiometers," Applied Optics, Vol. 7, No. 9, 1705-1710, September, 1968.

Maul, George A. "Infrared Sensing of Ocean Surface Temperature," Reprinted from The Second Fifteen Years in Space, Vol. 31, Science and Technology, American Astronautical Society Publication, 1973.

Maul, George A. and Hansen, Donald V. "An Observation of the Gulf Stream Surface Front Structure by Ship, Aircraft, and Satellite," Remote Sensing of Environment, Vol. 2, 109-116, 1972.

Maul, George A. and Sidran, Miriam. "Comment on 'Estimation of Sea Surface Temperature from Space' by D. Anding and R. Kouth, Remote Sensing of Environment, Vol. 2, 165-169, 1972.

Maul, George A. "Atmospheric Effects on Ocean Surface Temperature Sensing from the NOAA Satellite Scanning Radiometer," J. Geophysical Research, Vol. 78, No. 12, 1909-1916, April 1973.

Richards, T. L. "Great Lakes Water Temperatures by Aerial Survey," I.A.S.H. Symposium of Garda, No. 70, 406-419.

Richards, T. L. and Massey, D. G. "An Evaluation of the Infrared Thermometer as an Airborne Indicator of Surface Water Temperature," Canada Department of Transport TEC, 592, January 1966.

Saunders, Peter M. "Corrections for Airborne Radiation Thermometry," J. Geophysical Research, Vol. 75, No. 36, 7596-7601, December 1970.

Shaw, R. W. and Irbe, J. G. "Environmental Adjustments for the Airborne Radiation Thermometer Measuring Water Surface Temperature," Water Resources Research, Vol. 8, No. 5, 1214-1224, October 1972.

Smith, W. L., et al. "A Regression Method for Obtaining Real-Time Temperature and Geophysical Height Profiles from Satellite Spectrometer Measurements and its Application to NIMBUS 3 'SIRS' Observations," Monthly Weather Review, Vol. 98, No. 8, 582-603, August 1970.

Smith, W. L., et al. "The Determination of Sea-Surface Temperature from Satellite High Resolution Infrared Window Radiation Measurements," Monthly Weather Review, Vol. 98, No. 8, 604-611, August 1970.

Summers, W. R., et al. "Infra-Red Scanning Techniques Applied to Quantitative Analysis of Thermal Plumes," Proceedings of Thermo-Fluids Conference, 1972.

Tien, C. L. "Atmospheric Corrections for Airborne Measurements of Water Surface Temperature," Applied Optics, Vol. 13, No. 8, August 1974.

Wark, D. Z., et al. "Methods of Estimating Infrared Flux and Surface Temperature from Meteorological Satellites," J. Atmospheric Science, Vol. 11, 369-384, September 1962.

Weinreh, Michael P. and Crosby, David S. "Estimation of Atmospheric Moisture Profiles from Satellite Measurements by a Combination of Linear and Non-Linear Methods," *Proceedings of AMS Third Conference on Probability and Statistics in Atmospheric Science*, 231-235, June 1973.

Wormser, Eric M. "Sensing the Invisible World," *Applied Optics*, Vol. 7, No. 9, 1667-1671, September 1968.

Wyatt, Philip J., et al. "The Infrared Transmittance of Water Vapor," *Applied Optics*, Vol. 3, No. 2, 229-241, February 1964.

REMOTE SENSING APPLIED TO NUMERICAL MODELLING

S. SENGUPTA, S. S. LEE, T. N. VEZIROGLU
Clean Energy Research Institute
School of Engineering and Enviromental Design
University of Miami, Coral Gables, Florida, U.S.A.

R. BLAND
Earth Resources Branch
NASA Kennedy Space Center, Florida, U.S.A.

INTRODUCTION

The finite nature of earth's water resources have become shockingly apparent in recent years. No longer can rivers, lakes or even oceans be treated as infinite sinks or sources. Whether it be thermal discharges from power plants or pollutants from chemical plants and sewage systems, the effects on the environment are often quite severe and long term. Changes in temperature and chemical composition lead to disturbances in the delicate balance of an ecosystem. It is estimated that 50 trillion gallons of heated water is being discharged into the waterways by electric power plants [1]. The doubling rate of ten years in power generation is expected to accelerate in coming years [2], [3]. The need to understand, monitor and predict environmental conditions is therefore imperative.

This need is further emphasized by our need to understand the flows in oceans in order to relate ocean conditions to atmospheric flows. For more meaningful meteorological predictions ocean temperatures and currents must be considered as important boundary conditions to the atmospheric problem.

The sudden focus of attention on energy needs has made the extraction of power from ocean currents and ocean temperature variations more attractive [4]. One of the important concerns of ocean energy systems is site selection. For economical utilization of ocean power, areas of fairly constant strong currents or large temperature differences have to be identified. Proper mapping of the velocities and temperatures as well as wave height seems imperative as a first step towards ultimate commercial utilization of ocean energy.

To obtain the understanding of the hydrodynamics and thermodynamics of earth's water basins requires extensive field data collection. Even the most rudimentary data acquisition field trip is extremely expensive. Experimental observations from physical models are often unsatisfactory owing to the impossibility of obtaining appropriate similarity. Even when similarity parameters are approximated it is impossible to reproduce the actual turbulent scales in laboratory facilities. Consequently, there is always the possibility of getting observations from laboratory facilities which do not correspond to flows on geophysical scales even qualitatively.

Developing theoretical predictive ability is the only way to obtain the large volume of information required to understand the environment. Besides, field data can merely determine base states or "after the fact" conditions. Models can be used predictively to arrive at meaningful decision parameters a priori for site selection for power plants and other modifying factors.

Analytical solutions for wind driven circulation were first obtained by Ekman [5]. He developed a theory for sea-level fluctuations produced in a deep sea by steady wind stress. The steady-state and time dependent, finite depth basin was modelled by Welander [6]. Important mechanisms and characteristics of wind driven circulation were understood by using analytical models. Detailed investigations of circulations and temperature fields, taking into account the effects of bottom topography and non-linear terms, however, require numerical modelling. A numerical model needs to be compared with field data. Collection of field data in a systematic, exhaustive manner for ecosystems of general interest like the Great Lakes, coastal regions or the oceans is expensive and time consuming. A large amount of data at different locations have to be gathered, over a period of time to form a comprehensive idea of the transport processes in the system.

A major disadvantage of data gathered on field trips is the difficulty of obtaining data over large areas at the same instant of time. For numerical models data at a particular time over a large domain is often necessary to set initial and boundary conditions. Such data sets are also required to check values predicted by numerical models. The unique characteristics of remote sensing become advantages for these reasons. Aircraft and satellite remote sensing devices can collect information over large areas with very little lapse time between locations. Temperature, velocity, topography and other variables measured by remote sensing techniques will form the basis for setting initial and boundary conditions for numerical models in the coming years. Model reliability will also be greatly improved by calibration and comparison with remotely sensed data.

FUNDAMENTALS OF NUMERICAL MODELLING

The characteristics of a body of water is affected by natural influences originating from the geophysical, meteorological and hydrological characteristics of the area under consideration, and also man-made disturbances caused by industry, agriculture and urban areas. The complex interaction of these parameters may be considered as a system comprised of interrelated subsystems.

The essential technique of modelling is to describe the physical interactions in terms of mathematical relations which satisfy the basic laws governing the system. The complete set of mathematical relations together with the proper initial and boundary conditions constitute a mathematical model that describes the physical system. Given a set of forcing functions, initial and boundary conditions the mathematical model responds to give values of variables in the equations. These values describe the predicted response of the physical system. If the mathematical system, the inputs and the solution were "exact" the model would describe the physical condition on a one-to-one basis. More often then not this condition of "exactness" cannot be achieved. The essence of modelling therefore becomes an attempt to formulate an approximate system of equations within the powers of available mathematics and physical understanding to arrive at predicted behavior not grossly different from reality.

The governing equations which describe environmental systems, even in their approximate forms are usually not amenable to analytical solution techniques. They are often, coupled, time-dependent, non-linear, multi-dimensional partial differential equations. With the advent of high-speed, large core computers numerical techniques have become the usual methods for solving these equations. For this reason most complex ecosystem models are numerical models. The present paper will therefore describe the state of the art in numerical modelling, assuming for the most part that analytical solutions relate to remote sensing in a similar manner.

GOVERNING EQUATIONS

The state at a point in a flow field is described by the solution of a system of coupled, non-linear, second-order, three dimensional partial differential equations which satisfy local conservation laws for total mass, species mass, momentum and energy. The system of equations is completed by the constitutive equations. In case of laminar flow the molecular transport properties for heat and momentum can be used. For most environmental situations the flows are turbulent. Therefore the time averaged Navier-Stokes equations are used and they are similar to the laminar equations when the Reynolds stress terms are approximated by eddy transport coefficients. The conservation laws in non-dimensional form under these assumptions for an incompressible fluid are:

<u>Continuity:</u>

$$\frac{\partial u}{\partial x} + \frac{\partial v}{\partial y} + \frac{\partial w}{\partial z} = 0$$

<u>Momentum:</u>

$$\frac{\partial u}{\partial t} + \frac{\partial (uu)}{\partial x} + \frac{\partial (uv)}{\partial y} + \frac{\partial (uw)}{\partial z} - \frac{1}{R_B} v = - \frac{\partial p}{\partial x}$$
$$+ \frac{1}{R_e} \frac{\partial}{\partial x} (A_H^* \frac{\partial u}{\partial x}) + \frac{\partial}{\partial y} (A_H^* \frac{\partial u}{\partial x}) + \frac{1}{R_e \varepsilon^2} \frac{\partial}{\partial z} (A_V^* \frac{\partial u}{\partial z})$$

$$\frac{\partial v}{\partial t} + \frac{\partial (uv)}{\partial x} + \frac{\partial (vv)}{\partial y} + \frac{\partial (wv)}{\partial z} + \frac{1}{R_B} u = - \frac{\partial p}{\partial y}$$
$$+ \frac{1}{R_e} \frac{\partial}{\partial} (A_H^* \frac{\partial v}{\partial x}) + \frac{\partial}{\partial y} (A_H^* \frac{\partial v}{\partial y}) + \frac{1}{R_e \varepsilon^2} \frac{\partial}{\partial z} (A_V^* \frac{\partial u}{\partial z})$$

<u>Hydrostatic Equation:</u>

$$\frac{\partial p}{\partial z} = Eu\ (1 + \rho)$$

<u>Energy:</u>

$$\frac{\partial T}{\partial t} + \frac{\partial (uT)}{\partial x} + \frac{\partial (vT)}{\partial y} + \frac{\partial (wT)}{\partial z} = \frac{1}{P_e} \frac{\partial}{\partial x} (B_H^* \frac{\partial T}{\partial x}) + \frac{\partial}{\partial y} (B_H^* \frac{\partial T}{\partial y})$$
$$+ \frac{1}{P_e \varepsilon^2} \frac{\partial}{\partial z} (B_V^* \frac{\partial T}{\partial Z}) + Q^*$$

<u>Species:</u>

$$\frac{\partial Ci}{\partial t} + \frac{\partial (uCi)}{\partial x} + \frac{\partial (vCi)}{\partial y} + \frac{\partial (wCi)}{\partial z} = \frac{1}{Pe_i} \frac{\partial}{\partial x} (B_{Hi}^* \frac{\partial Ci}{\partial x}) + \frac{\partial}{\partial y} (B_{Hi}^* \frac{\partial Ci}{\partial y})$$
$$+ \frac{1}{P_{ei} \varepsilon^2} \frac{\partial}{\partial z} (B_{vi}^* \frac{\partial Ci}{\partial z}) + S_i^* \qquad i = 1, 2 \ldots n$$

<u>Equation of State:</u>

$$\rho = (T)$$

where

$$u = \tilde{u}/U_{ref}\ ;\ \tilde{v} = v/V_{ref}\ ;\ w = \tilde{w}/\varepsilon U_{ref}\ ;\ t = \tilde{t}/t_{ref}$$
$$x = \tilde{x}/L\ ;\ y = \tilde{y}/L\ ;\ z = \tilde{z}/H\ ;\ \varepsilon = H/L$$
$$P = \tilde{P}/\rho_{ref} U^2_{ref}\ ;\ T = \frac{\tilde{T} - T_{ref}}{T_{ref}}\ ;\ \rho = \frac{\rho - \tilde{\rho}_{ref}}{\rho_{ref}}$$

$$A_H^* = A_H/A_{ref} \quad ; \quad A_v^* = A_v/A_{ref} \quad ; \quad B_H^* = B_H/B_{ref} \quad ; \quad B_v^* = B_v/B_{ref}$$

$$t_{ref} = L/U_{ref}$$

Quantities with subscript 'ref' are reference quantities, H and L are vertical and horizontal length scales. The variables .($\sim$) are dimensional quantities. u, v, w, are the velocity components, p the pressure, T the temperature, ρ the density, A_H, A_V are the horizontal and vertical eddy viscosities, B_H and B_V are the horizontal and vertical eddy diffusivites; Q^* is the heat source and s_i^* the source term for species 'i'. The non-dimensional parameters are defined as,

$$Re = \frac{U_{ref}L}{A_{nef}} \quad \text{Reynolds Number} \quad \frac{\text{Inertia}}{\text{Viscous}}$$

$$R_B = \frac{U_{ref}}{fL} \quad \text{Rossby Number} \quad \frac{\text{Inertia}}{\text{Coriolis}}$$

$$Pr = \frac{A_{ref}}{B_{ref}} \quad \text{Prandtl Number} \quad \frac{\text{Viscous Diffusion}}{\text{Thermal Diffusion}}$$

$$Pe = \frac{U_{ref}L}{B_{ref}} \quad \text{Peclet Number} \quad \frac{\text{Convection}}{\text{Diffusion}}$$

$$Eu = \frac{gH}{U_{ref}} \quad \text{Euler Number} \quad \frac{\text{Body Force}}{\text{Inertia Force}}$$

The above equations together with the appropriate initial and boundary conditions form the mathematical model.

INITIAL AND BOUNDARY CONDITIONS

The time dependent nature of the primitive equations require conditions to be specified at a given time. This initial condition specifies the values of dependent variables at sometime so that the equation may be integrated to find the values after the desired prediction interval. The velocities, surface elevations, temperatures, water quality parameters and properties have to be specified. Various driving mechanisms like winds, tides and air-water temperature difference have to be specified.

In most cases the initial condition is not easily defined. The data base necessary for a complete specification of initial conditions is too large and requires extensive in situ measurements resulting in unjustifiable cost. Even where in situ measurements can be made the simultaneous measurement of all variables at all locations is impossible. For most cases therefore the initial conditions are applied using insufficient data with the expectation that the effects of erroneous initial conditions are negligible for long term integration. Certain initial conditions can be specified accurately. For example, at pollution sources, the temperature, velocity and species mass fractions are easily determined.

The boundary conditions are specified for the air-water interface, horizontal boundaries of the domain, the bottom of the basin and at points of injection.

At the air-water interface the boundary conditions for the momentum equations are obtained from the surface shear stress owing to wind. Experimental correlations by Nilson [7] relate wind velocities 6 meters above the water surface and surface shear-stress. This stress can be related to the vertical gradients of the horizontal velocities using the vertical eddy viscosity. Defant [8] has given a list of values for vertical eddy viscosity A_z for various wind velocities.

At the air-water interface the boundary condition on the energy equation is derived from a heat flux condition. Edinger and Geyer [9] have determined the heat flux relations at this interface for various meteorological conditions. They derive heat flux as a product of a heat transfer coefficient and the temperature difference between the water surface temperature and the equilibrium temperature. The equilibrium temperature is defined as that surface temperature at which no heat flux takes place at the surface. The equilibrium temperature is a function of the ambient air temperature and meteorological conditions. At the interface a condition for the free surface has to be specified.

The conditions on the side walls are no-slip and no normal velocity for the momentum equations. The heat flux is specified (often adiabatic). A species flux is specified for the species equation. At the bottom, the conditions of no-slip and no normal velocity are applicable. The energy equation has a heat flux boundary condition. Species fluxes are specified for the species equation. At points of injection velocities, temperatures and species mass fraction are specified.

Boundary conditions are more difficult to specify at open boundaries. For example, for open-ocean boundary conditions, velocities, temperatures, species fluxes and free-surface heights must be specified. Often these are not well known and require considerable ingenuity on the part of the modeller to specify. The boundary conditions in summary, therefore are (refer to Fig. 1).

At the air-water interface:
   at $z = 0$ $w = 0$ (Rigid Lid)
              or Kinematic condition (Free Surface)

$$\frac{\partial u}{\partial z} = \frac{H}{U_{ref} A_v} \tau_{zx}$$

$$\frac{\partial v}{\partial z} = \frac{H}{U_{ref} A_v} \tau_{zy}$$

$$\frac{\partial T}{\partial z} = \frac{H K_s}{K_s} (T_E - T_S) \; ; \quad T_E: \text{Equilibrium temperature}; \; T_s: \text{Surface temperature}; \; K_s: \text{Surface heat transfer coefficient}$$

$$\frac{\partial C_i}{\partial z} = 0 \; ; \quad i = 1, 2 \ldots n$$

Figure 1. Boundary Conditions.

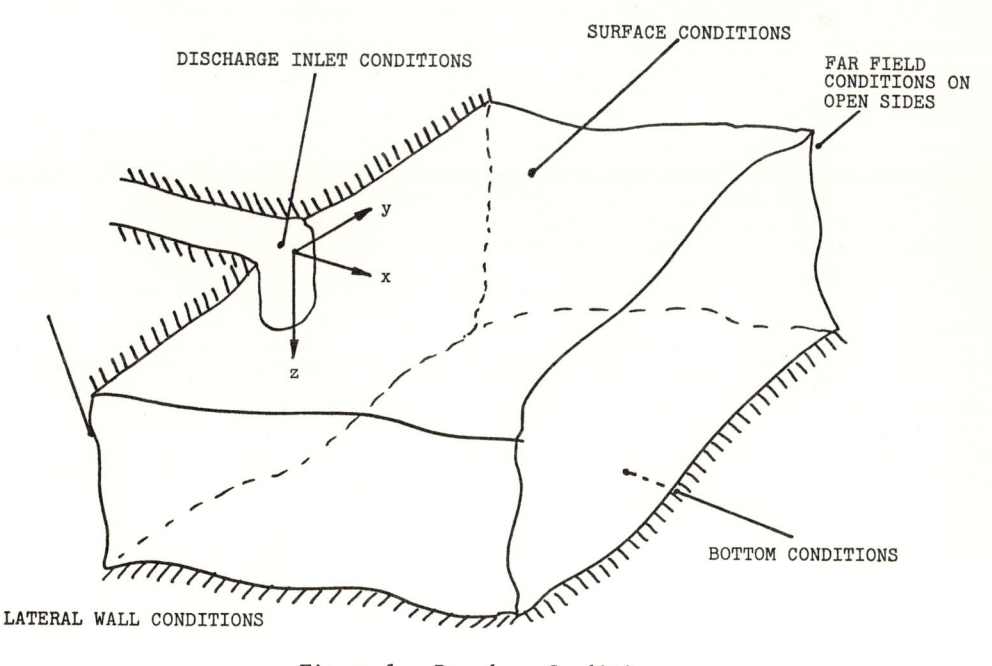

At the bottom of basin:
  z = h   w = 0
           u = 0
           v = 0

$\dfrac{\partial T}{\partial z}$ = Specified value, often adiabatic

$\dfrac{\partial Ci}{\partial z}$ = Specified value.   i = 1, 2 ....n

On lateral walls:
  u = 0
  v = 0
  w = 0 (owing to hydrostatic assumption)

$\dfrac{\partial T}{\partial x}$ or $\dfrac{\partial T}{\partial y}$ = Specified, often adiabatic

$\dfrac{\partial Ci}{\partial x}$ or $\dfrac{\partial Ci}{\partial y}$ = Specified value   i = 1, 2 ... n

At open boundaries:
Depending on formulation u, v, w, T, Ci or their derivatives are specified. For free-surface formulation, surface height variations have to be known.

At points of Injection:
  u, v, w, T, Ci are specified.

The governing equations of conservation of mass, momentum, energy and species with the boundary conditions described form the mathematical model.

SOLUTION TECHNIQUES

The system of equations described is too complicated to lend itself to analytical solutions. Only in special cases with confining assumptions can the system be sufficiently simplified for analytical solutions. Therefore, most environmental models are numerical. The mathematical theory of numerical analyses of non-linear partial differential equations is still in its infancy. Rigorous methods for obtaining stability criteria, error estimates, or convergence proofs are non-existant. Therefore the trend has been to analyze simpler, linearized problems of related nature to obtain insight into complex systems by heuristics, physical intuition and computational experience.

The two general classes of numerical techniques that have been used in solving hydrodynamic problems are the finite element method and the finite difference method. The finite element method though relatively new has been well developed for structural analysis. The basic concept was developed by Argyris [10] in a series of papers. Since then the application of finite element methods to solid mechanics and structures problems has become a sophisticated technique. In the area of fluid mechanics finite element techniques are still in the exploratory stages. Baker [11], [12] in a recent two part paper has described the theoretical development and applications of finite element methods to environmental hydrodynamics. One of the major advantages of finite element methods over finite difference techniques is the easier modelling of irregular domains by using non-uniform mesh sizes and shapes. However, a careful comparison of the two techniques in a large scale hydrodynamic problem is yet to be made. Finite difference methods, on the other hand, have been widely applied to numerical modelling of the environment.

Roache [13] in his book on computational fluid dynamics has discussed the existing state of the art in the application of finite difference techniques. He has discussed several methods for integration of the governing equations for incompressible flow. The methods may be explicit or implicit. The explicit methods use known values of variables, i.e., at n, n-1, . . ., to calculate values of variables at n+1. The implicit methods use values at the advanced time level (n+1) to calculate spatial derivatives. This requires the solution of a set of simultaneous equations at time level (n+1). The use of implicit methods usually allow larger integration time steps but the solution of the simultaneous equations is often time consuming. For problems which are one-dimensional in space, implicit methods are useful since the system of simultaneous equations is tri-diagonal and direct methods of solution exist. Alternating Direction Implicit methods may be used to obtain tri-diagonal matrices even for multidimensional equations. However, irregular boundaries make ADI methods cumbersome.

The system of equations for most environmental models is three dimensional and are to be solved for irregular domains. Therefore, explicit methods are widely used in spite of the stringent stability criteria.

Explicit integration of time dependent equations over long time periods may result in leakage of mass, momentum and energy even in non-dissipative systems. The equations are therefore often put in a conservative form suggested by Arakawa [14]. Special care has also to be taken at the boundaries to insure that flow quantities are conserved.

ASSUMPTIONS AND SIMPLIFICATIONS

For many environmental investigations the information required allows certain assumptions and consequent simplifications. In situations where variations with time are slow or the effect of input changes causes the system to reach steady-state relatively rapidly the time-dependence of the governing equations may be eliminated. This leads to mathematically simpler models and savings in computation time. However, in complicated three-dimensional models sometimes, even when only steady-state results are desired, it is easier to integrate the time dependent equations to reach the steady state.

Reduction in spatial dimensions is a powerful means of simplifying the general system. In river models often the well-mixed approximation is made leaving only one spatial dimension along the river. Models investigating the vertical structure of lakes may include only the vertical dimension resulting in a one-dimensional energy equation to be solved. Vertically integrated models for circulation are used where horizontal mass flux rather than the vertical distribution is important. Layered models are effective in saving computation time by dividing the domain into horizontal layers with each layer having no vertical structure. The relevant variables are matched at layer boundaries. Layered models have been extensively used in atmospheric, and oceanic investigations. However, they are not truly three-dimensional and with increasing number of layers the computation time savings are minimal compared to three-dimensional models.

There are various other simplifications that can be introduced depending on the particular situation that is being modelled. The vertical velocity at the surface can be equated to zero, thereby eliminating surface gravity waves. This rigid-lid assumption allows larger integration time steps. Crowley [15], [16], [17] and Berdahl [18] have made comparisons of rigid-lid and free-surface models for oceanic circulation. They found that the circulation is approximately the same. However, the Rossby wave speeds are erroneously predicted by the rigid-lid model as may be expected from the nature of the approximation. For most flow situations the non-dimensional numbers are such that the vertical momentum equation may be reduced to the hydrostatic equation, i.e., a balance between the body force and the pressure gradient terms. Except in regions of severe upwelling or downwelling this is an excellent approximation. The effects of turbulence are modelled by replacing the Reynolds stresses by eddy transport coefficients. Sometimes the eddy transport coefficients are made empirical functions of the Richardson number to reproduce the influence of stratification and velocity shear. Munk and Anderson [19] have investigated the effects of variable eddy viscosity. The effects of density changes are usually small except in the buoyancy term. Therefore the Boussinesq approximation is made whereby density is assumed constant in the equations with the effects of variation included only in the buoyancy term. For environmental models the fluid is usually considered incompressible and pressure work terms are neglected in the energy equation.

APPLICATION AND VERIFICATION OF MODELS

The numerous assumptions and approximations that are usually present in a model demand that the solutions obtained from the model be carefully tested against observed data. The prevalent procedure is to first check for limiting cases, namely simple geometries and conditions where analytical solutions exist. If comparisons are valid then the model is calibrated with data from an ecological system where extensive measurements have been made. This process allows determination of various parameters like the transport properties more accurately. However, it is imperative to test the model in a physical situation against which the model has not been calibrated. If the evaluated distributions of velocity, concentrations etc. compare well with this physical situation the model may be considered verified and is ready for application in new situations where experimental or field data is scarce.

During the process of application of the model there is a constant need for modifications if necessary. This insures a self-improving model which "learns from experience." The interaction between modelling and observations is necessary to insure useful, physically realistic models.

DATA REQUIREMENTS AND REMOTE SENSING CAPABILITIES

The need for data acquisition for calibration and verification of environmental models has been established. Topography, surface winds, gravity waves, temperature and other variables have to be measured extensively even when models are used. The simultaneity of measurements over large areas make remote sensing an excellent tool.

Table I from a paper by Nagler and Loomis [20] shows the data requirements for oceanographic investigation. The same data base is required also for model development. The SEASAT Program initiated by NASA and SEASAT-A, to be launched in the beginning of 1978, is designed for oceanographic data acquisition. The payload sensors will be all microwave except for a visible and IR scanner. Table II shows the SEASAT-A capabilities discussed by Apel [21]. This satellite will collect data on wave height, directional wave spectrum, surface wind field, surface temperature field, Geoidal Heights, sea surface topography and oceanic and atmospheric features. It should be noted that even this advanced satellite remote sensing system will not include measurements of ocean-surface velocities. Laser-doppler anemometry techniques are still not developed enough for remote sensing uses.

An important characteristic of a body of water is its thermal stratification. Vertical profiles of temperature is at present determined by in-situ measurements. This temperature data is essential for initial and boundary conditions for most environmental models. However, remote sensing of sub-surface temperature is not feasible at present. Hirschberg et al [22] have discussed a method of using Brillouin scattering and the circular depolarization of Raman scattering to determine sub-surface temperature and salinity profiles. The technology is yet to be developed.

At present surface temperature is the only variable that can be measured accurately and reliably by remote sensing techniques. Lee, Veziroglu, Hiser, Weinberg, and Sengupta from the University of Miami, in a series of papers

TABLE I. OCEANOGRAPHIC MEASUREMENT REQUIREMENTS IN GEOPHYSICAL TERMS

Measurement			Range	Precision/Accuracy	Resolution	Spatial Grid	Temporal Grid
Topography	Geoid		5 cm - 200 m	<±10 cm	<10 km	<20 km	Weekly to Monthly
	Currents, Surges, etc.		10 cm - 10 m 5 - 500 cm/sec	<±10 cm ± 5 cm sec	10 - 1000 m	<10 km	Twice a day to Weekly
Surface Winds	Amplitude	Open Ocean	3 - 50 m/s	± 2 m/s OR±10 - 25%	10 - 50 km	50-100 km	2 - 8/day
		Closed Sea			5 - 25 km	25 km	
		Coastal			1 - 5 km	5 km	Hourly
	Direction		0 - 360°	±20%			
Gravity Waves	Height		0.5 - 20 m	±0.5m OR ±10 - 25%	<20 km	<50 km	2 - 8/day
	Length		6 - 1000 m	±10 - 25%	3 - 50 km		2 - 4/day
	Direction		0 - 360°	±10 - 30°			
Surface Temperature	Open Ocean		-2 - 35°C	0.1 0 2°Relative 0.5 - 2°Absolute	25 - 100 km	100 km	Daily to Weekly with Spectrum of Times of Day and Times of yr.
	Closed Sea				5 - 25 km	25 km	
	Coastal				0.1 - 5 km	5 km	
Sea Ice	Extent and Age			1 - 5 km	1 - 5 km	1 - 5 km	Weekly
	Leads		>50 m	25 m	25 m	25 m	
	Icebergs		>10 m	1 - 50 m	1 - 50 m		2 - 4 day
Ocean Features	Open Ocean			50 - 500 m			Twice Daily to Daily
	Coastal			10 - 100 m			
Salinity			0 - 30 ppt	±0.1 - 1 ppt	1 - 10 km	100 km	Weekly
Surface Pressure			930 - 1030 mb	±2 - 4 mb	1 - 10 km	1 - 10 km	Hourly

TABLE II

CAPABILITY OF SEASAT-A IN MEETING USER REQUIREMENTS

PHYSICAL PARAMETER	INSTRUMENTS	RANGE	PRECISION	RESOLUTION OR IFOV	TOTAL FOV	COMMENTS
Wave Height	Pulse Altimeter Coherent Alt.	1.0 - 20 m	±0.5 m or ±10%	2x7 km spot	2-km swath	along subsatellite track only
Directional Wave Spectrum $S(\lambda, \theta, x, y)$	Imaging Radar (2-D transform)	S: unknown λ: 50-1000 m θ: 0-360°	S: ---- λ: ±10% θ: ±10°	50-m resolution	20x20 km squares	global samples at 250-km intervals
	2-f Wave Spectrometer	S: unknown λ: 6-500 m θ: 90° sector	S: ---- λ: ±10% θ: ±90°	8 x 25 km spot	300-km swath about nadir	global samples at 150-km intervals
Surface Wind Field, $\vec{U}(x,y)$	Scatterometer	U: 3-25 m/s θ: 0-360°	±2 m/s, ±10% ±20°	≤50 km spot	two 450-km swaths	global, 36 hrs (low speeds)
	µW Radiometer	U: 10-50 m/s θ: unknown	±2 m/s, ±10% ----	≤100 km spot	900-km swath about nadir	global, 36 hrs (high speeds)
Surface Temperature Field $\mathcal{T}(x,y)$	IR Radiometer	-2° to +35°C	±¼° - 1°C	107 km IFOV	1500-km swath about nadir	global, 36 hrs (clear air only)
	µW Radiometer	0° to 35°C	±1.5°C	100 km spot	900-km swath about nadir	global, 36 hrs (clouds & lt. Rain)
Geoidal Heights, h(x,y) (above reference ellipsoid)	Pulse Altimeter Coherent Alt.	7cm - 200 m	±7 cm	2x7 km spot	18-km spacing along equator	sampled throughout one year
Sea Surface Topography (x,y) (departures from geoid)	Pulse Altimeter Coherent Alt.	7 cm - 10 m	±7 cm	2x7 km spot	2-km swath	along subsatellite track only
Oceanic, Coastal, & Atmospheric Features (Patterns of waves, temp., currents, ice, oil, land clouds, atmospheric water content)	Imaging Radar	high resolution	all weather	25 or 100m	100 or 200 km	sampled direct or stored images
	IR Radiometer	high resolution	clear air	1-7 km	1500-km swath	broadly sampled images
	µW Radiometer	low resolution	all weather	15-100 km	900-km swath	global images

[23], [24], [25], [26] have discussed the application of remote sensing and modelling to the detection and monitoring of thermal pollution. They have used both airborne sensors and satellite sensors to get IR mapping of Biscayne Bay to be coupled with a numerical modelling effort. Data from Defense Meteorological Satellite Program (DMSP), National Oceanic and Atmospheric Administration NOAA Series and NASA-6 aircraft has been used to establish an extensive data base.

The capability of the present DMSP provides the following:

Data Type	Spectral Interval	Resolution
VHR Data	.4 - 1.1M	0.3 nautical mile
WHR Data	8 - 13M	0.3 nautical mile
HR Data	.4 - 1.1M	2.0 nautical mile
IR Data	8 - 13M	2.0 nautical mile

The infrared photographs (WHR and IR) receive thermal energy in the 210°K to 310°K range. Sixteen grey shades are available in increments of 1°K or a 25°K span can be divided into 16 grey shades.

The NOAA-2 and NOAA-3, ITOS series of satellites provide thermal IR data in the 10.5 to 12.5 micron region. These satellites are sun-synchronous, 790 n.m. polar orbit with a period of 115.14 minutes. The resolution is 0.5 nautical miles.

Among the variety of airborne infrared systems the Bendix Thermal Mapper (BTM), Barnes Airborne Radiation Thermometer (ART) and the Daedalus Scanner Systems are the most commonly used. NASA-6 aircraft carries the Daedalus Scanner Systems. This system has a scan angle of 120° centered above the vertical. Broad band coverage from 0.38 micron to 14 micron is attainable with this system by selecting the appropriate detectors. In this study, a Hg : Cd : Te detector is used for sensing 8-14 micron radiation. The resolution of IR data obtained from NASA-6 is sufficient to be used for thermal plume mapping and modelling.

APPLICATION OF ENVIRONMENTAL MODELLING AND REMOTE SENSING TO THERMAL POLLUTION

One of the important areas of application of environmental modelling is thermal pollution. Thermal pollution caused by power plant discharges or other industrial effluents can cause serious changes in an ecosystem. Considerable amount of work has been done on this topic. A three dimensional model including buoyancy, topography and other parameters has not been developed yet. Simpler models are in existence. Akers [27] discussed some of the models which are being used. Policastro [28], [29], [30] in a series of review papers has compared the existing plume models with field data. He considered both analytical and numerical models. Very few of the models are adequate. Harleman's [31], [32], [33] pioneering work led to a numerical model with Stolzenbach [32] which has been widely used in plume analysis.

Investigations in mathematical modelling of thermal discharges including ocean currents, cyclic tidal flushing in bays and estuaries, variable winds and realistic bottom topography have been few. Veziroglu, et al [34] at the University of Miami together with NASA Kennedy Space Center is developing a three-dimensional thermal pollution model which will be verified and calibrated using ground truth and remotely sensed data. The initial site of application is Biscayne Bay. The model will then be applied to Hutchinson Island Nuclear Power Plant of the Florida Power and Light Company.

Thermal anomalies caused by a heated discharge usually effect areas of about a few miles extent. Initially the jet motion is governed by the momentum and buoyancy of the discharge. Then the jet gradually spreads at the surface with vertical stratification. Finally the flow is dominated by far field conditions and the ambient meteorological state. The domain of interest can be classified into a near field where effects of the discharge are measurable and a far field which affects the plume but is appreciably unaffected by the plume. The numerical characteristics of these two domains are quite different. The procedure therefore is to obtain a far field solution with coarse finite difference grid and use this to obtain the near field using a finer mesh size.

The problem has therefore been divided into five separate parts. The combination of these subsystems constitute the complete thermal pollution model.

a) A one-dimensional study without convective or horizontal diffusion terms is being made. The results of this model when compared to vertical temperature profiles from field data give measures of vertical eddy transport quantities, solar radiation absorption characteristics and boundary conditions.
b) A rigid-lid model for the Bay with no tidal effects has been used to obtain the wind driven circulation in the Bay. The effects of oceanic inflow and outflow owing to tides is also included.
c) A free surface model which models tidal boundary conditions accurately is being used to obtain the velocity and temperature fields for the far field.
d) A near field rigid-lid model is being used to obtain the thermal anomalies owing to the plume.
e) A free surface near field model is being used in cases where tidal influences greatly affect the near field.

These models together with field data, satellite and NASA-6 airborne data collection systems are being used to develop a predictive computer package for thermal pollution for widely different topographical and meteorological conditions. The inter-relationships of the subsystems are shown in Figure 2.

All these programs have been discussed by Lee, et al [34]. The solution techniques and some results for the rigid-lid far field circulation model will be discussed here. Some of the remotely sensed data will be presented to show their applicability to modelling.

To obtain the general circulation in Biscayne Bay under the rigid-lid assumption the Sengupta-Lick [35] three-dimensional model has been applied. In this model the horizontal momentum equations each include the nonsteady, non-linear inertia, Coriolis, pressure gradient and all three viscous terms. The energy equation includes the non-steady, convective and all three diffusion terms. The hydrostatic and Boussinesq approximations are used. A Poisson equation is derived from the vertically integrated horizontal momentum equation for surface pressure. An iterative scheme with normalization is used to solve the Poisson equation for pressure with Neumann boundary conditions. A vertically stretched system of equations which may variable depth domains to constant is used.

METHOD OF SOLUTION

Fig. 3 shows the grid system and the arrangement of variables. u and v are located on integral modes (i, j, k) on the horizontal or x,y plane. P, w, ρ and T are located at half grid points (i + 1/2, j + 1/2, k). The arrangement is repeated in the z direction, represented by K.

The finite difference equation for the momentum and energy equation may be represented in the general form.

$$\frac{u^{n+1} - u^n}{\Delta t} = (\text{convection})^n + (\text{pressure})^n + (\text{viscous})^n$$

Where u may be replaced by v or T (for the T equation the pressure term is not there). The spatial derivatives are central differenced.

The pressure equation is approximated by a five-point scheme and solved by Liebmann relaxation procedure.

Figure 4 shows the flow chart for the steps in the solution algorithm summarized below:
- a) Using values at time step n, calculate the forcing term for the pressure equation.
- b) Solve the pressure equation iteratively.
- c) Calculate u, v from the momentum equations.
- d) Calculate w from the continuity equation using u, v at n + 1.
- e) Calculate T from the energy equation.
- f) Calculate ρ from the empirical equation of state.

The values at (n + 1) have now been obtained. Repeat the procedure for (n + 2) using value at (n + 1).

SOME RIGID-LID RESULTS

The rigid-lid model will be used to obtain some preliminary understanding of the general wind driven circulation in Biscayne Bay.

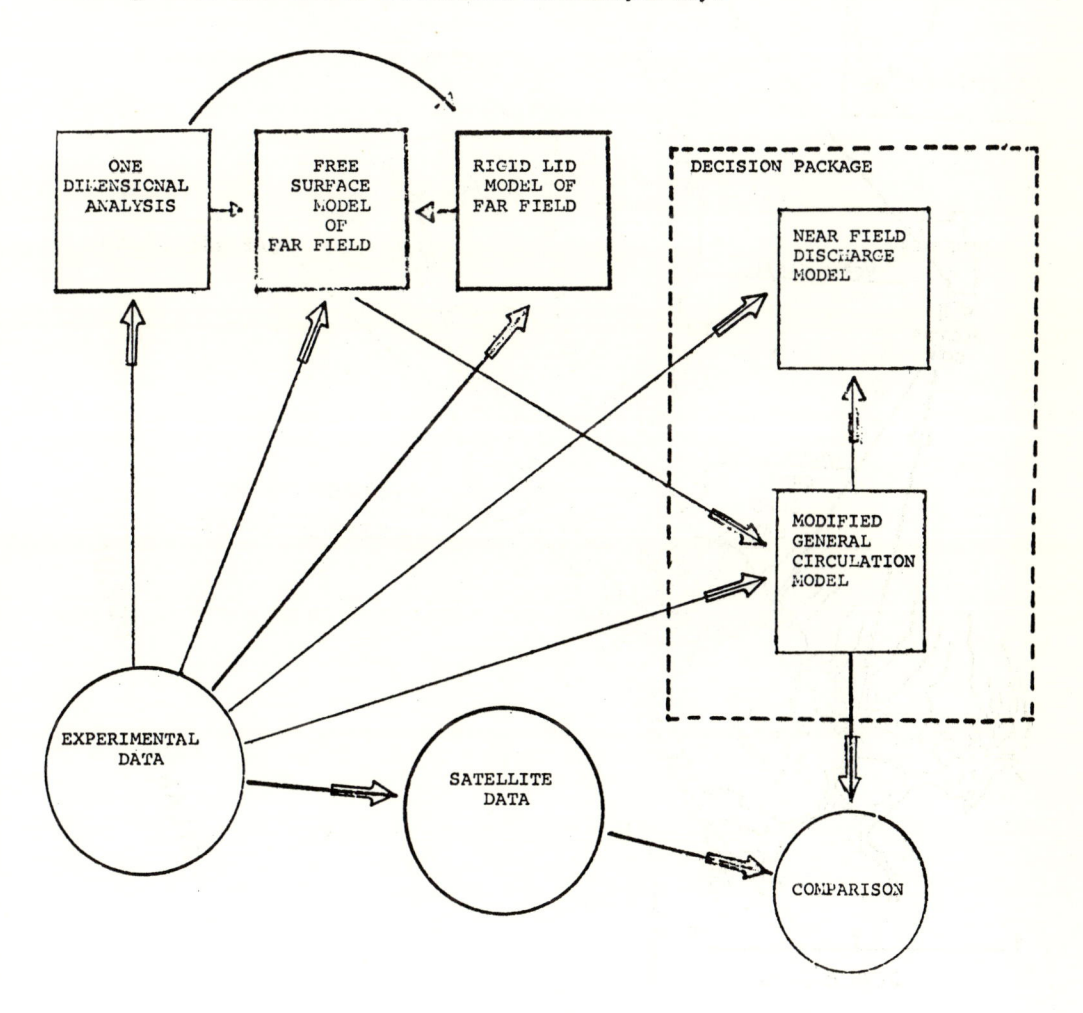

Figure 2. Interrelationships of sub-programs.

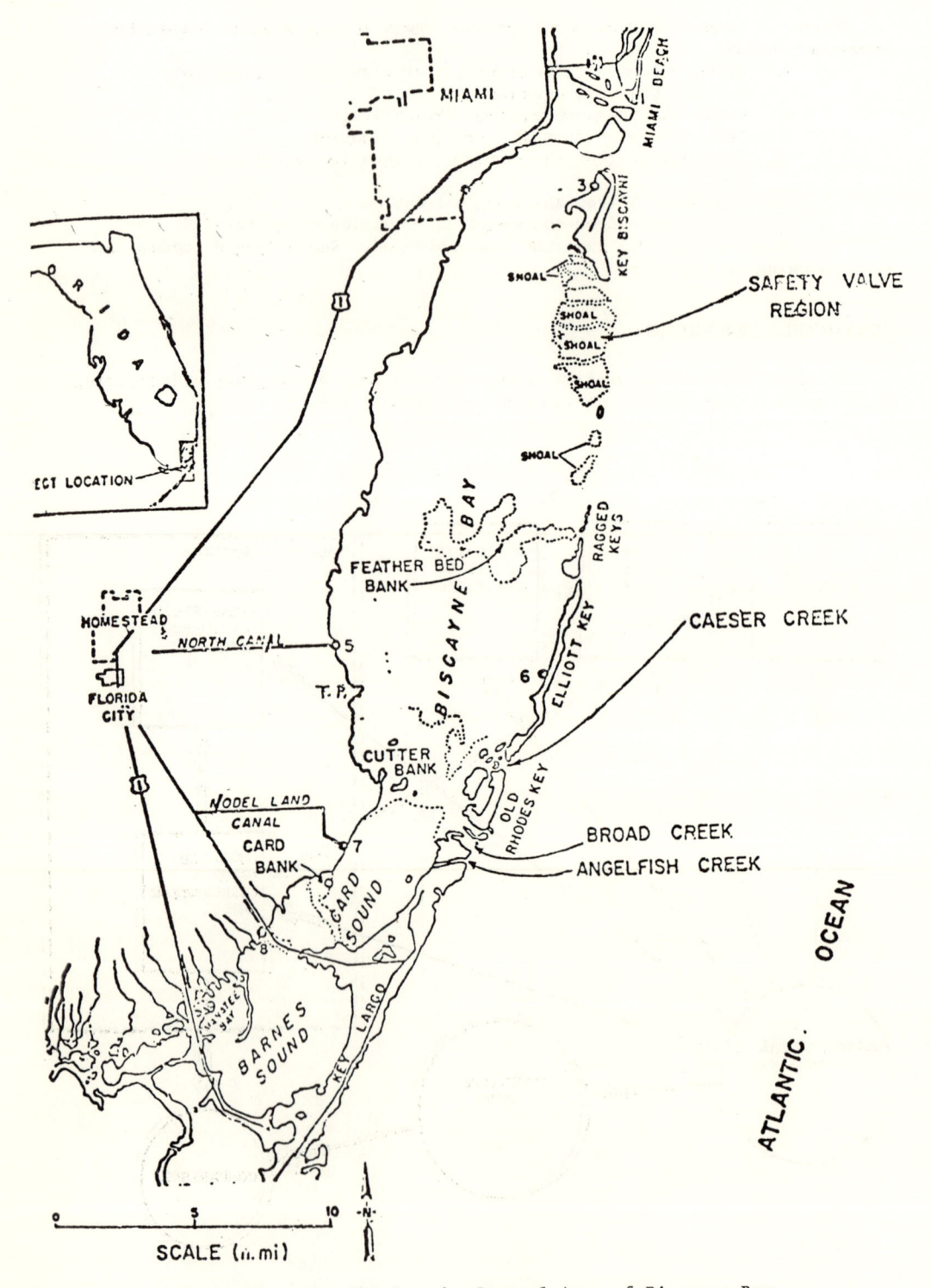

Figure 2a. Map Showing the General Area of Biscayne Bay.

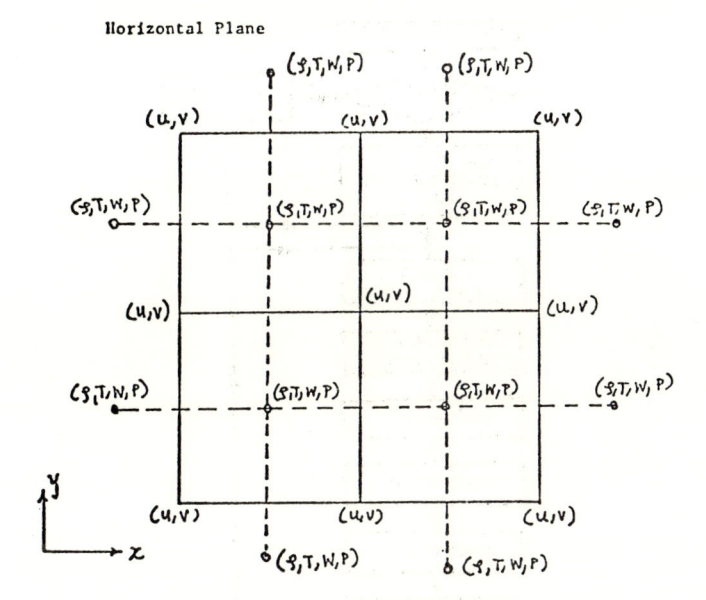

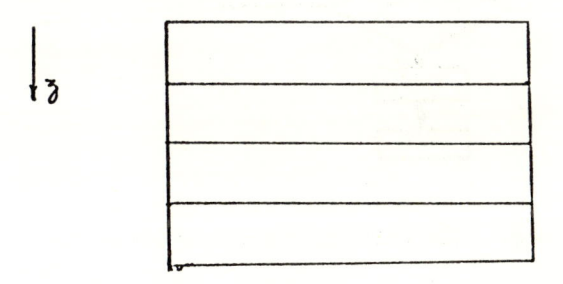

Figure 3. Grid-System and arrangement of variables.

Figure 5 shows the horizontal grid system used to approximate the shore line of the Bay. No such rectangular fitting is necessary in the vertical direction owing to the mapping of the Bay to a constant depth basin. This mapping is obtained by using a transformation of the form $\gamma = Z/h\ (x,y)$.

The basin was assumed closed and tidal effects were ignored. The purpose of the investigation here is to find the wind-driven circulation pattern in the Bay. Later tidal fluxes will be introduced to study the modifications caused by ocean tide.

A vertical eddy viscosity of 5 cm^2/sec. was assumed. The wind stress magnitude was taken to be .1 dynes/cm^2. Two cases have been solved. One for wind from the north, the other for wind from the southeast. The average wind during summer and spring in Miami is from the southeast and is at 5 miles per hour. Qualitative assessment of circulation for intermediate wind directions can be made from the results of these two cases.

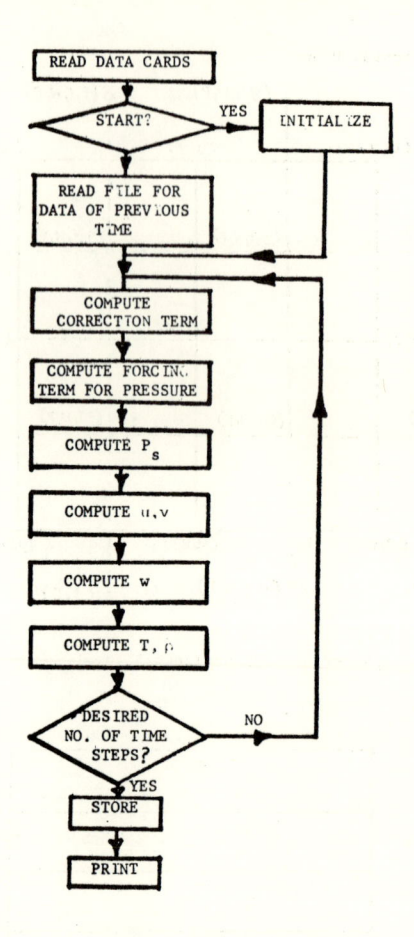

Figure 4. Flow Chart for Rigid-Lid Computation.

Figure 6 shows some of the representative non-dimensional numbers for the Bay. The relative importance of various terms in the governing equation is thereby illustrated.

WIND FROM NORTH

Figure 7 shows the surface velocities at steady state with wind from the north at about 5 miles per hour. The velocities vary significantly in the domain. The variation of velocity direction and magnitude is as a result of a number of effects. The outline of the Bay guides the current near the shoreline. The bottom slopes tend to influence the velocity, and the local depth of the Bay is significant also. The shallower regions have smaller velocities. This is seen in the shoal areas and near the shore. The reduction of velocity near the shore is caused mainly by the smaller depth rather than by the shore boundary-layers. Estimates by Sengupta and Lick [35] have indicated that the sidewall boundary layers are quite thin for similar situations and do not extend to as far as the nearest interior node (.8 kilometers from the shore). Figure 8 shows the velocities at a depth of 2 meters from the surface. There are no

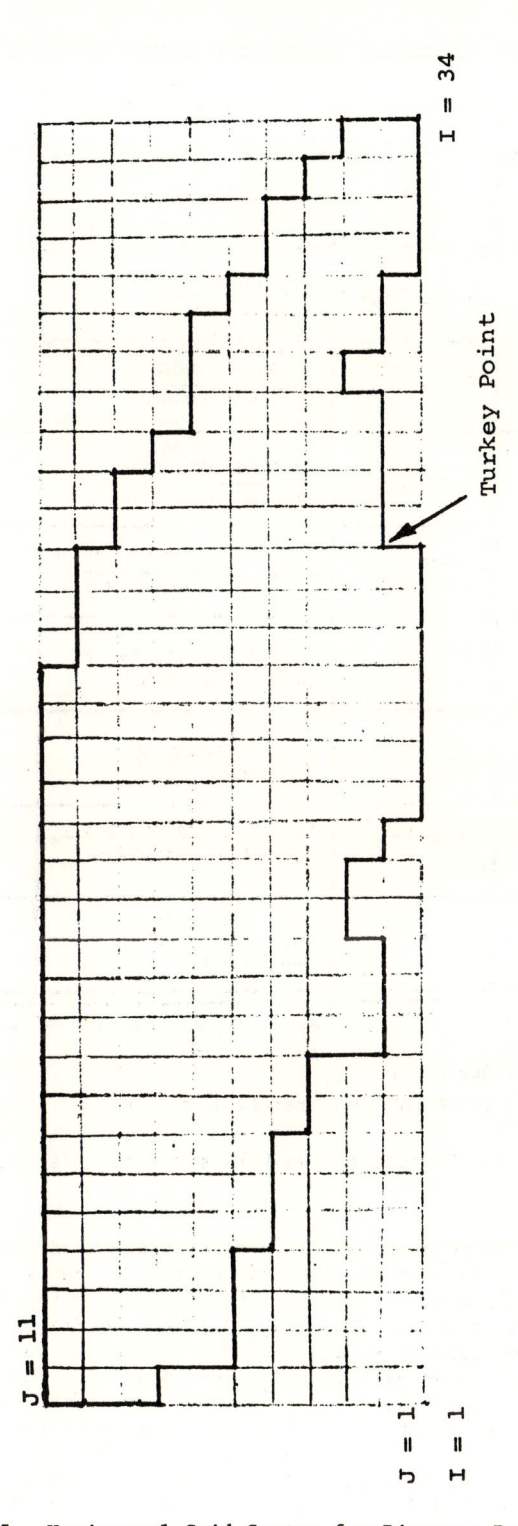

Figure 5. Horizontal Grid System for Biscayne Bay.

PARAMETER	EXPRESSION	SIGNIFICANCE	ESTIMATES * FOR BISCAYNE BAY
ASPECT RATIO	H/L	COMPARISON OF HORIZONTAL AND VERTICAL SCALES	$.5 \times 10^{-3}$
Re; REYNOLDS NO.	$U_{ref} \cdot L / A_{ref}$	$\dfrac{\text{INERTIA FORCES}}{\text{VISCOUS FORCES}}$	800
R_B ROSSBY NO.	U_{ref}/fL	$\dfrac{\text{INERTIA FORCES}}{\text{CORIOLIS FORCES}}$	.238
Pr PRANDTL NO.	A_{ref}/B_{ref}	$\dfrac{\text{VISCOUS DIFFUSION}}{\text{THERMAL DIFFUSION}}$	1
Pe PECLET NO.	$\dfrac{U_{ref} \cdot L}{B_{ref}}$	$Re \cdot Pr \sim \dfrac{\text{CONVECTION}}{\text{DIFFUSION}}$	800
Eu EULER NO.	$\dfrac{gH}{U_{ref}^2}$	$\dfrac{\text{HYDROSTATIC PRESSURE}}{\text{DYNAMIC PRESSURE}}$	4000
Fr FROUDE NO.	$\dfrac{U_{ref}}{gH}$	INERTIA/BODY OR BUOYANCY	.0159
VISCOUS TIME SCALES	L^2/A_H and H^2/A_V	VERTICAL VISCOUS SCALE, HORIZONTAL VISCOUS SCALE	64×10^6 secs. and 32,000 secs
CONVECTIVE TIME SCALE	L/U_{ref}	TIME TAKEN FOR PARTICLE TO COVER LENGTH SCALE	80,000 secs.

* BASED ON $L = 8$ Km. $A_H = 10,000$ cm^2/sec, $A_V = 5$ cm^2/sec.
$H = 4$M $f = .6 \times 10^{-4}$/sec.
$U_{ref} = 10$ cm/sec.
$A_{ref} = 10,000$ cm^2/sec; $B_{ref} = 10,000$ cm^2/sec.

Figure 6. Representative Parameters for Biscayne Bay.

velocity vectors in large parts of the domain since the Bay is less than 2 meters deep in those parts. The return flow can be clearly seen.

A cross-sectional view of the velocity field indicates a cell like structure. Figure 9 shows a section through J=7 (refer to Figure 5). It can be seen that the surface velocities are in the direction of the wind stress. Except for the top third at each location the flow is in the opposite direction indicating a return flow. There is sharp upwelling at the near shore region. Wherever there is sharp change of bottom depth, terms of the form $\partial u/\partial x$ become significant resulting in vertical velocities to satisfy continuity.

It is very important to note that at point A in Figure 9 there is close proximity of upwelling and downwelling. The two cell-like circulation patterns divide the Bay.

Wind Stress = .1 dynes/cm^2 .8 kilometers

2 cm/sec

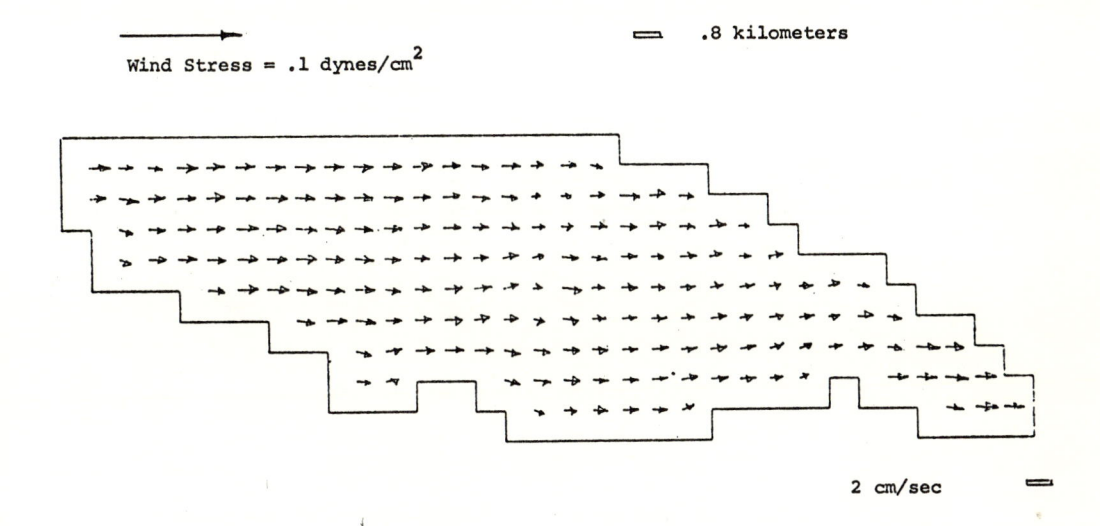

Figure 7. Surface Velocities in Biscayne Bay with wind from the north.

Wind Stress .1 dynes/cm^2 .8 Kilometers

2 cm/sec

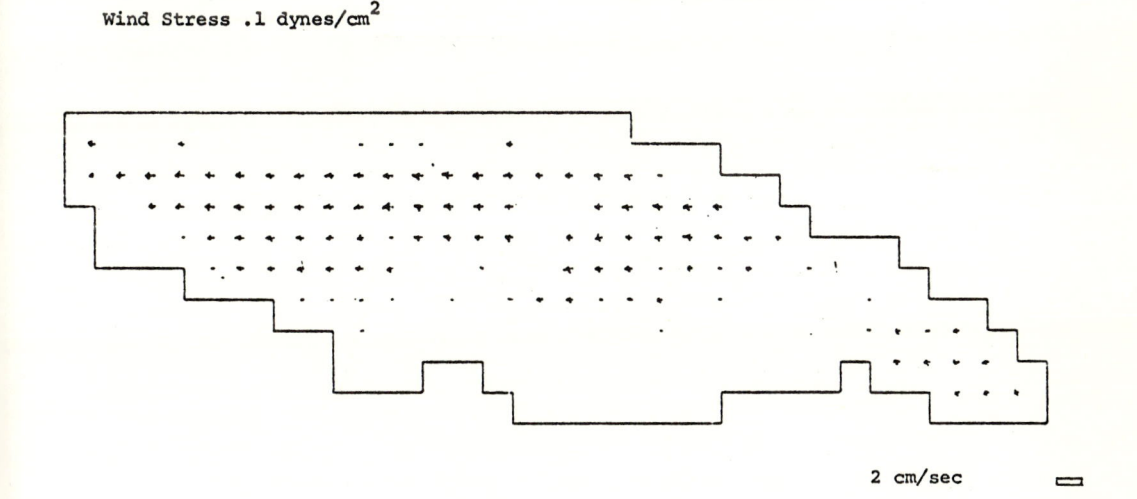

Figure 8. Velocities at a depth of 2 meters with wind from the north.

This indicates that fluid in the region left of A is not exchanged with fluid to the right of A with any degree of facility.

Figure 10 shows a section through J=9. A similar kind of flow as was discussed above may also be seen here with the basin being divided into two parts at point A.

It is evident that shoals are regions which may divide the bay into separate circulation regions under proper wind conditions.

Figure 9. Velocities at section J = 7 with wind from the north.

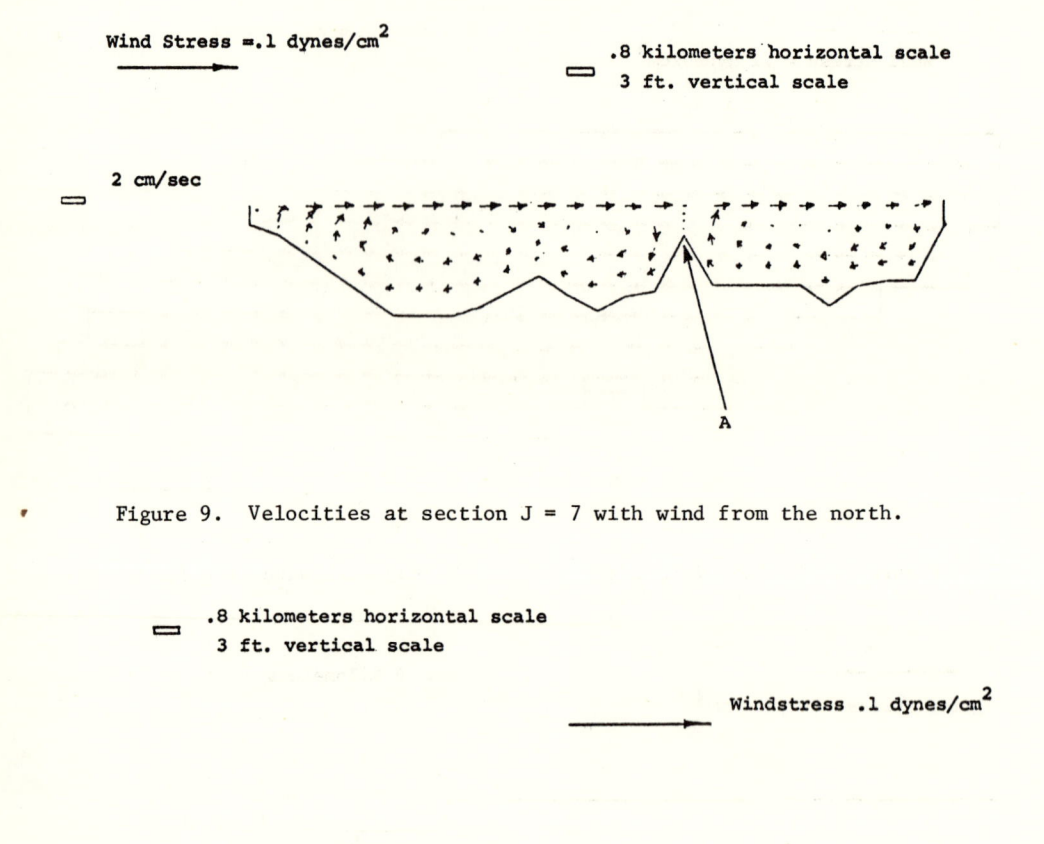

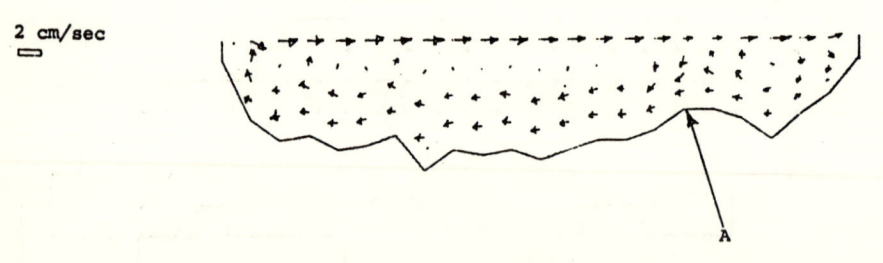

Figure 10. Velocities at section J = 9 with wind from the north.

WIND FROM SOUTHEAST

The prevailing winds during spring and summer are from the southeast. The surface velocities for this case are shown in Figure 11. The velocities are approximately in the direction of the wind. The magnitude of the velocities vary with depth and are smaller near-shore and in shoal regions. It is interesting to note the effect of bottom slope. In some parts near the mainland shore the velocities are almost parallel to the shore and not in the direction of the wind. The return flow can be seen in Figure 12 which shows the velocities at a depth of 2 meters.

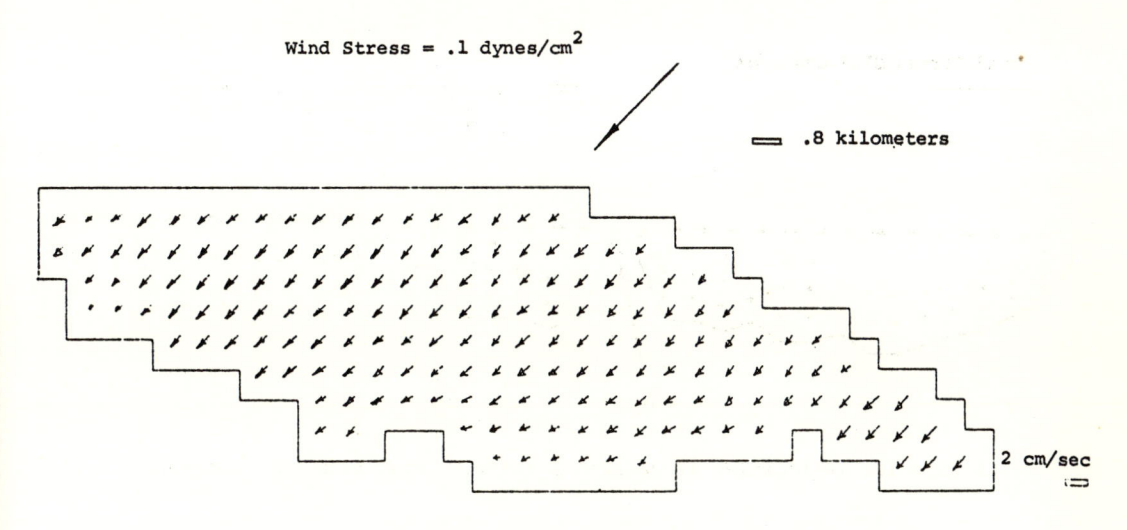

Figure 11. Surface Velocities with wind from southeast.

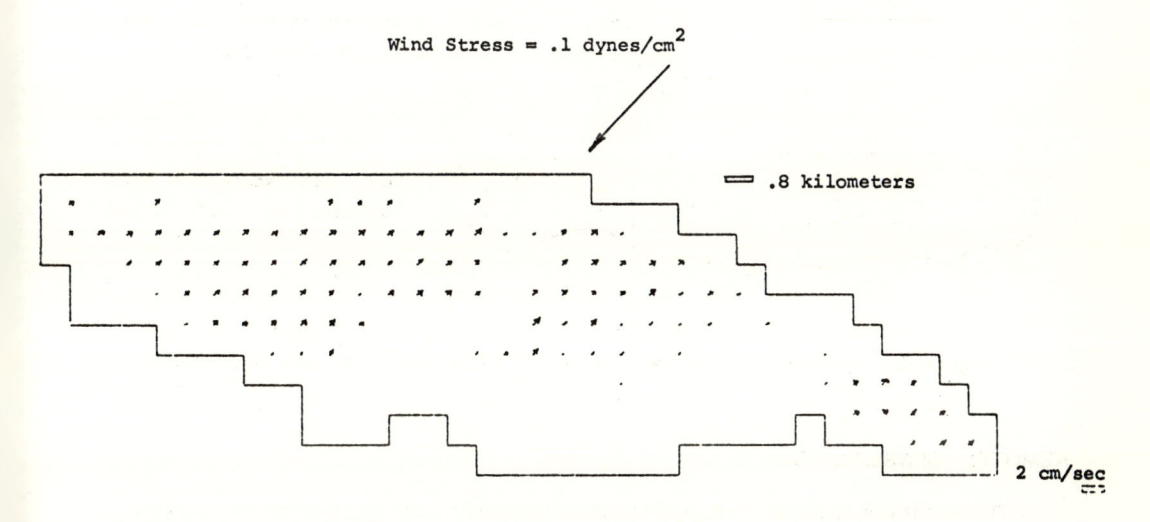

Figure 12. Velocities at a depth of 2 meters with wind from the Southeast.

    Figure 13 shows a vertical cross-section of the Bay along J=8. The cell-like circulation pattern is quite clear with a demarcation point between the cells at point A. The effect of shoals is thus quite clear. Figure 14 shows the velocities at section J=9. Here again it is seen that the top quarter of the bay has velocities in the direction of the wind component with the deeper regions being a region of return flow. Vertical velocities are seen in regions of large bottom slopes. Figure 15 shows vertical sections at I=7 and I=11. These east-west sections also show a cell-like circulation pattern. The comments regarding the regions of upwelling and downwelling are the same as before.
    The circulation patterns discussed are typical and represent the basic nature of the wind driven circulation in Biscayne Bay. It is important to note that the ocean tides will greatly alter the flow patterns computed by the present study. Inclusion of tidal effects and free-surface modelling will result in more realistic flow patterns.

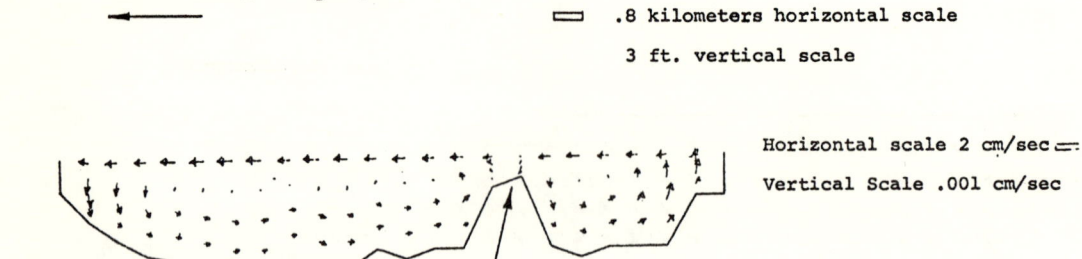

Figure 13. Velocities at section J = 8 with wind from the Southeast.

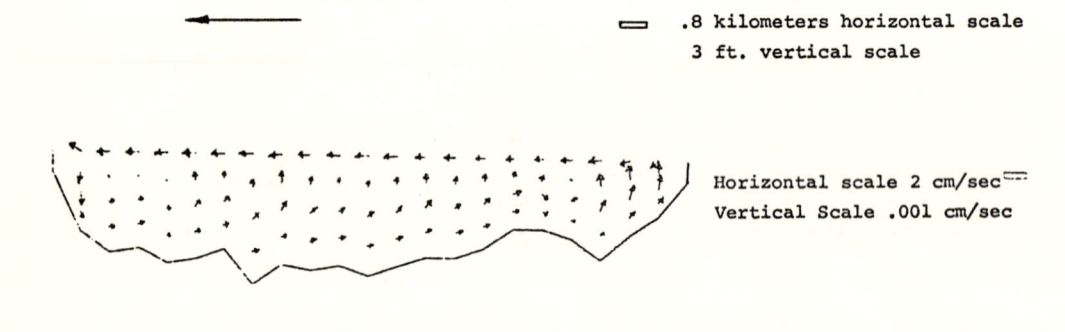

Figure 14. Velocities at section J = 9 with wind from the Southeast.

REMOTELY SENSED DATA

The relationships of the remote sensing study with ground truth measurements and the mathematical model is shown in Figure 16. The satellite and aircraft data are calibrated by ground truth measurements after correction for atmospheric effects using Radiosonde data. This data is then used for verification and calibration of the model.

On the morning of 29 July 1974 an experiment was conducted on Biscayne Bay. Temperature and velocities were measured from a boat. Satellite thermal I-R data was obtained from the NOAA-2 satellite. The NASA-6 aircraft flew several times on an east-west path near the Cutler Ridge power plant. Flights were made along the discharge canal from the power plant. Temperatures were interpreted from the I-R photograph using the DIGICOLOR system which shows false colors for different temperature ranges. Figures 18, 19, 20, 21 show isotherms for the power plant discharge at different times. The plume spreading can be distinctly seen. Since the initial and boundary conditions in terms of winds, ambient temperature, far field temperature, discharge temperature, etc. are known the near field plume model can be used to predict the surface temperature variation with time. The remote sensed data can therefore be directly used to verify the model. This modelling effort is now in progress.

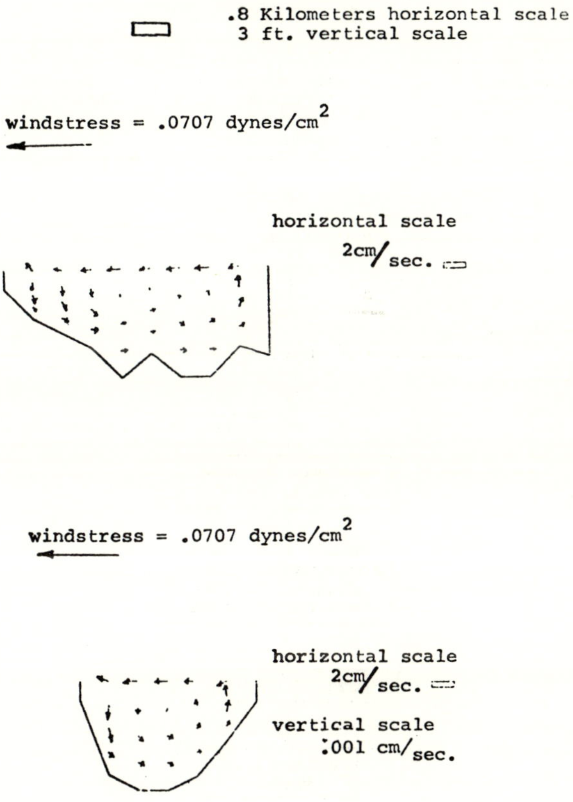

.8 Kilometers horizontal scale
3 ft. vertical scale

windstress = .0707 dynes/cm^2

horizontal scale
2cm/sec.

windstress = .0707 dynes/cm^2

horizontal scale
2cm/sec.
vertical scale
.001 cm/sec.

Figure 15. Velocities at section I = 7 with wind from the Southeast.

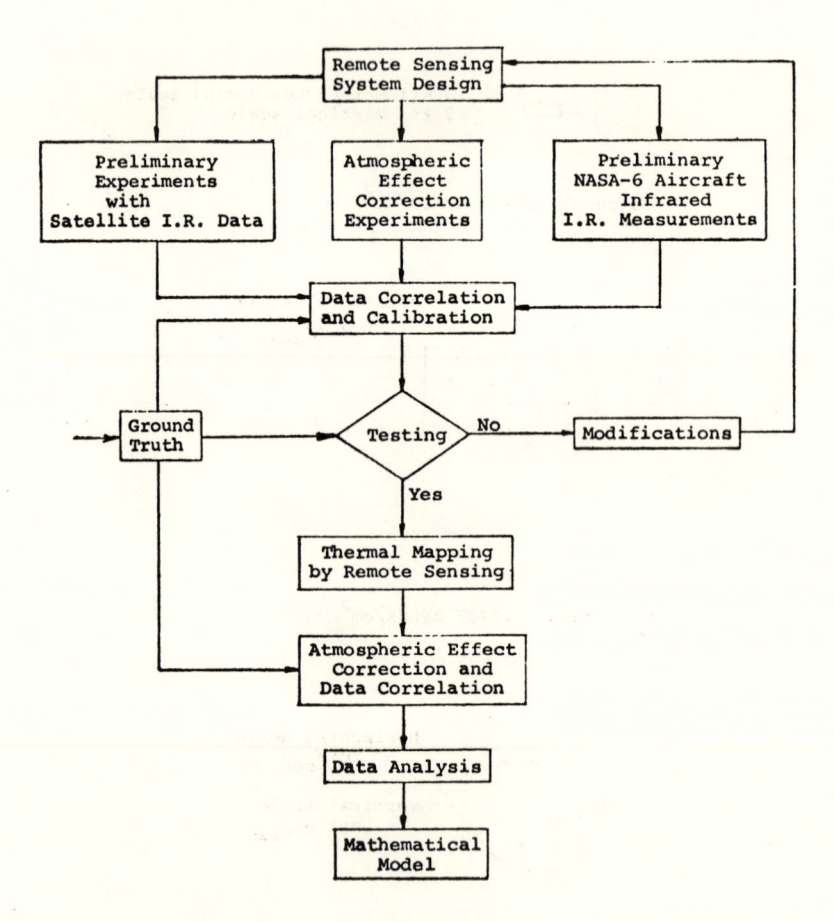

Figure 16. Flow Chart for Remote Sensing Study.

Fig. 17 Flight Path of NASA-6 Beechcraft

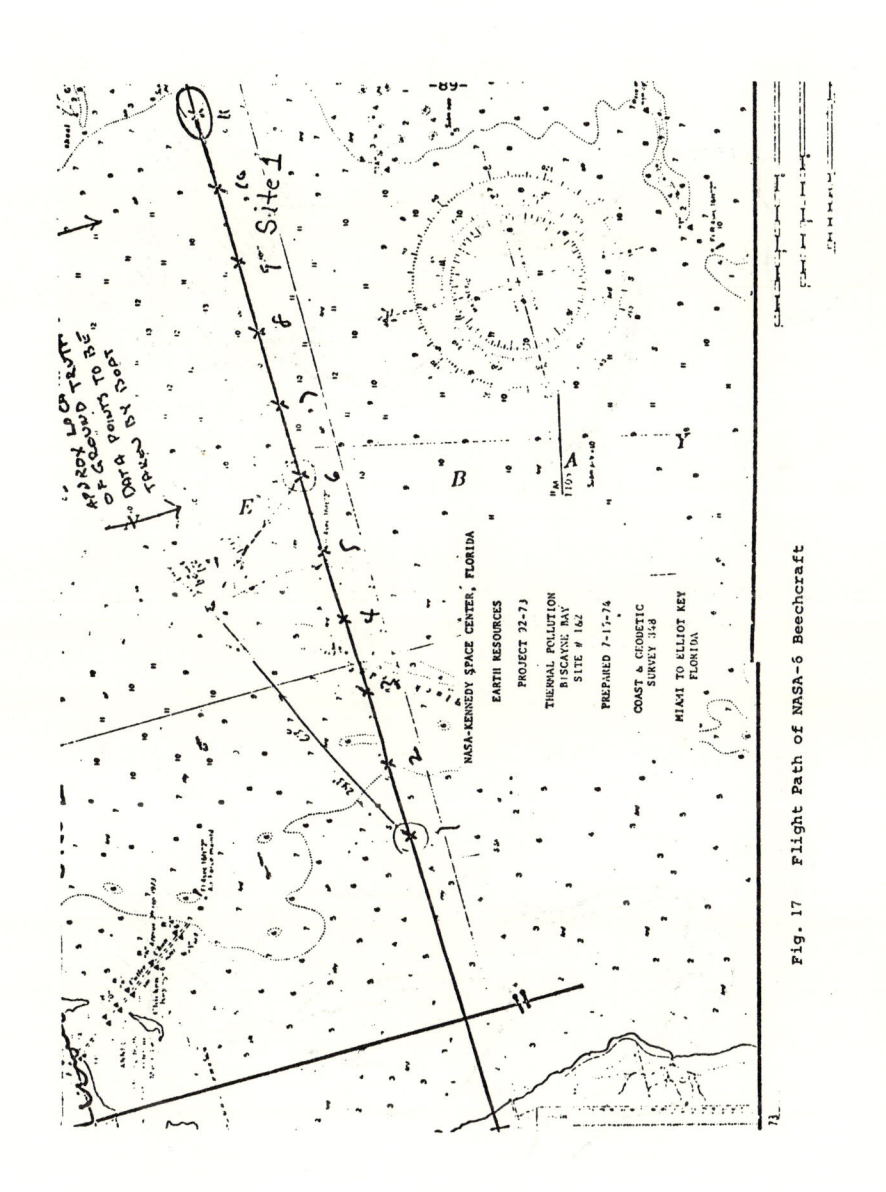

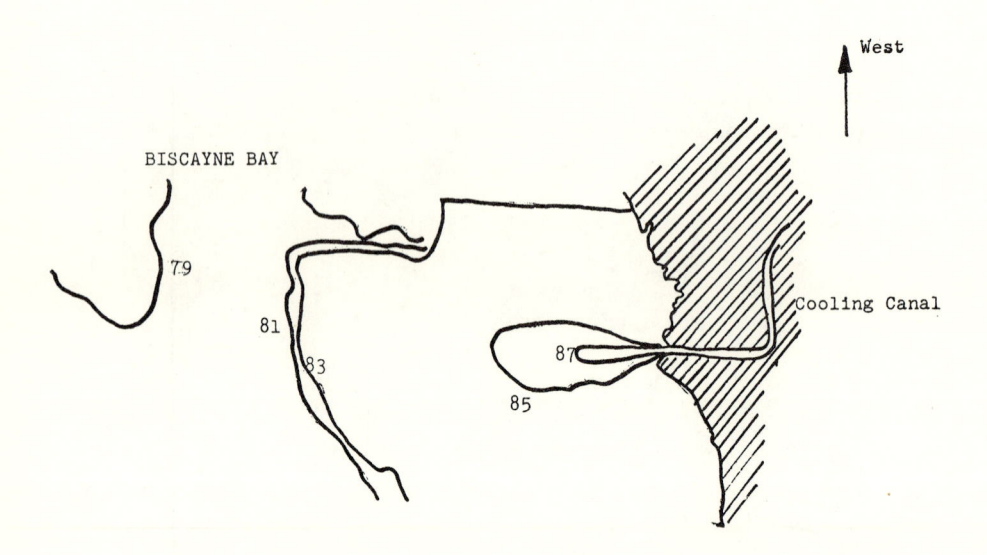

Figure 18. Surface Isotherms (°F) for Thermal Plume from Cutler Ridge Power Plant at 9:30 a.m.

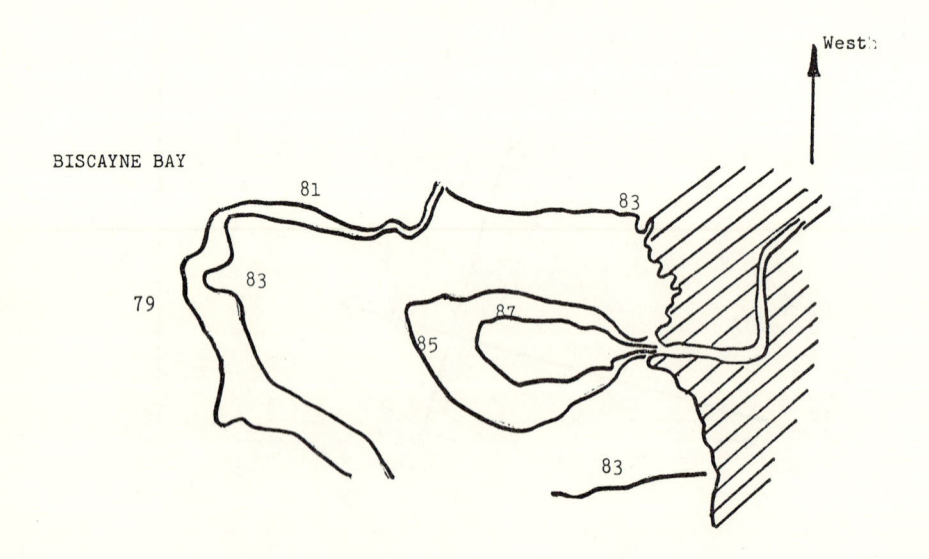

Figure 19. Surface Isotherms (°F) for Thermal Plume from Cutler Ridge Power Plant at 10:30 a.m.

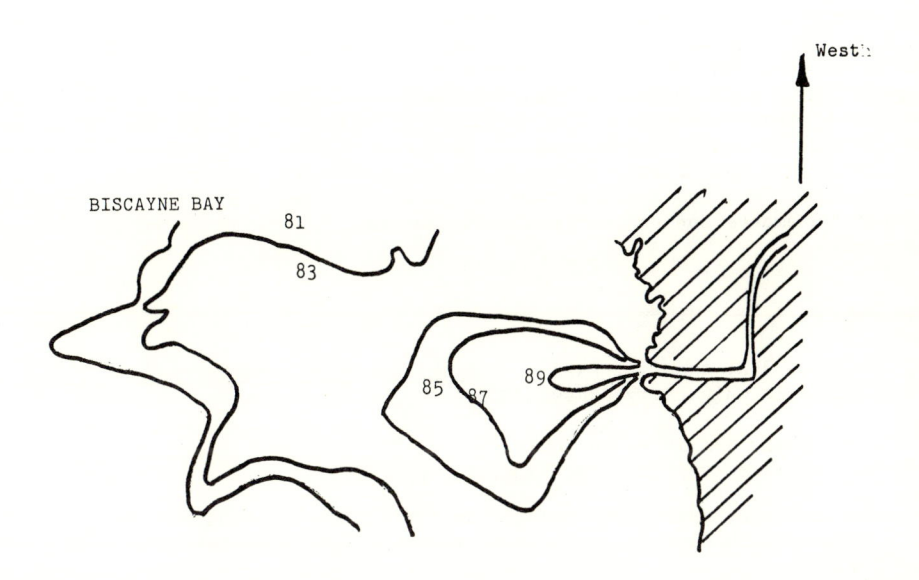

Figure 20. Surface Isotherms (°F) for Thermal Plume from Cutler Ridge Power Plant at 11:30 a.m.

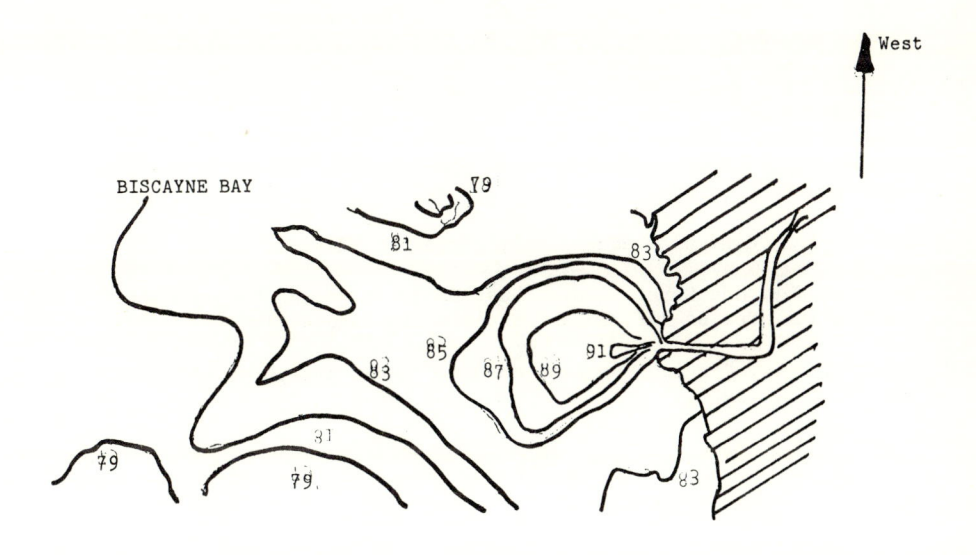

Figure 21. Surface Isotherms (°F) for Thermal Plume from Cutler Ridge Power Plant at 12:30 p.m.

CONCLUSIONS

Predictive studies of ecosystems can only be done by mathematical models. The need for large data bases to calibrate and verify models has been demonstrated. Though the SEASAT program promises to supply information on a wide range of ocean surface parameters, at present only temperature can be reliably obtained by remote sensing. The satellite I-R data has resolution sufficient for macro-scale models for oceans and the Great Lakes. For meso-scale analysis on Biscayne Bay and thermal plumes airborne radiometers are suitable at the present time. A total systems approach to environmental analysis using ground truth measurements and remote sensing is imperative for the development of useful models.

REFERENCES

[1] Krenkel, R.A. and F.L. Parker, Biological Aspects of Thermal Pollution, Vanderbilt University Press, 1969.

[2] U.S. Senate Committee on Public Works, "A Study of Pollution: Water," U.S. Government Printing Office, 100 pp. 1963.

[3] Federal Water Pollution Control Administration, "Industrial Waste Guide Thermal Pollution," U.S. Department of the Interior, September 1968, Revised.

[4] Proceedings of the MacArthur Workshop on the Feasibility of Extracting Usable Energy from the Florida Current, Palm Beach Shores, Florida, 197

[5] Ekman, V.X., Uber Horizontal Cirkulation Winderzengten Meeresstromungen, Arkue, Mat. Astr. Fysik. 17, No. 26, 1923.

[6] Welander, P., Wind Action on a Shallow Sea: Some Generalizations of Ekman's Theory," Tellus, Vol. 9, No. 1, February 1957.

[7] Nilson, B.W., "Note on Surface Wind Stresses over Water at Low and High Wind Speeds," Journal of Geophysical Research, Vol. 65, No. 10, 1960.

[8] Defant, A., Physical Oceanography, Pergamon Press, 1961.

[9] Edinger, J.E. and Geyer, J.C., "Heat Exchange in the Environment," E.E.I. Publication, No. 65-902, Edison Electric Institute, 1971.

[10] Argyris, J.H., "Energy Theorems and Structural Analysis," Aircraft Eng. 26, 1954; 27, 1955.

[11] Baker, A.J., "Predictions in Environmental Hydrodynamics Using the Finite Element Method, 1," Theoretical Development, AIAA Journal, Vol. 13, No. 1, 1975.

[12] Baker, A.J., and S.W. Zelany, "Predictions in Environmental Hydrodynamics Using the Finite Element Method, II," Applications, AIAA Journal, Vol. 13, No. 1, 1975.

[13] Roache, J.P., Computational Fluid Dynamics, Hermosa Publishers, 1972.

[14] Arakawa, A, "Computational Design for Long-Term Numerical Intergration of the Equations of Fluid-Motion: Two-Dimensional Flow, Part I," J. Comp. Phys. 1, 1966.

[15] Crowley, W.P., "A Global Numerical Ocean Model: Part I," J. Comp. Phys. 3, 1968.

[16] Crowley, W.P., "A Numerical Model for Viscous Free Surface Barotopic Wind-Driven Ocean Circulation," J. Comp. Phys. 5, 1969.

[17] Crowley, W.P., "A Numerical Model for Viscous Non-Divergent Barotopic, Wind-Driven, Ocean Circulations," J. Comp. Phys., 6, 1970.

[18] Berdahl, P., Oceanic Rossby Waves: A Numerical Rigid-Lid Model, ITD-4500, UC-34, Lawrence Radiation Laboratory, University of California, Livermore.

[19] Munk, W.H. and E.R. Anderson, "Notes on a Theory of the Thermocline," Journal of Marine Research, Vol. 7, 1948.

[20] Nagler, R.G., and A.A. Loomis, "SEASAT-A Monitoring the Oceans From Space," Presented at IEEE Electronic and Aerospace System Convention, EASCON 74, Oct. 9, 1974.

[21] Apel, J.R., SEASAT: A Spacecraft Views the Marine Environment with Microwave Sensors, Presented at Remote Sensing Applied to Energy Related Problems Symposium, Miami, 1974.

[22] Hirschberg, J.G., A.W. Wouters, and J.D. Byrne, Laser Measurements of Sea, Salinity, Temperature and Turbidity in Depth, Presented at Remote Sensing Applied to Energy Related Problems Symposium, Miami, 1974.

[23] Lee, S., T.N. Veziroglu, S. Sengupta, and N.L. Weinberg, Remote Sensing Applied to Thermal Pollution, Proceedings of the Symposium on Remote Sensing Applied to Energy Related Problems, 1974.

[24] Lee, S., T.N. Veziroglu, S. Sengupta, and C.F. Tsai, Mathematical Modelling of Thermal Pollution in Coastal Regions, Accepted for presentation at the Southeastern Seminar on Thermal Sciences, 1975.

[25] Hiser, H.W., S. Lee, T.N. Veziroglu, and S. Sengupta, Application of Remote Sensing to Thermal Pollution Analysis, Presented at Fourth Annual Remote Sensing of Earth Resources Conference, University of Tennessee Space Institute, Tullahoma, Tennessee, 1975.

[26] Veziroglu, T.N., S. Lee, and H.W. Hiser, Monitoring of Thermal Discharges into Biscayne Bay, Accepted for presentation at the Southeastern Seminar on Thermal Sciences, 1975.

[27] Akers, P., Modelling of Heated Discharges in Engineering Aspects of Thermal Pollution, Krenkil and Parker (Ed.) Vanderbilt Univ. Press, 1969.

[28] Policastro, A.J., Heated Effluent Dispersion in Large Lakes, Presented at the Topical Conference, Water Quality Consideration Siting and Operating of Nuclear Power Plants, Atomic Industrial Form Inc., 1972.

[29] Policastro, A.J. and J.V. Tokar, Heated Effluent Dispersion in Large Lakes, Report No. ANL/ES-11, Argonne National Laboratory, Argonne, Illinois, 1972.

[30] Policastro, A.J. and R.A. Paddock, Analytical Modelling of Heated Surface Discharges with Comparisons to Experimental Data, Interim Report No. I, Presented at the 1972 Annual Meeting of the A.I. Ch. E.

[31] Stolzenbach, K.D. and D.R.F. Harleman, "An Analytical and Experimental Investigation of Surface Discharges of Heated Water," R.M. Parsons Laboratory for Water Resources and Hydrodynamics, M.I.T., Cambridge, Massachusetts, Tech. Report No. 135, 1971.

[32] Stolzenbach, K.D. and D.R.F. Harleman, "Three-Dimensional Heated Surface Jets," Water Resources Research, Vol. 9, No. 1, 1973.

[33] Jirka, G.H. and D.R.F. Harleman, "The Mechanics of Submerged and Multiport Diffusers for Buoyant Discharges in Shallow Water," R.M. Parsons Laboratory for Water Resources and Hydrodynamics, M.I.T., Cambridge, Massachusetts, Tech. Report No. 169, 1973.

[34] Veziroglu, T.N., S.S. Lee, N.L. Weinberg, S. Sengupta, et.al., Application of Remote Sensing for Prediction and Detection of Thermal Pollution, NASA-CR-139182, 1974.

[35] Sengupta, S. and W. Lick, A Numerical Model for Wind Driven Circulation and Temperature Fields in Lakes and Ponds, FTAS/TR-74-99, Case Western Reserve University, 1974.

PART V

SPECIAL TOPICS

THE SATELLITE SOLAR POWER STATION OPTION

PETER E. GLASER

*Arthur D. Little, Inc.,
Cambridge, Massachusetts, U.S.A.*

ABSTRACT

The possibilities for using satellite solar power stations for large-scale power generation on Earth, converting solar energy into microwave energy, transmitting it to the Earth's surface, and transforming it into electricity have recently been explored. The current state of technology and the necessary developments for accomplishing the four functions — i.e., collection of solar energy, conversion to and transmission of microwaves and rectification to DC on the ground — are reviewed. The requirements for flight control, Earth to orbit transportation and orbital assembly are discussed. Considerations are given to cost projections, resource use and economics. Environmental issues, including impact of waste heat release, water injection into the upper atmosphere by space vapor exhaust, and location of antenna sites are listed. Biological effects and radio frequency interference are explored. The time frame for accomplishing the operational system is outlined.

INTRODUCTION

Among the different sources of energy, whether they be non-renewable such as fossil or nuclear fuels or continuous such as tidal or geothermal, none have a greater potential than solar energy. The question that needs to be answered is: Has society reached the level of sophistication to apply solar energy for its overall long-term benefit consistent with the balance of nature? The answer is not yet obvious. However, efforts required to provide the answer are beginning to be made.

The total influx from solar, geothermal, and tidal energy into the Earth's surface environment is estimated to be $173,000 \times 10^{12}$ watts. Solar radiation accounts for 99.98% of it. The sun's contribution to the energy budget of the Earth is 5,000 times the energy input of other sources combined.

The vast quantities of energy radiated by the the sun reach the Earth in very dilute form. Thus, any attempts to convert solar energy to power on a significant scale will require devices which occupy a large land area as well as locations that receive a copious supply of sunlight. These requirements restrict Earth-based solar energy conversion devices which could produce power to a few favorable geographical locations. Even for these locations energy storage must be provided to compensate for the day-night cycle and cloudy weather.

One way to harness solar energy effectively would be to move the solar energy conversion devices off the surface of the Earth and place them in orbit away from the Earth's active environment and influence and resulting erosive forces.[1,2] The most favorable orbit would be one around the sun, but as a first approximation toward this very long-term goal, an orbit around the Earth where solar energy is available nearly 24 hours of every day could be used.

Since the concept of a satellite solar power station (SSPS) (See Figure 1) was presented as an alternative energy production method,[3] the impending energy crisis has been recognized as one of the major issues facing the nation. Various options to meet the crisis are now being explored.

An assessment of the feasibility of the SSPS concept has shown that it is worthy of consideration as an alternative energy production method.[4] Its development can be realized by building on scientific realities, on an existing industrial capacity for mass production, and on demonstrated technological achievement particularly those associated with remote sensing of the earth's environment and resources.

A. PRINCIPLES OF A SATELLITE SOLAR POWER STATION

Figure 2 shows the design principles for an SSPS. Two symmetrically arranged solar collectors convert solar energy directly to electricity by the photovoltaic process while the satellite is maintained in synchronous orbit around the Earth. The electricity is fed to microwave generators incorporated in a transmitting antenna located between the two solar collectors. The antenna directs the microwave beam to a receiving antenna on Earth where the microwave energy is efficiently and safely converted back to electricity.

An SSPS can be designed to generate electrical power on Earth at any specific level. However, for a power output ranging from about 3,000 to 15,000 megawatts (MW), the orbiting portion of the SSPS exhibits the best power-to-weight characteristics. Additional solar collectors and antennas could be added to establish an SSPS system at a desired orbital location. Power can be delivered to most desired geographic locations with the receiving antenna placed either on land or on platforms over water near major load centers and tied into a power transmission grid. The status of technology and the advances which will be required to achieve effective operation for an SSPS are described in the following sections.

1. LOCATION OF ORBIT

The preferred locations for the SSPS are the Earth's equatorial synchronous orbit stable nodes which occur near the minor axes, at longitudes of about 123° West and 57° East. The SSPS would be positioned so its solar collectors

FIGURE 1 DESIGN CONCEPT FOR A SATELLITE SOLAR POWER STATION

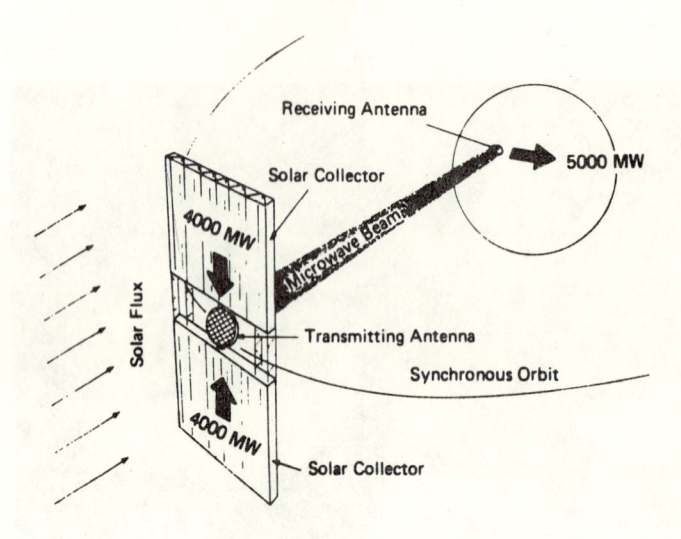

FIGURE 2   DESIGN PRINCIPLES FOR A SATELLITE SOLAR POWER STATION

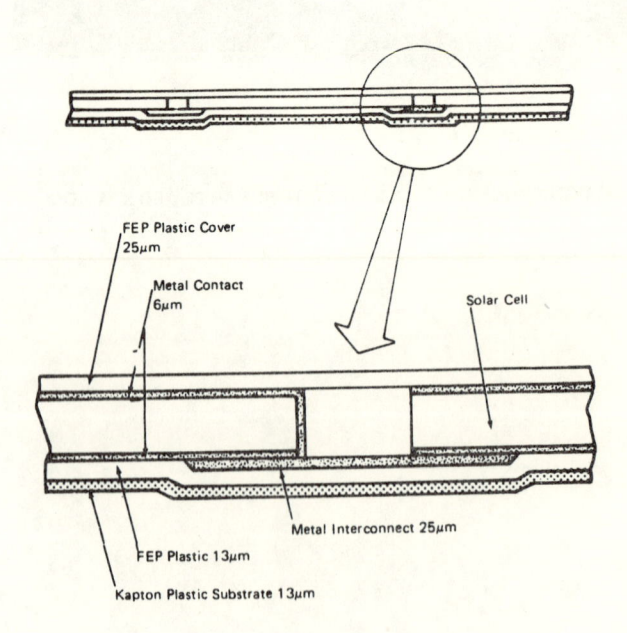

FIGURE 3 SOLAR CELL BLANKET CONSTRUCTION

always faced the sun, while the antenna directed a microwave beam to a receiving antenna on Earth. In an equatorial, synchronous orbit, the satellite can be maintained stationary with respect to any desired location on Earth. About 30,000 pounds per year of propellants for attitude control would be required to overcome orbit-disturbing influences, such as the gravitational effects of the sun and moon, solar pressure, and the eccentricity of the Earth. The microwave beam would permit all-weather transmission so that full use could be made of the nearly 24 hours of available solar energy.

An SSPS in synchronous equatorial orbit would pass through the Earth's shadow around the time of the equinoxes, when it would be eclipsed for a maximum of 72 minutes a day (near midnight at the receiving site). This orbit provides a 6- to 15-fold time advantage over solar energy conversion on Earth. A comparison of the maximum allowable costs of solar cells indicates that for terrestrial solar power applications these devices are competitive with other energy conversion methods if they cost about $2.30 per square meter. Because of the favorable conditions for energy conversion that exist in space, these cells would be competitive if they cost about $45 per square meter in an Earth-orbit application.[5]

2. SOLAR ENERGY CONVERSION

The photovoltaic conversion of solar energy into electricity is ideally suited to the purposes of an SSPS. In contrast to any process based on thermodynamic energy conversion, there are no moving parts, fluid does not circulate, no material is consumed, and a photovoltaic solar cell can operate for long periods without maintenance. Photovoltaic energy conversion has been developed substantially since the first laboratory demonstration of the silicon solar cell in 1953. Today, such cells are a necessary part of the power supply system of nearly every unmanned spacecraft, and considerable experience has been accumulated to achieve long-term and reliable operations under the conditions existing in space. As a result of many years of operational experience, a substantial technological base exists on which further developments can be based.[6] These developments will be directed toward increasing the efficiency of solar cells, reducing their weight and cost, and maintaining their operation over extended periods.

A. EFFICIENCY INCREASE

The maximum theoretical efficiency of a silicon solar cell is about 22%. The most widely used single-crystal silicon solar cells can routinely reach an efficiency of 11% and efficiencies of up to 16% have been reported.[7]

The silicon solar cell which is produced from single-crystal silicon is typically arranged with the P-N junctions positioned horizontally. More recently, vertically illuminated, multi-junction silicon solar cells have been investigated.[8] These have the potential of reaching higher efficiencies and of being more resistant to solar radiation damage.

Solar cells made from single-crystal gallium arsenide exhibit an efficiency of about 14% with a theoretical limit of about 26%. A modified gallium arsenide solar cell was reported to have reached an efficiency of 18%.[9] This substantial increase in efficiency is particularly significant, because these cells can operate at higher temperatures than silicon solar cells, are more radiation-resistant, and can be prepared in thickness about one-tenth that of a silicon solar cell.

Several other materials may be suitable for photovoltaic solar energy conversion. Among these are various combinations of inorganic semiconductors which have only partially been investigated. Organic semiconductors which exhibit the photovoltaic effect and which do not have known boundaries to the theoretical efficiency also remain to be explored, so that their potential for photovoltaic solar-energy conversion can be established.

B. WEIGHT

Single-crystal silicon solar cells are presently 500 to 1000 microns thick, although their thickness could be reduced to about 50 microns with acceptable efficiency. However, gallium arsenide cells need be only a few microns thick.

The individual solar cells have to be assembled to form the solar collector. The weight of a solar cell array can be reduced by assembling the solar cells in a blanket between thin plastic films, with electrical interconnections between individual cells obtained by vacuum-depositing metal alloy contact materials (see Figure 3). The collector weight can be further reduced when solar energy concentrating mirrors arranged to form flat-plate channels are used so that a smaller area of solar cells is required for the same electrical power output (see Figure 4). The weight and cost of a given area of a reflecting mirror used to concentrate solar energy are considerably less than those for the same area of solar cells.

Suitable coatings on mirrors to reflect only the component of the solar spectrum most useful for photovoltaic conversion can reduce heating of the solar cells and thus increase efficiency. There is a balance between concentrating the solar radiation per unit area of the cell, which may lead to a rise in temperature and a consequent decrease in solar cell efficiency, and the desire to maximize the collection of solar energy. An array configuration that includes mirrors with a concentration factor of about 2 has been chosen for the solar collector as a baseline.

Since 1965, the weight of solar cells has decreased substantially. With the development of blanket-type construction for solar cell arrays, this weight is projected to drop to 8 lb/kW by 1975. The use of solar energy-concentrating mirrors can reduce the solar collector weight to about 2 lb/kW if 50-micron thick silicon solar cells are used, and even less if gallium arsenide solar cells are used.

C. COST

The present cost of silicon solar cells for use in spacecraft — about $80/W — is prohibitive. New methods for producing single-crystal silicon and mass-production assembly techniques will have to be developed to reach the goal of less than $1/W. Based on the experience of present manufacturers, there is a high probability that low-cost, high-volume silicon single crystals of the desired thickness can be produced once there is a large enough market. Hand-assembly techniques of solar cells, which are adequate to meet the present very small demand, will have to be replaced by automated methods similar to those that have been perfected for the production of other semiconductors devices. For example, recent work with edge-defined, film-fed growth (EFG) of silicon ribbons[10] has demonstrated that a one-inch-wide, six-feet-long ribbon can be grown which results in 10% efficient silicon solar cells.

Cost projections based on actual experience in the production of single-crystal silicon solar cells are presented in Figure 5.[11] The 2x2 cm cells are widely used on unmanned spacecraft. The 2x6 cm cells were produced for the Apollo telescope mount portion of the "Skylab" spacecraft. The 2-inch diameter cell represents the costs of solar cells for terrestrial applications produced from 2-inch diameter wafers of single-crystal silicon boules. The SSPS solar collector cost projections are based on the EFG ribbon-grown solar cell process.

The industry perspective on experience with semiconductor devices can be used to test the reasonableness of the SSPS solar collector cost projections. The cost projections correspond to reasonable slopes, since experience indicates that typical cost reductions follow about a 70% slope.[12] These cost reductions could be approached as markets develop for the different solar cell applications.

D. EFFECTS OF THE SPACE ENVIRONMENT

The state-of-the-art of solar cells is now at a level where lifetimes of 10 years are achievable. For example, the effective life expectancy of the Intelsat IV satellite is eight years. But the operations of the solar cell arrays will be influenced by the space environment.

One influencing factor in space will be solar radiation, damage from which will cause a logarithmic decay of solar-cell effectiveness. However, improvements in radiation-resistant solar cells are expected to result in a 30-year minimum operational lifetime for the SSPS, after which normal SSPS effectiveness can be restored by adding a small area of new solar cells. Thus, there will be no absolute time limit on effective SSPS operation.

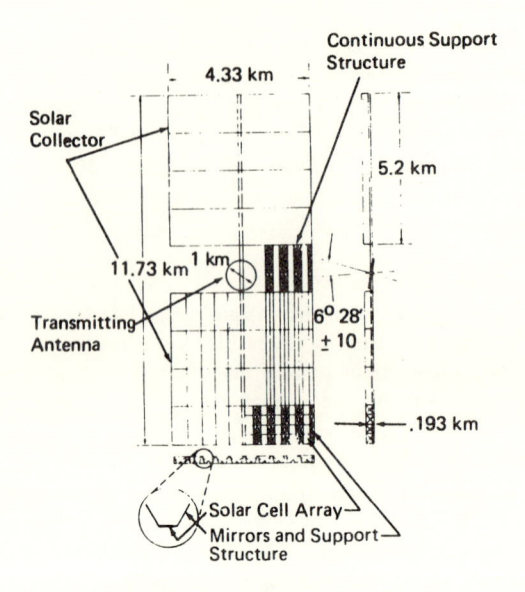

FIGURE 4 SSPS DIMENSIONS

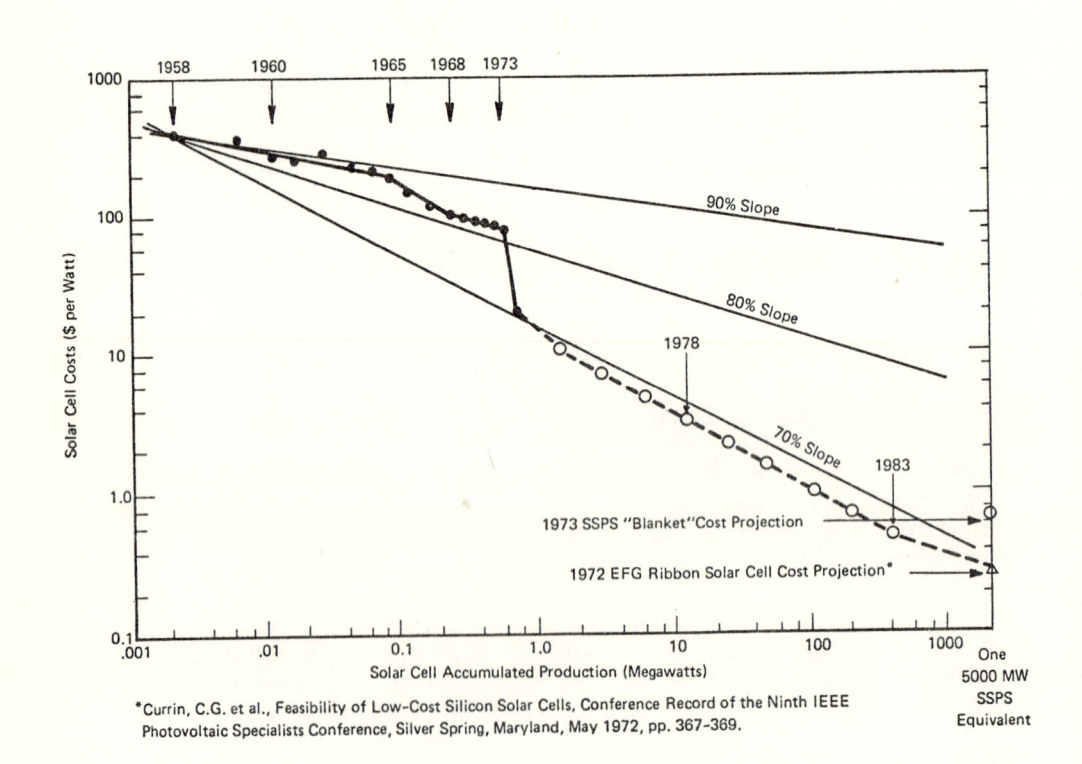

*Currin, C.G. et al., Feasibility of Low-Cost Silicon Solar Cells, Conference Record of the Ninth IEEE Photovoltaic Specialists Conference, Silver Spring, Maryland, May 1972, pp. 367-369.

FIGURE 5 SILICON SOLAR CELL COST PROJECTIONS

NORTHWEST

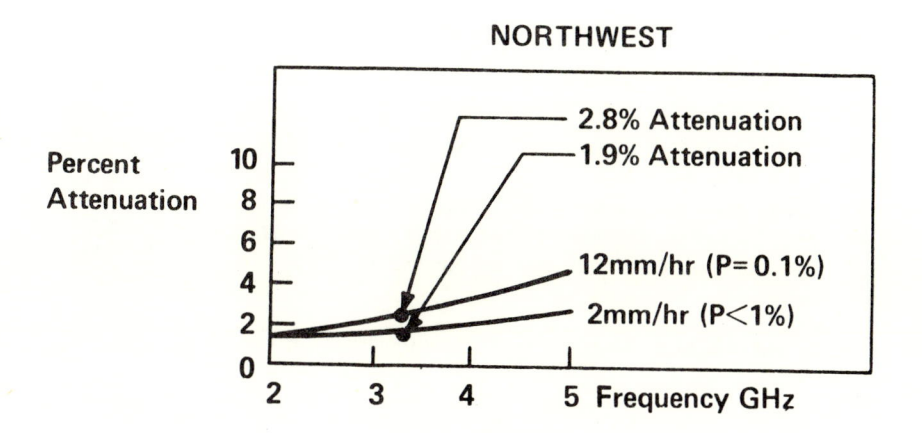

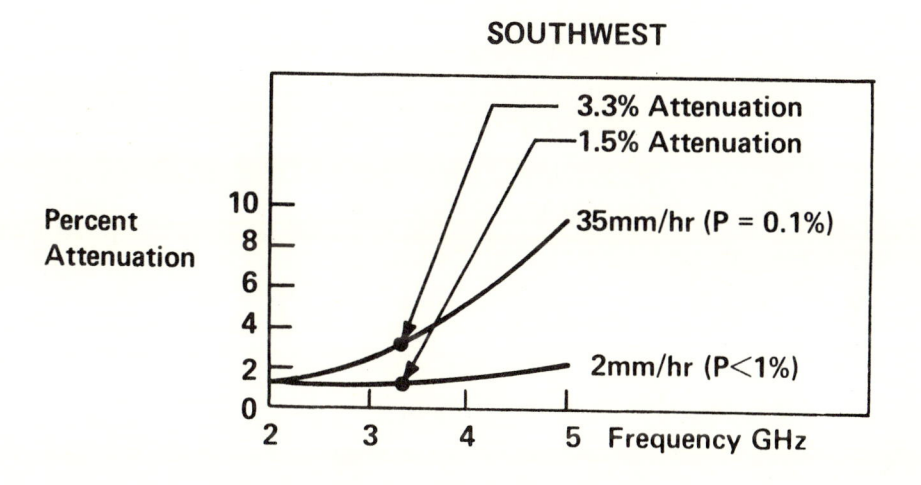

SOUTHWEST

FIGURE 6 ATMOSPHERIC ATTENUATION OF MICROWAVES IN TWO UNITED STATES LOCATIONS

Another aspect of the space environment that will influence SSPS operations is the impact of micrometeoroids. In synchronous orbit, the SSPS is expected to suffer a 1% loss of solar cells, based on the probability of damage-causing impacts by micrometeoroids during a 30-year period.

The benign nature of the space environment and the absence of significant gravitational forces, however, permit the design of solar collector arrays which have a minimum material mass. In addition, their performance would be much more predictable than that of an Earth-based solar energy conversion device because of the absence of the vagaries of the Earth environment.

Based on these considerations, key issues and performance goals can be established as follows:

Key Issue	Performance Goal
Solar cell performance	18% efficiency, 2-mill-thick cell
Solar cell cost	$0.28 per cm^2
Blanket weight	950 W/kg (430 W/lb)
Blanket cost	$0.68 per cm^2
Solar collector cost and weight	$310/kW, 3 lb/kW
Operational life	30-year-life, 6% degradation in 5 years
Energy input to process	3 months of SSPS operation
High-voltage circuit control	40 kV – 5% loss

The detailed technology which can be identified for this portion of the SSPS system indicates the advanced state-of-the-art of photovoltaic solar energy conversion.

3. MICROWAVE POWER GENERATION, TRANSMISSION, AND RECTIFICATION

The power generated by the SSPS in synchronous orbit must be transmitted to a receiving antenna on the surface of the Earth and then rectified. The power must be in a form suitable for efficient transmission in large amounts across long distances with minimum losses and without affecting the ionosphere and atmosphere. The power flux densities received on Earth must also be at levels which will not produce undesirable environmental or biological effects. Finally, the power must be in a form that can be converted, transmitted, and rectified with very high efficiency by known devices.

All these conditions can best be met by a beam link in the microwave part of the spectrum. In this part of the spectrum a frequency of 3.3 GHz was selected because induced radio frequency interference can be limited so that an appropriate internationally agreed upon frequency could be assigned to an SSPS.

As early as 1963, Brown[13] demonstrated that large amounts of power could be transmitted by microwaves. The efficiency of microwave power transmission will be high when the transmitting antenna in the SSPS and the receiving antenna on Earth are large. The dimensions of the transmitting antenna and the receiving antenna on Earth are governed by the distance between them and the choice of wavelength.[14]

The size of the transmitting antenna is also influenced by the inefficiency of the microwave generators due to the area required for passive radiators to reject waste heat to space and the structural considerations as determined by the arrangement of the individual microwave generators. The size and weight of the transmitting antenna will be reduced as the average microwave power flux density on the ground is reduced by increasing the size of the receiving antenna and as higher-frequency microwave transmission is used. The size of the receiving antenna will be influenced by the choice of the acceptable microwave power flux density, the illumination pattern across the antenna face, and the minimum microwave power flux density required for efficient microwave rectification.

Several power distributions for the transmitting antenna can be identified which concentrate at least 90% of the power at the ground within the main lobe, along with an indication of the nature of side lobes to be expected. Several approaches are available to control the side lobes. The ground site selection criteria will be greatly influenced by results of projected Earth resources studies, as well as social and political considerations. System aspects of site selection lend themselves to relatively simple and known analysis techniques.

Highly efficient ground power distribution systems, such as the underground, cryogenic, high-voltage dc system, are anticipated to be well developed in the operation time period of the SSPS. Such systems will permit a rationale for location of power consumption centers, such that they may not necessarily be close to the power supply station. The low power flux density and the efficient receiving antenna will permit a full utilization of land under the antenna. The biological effects of the non-ionizing radiation, which are the subject of investigation now, can be anticipated to become sufficiently well understood so that a straightforward SSPS biological effects program can be implemented to define criteria and limitations for several ground selection and multiple utilization options.

A. MICROWAVE ATTENUATION[15]

Ionospheric attenuation of microwaves is low (less than 0.1%) for wavelengths between 3 and 30 cm and for the microwave power flux densities occurring within the beam. Tropospheric attenuation is low for wavelengths near and above 10 cm, but attenuation will increase as wavelengths are reduced. Moderate rainfall attenuates microwaves approximately 3% at a wavelength of 10 cm at a nadir angle of 60 deg. The efficiency of transmission through the atmosphere in temperate latitudes, including some rain (2 mm/hr), is approximately 98% and decreases to 94% for moderate (33 mm/hr) rainfall, depending on location (see Figure 6).

B. MICROWAVE TRANSMISSION SYSTEM EFFICIENCY

Table 1 indicates the efficiencies that have been demonstrated in the three major functional categories of the microwave transmission system and the projected efficiencies which should be achievable with further development.[16] In recent tests at the Goldstone antenna site a 75% microwave to DC rectification efficiency was demonstrated. Further tests to transmit 15 kW of microwave power over a one mile distance are planned for early 1975 at the site.

Including microwave attenuation, the overall efficiency of microwave transmission from dc in the SSPS to dc on the ground is projected to be about 70%.

C. MICROWAVE GENERATION

The microwave generator design is based on the principle of a crossed-field device which has the potential of achieving a high reliability and a very long life.[17] A pure-metal, self-starting, secondary emitting, cold cathode is employed in a non-reentrant circuit matched to an input and output circuit so as to provide a broad-band gain device. The device is designed to be capable of automatically self-regulating its power output. The use of samarium-cobalt permanent magnet material leads to substantial weight reduction compared to previously available magnet material. The vacuum in space obviates the glass envelope required on Earth (see Figure 7). The cathode and anode of the microwave generator are designed to reject waste heat with passive extended-surface radiators which radiate to space (see Figure 8). The output of an individual microwave generator weighing a fraction of a kilogram per kilowatt can range from 2 to 5 kW. The cost of present commercially produced microwave generators is about $5/kW.

Table 1. Microwave Power Transmission Efficiencies

	Efficiency Presently Demonstrated[a]	Efficiency Expected with Present Technology[a]	Efficiency Expected with Additional Development[a]
Microwave Power Generation Efficiency	76.7[b]	85.0	90.0
Transmission Efficiency from Output of Generator-to-Collector Aperture	94.0	94.0	95.0
Collection and Rectification Efficiency (Rectenna)	64.0	75.0	90.0
Transmission, Collection, and Rectification Efficiency	60.2	70.5	85.0
Overall Efficiency	26.5[c]	60.0	77.0

a. Frequency of 2450 MHz (12.2-cm wavelength).

b. This efficiency was demonstrated at 3000 MHz and a power level of 300 kW CW.

c. This value could be immediately increased to 45% if an efficient generator were available at the same power level at which the efficiency of 60.2% was obtained.

THE SATELLITE SOLAR POWER STATION OPTION

FIGURE 7 COMPARISON BETWEEN A COMMERCIALLY AVAILABLE MICROWAVE DEVICE WITH THE MICROWAVE GENERATOR COMPONENTS DESIGNED FOR USE IN SPACE

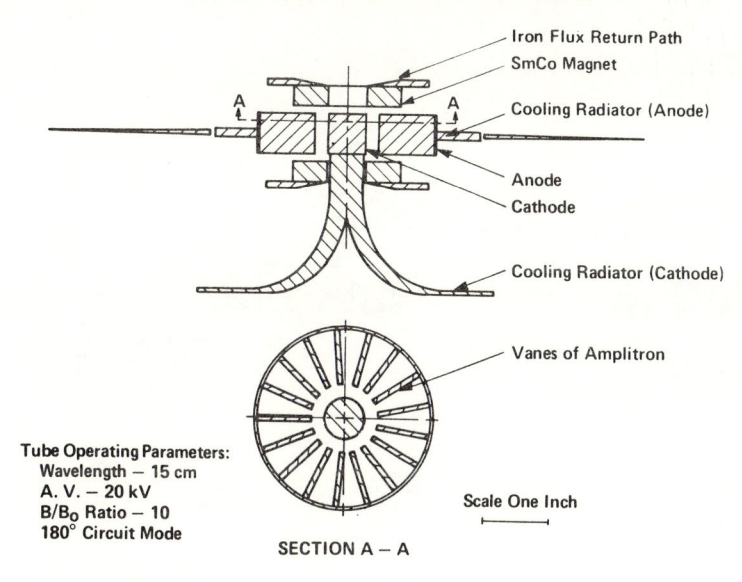

Tube Operating Parameters:
 Wavelength — 15 cm
 A. V. — 20 kV
 B/B_0 Ratio — 10
 180° Circuit Mode

FIGURE 8 ARRANGEMENT OF SPACE RADIATORS TO REJECT WASTE HEAT

FIGURE 9 ASSEMBLY OF MICROWAVE GENERATORS WITH SLOTTED WAVEGUIDE, PHASED-ARRAY TRANSMITTING ANTENNA

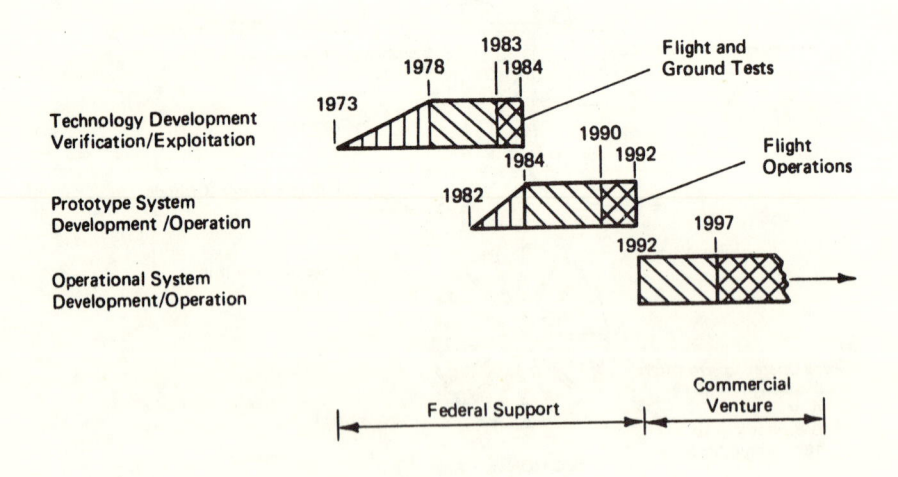

FIGURE 10   PROGRAM PHASING

The quantity of 1 to 2 million tubes that would be needed for each SSPS is large enough to warrant large-scale, highly efficient mass production. There is substantial production experience on magnetrons, similar in many respects to the Amplitron device projected for use in the SSPS. These magnetrons are produced in quantities of 250,000 per year for less than $30.00 each, with production volume in 1974 projected to reach 1,250,000. The Amplitron is projected to cost about $125 for a 5-kW tube, or $25/kW, after making allowance for the higher material costs in their construction and their higher power rating.

D. MICROWAVE TRANSMISSION

A series of microwave generators will be combined in a subarray (e.g., about 10 m by 10 m) which forms part of the antenna. Each subarray must be provided with an automatic phasing system so the individual antenna radiating elements will be in phase. These subarrays will be assembled into a slotted waveguide, phased-array, transmitting antenna about 1 km in diameter to obtain a microwave beam of a desired distribution (see Figure 9). The distribution can be designed to range from uniform to gaussian.

For this diameter antenna, the diameter of the receiving antenna on Earth would have to about 7 km for an ideal gaussian distribution in the beam within which 90% of the transmitted energy is intercepted. The use of such a large receiving antenna area will reduce the microwave power flux density on the Earth to a value low enough so that the flux density at the edges of the receiving antenna would be substantially less than the continuous microwave exposure standard presently accepted in the United States (i.e., 10 mW/cm^2). Within several kilometers beyond the receiving antenna, the microwave density levels drop to about 10 μW/cm^2 a value low enough to meet even the most stringent international standards for microwave exposure.

To achieve the desired high efficiency for microwave transmission, the phased-array antenna will be pointed by electronic phase shifters. Proper phase setting for each subarray must be established to form and maintain the desired phase front. Deviations can be detected and appropriate phase shifts made to minimize microwave beam scattering. A master phase control in the antenna can be developed if the microwaves are to be transmitted efficiently and the microwave beam always directed toward the receiving antenna.[18] The master phase front control system can be designed to compensate for the tolerance and position differences between the subarrays by sensing the phase of a pilot signal beamed from the center of the area occupied by the receiving antenna to control the phase of the microwaves transmitted by each subarray. The pilot signal will be of a substantially different frequency than that of the microwave power beam, so wave filters could be used to separate them.

Precision pointing of the receiving antenna is not necessary to the operation of the SSPS and inhomogeneities in the propagation path are not significant. Any deviation of the microwave beam beyond allowable limits would preclude acquisition of the pilot signal. Without the signal, the coherence of the microwave beam would be lost, the energy dissipated, and the beam spread out so the microwave power density would approach communication signal levels. This phase-control approach would assure that the beam could not be directed either accidentally or deliberately toward any other location but the receiving antenna. This inherent fail-safe feature of the microwave transmission system is backed up by the operation of the switching devices, which would open-circuit the solar cell arrays to interrupt the power supply to the microwave gnerators. The microwave generation and transmission system is projected to cost $130/kW.

E. MICROWAVE RECTIFICATION

The receiving antenna is designed to intercept, collect, and rectify the microwave beam into dc which can then be fed into a high-voltage dc transmission network or converted into 60-Hz ac. Half-wave dipoles distributed throughout the receiving antenna capture the microwave power and deliver it to solid-state microwave rectifiers.[19] Schottky barrier diodes have already been demonstrated to have a 80% microwave rectification efficiency at 5 watts of power output. With improved circuits and diodes, a rectification efficiency of about 90% will be achievable.

The diodes combined with circuit elements which act as half-wave dipoles are uniformly distributed throughout the receiving antenna, so the microwave beam intercepted in a local region of the receiving antenna is immediately converted into dc. The collection and rectification of microwaves with a receiving antenna based on this principle have the advantages that the receiving antenna is fixed and does not have to be pointed precisely at the transmitting antenna. Thus, the mechanical tolerance in the construction of the receiving antenna can be relaxed. Furthermore, the illumination distribution of the incoming microwave radiation need not be matched to the radiation pattern of the receiving antenna; therefore, a distorted incoming wavefront caused by non-uniform atmospheric conditions across the antenna will not reduce efficiency.

The amount of microwave power that is received in local regions of the receiving antenna can be matched to the power-handling capability of the solid-state diode microwave rectifiers. Any heat resulting from inefficient rectification in the diode circuits can be convected by the ambient air in the local region of the receiving antenna with atmospheric heating similar to that over urban areas. Only about 10% of the incoming microwave beam would be lost as waste heat. The low thermal pollution achievable by this process of rectification cannot be equaled in any known thermodynamic conversion process. The cost of the diodes, circuits, and supporting structure is projected to be about $11/m^2. The projected cost of the receiving antenna required for a 5000-MW output on Earth is $100/kW and for a 10,000-MW output on Earth $50/kW.

The rectifying elements in the receiving antenna can be exposed to local weather conditions. The antenna can be designed so that sunlight would still reach the land beneath it, with only a fraction lost due to shadowing. Thus the land could be put to productive use. The successful demonstration of high-efficiency rectification of microwaves to dc shows that the SSPS will be capable of generating power on Earth with an efficiency which has not yet been equaled by any known power generation method.

B. SSPS FLIGHT CONTROL

Although the SSPS is orders of magnitude larger than any spacecraft yet designed, its overall design is based on present principles of technology. Thus, its construction and the attainment of a 30-year operating life require not new technology, but substantial advances in the state of the art.

The SSPS structure is composed of high-current-carrying structural elements, the electromagnetic interactions of which will induce loads or forces into the structure. Current stabilization and control techniques are capable of meeting the requirements of spacecraft now under development. Most of these spacecraft have comparatively rigid structures and are amenable to control as a single entity by reaction jets or momentum storage devices. But the large size of an SSPS at first suggests that new structural and control system design approaches may be needed to satisfy orientation requirements. However, results of analyses indicate that present structural analytical techniques and tools are adequate and that an SSPS can be controlled such that the solar collector will point at the sun to ± 1 degree, the transmitting antenna can be mechanically controlled to ± 1 arc min and the direction of the microwave beam electronically maintained to ± 1 arc sec.[4] Low-thrust, ion propulsion systems appear promising for SSPS control because their performance characteristics are compatible with the potential lifetime required of the SSPS.

C. EARTH-TO-ORBIT TRANSPORTATION

A high-volume, two-stage transportation system will be required for an SSPS: (1) a low-cost stage capable of carrying high-volume payloads to low-Earth orbit (LEO); and (2) a high-performance stage capable of delivering partially assembled elements to synchronous or some intermediate orbit altitude for final assembly and deployment. The factors affecting flight mode selection include payload element size, payload assembly techniques, desirable orbit locations for assembly, time constraints, and requirements for man's participation in the assembly. The choice of transportation system elements includes currently planned propulsion stages and advanced concepts optimized for an operational SSPS system. Minimum cost transportation combinations will have to be identified which can fulfill the requirements for SSPS delivery, assembly, and maintenance for an operational system. The challenge of an SSPS capable of generating 50000 MW of power on Earth is to place into orbit a payload of about 25 million pounds and propellant supplies for station-keeping purposes of about 30,000 pounds per year.

1. GROUND TO LOW-EARTH ORBIT TRANSPORTATION

The space shuttle now under development provides the necessary first steps toward a very high-volume, low-cost, Earth-to-orbit transportation system. This shuttle can be used for SSPS technology verification and flight demonstration activities and for transporting elements of a prototype SSPS into LEO. Operational experience with the prototype SSPS will be essential to permit an orderly evolution to the very high-volume Earth-to-orbit-to-synchronous transportation system needed for an operational SSPS.

The nominal building block sizes for the solar cell arrays and microwave antenna element payloads would be established from representative SSPS configurations. Nominal component sizes and orbital assembly factors inherent in the large orbital configuration can then be identified. These SSPS payload sizing requirements and orbital assembly considerations can then be combined with an appropriate LEO-to-synchronous orbit transportation mode, including the most desirable propulsion elements, to identify the combined total payload size with which an advanced shuttle must contend, so that a desirable low-operating-cost transportation system can be evolved.

2. LEO-TO-SYNCHRONOUS ORBIT TRANSPORTATION

The LEO-to-synchronous orbit transportation system under consideration by NASA includes reusable space tugs using cryogenic propellants or storable propellants. None of these chemically powered upper-stage systems has sufficiently high performance for an SSPS. For an SSPS it appears necessary to develop advanced high-performance propulsion systems for exclusive operation in the space environment.

The ion propulsion system for a space tug can be designed so it will interface with payloads delivered to LEO by the space shuttle. The payloads would be transferred to a space tug which, over a period of several months, would follow a spiral trajectory to synchronous orbit.

Preliminary flight profiles and propulsion vehicle combinations have been investigated to expose the type of environmental and gross operational factors that must be considered in evolving a space transportation system for an SSPS. It is evident that both orbital assembly considerations and alternative transportation modes are involved in the definition and development of a low-cost, high-volume, Earth-to-synchronous orbit advanced space transportation system. This definition effort will involve considerable system engineering and technology development which will greatly benefit from the activities currently under way in support of the present space shuttle development. Cost projections for the LEO-to-synchronous orbit transportation system range from $100 per pound for an advanced space transportation system to $300 per pound for a modified space shuttle vehicle.

D. ORBITAL ASSEMBLY

Operations in space involving assembly have been limited to docking two actively maneuvering vehicles together. However, studies of modular space stations which dock a number of similar masses to form a large complex in orbit have been performed. In general, to effect a mating, attitude control is required on both target and docking vehicles.

After a docking vehicle contacts a space assembly, several things happen: (1) the assembly experiences loads; (2) modular elements deflect relative to each other, and (3) possibly disruptive control forces and G loads are transmitted throughout the space assembly. In general, the weak link in a space assembly is the docking interface, since it has a smaller cross-sectional area than the prime construction element. Relatively large space assemblies have been analyzed in modular space station studies and found to be controllable during assembly, providing that a prescribed build-up sequence is followed and that grossly asymmetric configurations are avoided. In addition, there is a significant interplay between docking contact velocities, target and docking vehicle flight control, and docking mechanism characteristics. In general, direct vehicle docking contact velocities of about 0.5 fps and manipulator docking contact velocities of 0.1 fps are compatible with currently planned spacecraft systems. The large, more flexible SSPS type spacecraft will require significantly lower contact velocities because of the size and flexibility of the system and its potential damping characteristics. Zero or near-zero contact velocities may be required, together with appropriate docking or joining mechanisms and control techniques during assembly. Because a considerable portion of the SSPS structure may be part of the power distribution system, new joining and assembly techniques may be required.

Assembly sequences and modes and desirable assembly altitudes have to be identified to define the assembly requirements for the SSPS's large area and light-weight structure. Once the framework for the basic SSPS structure has been developed, including the potential sizes and sub-element characteristics of major components, such as the solar collector arrays, the transmitting antenna and structural members, alternative assembly sequences modes and attitudes can be evaluated. These could include automatic or operator assembly options, and assessments as to the portions of the assembly process to be carried out at low-Earth orbit altitudes, at intermediate or at synchronous altitudes. With this knowledge, the appropriate control techniques during the assembly process can then be identified and developed.

E. ECONOMIC CONSIDERATIONS

1. COST PROJECTIONS

Workable versions of each component for the SSPS exist today or can be built, although some entail considerable development. The costs of the major components, such as solar cells, microwave generators, and rectifiers can be drastically reduced. Programs are already under way to achieve cost reductions. Further reductions can be achieved by mass production as demonstrated by the successful transition from development to production.

Allowing for the potential cost reductions, but using the present design approaches, and a modified space shuttle vehicle, the major SSPS components capital costs for a prototype SSPS designed to generate seven hundred MW on Earth are summarized below:

	Prototype $/kW
Solar cells and solar collector array	350
Microwave generators and transmitting antenna	150
Rectifiers and receiving antenna	100
Transportation to orbit and assembly	1000
	1600

The dominant cost element for a prototype SSPS is space transportation. This underscores the need for considerable systems engineering to evolve and identify a low operating cost transportation system for an operational SSPS. Design improvements and weight reduction potentials indicate that the capital costs for an operational SSPS could be reduced to about $1000/kW. At this cost level the SSPS can be projected to be competitive with other energy production systems which are planned to be operational by the year 2000.

2. RESOURCE USE

The SSPS represents an approach to power generation which does not use a terrestrial energy source. Thus, the environmental degradation associated with mining, transportation, or refining of natural energy sources is absent. Natural resources will have to be used to produce the components for the SSPS and the propellants for transportation to orbit. Nearly all the materials to be used for the components are abundant. For each SSPS, the rare materials required, such as platinum or gallium, would be less than 2% of the supply available to the United States per year.

In addition, energy resource requirements for the SSPS will be minimal. Thus, the time required for one operational SSPS to pay back the direct energy expended during the construction phases, including raw materials, manufacturing processes, component assembly, space transportation, and ground support facilities will be:

Propellants	6 months
Solar cells	3 months
Ground support equipment	3 months

3. ECONOMIC COMPARISONS

The three issues that will have to be addressed when making an economic comparison of the SSPS with other means of generating power and the methodology include the following:

- The costs and benefits associated with the SSPS which have to be evaluated to determine the economic feasibility of an investment of this type. Such an evaluation has to answer the question whether the expenditure of funds on this project will result in a product which will earn a satisfactory rate of return and thereby be considered as a worthy undertaking. This analysis must assess the social and environmental side effects of both the old and the new technology needed and establish a standard of comparison with other power generation and distribution systems.

- The macro-economic interindustry effects produced by the SSPS which have to be examined to analyze the effects such an investment might have on the structure of the economy as a whole. This analysis can be accomplished by using an input-output model of the economy to test the industry effects and to establish demand patterns for capital, material, and labor resources.

- The consumption effects created by the SSPS which will be reflected in both the cost/benefit analysis and the analysis of macro-economic interindustry effects. Because of its importance, this is an issue worthy of separate consideration and attention. It is concerned with total energy requirements, the impact on the capital management, and labor resources of the U.S. economy, and the demand for raw material, the effect on prices of those materials in the face of a massive demand, and the resulting potential political implications.

These issues must be related not only to an efficiency criterion, where an optimal allocation of scarce resources is the sole objective, but must also indicate the effects associated with large-scale changes that might be created in the economic structure if the projected benefits of an SSPS are to be realized. An analysis of industry and consumption effects can quantify the impact of an SSPS as a necessary component of a technology assessment which will be required for a decision-making process among different energy-production systems required to meet future energy demands.

F. ENVIRONMENTAL ISSUES

1. ENVIRONMENTAL/ECOLOGICAL IMPACT

The SSPS appears to have limited environmental and ecological impacts at the receiving antenna. The following are the potential environmental impacts:

- Waste heat released by natural convection at the receiving antenna does not constitute a significant thermal effect on the atmosphere. If the antenna is located in desert regions where water is limited, there may be a slight modification of the plant community at and near the antenna site.

- With RF shielding incorporated below the rectifying elements, the receiving antenna operation can be compatible with other land uses, because there is only a small degree of reduction of solar radiation received on the ground below the antenna. However, installation and maintenance of the antenna has to be planned because extensive activities may be damaging to some ecologically important systems.

- Injection of NO and water into the stratosphere and upper atmosphere by space vehicle exhausts would be small in contrast to the natural abundance, and does not appear to constitute a significant environmental effect. However, a detailed assessment of these effects will have to await a better definition of the nature of the upper atmosphere.

- Noise pollution from the launch operations would be of concern in the immediate vicinity of the launch facility and would have to be reduced by suitable design techniques or location of the launch facilities.

- There is substantial flexibility in choosing a suitable location for the receiving antenna. The area has to be contiguous but need not be flat terrain. The location can be in a region where the land is not suitable for other uses (e.g., desert areas, previously strip-mined land), or near major electrical power users (e.g., aluminum smelters).

2. BIOLOGICAL EFFECTS

There exist conflicting interpretations of the effects of microwave exposure throughout the scientific community. Because of the lack of internationally accepted standards, based on experimental data, to place a specific and allowable level on microwave exposure, the SSPS will have to be designed to accommodate a wide range of frequencies and microwave power flux densities. Precise pointing of the microwave beam must be achievable with attitude stabilization and automatic phase control to assure efficient transmission of the power to the receiving antenna. The design approaches already identified indicate that this objective can be met. The transmitting antenna size, the shape of the microwave power distribution across the antenna, and the total power transmitted will determine the level of microwave power flux densities in the beam reaching the Earth.

The effects on birds exposed to microwave power flux densities within the beam at the receiving antenna and the effects on aircraft accidentally flying through the beam, even though projected to be negligible, will have to be determined experimentally. The microwave system design requirements of the SSPS must be established so as to result in low microwave exposure levels to assure safe operations which will be acceptable on an international basis.

3. RADIO FREQUENCY INTERFERENCE

Design and development of the microwave generation device with its filters and controls are required to further define criteria for radio frequency spectrum contributions associated with noise and harmonics. High overall efficiency is attributed to circuit efficiency and internal dc-to-rf conversion efficiency which depend on achievable values of magnetic field and low but achievable power losses. The RF spectrum effects are associated with the fundamental frequency and its harmonics, turn-on and shut-down sequence, random background energy, and other superfluous signals resulting from the specific design.

Further studies are required in the areas of filtering technology and the design of the device integrated with its power input and control, and integrated in the higher level of assembly with the transmitting antenna; narrowband operation permitted by the SSPS application in determining and achieving near-optimum gain; and noise measurements specifically associated with the microwave tubes.

The effects of SSPS radio frequency interference on other users will be substantial, unless the frequency is selected so as to avoid the more sensitive services close to the fundamental frequency. It is very likely that current users would be required to relinquish their frequency allocation to the SSPS. Frequency allocation priority may be achieved if the SSPS can be shown to contribute significantly to fulfilling the nation's and the world's power needs.

Most sensitive radio astronomy services require fixed narrow bands for operation spread throughout the spectrum. The 3.3 GHz for SSPS fundamental was selected primarily to maximize the capability of filtering SSPS noise at the existing radio astronomy frequencies because filters can be designed and developed to achieve this objective.

Potential interference with shipborne radar must be investigated unless the advanced technologies in Amplitron device design and filtering are achieved.

Amateur sharing, state police radar, and radio location from high-power defense radar in the 3.23- to 3.37-GHz band will suffer interference. Thus, specific allocation for the SSPS would have to be negotiated with such users in mind. This is not to infer that interference and re-allocation are recommended as acceptable. Rather, detailed and specific effects and impacts on these users in this small band must be determined before the acceptability of specific frequency allocations can be established.

G. TIME FRAME

Based on an assessment of the steps required to develop the various technologies, it is reasonable to conclude that a prototype SSPS can be demonstrated in the early 1990's (see Figure 10). In parallel with the development of technology, environmental effects, economic constraints and social impacts should be assessed so that the overall desirability of an SSPS can be compared with other energy production methods.

The activities and developments specifically related to an operational SSPS system consisting of a network of satellites are assumed to be directed towards a commercial venture with potential for international participation. Such participation would assure that the benefits of the development of the SSPS on a scale which can be of benefit to many nations could be modeled on the successful introduction of communication satellites. The research, development, and verification program is aimed at providing an option for power generation on a time scale which is meaningful in terms of the development of other options whether they be based on solar energy or other energy sources while providing tangible returns even if the SSPS option is not exercised.

CONCLUSIONS

The realization of the concept of a Satellite Solar Power Station not only represents a major challenge to technology but also an unparalleled opportunity to apply space technology for the benefit of mankind. No fundamental breakthroughs are required to achieve the objectives identified with the SSPS but rather major advances and improvements over existing technology. Just as only 15 years ago it was unimaginable that a space shuttle could be developed to transport massive payloads into orbit and repeat the mission 100 times, so today the mission to orbit a satellite as large as an SSPS appears to be a formidable undertaking. The questions that need to be answered are not only whether the required technology can be developed — because the answer most likely will be yes — but rather whether the development of an SSPS to meet future energy demands is an option that should be pursued.

In addition to solar energy, several other energy sources have the potential to meet future energy requirements, but only very few are truly in harmony with the environment and conserving the finite resources of the Earth. Solar energy applications, such as the SSPS, are still in an early stage of development. Thus, it is too early to tell which of the approaches now being studied will be judged to have the greatest potential to be of overall benefit to society. As more is learned about the operating characteristics of potentially competitive electrical energy-generating systems, the views on what best performance represents in terms of particular operating parameters will continue to evolve. Thus, the criteria for decision-making, whether based on cost, resource conservation, or environmental enhancement, may be quite different in the future, and will continue to change as long as technical developments continue actively on the various energy-production methods.

Alternative energy production methods, such as an SSPS, if successfully developed over the next few decades, will permit society to look beyond the year 2000 with the assurance that future energy requirements could be met without endangering the planet Earth. But even successful development of solar energy conversion alternatives will still require approaches to reduce energy consumption, and to find the balance with nature essential to the flourishing of human civilization.

REFERENCES

1. Glaser, P.E., Solar Power via Satellite, Astronautics and Aeronautics, Vol. 11 (eleven) August 1973, pp. 60-68.

2. Glaser, P.E., "Power from the Sun: Its Future," SCIENCE, Vol. 162 (November 1968), pp. 857-886.

3. Glaser, P.E., The Future of Power from the Sun, IECEC 1968 Record; IEEE Publication 68C21-Energy, pp. 98-103.

4. Glaser, P.E., et al., Feasibility Study of a Satellite Solar Power Station, NASA CR-2357, February 1974.

5. Wolf, M., Cost Goals for Silicon Solar Arrays for Large-scale Terrestrial Applications, Conference Record of the Ninth IEEE Photovoltaic Specialists Conference, Silver Spring, Maryland, May 1972, pp. 342-350.

6. JPL — California Institute of Technology, Photovoltaic Conversion of Solar Energy for Terrestrial Applications NSF-RAN-74-013, Workshop Proceedings October 23-25, 1973, Cherry Hill, New Jersey.

7. Lindmayer, I., and Allison, J., An Improved Silicon Solar Cell — The Violet Cell, Conference Record of the Ninth IEEE Photovoltaic Specialists Conference, Silver Spring, Maryland, May 1972, pp. 83-84.

8. Chadda, T.B.S., and M. Wolf, "The Effect of Surface Recombination Velocity on the Performance of Vertical Multi-Junction Solar Cell," Conference Record of the Ninth IEEE Photovoltaic Specialists Conference, Silver Spring, Maryland, May 1972, pp. 87-90.

9. Woodall, J.M., "Conversion of Electromagnetic Radiation to Electrical Power," Pat. 3,675,026, issued July 4, 1972.

10. Bates, H.E., et al., Thick Film Silicon Growth Techniques, NAS 7-100, Tyco Laboratories, Inc., Waltham, Mass. May 31, 1974.

11. Ralph, E.L., Spectrolab Division, Textron Inc., Private Communication.

12. *Perspectives on Experience,* The Boston Consulting Group, Inc., Boston, Massachusetts, 1968.

13. Brown, W.C., Experiments in the Transportation of Energy by Microwave Beams, 1964 IEEE Intersociety Conference Record, Vol. 12, Pt. 2, 1964, pp. 8-17.

14. Goubau, G., Microwave Power Transmission from an Orbiting Solar Power Station, *Microwave Power,* Vol. 5, No. 4, December 1970, pp. 223-231.

15. Falcone, V.J., Jr., Atmosphere Attenuation of Microwave Power, *Microwave Power,* Vol. 5, No. 4, December 1970, pp. 269-278.

16. Brown, W.C., The Satellite Solar Power Station, IEEE Spectrum, March 1973, pp. 38-47.

17. Brown, W.C., High-Power Microwave Generators of the Crossed-Field Type, *Microwave Power,* Vol. 5, No. 4, December 1970, pp. 245-259.

18. Special Issue on Active and Adaptive Antennae, IEEE Transactions of Professional Group of Antennae and Propagation, March 1964.

19. Brown, W.C., The Receiving Antenna and Microwave Power Rectification, *Microwave Power,* Vol. 5, No. 4, December 1970, pp. 279-292.

SPACE ACQUIRED IMAGERY, A VERSATILE TOOL IN THE DEVELOPMENT OF ENERGY SOURCES

D.L. AMSBURY

Johnson Space Center, Houston, Texas, U.S.A.

INTRODUCTION

Earth-orbiting satellites have been returning geologically useful imagery for more than a decade [1, 7]. Most geologists aware of this imagery will agree for the sake of argument that space photographs and high-altitude (10,000-meter or greater) aerial photographs might be useful for geological reconnaissance in arid and little-known portions of the world. We believe that small-scale imagery is not merely useful but indispensable for gaining an understanding of regional geological features anywhere, and that plenty of unknown or poorly-known geological features exist in "mature petroleum provinces" such as the Texas Gulf Coastal Plain.

It seems to us much more rational to begin a subsurface study or geophysical study by making a map of the surface geology, than to ignore the two dimensions readily available for study using small-scale imagery. Well data comes from a zone around the well bore a few meters in diameter at most, with data points separated by hundreds of meters to kilometers; geophysical survey lines are separated by kilometers. Data stored on imagery is essentially continuous in two dimensions and may suggest the existence of geological features between subsurface and geophysical data points. Furthermore, use of this imagery can cut down the field work required for a surface survey by an order of magnitude.

APOLLO 9

During the Apollo 9 mission in March, 1969 a systematic effort was made by the crew and ground support personnel to photograph selected sites in the United States using a cluster of multiband cameras [3]. The four cameras contained color-infrared film, panchromatic film filtered to image the green band, panchromatic film filtered to image the red band, and black-and-white infrared film. In the southwestern United States, the red band is the best for geological interpretation, but for the Texas Gulf Coast color-infrared and black-and-white infrared are more useful because they show greater tonal contrasts caused by vegetation than do the other bands.

Figure 1 is a black-and-white infrared photograph taken during the Apollo 9 mission, and Figure 2 is a geological interpretation with geographic annotations. Two types of very large geological features stand out on photographs such as this: circular to elliptical features and linear ones. The large elliptical anomaly in Fort Bend County, outlined partially by the Brazos River, is particularly striking. At least 7 salt domes occur around the periphery of this feature; regional growth faults cross it but there is some suggestion of doming in this area on the regional geological map published [4].

Fig. 1. - Black-and-White Infrared Photograph (AS9-26C-3727C) from the S-065 Experiment on the Apollo 9 Mission. The Taking Scale was about 1:4,000,000, and the Area Shown Covers About 10,000 Square Miles of Southeastern Texas.

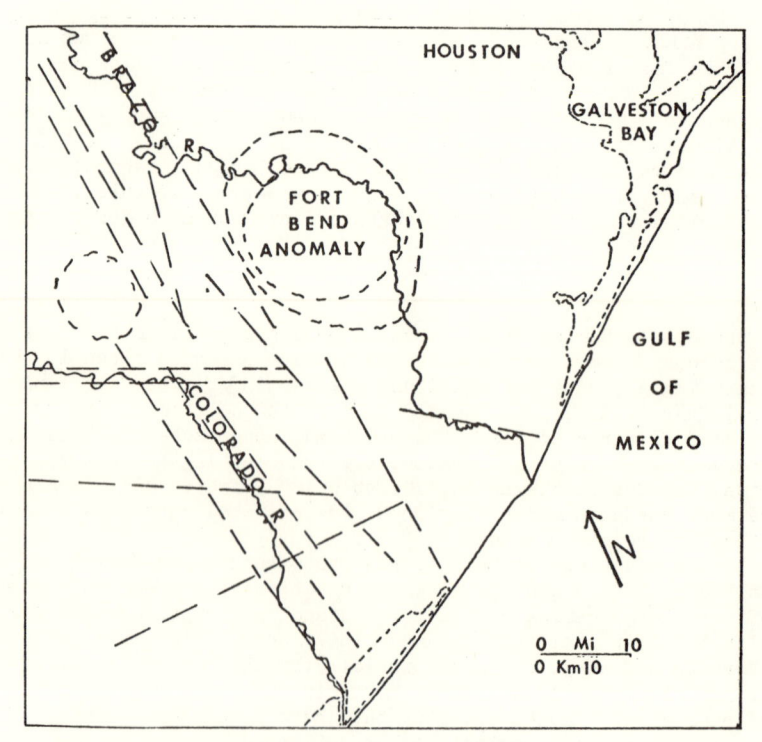

Fig. 2. - Annotation of Figure 1 Showing Geographic Features, Photolinears, and Elliptical Drainage and Tonal Anomalies.

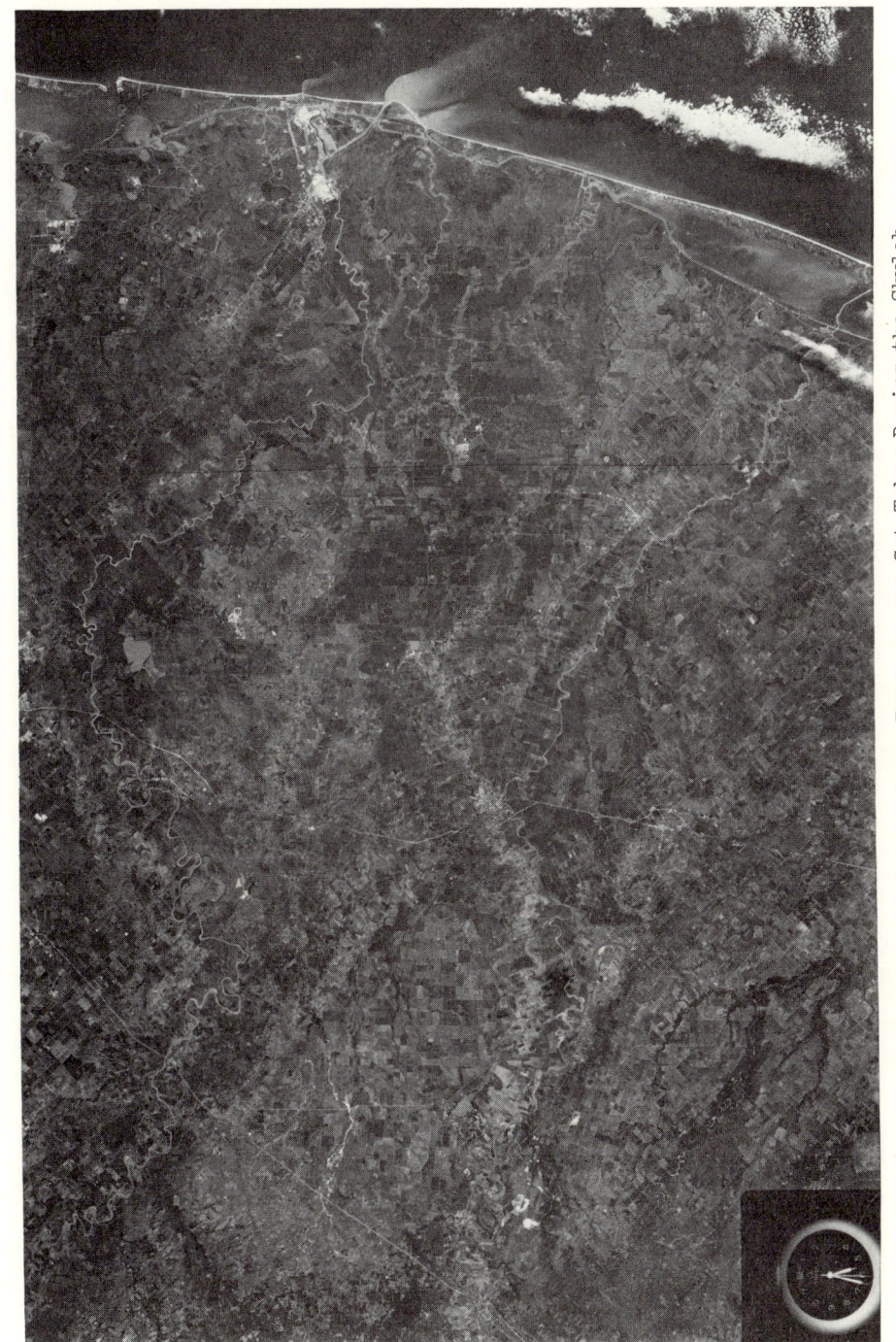

Fig. 3. — Print From an Original Color-Transparency Set Taken During the Skylab 4 Mission (SL4-94-128, 130). The Taking Scale was About 1:1,000,000.

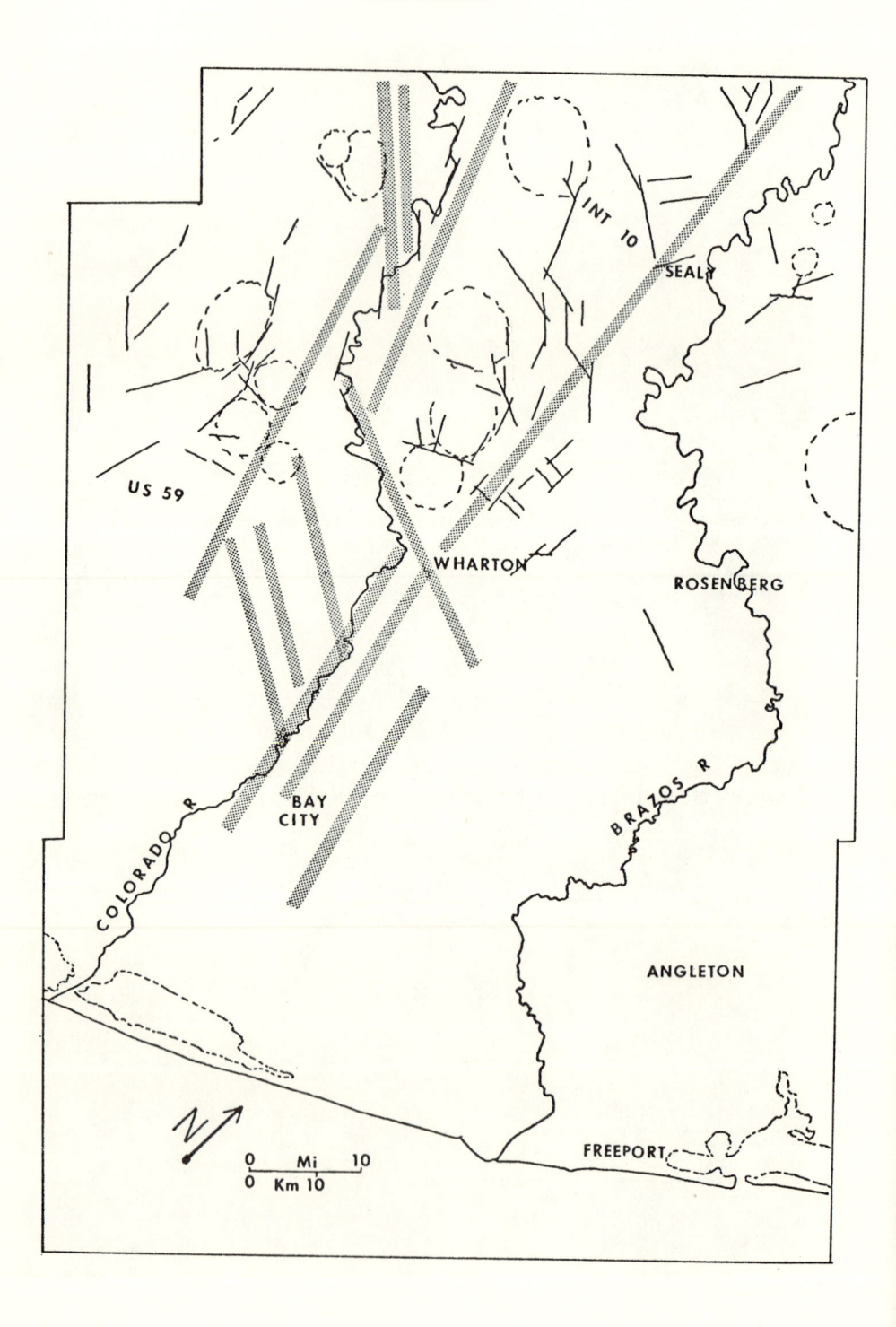

Fig. 4. - Annotation of Figure 2 Showing Linear Zones (Pattern), Straight Stream Segments, and Elliptical Anomalies.

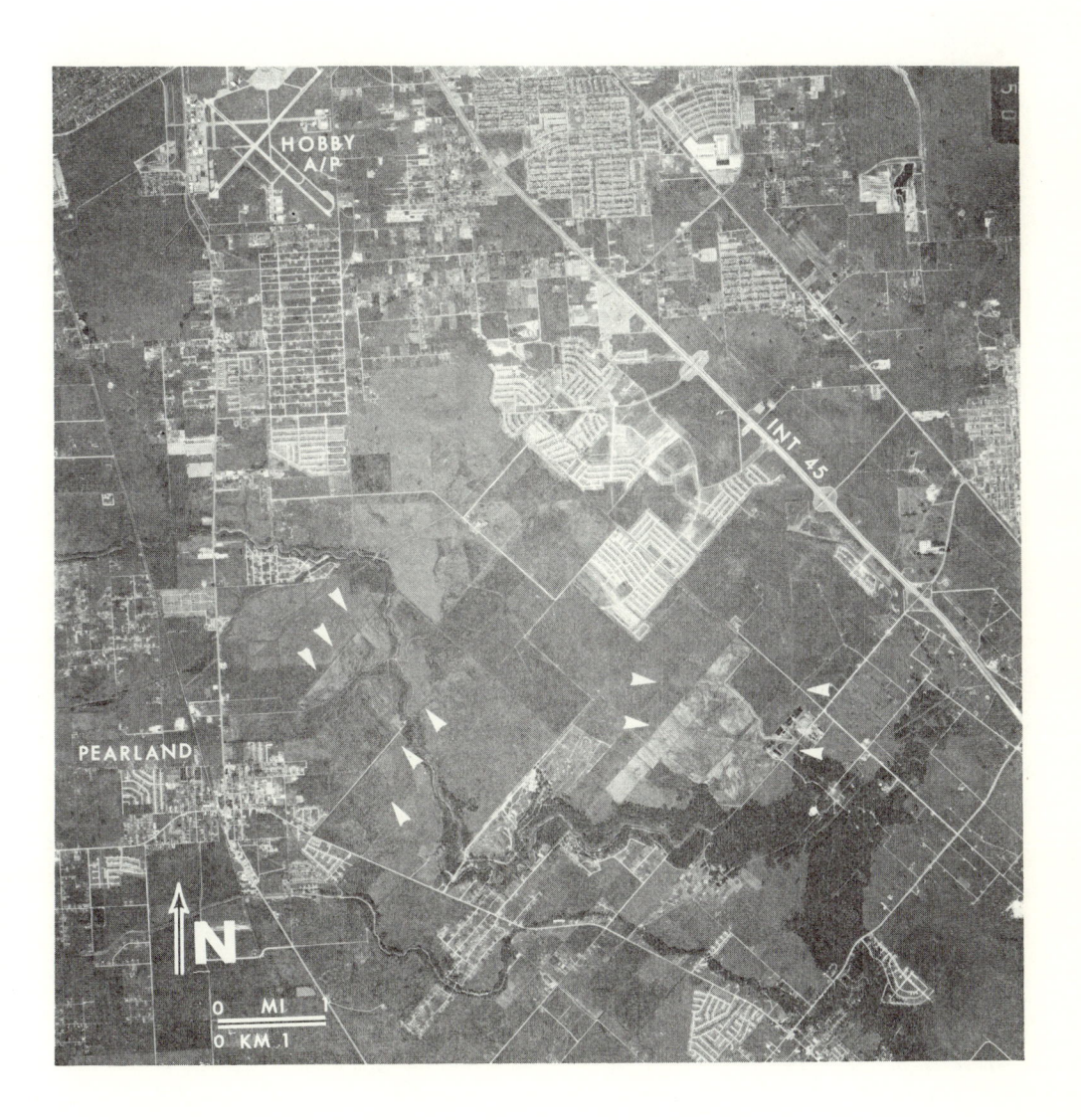

A VERSATILE TOOL IN THE DEVELOPMENT OF ENERGY SOURCES

Fig. 5. - Contact Print From a Positive Transparency, Color Infrared Film, of Frame 150, Roll 69, RB-57F Mission 145 on November 3, 1970. The Contact Scale was 1:120,000 and the Area Shown Includes About 225 Square Miles.

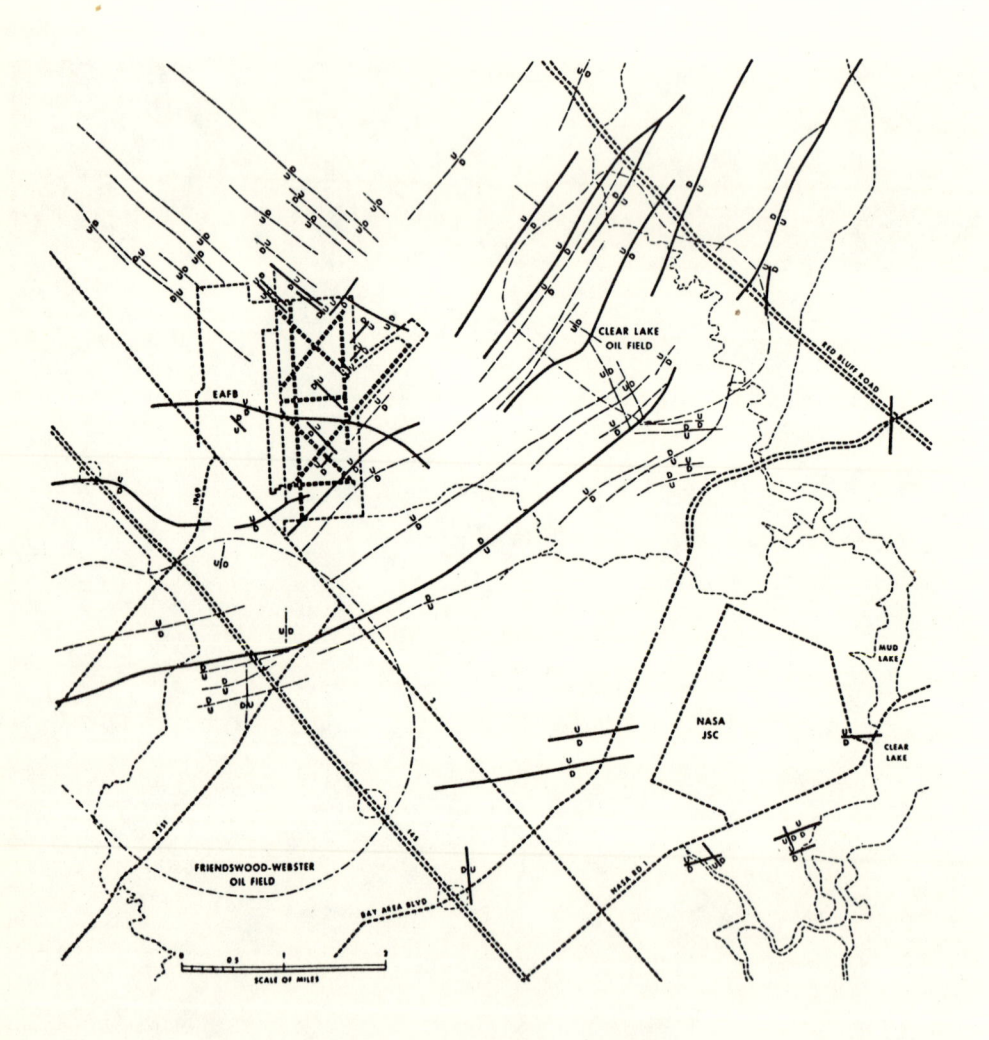

Fig. 6. - Map of the Ellington Air Force Base - NASA Johnson Space Center Area, Showing Active Faults (Heavy Lines) and Photolinears Having Topographic Expression (Light Lines). This Map was Prepared as an Overlay on Color-Infrared Photographs Whose Taking Scale was 1:60,000, Enlarged to 1:20,00 Scale.

A zone of photo-linears trending more-or-less north-south is well displayed on the infrared photograph (Figure 1). The linears show up because of tonal differences caused by differences in vegetation across them. They evidently control the location of major rivers on the Gulf Coast, and may have done so in the past. The linears coincide with the western edge of the Houston Embayment Salt Basin and the eastern flank of the San Marcos Arch. It is possible that they represent the surface traces of deep-seated, perhaps old, fracture systems. The most prominent east-west linear is aligned with the Wharton Fault, mapped farther west [4].

SKYLAB

Figure 3 is a strip mosaic of two frames of Ektachrome, S-190B photographs taken in January, 1974 during the third manned Skylab mission. Tonal contrasts between different types of vegetation are less for the Ektachrome film than for infrared film, so that some geological features do not show up as well in color film. On the other hand, the increased resolution of the 45-cm focal-length lens used on the S-190B camera, compared to the resolution of the 7.5 and 15-cm lenses used on other NASA cameras, allows one to map short, straight stream segments as well as circular tonal anomalies (Figure 4). The straight stream segments seem to be the surface traces of fracture systems, either faults or joint swarms; some of the circular anomalies, at least, are the surface traces of salt domes or "deep-seated domes."

RB-57

High-altitude aerial photographs were taken for NASA during the Apollo 7 mission in 1968 and the Apollo 9 mission in 1969. Since then high-altitude aircraft such as the RB-57F and U-2 have formed an important link between low-altitude observations and field work on the one hand, and space experiments on the other. Large portions of the United States have been photographed using color, color-infrared, and multiband black-and-white film at scales of 1:60,000, 1:120,000, and 1:400,000. Figure 5 is an example of this high-altitude photography, taken south of Houston, Texas in 1970. When studied stereoscopically, a large number of photolinears can be mapped, especially on color-infrared film at a scale of 1:60,000. Some of the photolinears coincide with surface traces of active faults that were mapped previously [5,6,8]. Field work 2 has demonstrated that most of the photolinears have some surface expression, in topographic scarps and vegetation boundaries as well as in soil boundaries. Figure 6 is a map prepared by photointerpretation of high-altitude photographs coupled with about 2 man-weeks of field checking, in the area just east of Figure 5. We conclude that the use of high-altitude photographs can shorten the time required to investigate an area for potential fault hazards; perhaps more importantly, use of the photographs leads to evidence that otherwise might be ignored, and increases confidence in the conclusions drawn from a survey.

SUMMARY

Essentially all of the United States has been imaged more than once by the multiband scanner on board Landsat I or Landsat II; for most places cloud-free imagery taken at several different seasons is available. Much of the United States was photographed by higher-resolution cameras on board Skylab. NASA aircraft (RB-57 and U-2P) have photographed large areas using color, color-infrared, and multiband black-and-white film. Every part of the United States was photographed at a scale of 1:60,000 to make the 1:250,000 AMS map sheets, and for many areas multiple government and commercial coverage is available. We believe that every regional geological study should begin with a study of as much as possible of the available small-scale imagery.

The most important factor in a successful study is unity; the geologist who makes the photo study must be the person who is responsible for the final conclusions, for two reasons. A regional image analysis is the best and quickest way to learn the geography and geology of an area, and there is a continual need for interaction between image analysis and the synthesis of surface, subsurface, and geophysical data.

REFERENCES

1. Amsbury, D.L., 1969, Geological Comparison of Spacecraft and Aircraft Photographs of the Potrillo Mountains, New Mexico and the Franklin Mountains, Texas: University of Michigan Proc. 6th Symp. on Remote Sensing of Env., p. 394-515.

2. Clanton, U.S., and Amsbury, D. L., 1975, Active Faults in Southeastern Harris County, Texas: Environmental Geology, in press.

3. Colwell, R.N., 1972, Monitoring Earth Resources from Aircraft and Spacecraft: NASA SP-275, U.S. Government Printing Office.

4. Hickey, H. N., et al., 1972, Tectonic Map of Gulf Coast Region U.S.A.: American Association of Petroleum Geologists, Tulsa, Oklahoma.

5. Reid, W. M., 1973, Active Faults in Houston, Texas: Dissertation, the University of Texas at Austin, Texas.

6. Sheets, M. M., 1971, Active Surface Faulting in the Houston Area, Texas: Houston Geological Society Bulletin, Vol. 13, p. 24-33.

7. Short, N. M., and Lowman, P. D., Jr., 1973, Earth Observations From Space: Outlook for the Geological Sciences: NASA Goddard Space Flight Center Document X-650-73-316.

8. Van Siclen, D., 1967, The Houston Fault Problem: Proc. 3rd Annual Meeting Texas Section Inst. Prof. Geol., p. 9-31.

MULTISPECTRAL DATA SYSTEMS FOR ENERGY RELATED PROBLEMS

C. L. WILSON and R. H. ROGERS

Bendix Aerospace Systems Division, Ann Arbor, Michigan, U.S.A.

ABSTRACT

A multispectral data system consists of data collection, data processing, data analysis and interpretation, and information dissemination (application) subsystems. The ERTS MSS and the data processing facility at Goddard Space Flight Center are examples of the first two subsystems respectively. The characteristics of airborne and satellite digital multispectral scanner data will be described, as will the digital data processing and analysis techniques which have been developed to automatically analyze and interpret the data. Techniques for output product generation and information dissemination will be illustrated using strip mine monitoring and environmental impact assessment for power plant siting and transmission line routing as examples. The problems associated with the development of semiautomated performance of the cited applications will be discussed, as will current status of the development programs. Expected trends in the development of operational application subsystems will be presented.

INTRODUCTION

The current surge in the use of remote sensing techniques, due in large part to the availability of data from the ERTS satellite, has created interest and anticipation both among resource managers and the general public.

Applied research in the use of remote sensing techniques has shown that many of the applications are indeed feasible and some are in use on a limited scale. However, much of the feasibility demonstration work has been done using manual interpretation of aerial photography, and the techniques are useful on only a limited basis unless large amounts of manpower and other resources are committed to data interpretation. The problems occur when techniques used on a small area are extended to cover large areas of terrain. An example is the ERTS-1 satellite. During its first year of operation, ERTS-1 covered the entire earth 20 times. 56,000 scenes were taken, 771,800 black and white film products were generated, and 21,000 color film products were generated. Most of these film products were restricted to North American coverage, otherwise, the film volumes would have been much larger. The data base exists to extract large amounts of information about the earth on which we live, yet the new information has been but a trickle.

One of the major reasons for the trickle is the lack of skilled interpreters to analyze the data or the funds to pay them if they were available. What will be required to fully utilize the capabilities of remote sensing techniques is the development of automatic processing and analysis techniques to alleviate the manual data interpretation workload. It is difficult to implement automatic processing and interpretation of photographic data. However, there is a new class of remote sensing instrumentation which was designed specifically to use automatic data processing techniques. These instruments are multispectral scanners. The most prominent example of this device is the ERTS-1 Multispectral Scanner (MSS). There are also a number of airborne multispectral scanners in existence. Unfortunately, because of the failure of the return beam vidicon package on the ERTS satellite, the ERTS MSS data is being used primarily with manual data interpretation as if it were a multiband or multispectral camera and very few users are using the data as it was designed to be used.

It is common to think of and discuss remote sensing technology only in terms of the data collection instrumentation and interpretation techniques. However, the use of automated interpretation of multispectral scanner data requires a system approach.

The purpose of this paper is to describe the characteristics of the subsystems of which a multispectral data system consists, and to illustrate their functions through examples of application to energy related problems.

2. MULTISPECTRAL DATA SYSTEMS

Multispectral Data Systems consist of four modules or subsystems, as shown in Figure 1. The data collection subsystem is a digital multispectral scanner such as the ERTS MSS, the SKYLAB S192 multispectral scanner, or the Bendix airborne M^2S (modular multispectral scanner). The data collection system also includes data recording, either on board the carrier vehicle, or after transmitting the data to the earth. The data processing subsystem performs reformatting, geometric and radiometric corrections, and otherwise prepares the data for interpretation and analysis. The NASA Data Processing Facility at the Goddard Space Flight Center is an example, and as the EROS Data Center at Souix Falls will be when the planned expansion is completed. Equipment such as the Bendix MDAS (Multispectral Data Analysis Station) performs this function for airborne multispectral scanners when the high density magnetic tape option is included. The data interpretation and analysis subsystem extracts the desired information from the data. Examples are the Laboratory for Application of Remote Sensing (LARS) facility at Purdue University, the Bendix MDAS facility, the facility at the Environmental Research Institute of Michigan (ERIM), and the General Electric System 100.

The listings of facilities given above are not intended to be all inclusive, but representative.

The final subsystem, information dissemination, accepts the information provided by the previous subsystem, places the information into a format suitable for the needs of the end user, and disseminates the information to the end user.

Most of the remote sensing applications research and development effort has been devoted to the first three subsystems described. Little effort has been devoted to the fourth. Consequently, several examples, known to most

MULTISPECTRAL DATA SYSTEMS FOR ENERGY RELATED PROBLEMS

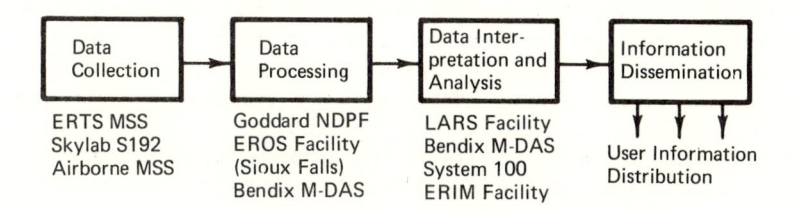

Fig. 1. – Multispectral Data System.

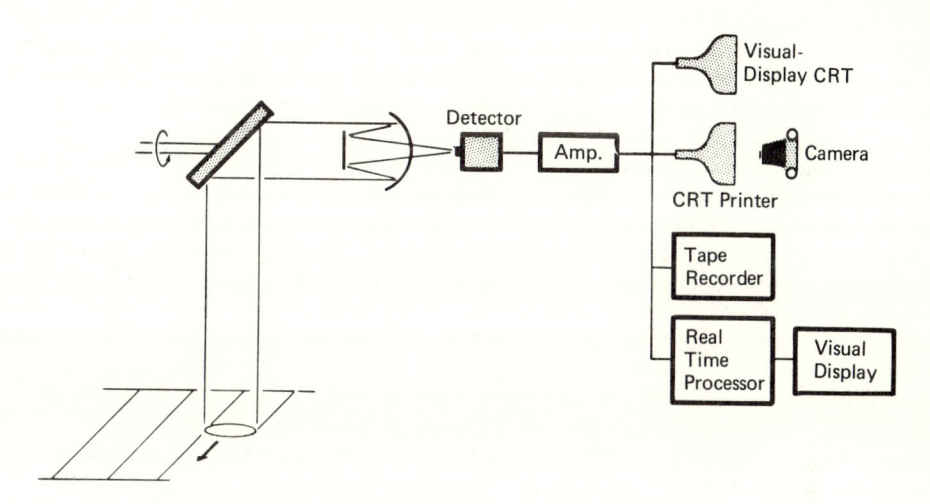

Fig. 2. – Optical Schematic Airborne Line Scanner.

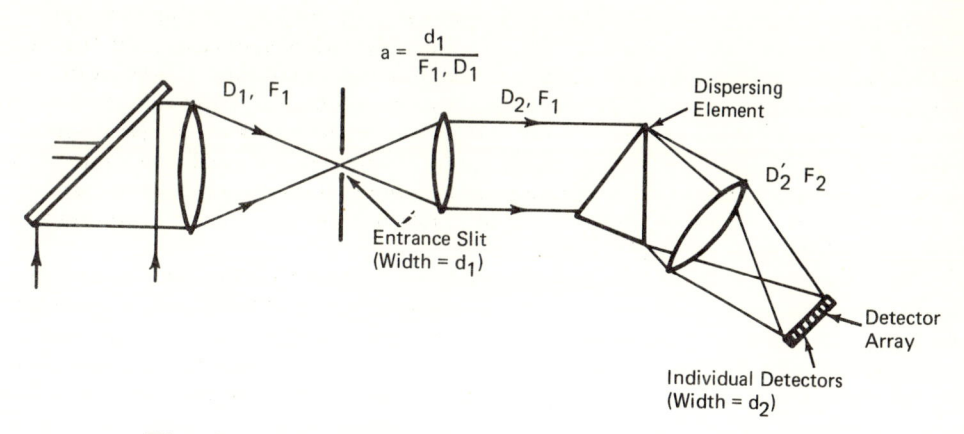

Fig. 3. – Optical Schematic - Multispectral Scanner.

users of remote sensing data, can be cited of the first three subsystems but no commonly known example of the fourth can be cited.

This section of this paper describes the operations and functions of the first three of these subsystems. The current development effort at Bendix related to the fourth subsystem will be described concurrently with the prospective applications in the next section, since the information dissemination function is closely related to the specific application, as will be seen.

2.1 DATA COLLECTION

Multispectral scanners are remote sensing devices which collect data concerning the terrain below the sensing vehicle in multiple wavelength regions. They are not to be confused with multiband or multispectral cameras, which use conventional photographic techniques and various combinations of film and filters to collect photographs in several different wavelength regions. With multispectral photography, data analysis is performed using special film viewing devices and conventional photointerpretation techniques. With multispectral scanners, data analysis is performed in a computer.

Multispectral scanners can be regarded as a combination of a line scanner and a spectrometer. A typical airborne line scanner is shown in schematic form in Figure 2. A rotating scan mirror provides a method of scanning the field of view of a telescope across the terrain. A single detector in the focal plane of the telescope receives the energy from the terrain. The size of the detector and the focal length of the telescope determines the size of the resolution cell or "pixel" scanned on the ground. The forward motion of the aircraft provides the second dimension of the scanned image. The signal from the detector can be tape recorded or directly printed on film to yield the familiar "strip map" produced by an infrared line scanner or thermal mapper.

If the single detector element is replaced by a spectrometer, then multiple channels or bands of data can be achieved instead of one band. This is illustrated schematically in Figure 3. The entrance aperture of a prism spectrometer disperses the received energy into a spectrum. A detector array is located in the exit plane of the spectrometer. The number of spectral bands in the multispectral scanner is determined by the number of detector elements in the detector array. The wavelength response of each band is determined by the size of each detector element and its location in the spectrum of the spectrometer. Each resolution cell, or "pixel" of the scanner, instead of being a single reading, can be regarded as a spectrum. If the detector array has ten elements, then ten samples of each "pixel" are being received simultaneously. If the ten signals are recorded, then ten strip maps of the same terrain can be generated, each in a different wavelength region. These ten images could be photointerpreted, which is done with multispectral photography. However, the multispectral scanner data is interpreted in a digital computer or multispectral ground station where the desired information is extracted and presented to the user.

A schematic of the Bendix M^2S scanner is shown in Figure 4. The optical system contains the scan mirror, the telescope, and a reflection grating spectrometer which provides ten bands in the near ultraviolet, visible, and near infrared portions of the spectrum. There is also a thermal infrared detector which receives its energy from a dichroic mirror located in front of the spectrometer entrance slit. There are two calibration sources for each band which are injected into the scanning system each revolution of the mirror.

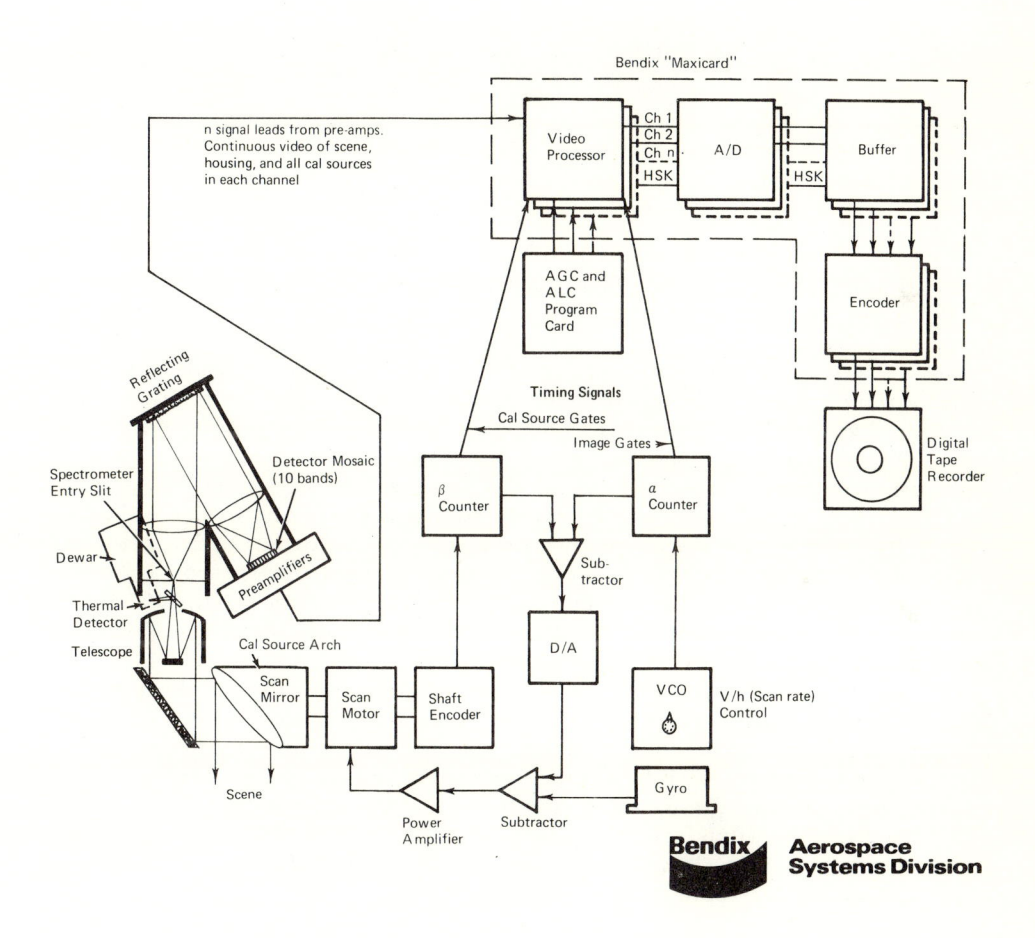

Fig. 4. — Bendix Modular Multispectral Scanner - Block Diagram.

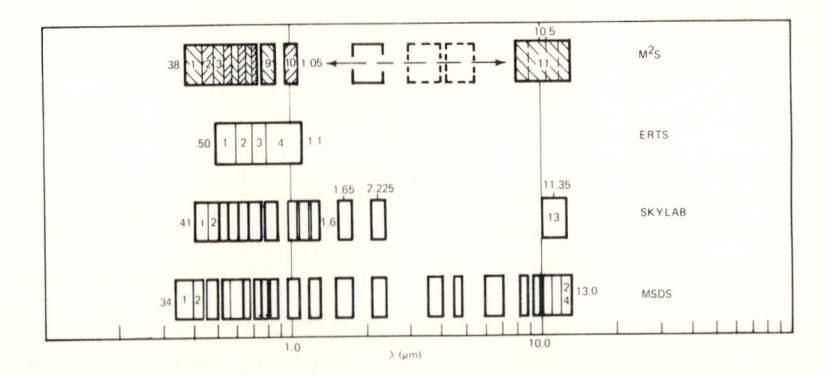

Fig. 5. — Band Locations - Digital Multispectral Scanners.

The output from each detector preamplifier proceeds to a video processor for each band. The video processor provides dc stabilization and gain calibration automatically for each band. The signals from the calibration sources as they appear in the scanner video are used as the stabilization and calibration references. Gating signals from the shaft encoder are provided by the master timing system to the video processors for location of the calibration source signals in the video. The master timing also provides clock signals to the analog to digital (A/D) converters, the data buffers, and the data encoders. The A/D converters digitize the calibrated video signal with an accuracy of eight bits once each resolution cell. The data buffers select the 100 degrees of active video signal and the calibration source signals and prepare these signals for encoding. The encoder drives the tape recorder with a modified pulse code modulated format. The tape recorder speed is controlled by the master timing to provide a constant data recording density of 10,000 bits per inch per tape recorder track. Maximum data rate of the Bendix M^2S system is approximately 7.5 million bits per second.

Nearly all multispectral scanners use digital recording techniques. This includes the ERTS and the SKYLAB multispectral scanners. The advantages of digital recording include greater stability, wider dynamic range, elimination of band to band data misregistration due to tape recorder skew and direct compatibility with digital computers for data processing.

The band locations for the Bendix M^2S, the ERTS MSS, the SKYLAB MSS, and the NASA 24 Channel scanner are shown in Figure 5. All of these scanners use digital recording of the data.

2.2 DATA PROCESSING

The data processing subsystem of a multispectral data system provides three basic functions: reformatting, radiometric correction, and geometric correction. The reformatting function is straightforward and involves conversion of the original high density digital tapes into either computer compatible magnetic tapes for later analysis, or film for direct interpretation. Manual photointerpretation of multispectral data products will not be discussed in this paper.

Radiometric and geometric correction of both space and airborne multispectral scanner data are presently experimental in nature. Correction of radiometric errors in the scanner is either conceptually simple or not required because of internal calibration in the scanner itself, as was mentioned in the description of the Bendix M^{2}S scanner. The major emphasis for radiometric calibration for operational usage of airborne and spaceborne scanners must be placed upon correction for atmospheric effects. At present, correction for atmospheric effects is not performed in the existing experimental data processing stations and either is performed as part of the analysis function, or is not performed at all.

For ERTS data, atmospheric changes from one scene to another are generally ignored because the effects vary little with the small scan angle changes, scenes to be analyzed are selected for their clarity, and the computer analysis techniques compensate automatically for static atmospheric effects. The analysis techniques used involve selection of training sets from the scene, correlated with ground truth, to develop algorithms for automated categorization of the data. Since both the training set data and the total set of data to be analyzed include the same atmospheric variants, the training process automatically compensates for the atmospheric effects that obtain at the time. However, for wide area operational usage of ERTS type data, the desirability of automated correction for scene to scene atmospheric variations must be considered. If correction techniques for atmospheric variations can be developed, then many scenes can be analyzed using analysis algorithms derived from standardized signature data banks modified for existing atmospheric effects. The factors which mitigate against this approach are the requirements for establishing ground instrumentation networks for measuring atmospheric parameters plus the necessity of modifying the signatures in the data bank to account for seasonal and geographic variations. Bendix has developed techniques for measuring atmospheric parameters in the field and correcting the analysis algorithms for the derived atmospheric effects. [1] However, a comprehensive signature data bank which permits selection of terrain signatures based upon desired categories, season, and geographic location does not exist and insufficient research has been performed to demonstrate feasibility. The presently accepted technique is to develop analysis algorithms for each scene to be interpreted, and to select scenes for interpretation which have little in the way of atmospheric variations from one location to another in the scene. If automated correction for atmospheric and terrain feature signature variations cannot be developed, then the present techniques must be used for operational applications and development emphasis must be placed on minimizing the cost of generating the analysis algorithms.

Airborne multispectral scanners are susceptible to the same atmospheric effects as spaceborne scanners, which are further complicated by the much wider field of view of most airborne scanners. Additionally, airborne scanners do not have the advantage of only minor solar angle variations that ERTS has, because the airborne scanner can be flown at any time of day with a flight path at any angle with respect to the sun.

Because of the wide field of view, there are considerable variations in atmospheric path length between the nadir and the extremes of the field of view. The changes in path length provide both attenuation variations and backscatter variations. Variations in solar angle can also provide both larger or smaller terrain radiance signals and assymetrical (with respect to the scanner imagery) backscatter.

There are two general approaches to minimizing atmospheric effects on airborne scanner data. The most common is control of data collection conditions by controlling data collection procedures. Common practice is to collect data with solar angles of thirty degrees or higher. Flight lines are usually laid out into or out of the sun to reduce assymetrical backscatter from the atmosphere and specular reflections from the terrain. Finally, data collection is delayed if necessary to obtain the clearest possible atmospheric conditions. Residual atmospheric effects are not corrected, but are coped with by expanding the training set selection during the analysis phase to include similar categories selected from several different scan angles in the data. In extreme cases, the data are analyzed separately for several ranges of scan angles.

Research is being conducted at several different facilities on processing techniques for atmospheric correction of airborne multispectral scanner data. The Environmental Research Institute of Michigan (ERIM) has implemented analog preprocessing electronics to correct airborne scanner data for solar and scan angle effects and is in the process of developing similar digital electronics. [2]

Geometric correction of multispectral scanner data is required to permit generation of map overlays or maps from the data, and to permit entry of the information extracted from the data into the data bank portion of the information dissemination system. At present, because of the experimental nature of the ERTS program, geometric correction of the ERTS MSS data is not performed at the NASA ERTS data processing facility. Plans are being laid for incorporation of geometric correction into future more operationally oriented processing stations.

Geometric correction of ERTS data is being performed at a number of facilities, either before the analysis step is performed or after analysis. A number of arguments can be advanced for either alternative. The principal argument for performing geometric correction before analysis is that all the analyzed data is then in a common geographical format and can be more easily entered into a data bank or output in a standard format output device. The principal argument against is that correction before analyses requires that the data be deconvoluted and rescanned before analysis, resulting in some degradation of the spectral signature information. If geometric correction is performed after analysis, the above cited advantage becomes a disadvantage and vice versa. If future satellite data processing stations include geometric correction as planned, the argument becomes moot. However, scale changes will still be required and perhaps rotation into a different geographical coordinate system than that in which the data are processed.

Geometric correction of airborne multispectral scanner data is not being routinely done at present except for correction of scan angle and aircraft roll distortions. Because of the wide field of view and the constant speed rotation of the scan angle, the scanner instantaneous field of view traverses the terrain at a much faster rate at the extremes of the field of view than it does at nadir. This is because the rate over the ground is a function of both scan mirror rotational rate and path length from the scan mirror to the terrain. The path length is much longer at scan angle extremes than at nadir. Scan distortion is usually corrected by programming the output device (film recorder, plotter) with the inverse of the distortion function with scan angle. Roll distortion is removed by stabilizing the data with respect to a vertical reference, either in the air or as a processing step on the ground.

Geometric processing to correct airborne multispectral scanner data to conform to a base map is still in the research phase. The problem is considerably more difficult than correction of spacecraft data due to the lower altitude and wider field of view (making terrain relief distortion significant), and the inexact knowledge of aircraft altitude, attitude, and flight path. It is likely that the operational approach eventually evolved will be a combination of manual and automated techniques, similar in concept to that of a stereoplotter, and the corrections will be performed after data analysis.

2.3 DATA ANALYSIS

The analysis portion of the multispectral data system is presently perhaps the most refined subsystem of all. The basic statistical analysis techniques have been developed over several years by a number of facilities [3, 4, 5] and will not be discussed here. The mathematical principles involved are described in the cited references. The major emphasis over the last several years has been to develop operational implementation of the basic principles to provide cost effective approaches to the initial analysis step and to provide the highest possible throughput rates for generation of output information. Several different approaches have been developed to address these operationally oriented problems. The one discussed will be that used at Bendix.

First, we will discuss the objectives of the computer analysis of the multispectral scanner data. As was mentioned earlier, utilization of remote sensing techniques has often been proven feasible on a limited area but encountered problems on larger areas. The problems are caused by the large amounts of data to be interpreted. The objectives of computer interpretation of the data are to utilize what is known from normal interpretation or actual ground survey of a limited area to 1) develop an analysis algorithm which correlates well with the results from the initial small area, and 2) apply the analysis algorithm to the entire data set to yield the same results over the large area. In effect, the computer is being taught by an interpreter to recognize target categories. Once taught, the computer applies the interpretation criteria to the rest of the data.

When used to achieve these objectives, three functions are performed in the analysis station. These are: 1) screening and editing, 2) data analysis, and 3) data interpretation. Each of these functions will be described. Figure 6 is a block diagram of the screening and editing function. The data is played back and displayed on a color display. Samples of data are extracted from the data tape by the interpreter using the edit commands. The samples selected

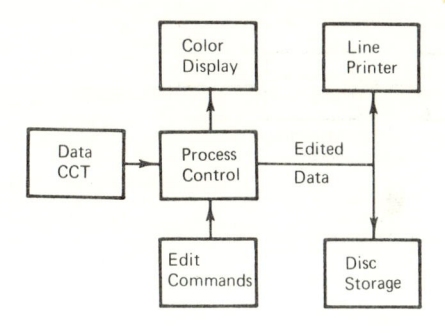

Fig. 6. – Screening and Editing.

are from areas of terrain of known characteristics. The characteristics are known either from a ground survey or from manual interpretation of data presented as imagery. The samples selected are called training samples and are stored on a magnetic disc and printed out on a line printer. An example is training selection for agricultural classification. If it is desired to automatically categorize wheat, oats and soybeans into respective categories then the interpreter will select sample spectra from several fields known to contain wheat, oats and soybeans, respectively. Training sets from the three categories will then be stored on the disc with identifying labels.

The second function is data analysis, shown in Figure 7. The training sets are submitted to the computer for statistical analysis. The outputs are a set of analysis algorithm coefficients which, when applied to the data, separate the readings or signals from the individual resolution cells into the classes or categories desired. In addition to the analysis coefficient output, several other outputs are available from the data analysis step. These outputs, such as scatter diagrams, histograms, categorization accuracy tables, etc., are designed as diagnostic tools to assist the interpreter in assessing the results and to notify the interpreter of the expected categorization accuracy and false alarm rate for the categories selected. Figure 8 is a categorization accuracy table, which shows the accuracy with which the training set data are categorized and, if mis-categorized, as what.

The third step is data interpretation shown in Figure 9. Once the computer has been "trained" to interpret the data, the data to be interpreted is read into the ground station from the data tape. The analysis algorithm developed during the data analysis step is applied to the data. The interpreted results may be viewed on the color display and presented in hard copy form either as imagery on film or as map overlays on a computer driven plotting table, or as "categorized CCTs" to be used in the information dissemination portion of the multispectral data system.

3. INFORMATION DISSEMINATION AND APPLICATION

The information dissemination portion of the system is the bridge between the high technology portion of the total system and the end user. It must take the interpreted data generated by the analysis subsystem, distribute the ex-

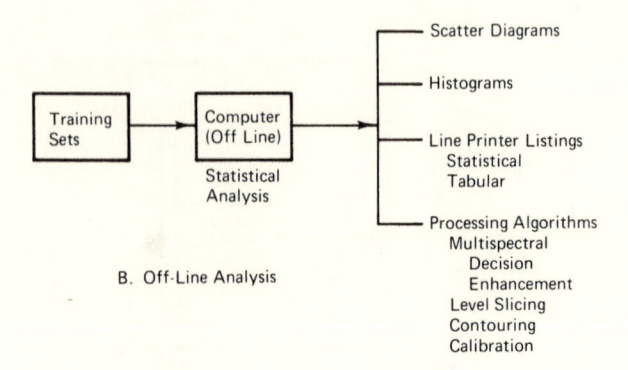

B. Off-Line Analysis

Fig. 7. – Off-Line Analysis.

```
CLASSIFICATION TABLE     12-SEP-73     11:02:50

REJECTION LEVEL =  0.000000    PER CENT
GROUP BIASES:    GROUP    BIAS
                   4     0.40000

TNG                         PER CENT CLASSIFIED AS GROUP
SET    0      1      2       3       4       5       6       7       8
 1   0.000  96.552  0.000   0.000   0.000   0.000   0.000   0.000   3.448
 2   0.000   0.000 96.552   0.000   0.000   3.448   0.000   0.000   0.000
 3   0.000   0.000  0.000 100.000   0.000   0.000   0.000   0.000   0.000
 4   0.000   0.769  0.000   0.769  90.000   5.385   0.000   0.000   3.077
 5   0.000   0.000  4.706   0.000   1.176  91.765   0.000   0.000   2.353
 6   0.000   0.000  0.000   0.000   0.000   0.000  98.182   1.818   0.000
 7   0.000   0.000  0.000   0.000   0.000   0.000   0.000 100.000   0.000
 8   0.000   1.562  1.562   0.000   0.000   3.125   0.000   0.000  93.750

PROGRAM RUN TIME = 00:00:36
```

0 Unclassified   5 Wetlands
1 Grass (tended)   6 Deep Water
2 Trees   7 Shallow Water
3 Barren Earth/Concrete   8 Grass (untended)
4 Urban

Classification accuracies, determined by computer integration of the probability density function determined by the computer with its decision on each of the areas.

Fig. 8. – Categorization Accuracy Table.

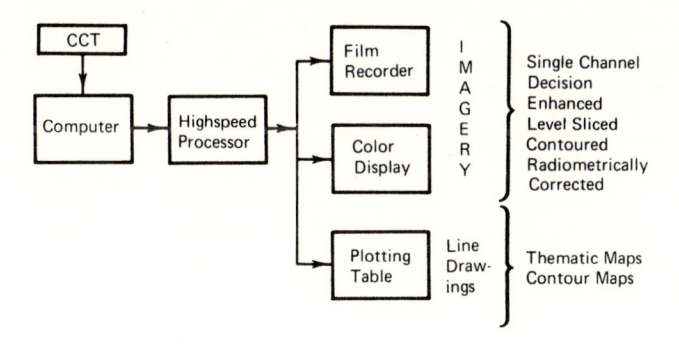

Fig. 9. – On-Line Interpretation.

tracted information, and present it to him in a manner in which he can utilize it. Generally speaking, this is in the form of maps and/or tabular data. The types of output products that can be generated and the manner in which they could be presented to users will be illustrated with two examples of energy related remote sensing data applications currently under investigation at Bendix.

The two examples chosen are at opposite ends of the application spectrum. The first, use of ERTS data for strip mine monitoring [6], is a feasibility demonstration of a potential application with little thought yet given to the problems of an operational system. The second, power plant site selection [7], has a potential information dissemination system already in existence if the application is proven technically feasible.

3.1 AUTOMATED STRIP MINE MAPPING FROM ERTS

For ERTS-1 Investigation MMC309, Bendix has been conducting an evaluation of the feasibility of using ERTS data to monitor the extent of strip mining for coal in southeastern Ohio and to catalog the progress and extent of reclamation activities. This experiment has been to determine the feasibility of such usage, not to set up an operational or pilot system for the routine performance of such monitoring.

The area encompassed by this investigation includes five counties in eastern Ohio that comprise nearly 3,000 square miles. The counties; Muskingum, Coshocton, Guernsey, Tuscarawas, and Belmont; have been disrupted by coal mining since the early 1800s. Strip mining, which generally began before the 1920s, has been practiced in all of them. The total area of stripping operations in each county was quite large during the period from 1914 to 1947, but was insignificant when compared to the area stripped from 1948 to the present time.

On-site examination of individual mines, and particularly older mines, is hindered by 1) a lack of adequate mine map coverage; 2) deeply eroded, non-existent, or blocked access roads; 3) lack of accurate or adequate records; 4) the great total size of the stripped area; 5) strip mine reclamation planting along roads that obscures adjacent barren land; and 6) dated aerial photographic coverage.

Local, state, and federal agencies must have repetitive coverage of mining areas and the capability for rapidly evaluating each situation. They also must be able to quickly determine areas of mining reclamation and progress or viability of replanted vegetation, at least on an annual basis. This, presently, cannot economically be done by ground teams, and aerial photographs rapidly become outdated.

Because of the more stringent 1972 Ohio strip-mine bill, reclamation now proceeds, in many areas, at the same rate as the mining. In fact, grading equipment may be operating just behind the giant mining machinery.

Various agencies within the Ohio state government collect certain types of coal-mining data. There is, however, little or no coordination between agencies; automatic data processing is non-existent; and various filing systems approach the chaotic. Consequently, reports available to the public are severely dated, commonly inaccurate, and difficult to acquire.

Although several specific areas have been examined, a very large mine in southeastern Muskingum County, owned and operated by the Ohio Power Company, was chosen for detailed examination. The mine (shown in the aerial

Fig. 10. — Aerial Photo-Mosaic, Showing Site A Ohio Power Company Strip Mine in Southeastern Muskingum County, Ohio. Acquired by NASA C-130 Aircraft Photography on 7 September 1973.

photograph of Figure 10) is very irregular in shape, nearly 14 km (9 miles) long, and as much as 8 km (5 miles) wide. Aerial photographs indicate that there had been no stripping in the area before 1950. By 1965, however, about 1.6×10^7 square meters (4,000 acres) had been disrupted and, by 1971, strip mining had claimed close to 4.5×10^7 square meters (11,000 acres). Reclamation, caused by the more stringent legislation enacted in 1972, is proceeding at a very rapid rate. Aerial photographs of the northern part of this mine were taken in May of 1972. The area was also examined in the course of field work in June of 1973. In several parts of the mine, there was no comparison between the landscapes that appeared on the 1972 photograph and the condition that existed only 13 months later. Many of the strip mine lakes had been filled, much of the area was graded, and various grasses had been planted as part of the reclamation program.

ERTS data from August of 1972 and from September of 1973 were interpreted, both to demonstrate that stripped earth and partially reclaimed areas could be automatically detected and tabulated, and to show that changes in the extent of stripping and extent of reclamation could be detected and documented.

For the site under discussion, the ERTS CCTs from the two scenes collected approximately one year apart were analyzed in the Bendix Earth Resources Data Center, which is the Bendix in-house facility for earth resource applications research. Aerial photography and the ERTS imagery itself was used to provide the ground truth information. The output products generated were tabular data from the computer line printer, and map overlays generated on a computer driven plotting table. The map overlays are the most interesting products, but since they are usually produced in color and depend on color for contrast, are less interesting when presented in black and white. Figure 11 shows examples of the stripped earth category computer generated map overlays, overlaid on an aerial photograph at a scale of approximately 1:40,000. The photograph used in both halves of the illustration is the same, and was collected in September of 1973. The ERTS stripped earth overlay on the left image was generated from ERTS data collected in August of 1972, while that of the right image was from September of 1973. The individual squares seen are the size of the ERTS resolution cell. As can be seen, the right hand ERTS data conforms closely to the area actually stripped shown in the photograph. The area shown as stripped in the 1972 data on the left conforms closely to that known of the situation at the time. The difference between the two, of course, is the additional area stripped in the intervening thirteen months. The other categories of the interpretation could also be overlaid over the photograph but are not too effective in a single color. Figure 12 is a photograph of a complete set of overlays at an original scale of 1:250,000 derived from the August 1972 data.

Tabular printouts available from the analysis system are summarized in Tables 1 through 3. Table 1 is a categorization accuracy table which is generated during the analysis phase. The computer uses the categorization algorithms which have been developed to categorize the training set data for an estimate of the categorization accuracy achieved. For all categories, the accuracy is greater than 90%. Table 2 is an area printout table listing the portion of the image covered by each category in percent, square kilometers, and acres. The total area is approximately that of the aerial photograph shown previously. Table 3 is a comparison of the same area from the August 1972 and September 1973 data, showing the changes for each category in the thirteen months between scenes.

Fig. 11. – Stripped Earth Category Mapped from 1972 and 1973 ERTS Data Overlaying NASA Photograph Obtained in September 1973. Approximate Scale 1:40,000. Site A, Ohio Power Company Mine in Muskingum County, Ohio.

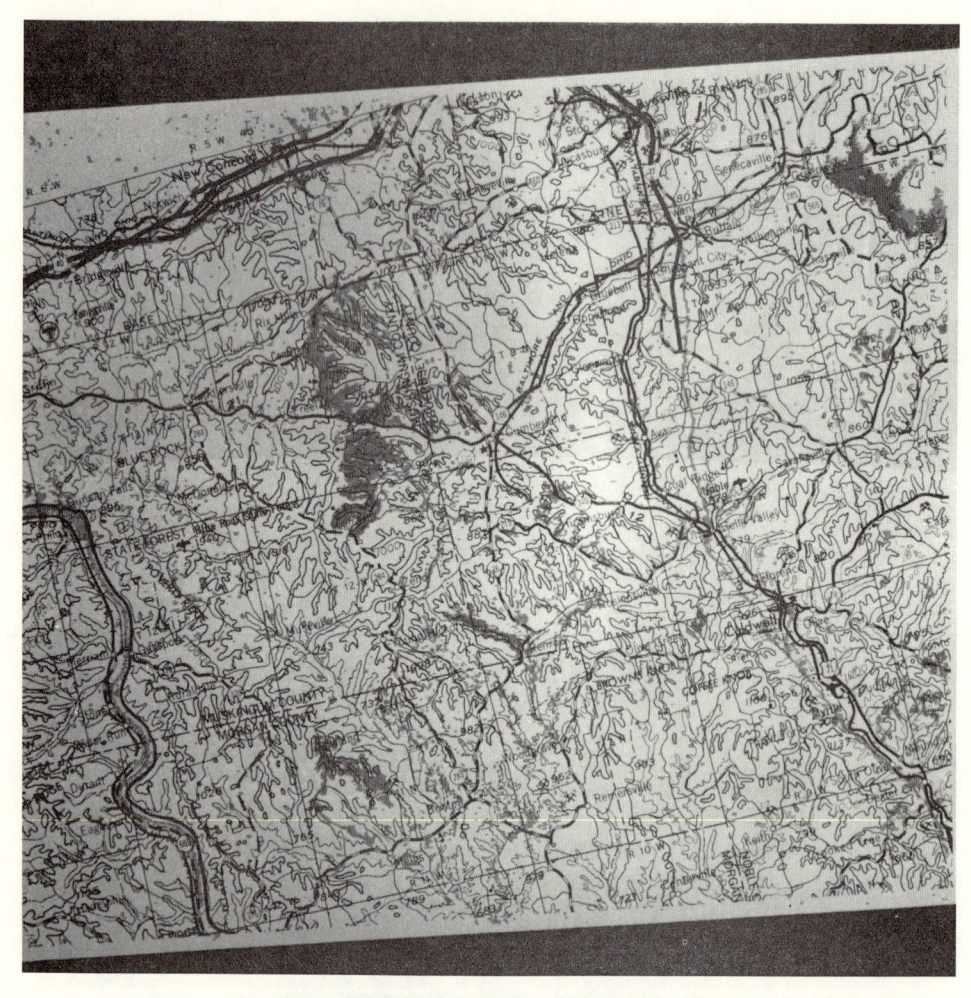

Dark Blue: Water without Sedimentation; Light Blue: Water
with Sedimentation (1:250,000 Scale Base Map)

Fig. 12. – Color-Coded Overlays Generated by Computer from ERTS Data Acquired 21 August 1972.

The overlay mapped from the 3 September 1973 ERTS overpass, when compared with the NASA photograph acquired at approximately the same time, shows good agreement in both geometric and categorization accuracy. Comparison in Figure 11 of the stripped earth overlays generated from the August 1972 and September 1973 ERTS tapes readily shows changes in stripping in the Ohio Power Company mine. Areas noted as Areas 1 and 2 in Figure 11, not stripped in 1972, are shown stripped in 1973. The area noted as Area 3 in Figure 11 was stripped in 1972 and partially reclaimed in 1973. Table 3, produced from area printout tables generated from 1972 and 1973 tapes, gives a quantitative measure of these changes.

Table 1. Categorization Accuracy Table (Units Percent).

	Category	1	2	3	4
1	Stripped Earth	96	4	0	0
2	Partially Reclaimed Earth	0	98	2	0
3	Natural Vegetation	0	8	92	0
4	Water	0	0	0	100

Table 2. Area Printout Table Test Area A, ERTS 3 September 1973.

Category	Percent of Total	Square Kilometers	Acres
Stripped Earth	15.54	15.44	3,814
Partially Reclaimed	11.86	11.79	2,913
Natural Vegetation	72.08	71.6	17,692
Water	0.53	0.52	129

Table 3. Site A Area and Area Changes.

Category	21 Aug 1972 Acres	3 Sept 1973 Acres	Difference (1973 - 1972) Acres
Stripped Earth	2,948	3,814	+868
Partially Reclaimed	2,512	2,913	+401
Natural Vegetation	18,657	17,692	-965
Water	433	129	-304

A brief analysis of the area and area change data in Table 3 indicates that, between 21 August 1972 and 3 September 1973, an additional 3.5 square kilometers (868 acres) was stripped at this test site. Partially reclaimed land also increased 1.6 square kilometers (401 acres) during this period. That the mine is still growing or spreading out is indicated by the loss of 3.9 square kilometers (965 acres) of natural vegetation and the fact that the new stripping is occurring at about twice the rate of the reclamation. The loss of 1.2

square kilometers (304 acres) in surface water was not considered significant since it only confirmed, in this case, that 1973 was a much drier year than 1972.

It is believed that the work done to date has shown that it is feasible to monitor strip mining activity using ERTS data. Further, experience in automatic analysis of ERTS data on other Bendix projects has shown that large areas can be categorized at costs significantly less than $0.50 per square mile, so the technique is cost effective compared to other techniques (typically $10 per square mile for land use maps from aircraft data). However, implementation of a large area surveillance system requires considerable additional development work. Feasibility demonstration is not enough. Before use of ERTS data for strip mine monitoring can move from a scientific curiosity to a practical application, the following additional steps must be taken:

1. Creation of a data bank to store categorized data from a large area. The data bank must be designed to retrieve data from specified geographical areas on command, both for reporting and for comparison.

2. Development of software to merge the categorized data from a number of ERTS scenes for entry into the data bank in a common geographical format.

3. User needs analysis to determine who and where the end users are, what their information needs are, on what time scale, with what frequency, and where.

4. Development of software and hardware to fulfill the requirements of the users.

5. Develop a distribution system to provide the desired information to the users on the required time scale.

Completion of the above steps would provide the "information dissemination" subsystem for the Multispectral Data System for strip mine monitoring. Review of the steps shows that the first two steps are probably common to many applications, but the last three are specific to the strip mine application. This would tend to lead to a conclusion that except for some commonality in a data bank function, implementation of the information dissemination function will be different for each application.

3.2 POWER PLANT SITE SELECTION

Bendix has been conducting a joint project addressing power plant site selection with Commonwealth Associates, Inc. of Jackson, Michigan over the last year. Commonwealth Associates is an engineering consulting firm specializing in energy systems such as power plants, transmission lines, natural gas and electrical distribution, etc. An important aspect of their service is evaluation and selection of power plant sites and transmission line routes.

Commonwealth Associates has designed a family of computer programs (named ENVIRO) for environmental impact analysis. These programs, using numerical rating techniques, have been developed to evaluate the complex considerations involved in large development projects.

Environmental impact analysis is a rapidly developing technology that assesses the complex and interrelated effects on the environment resulting from construction and operation of a large facility. It is a multi-dimensional concept embracing impacts on human activities as well as disruptions of the natural environment. Under provisions of the National Environmental Policy of 1969, an Environmental Impact Statement, including permits and licenses, is required for all major federal actions having a significant effect on the environment. Also, prudent management, more than ever, is developing a greater awareness of both the short and long term implications of large new projects.

A recent survey of electrical utility company siting practices performed by Commonwealth Associates, Inc. for the Atomic Industrial Forum has identified an underlying process consisting of three major stages involving a selection of candidate areas, the identification and preliminary evaluation of potential sites, and the comparative evaluation of candidate sites.

Numerical rating methods are most effective during the final stage of this process where the emphasis is on narrowing down the range of alternatives. However, it has applicability in any analysis where: (a) a specifically defined set of alternatives has been identified; (b) a comparative evaluation of the relative suitability of the various alternatives is required; and (c) a large number of inherently different kinds of considerations are involved. Numerical rating techniques are based on the concept that a number of widely different considerations can be combined into a single composite rating which is indicative of the total effective impact.

Commonwealth has found that an "Environmental Hierarchy" is a particularly effective way to insure that all significant effects are taken into account. The development of quantitative models for each specific impact within the hierarchy is a significant feature of the way in which Commonwealth has applied the numerical rating techniques. Finally, an adaptive philosophy in categorizing impacts, developing impact models and compiling data requirements has led to a high degree of refinement in the application of these techniques.

Probably the most controversial aspect of numerical rating methods is the way in which specific impacts are combined into a single composite rating. Since all specific impacts are not equally important, the significance of each impact relative to all other impacts is factored into the evaluation by the use of importance weighting factors. Although the structured nature of the numerical rating method helps to promote objectivity, the assignment of importance weights is, in the last analysis, a subjective evaluation. However, Commonwealth has devised several techniques for structuring these subjective aspects, thus fostering greater objectivity.

The ENVIRO family of programs provides a convenient tool to perform these functions. It is designed to facilitate the analysis of environmental impacts associated with any situation involving a wide range of distinctively different considerations or a large number of siting alternatives. The system provides great flexibility in handling all of the computations and bookkeeping tasks associated with numerical techniques. Input to the program consists of: (a) site specifications, (b) a set of specific impact values for each site, and (c) a set of relative weighting values. For each specific impact, in turn, the raw numerical value of impact magnitude is scaled to the site having the maximum impact rating. Thus, internal to the program all impact numbers are reduced to common scale. The specific impact ratings at each site are multiplied by the appropriate weighting values and summed for each site. The impact ratings at each site are thus a weighted sum of specific impacts. For ease in analysis, total impact ratings are further normalized by assigning an arbitrary impact rating of 1,000 to the site with the largest impact value; the ratings at other sites are proportioned accordingly.

The data bank used by the computer program is related to a 15 acre cell size geographical grid. The environmental criteria stored in the data bank and related to the grid includes vegetation, topography, water, transportation, population density, land use restraints, future land use, and soils. The weighting factors associated with the environmental criteria are a function of engineering considerations, cost, and potential environmental hazards. The weighting for environmental hazards is determined both from existing legal and administrative criteria, and from surveys conducted of the residents and local government and conservation agencies of the area. Figure 13 is a sample of a portion of a survey form for a proposed transmission line route. Figure 14 is an example of computer printouts summarizing values attached to environmental criteria for an environmental impact statement.

Land use, vegetation cover, and water are important input parameters to the computerized evaluation system. Present techniques utilized by Commonwealth Associates for deriving this information are primarily manual and are expensive and time consuming. If a land use map already exists, a 15 acre grid is overlaid on the map and each cell is aggregated to the predominant category. For each cell, an IBM card is keypunched with the cell category and the cell location. Twenty to thirty cubic feet of IBM cards can be accumulated in this fashion. A card to magnetic tape conversion then generates a tape suitable for entry into the data bank. If a land use map does not already exist, then aerial photography must be obtained, manually photointerpreted, and the interpretation transferred to a base map. The gridding and digitizing process can then be commenced. The entire process can consume three to four months and require several man years of effort.

The ERTS satellite data appeared a logical source for such data, if the land use information were accurate, since the data were already on magnetic tape and could be aggregated to achieve the desired 15 acre cell size. Geometric correction to a specified geographical coordinate system was also required, of course.

In July 1973 Bendix Aerospace Systems Division and Commonwealth Associates agreed to cooperate in a study to derive land use categories using spectral pattern recognition techniques from ERTS computer compatible tapes (CCT). These land use categories were to be used to study the feasibility of using ERTS MSS data in power plant siting and utility corridor selection. To avoid the possibility of unauthorized disclosure of future power plant location, two test sites, Kasota and Stearns Counties, were chosen in Minnesota. These sites had been part of an earlier site selection study by Commonwealth and were no longer being actively considered for power plants.

Bendix was to furnish Commonwealth with a color classified image of each test site and 1:250,000 scale map overlays for the Kasota test site of six land use categories. Also required were area measurement tables showing the amount of land falling under each of the six land use categories at both test sites. These measurements were to include a tabulation of land in each cate-

RANKING OF OBJECTIVES

Check The Box That Reflects Your Personal Value Judgement					
Objective	Very Important	Quite Important	Important	Indifferent	Unimportant
Minimize Damage To Natural Systems					
Minimize Conflict With Existing Land Uses					
Minimize Conflict With Proposed Land Uses					
Minimize Damage To Culturally Significant Features					
Minimize Visual Exposure					
Maximize Potential For Right-of-way Sharing					

Where Do You Live?
City ☐
Town ☐
Country ☐

Township:

Fig. 13. — Sample Local Survey Form.

environmental criteria

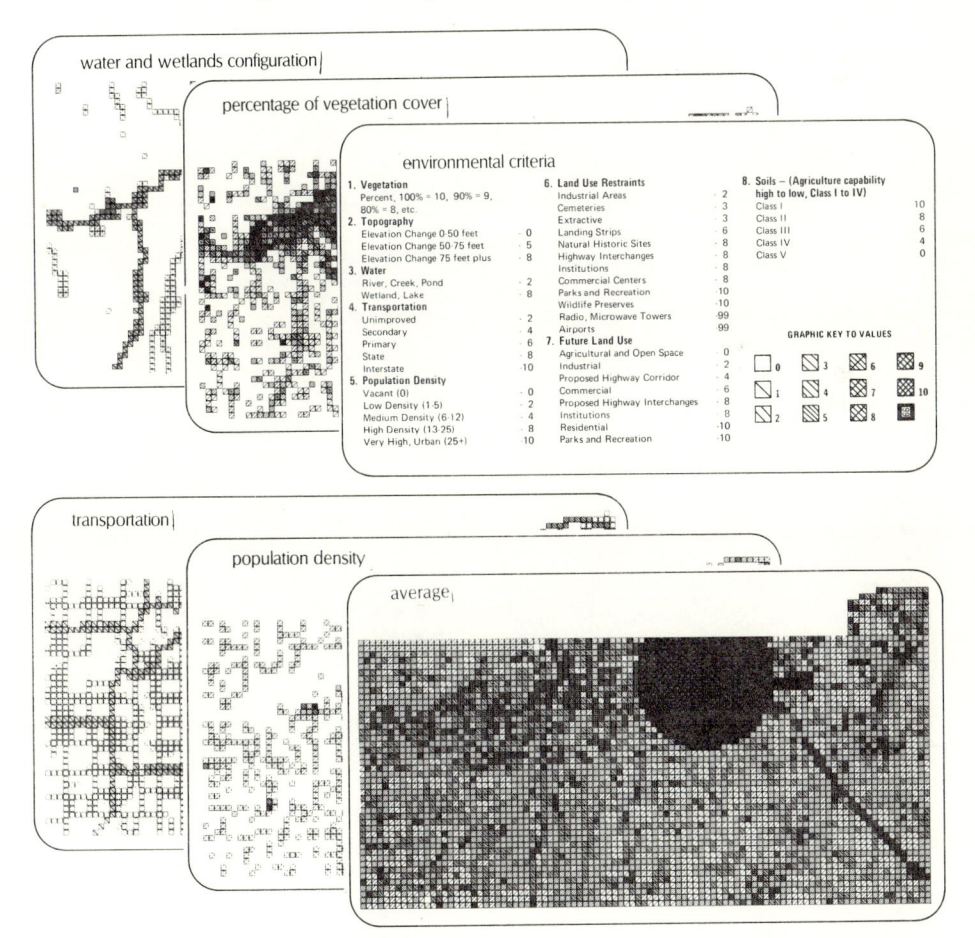

1. **Vegetation**
   Percent, 100% = 10, 90% = 9, 80% = 8, etc.
2. **Topography**
   Elevation Change 0-50 feet — 0
   Elevation Change 50-75 feet — 5
   Elevation Change 75 feet plus — 8
3. **Water**
   River, Creek, Pond — 2
   Wetland, Lake — 8
4. **Transportation**
   Unimproved — 2
   Secondary — 4
   Primary — 6
   State — 8
   Interstate — 10
5. **Population Density**
   Vacant (0) — 0
   Low Density (1-5) — 2
   Medium Density (6-12) — 4
   High Density (13-25) — 8
   Very High, Urban (25+) — 10
6. **Land Use Restraints**
   Industrial Areas — 2
   Cemeteries — 3
   Extractive — 3
   Landing Strips — 6
   Natural Historic Sites — 8
   Highway Interchanges — 8
   Institutions — 8
   Commercial Centers — 8
   Parks and Recreation — 10
   Wildlife Preserves — 10
   Radio, Microwave Towers — 99
   Airports — 99
7. **Future Land Use**
   Agricultural and Open Space — 0
   Industrial — 2
   Proposed Highway Corridor — 4
   Commercial — 6
   Proposed Highway Interchanges — 8
   Institutions — 8
   Residential — 10
   Parks and Recreation — 10
8. **Soils** — (Agriculture capability high to low, Class I to IV)
   Class I — 10
   Class II — 8
   Class III — 6
   Class IV — 4
   Class V — 0

GRAPHIC KEY TO VALUES

☐ 0   ◨ 3   ◩ 6   ■ 9
◪ 1   ◧ 4   ◫ 7   ■ 10
◩ 2   ◨ 5   ◪ 8

Fig. 14. — Computer Printouts - Weighted Environmental Criteria.

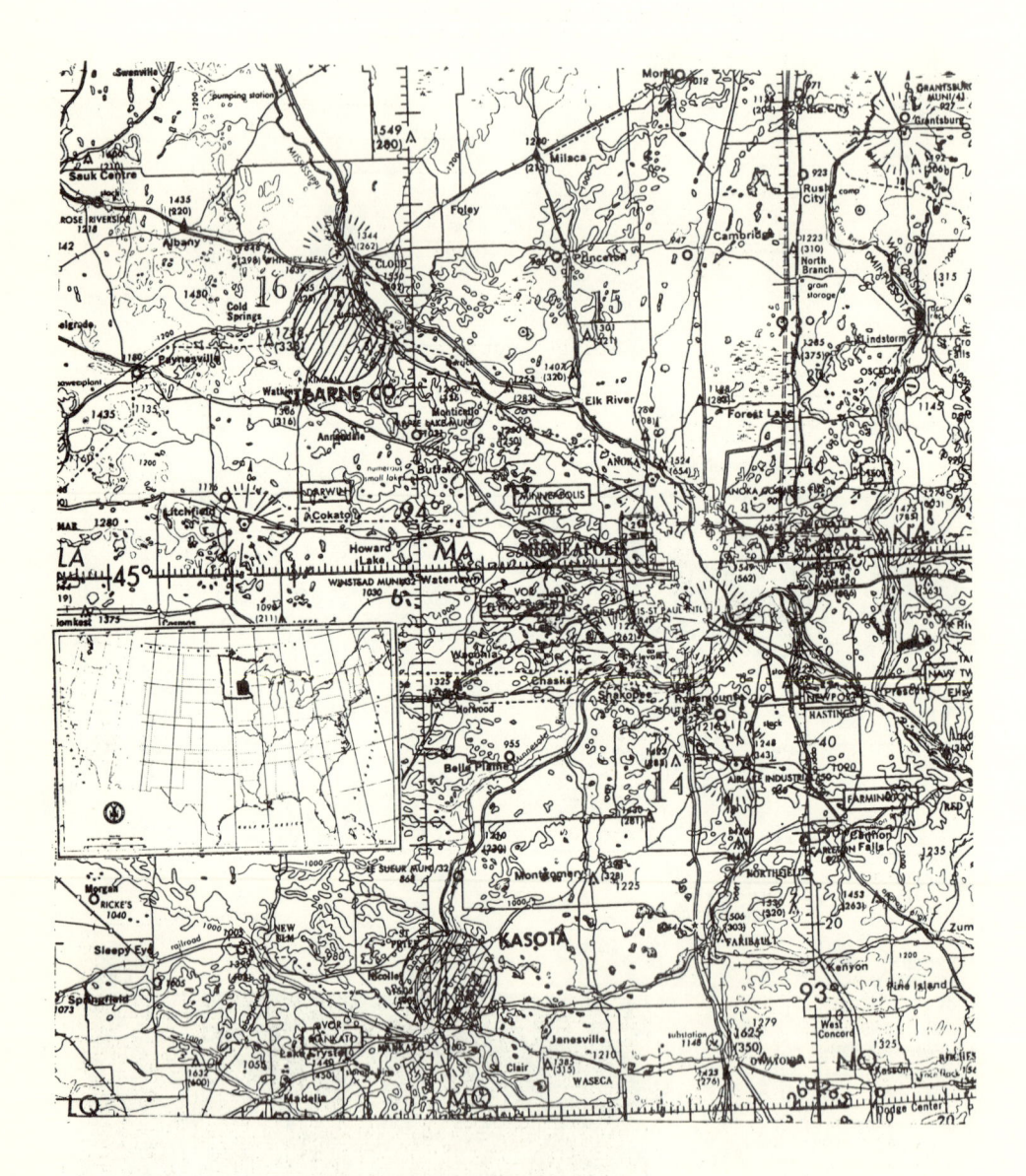

Fig. 15. – Minnesota Test Site Locations.

gory falling within the test site boundaries; and a tabulation of land falling within a five mile radius of the center of the test site, exclusive of the test site. The data produced were to be compared with data already derived through manual interpretation by Commonwealth of topographic maps and aerial photographs. Comparisons would include a discussion of sources of discrepancies, costs and schedule estimates of using ERTS data and manual methods.

Delivered to Bendix at the start of the project were enlargements of aerial photographs of the Kasota and Stearns Counties test sites produced in 1968, a photo mosaic of the Kasota site, and USGS topographic maps of both test sites. Delineated on both the maps and photography were the test site boundaries. It was felt that the older ground truth in conjunction with enlargements and color composites produced from the computer compatible tapes, would be more than sufficient for training set selection.

Two ERTS scenes, 1075-16314 and 1075-16321, of 6 October 1972, were needed to completely cover the test sites. Both scenes were classified as good quality with no clouds or haze noted in the target areas. The Kasota test site is located near the Minnesota River approximately 50 nautical miles southwest of Minneapolis and 8 nautical miles northeast of Mankota, Minnesota as shown in Figure 15. The Stearns County test site, also shown in Figure 15, is located near the Mississippi River, approximately 45 nautical miles northwest of Minneapolis and 8 nautical miles southeast of St. Cloud, Minnesota.

Both test sites are located in predominantly agricultural communities, characterized by numerous small shallow lakes and relatively level terrain. The large urban complex of Minneapolis lies within 50 miles of both areas but appears to have had little effect in terms of encroachment on the land use in those areas. The city of Mankota in the southern site and St. Cloud in the northern site do appear to be growing communities that might place new demands on the areas in the future.

The land use categories desired by Commonwealth for this study are listed below:
1. Open water
2. Wetland or swamp area
3. Forested areas
4. Tilled cropland or grassland
5. Quarries or gravel pits
6. Road, residential and all others

These categories are consistent with the data derived by manual methods.

In computer processing it is possible and often necessary to distinguish many more categories. For this particular study fourteen categories were distinguished and then combined into the required six.

All categories selected for study had categorization accuracies exceeding 90%. Table 4 is a combined categorization table for the six land use categories. It was formed by averaging the percentage of correct and incorrect categorizations in the fourteen categories used to build the final six categories. Combining of categories in this case improved categorization accuracy. The categorization table is only an internal check of the training sets and does not relate to the categorization accuracy of the test site. In this case the training sets were very carefully selected and more than 90% of the training cells were of the category selected.

Special software to derive area tables of the area contained within the test site and a five mile radius from the center of the test site, exclusive of the test site, was also developed for this study. This was accomplished by

Table 4. Integrated Categorization Table.

	0	1	2	3	4	5
1.	2.173	97.158	0.000	0.669	0.000	0.000
2.	0.252	0.000	99.748	0.000	0.000	0.000
3.	0.000	0.000	0.000	100.000	0.000	0.000
4.	2.778	2.778	0.000	0.000	94.444	0.000
5.	1.852	0.000	0.000	0.000	0.000	94.148

0 Unclassified
1 Tilled Cropland
2 Open Water
3 Trees/Brushland
4 Quarries/Gravel Pits
5 Wetlands or Swamp Areas

Table 5. Area Tables, Kasota Test Site.

Category	% of Total	Acres	Sq. Km
a. Test Site			
Road, residential and all other	3.01	124.09	0.50
Tilled cropland or grassland	85.82	3539.25	14.32
Open water	0.03	1.12	0.00
Trees/brushland	10.03	413.62	1.67
Wetlands or swamp area	0.95	39.13	0.16
Quarries/gravel pits	0.16	6.71	0.03
	100.00	4123.92	16.68
b. Five Mile Radius Exclusive of the Test Site			
Road, residential and all other	2.10	979.27	3.96
Tilled cropland or grassland	77.97	36339.43	147.06
Open water	6.14	2862.92	11.59
Trees/brushland	12.32	5740.39	23.23
Wetlands or swamp area	0.92	429.27	1.74
Quarries/gravel pits	0.55	255.88	1.03
	100.00	46606.16	188.61

Table 6. Area Tables, Stearns County Test Site.

Category	% of Total	Acres	Sq. Km
a. Test Site			
Road, residential and all other	0.27	21.24	0.09
Tilled cropland or grassland	82.11	6382.07	25.83
Open water	0.00	0.00	0.00
Trees/brushland	17.32	1345.95	5.45
Wetlands or swamp area	0.07	5.59	0.02
Quarries/gravel pits	0.23	17.89	0.07
	100.00	7772.74	31.46
b. Five Mile Radius Exclusive of the Test Site			
Road, residential and all other	0.79	343.19	1.38
Tilled cropland or grassland	68.11	29247.48	118.35
Open water	0.47	203.46	0.82
Trees/brushland	25.78	11077.22	44.82
Wetlands or swamp area	4.69	2016.68	8.16
Quarries/gravel pits	0.16	69.31	0.28
	100.00	42957.34	173.81

taking advantage of the categorized tape already produced. Computer line printouts were made of the test site and ERTS coordinates determined for the polygon corners and center of the test site. A special program was then implemented that corrected the data for certain geometric distortions, inherent in ERTS CCTs and then the pixels falling within each of the six groups were summed and multiplied by 1.15 acres, the actual area coverage of one pixel. The circle problem was accomplished in much the same manner. The coordinates for the center point of the test site and a radius in scan lines for the circle were input to the program, resulting in Tables 5 and 6.

Also produced but not included in this paper are geometrically corrected land use overlays at a scale of 1:250,000. These overlays are produced by editing the decision processed tape for a particular category and converting it to a tape format suitable for driving a computer driven plotting table. This results in an overlay (in color, if desired) that will fit over a map of any desired scale. In this study, six overlays were produced in different colors to represent the categories and overlaid onto a 1:250,000 AMS map. These overlays provide a direct comparison with a particular land use class and a map.

Preliminary comparisons between the computer derived and manually derived land use maps shows that the results are comparable in accuracy. Further, analysis of typical site selection and transmission line route selection problems has shown that costs for suitable output products of the type produced for this study will range from $3,000 to $13,000, and the products can be produced in six to nine weeks. The next step in the program is to develop software to reformat the output tape from the Bendix system into the input format of the Commonwealth Associates system.

CONCLUSIONS

Multispectral data systems, which involve computerized recognition of remote sensing parameters and digitally derived output products, have been shown to be potentially cost effective solutions to many energy related problems. The launching of the ERTS satellite has been a major contributor to decreasing the costs of multispectral data and has provided a considerable impetus to development of operationally oriented practical applications.

The data collection and data analysis/interpretation portions of the total multispectral data system can be regarded as quasi-operational. Present government plans for data processing, if carried to fruition, will also reduce the data processing function, as defined in this paper, to a quasi-operational status. The only major subsystem presently lacking definition is the information dissemination, or user information portion of the total system.

This latter subsystem tends to be peculiar for specific applications. Consequently, no single subsystem for performance of this function appears likely. However, several common factors will exist from application to application, such as geographically oriented data banks, and output product generating devices.

It has been shown that when a information dissemination or utilization structure is already in existence that can be adapted to the use of digital multispectral data (power plant site selection), cost/performance tradeoffs can readily be made and use of digital remote sensing data can be demonstrated to be a cost effective solution. Conversely, if an information dissemination network is not already in place (strip mine monitoring), then it is not obvious how to determine the path to convert a demonstrated feasible potential application into something in operational use.

The conclusions that can be drawn from these observations are obvious:

1. Practitioners of remote sensing technology must search for and find potential applications that are not only feasible, but also readily acceptable by the technology that exists at the time for utilizing the derived information.

2. Potential users of multispectral remote sensing information must recognize that cost effective utilization of remote sensing data may require modification of their existing methods of information dissemination and be prepared to make unbiased cost/performance tradeoffs concerning their true information needs, their current way of doing things, and the use of new technology.

3. Government agencies and other sources of research and development funds must recognize that the information dissemination and utilization subsystem will require support commensurate to that provided to the other portions of the total multispectral data system if effective operational usage of multispectral remote sensing technology is to be obtained.

REFERENCES

1. R. H. Rogers, K. Peacock and N. Shah; A Technique for Correcting ERTS Data for Solar and Atmospheric Effects, Paper I-7 presented at 3rd ERTS Symposium sponsored by NASA-GSFC at Washington, D. C. on 10 - 14 December 1973.

2. F. J. Krieger, et al; Preprocessing Transformations and Their Effects on Multispectral Recognition, Proceedings of the Sixth International Symposium on Remote Sensing of Environment, October 13 - 16, 1969.

3. R. H. Dye, 1974, Multivariate Categorical Analysis, Bendix Special Report, BSR 4149.

4. C. Connell, et al; Multivariate Interactive Digital Analysis System (MIDAS): A New Fast Multispectral Recognition System, Machine Processing of Remotely Sensed Data Conference, October 16 - 18, 1973, Purdue University.

5. D. A. Landgrebe, P. T. Min, P. H. Swain and K. S. Fu, 1968, The Application of Pattern Recognition Techniques to a Remote Sensing Problem, LARS Information Note 080568, Purdue University, also a paper presented at the Seventh Symposium on Adaptive Processes, UCLA, Los Angeles, California, December 16 - 18, 1968.

6. W. A. Pettyjohn, R. H. Rogers and L. E. Reed; Automated Strip Mine and Reclamation Mapping from ERTS, Paper E-3 presented at 3rd ERTS Symposium sponsored by NASA-GSFC at Washington, D. C. on 10 - 14 December 1973.

7. L. E. Reed; ERTS Land Use Mapping for Power Plant Site Selection, Bendix Report BSR 4160, October 1974.

LOCATING REMOTELY SENSED DATA ON THE GROUND

R. C. MALHOTRA and M. L. RADER
Lockheed Electronics Co., Inc., Houston, Texas, U.S.A.

ABSTRACT

This paper briefly discusses techniques for identifying the precise ground location represented by a specific set of remotely sensed data. Automated mathematical procedures using navigation (ephemeris) data and/or ground control points to identify the ground location of remotely sensed data sets by parametric modeling, "speculative" polynomial adjustment, and a method combining modeling and polynomial adjustment are described. Data from the NASA Skylab and the Earth Resources Technology Satellite (ERTS-1) projects are used in the examples given, but the basic methods apply to remotely sensed data collected from any aircraft or Earth-orbital satellite.

INTRODUCTION

The Earth location of data acquired by a remote sensing device may be determined by using navigation (ephemeris) and/or ground position information for specific data samples of the sensor. Each sample may then be assigned an earth location by a transformation derived from this navigation and/or ground position information and may be displayed in its correct position relative to other samples on a display device such as a film recorder, television screen, et cetera. The data may also be recorded sequentially on a magnetic medium (i.e., magnetic tape, disc) for additional analysis. The position of the data sample on the magnetic device would correspond to its relative location with respect to other data samples.

LOCATION OF REMOTELY SENSED DATA BY EPHEMERIS/NAVIGATIONAL DATA

Location of sensor data may be achieved by recording the sensor Earth location, its attitude and/or other data with respect to time (Table 1). With a time recorded for each data sample it is possible to locate this data by using the above ancilliary data in a dynamic math model (time study) of the sensor (internal) and vehicle motion. Examples of such time studies are given in Tables 1 and 2. These tables depict instantaneous field of view (FOV) of a sensor at certain times. This data is obtained by utilizing the SKYBET (Skylab Best Estimate Trajectory) data (Reference 1) in the time study.

A hard-copy of the remotely sensed data may be obtained by a data display (film recorder) of a channel or channels recorded by the sensor. Using navigation (ephemeris) data, certain geometric corrections may be applied to the imagery during the display to eliminate geometric distortions caused by factors such as sensor geometry, Earth's rotation, curvature of the Earth, et cetera.

Table 1. Skylab S190 Earth Terrain Camera Field of View (FOV) Tabulation

```
                Project          08-DPCA-0-31-31-C
                Mission    3     Orbit 31              Site 0
                Sensor S190B                           Flight Date 9 Sept. 1973

TRJGMT 253:20: 6:51.800      TMID 253:19:13:45.500   TSR 253:19:31:28.100   TSS 253:18:56: 0.900
TRJGET  44: 8:56: 1.800      ATM TIME 93: 4: 1:19.609 TESR 253:19:31:43.400 TESS 253:18:55:45.500
ACC.TIME  0: 0: 0: 0.000     REV NO. --------
```

---SKYBET EPHEMERIS DATA---

XECI	-5.6124258E 03	XECT	1.3056011E 03	MOONX	5.1125912E 01	ALPSI	2.4110041E 01	PHD1X	-1.3885263E-04
YECI	4.1034198E 02	YECT	5.4829761E 03	MOONY	-3.3083305E 01	BETSI	3.5208539E 02	PHD1Y	6.4167671E-02
ZECI	3.8134067E 03	ZECT	3.8002834E 03	MOONZ	-9.7166967E 00	GAMSI	3.4464157E 02	PHD1Z	5.4453837E-04
DXECI	2.2839687E 00	DXECT	-6.4042354E 00	SUNX	-9.7678196E-01	ALPZLV	3.5999164E 02	PHD2X	3.9363324E-04
DYECI	-6.1188688E 00	DYECT	-1.2649450E 00	SUNY	1.9654980E-01	BETZLV	1.4664196E-02	PHD2Y	6.3942745E-02
DZECI	4.0109010E 00	DZECT	4.0162272E 00	SUNZ	8.5235909E-02	GAMZLV	3.5980728E 02	PHD2Z	4.3419428E-04
ALTSPH	4.2441016E 02	R	6.7977749E 03	SMA	6.8102207E 03	ALPECI	2.9975870E 02	PHD3X	-3.8395790E-04
ALTOBL	4.2631702E 02	RPER	6.7969956E 03	ECC	1.9417877E-03	BETECI	3.0710474E 02	PHD3Y	6.4249612E-02
GCLAT	3.3989967E 01	RAPO	6.8234424E 03	INC	5.0039360E 01	GAMECI	3.5433121E 02	PHD3Z	2.5376436E-04
GDLAT	3.4157337E 01	HPER	4.2944669E 02	ARGPER	6.6653488E 01	SUNAZ	1.9898587E 02	EARTHAZ	4.5324508E-02
LON	-1.1529556E 02	HAPO	4.5589459E 02	LONAN	2.1030418E 02	SUNFL	1.1386707E 02	EARTHEL	8.9910263E 01
V	7.6644902E 00	ORBRAT	3.5758392E-04	TRUEANOM	3.4017953E 02	BETA	-1.7210472E 01	B	4.1700232E-01
FPA	-3.7655037E-02	ORBPER	1.5536363E 04	ECCANOM	3.4021716E 02	ESLAN	2.7582523E 01	L	1.8425816E 00
AZ	5.0770187E 01	LONANVE	1.4133673E 02	MEANANOM	3.4025500E 02	SEA	5.9850750E 01		

---CAMERA AND EXPOSURE DATA---

FRAME NUMBER	FILM TYPE	FILTER	EFF. F-NUMBER	EXPOSURE TIME
85-378	SO 242	1	F/ 4.0	5.0

---FIELD OF VIEW COORDINATES---

	CENTER OF FRAME	UPPER LEFT CORNER OF FRAME	UPPER RIGHT CORNER OF FRAME	LOWER LEFT CORNER OF FRAME	UPPER LEFT CORNER OF INSET	UPPER RIGHT CORNER OF INSET	LOWER LEFT CORNER OF INSET
LATITUDE (DEG:MIN)	34: 9.8	34:50.4	34: 5.4	34:13.8	33:42.8	33:35.3	33:36.7
LONGITUDE (DEG:MIN)	-115:17.5	-115:12.4	-114:28.7	-116: 6.3	-115:20.8	-115:13.6	-115:29.7

Table 2. Skylab S192 Conical Scanner Field of View (FOV) Tabulation

```
                Project          2-DPCA-0-47-52-7
                Mission    3     Orbit 47              Site 0
                Sensor S192      Recording Format 7    Flight Date 10 Sept. 1973
```

	LEFT		CENTER		RIGHT	
GMT	LONG	LAT	LONG	LAT	LONG	LAT
0:25: 6.338	135 32 6.29	33 51 22.70	135 56 36.09	33 43 36.96	136 1 11.03	33 22 4.09
0:25: 7.393	135 35 36.95	33 54 0.05	136 0 7.20	33 46 13.54	136 4 41.30	33 24 40.52
0:25: 8.448	135 39 7.72	33 56 37.22	136 3 30.46	33 48 49.47	136 8 11.75	33 27 16.81
0:25: 9.503	135 42 38.77	33 59 14.32	136 7 9.95	33 51 26.31	136 11 42.41	33 29 53.00
0:25:10.550	135 46 10.04	34 1 10.04	136 10 41.60	33 54 2.52	136 15 13.24	33 32 29.08
0:25:11.612	135 49 41.52	34 4 28.14	136 14 13.47	33 56 38.61	136 18 44.34	33 35 5.04
0:25:12.667	135 53 13.23	34 7 4.91	136 17 45.56	33 59 14.60	136 22 15.61	33 37
0:25:13.722	135 56 45.16	34 9 41.51	136 21 17.87	34 1 50.47	25 47.15	
	136 0 17.25	34 12 18.02	136 24 50.41	34 4 26		
	61	34 14 54.41	136 28 23.27	34		
		34 17 30.68	136 31			
		6.85	136			

Greenwich Mean Time, latitude and longitude of scene, frame number, et cetera, may also be displayed alongside the imagery on the film converter. Such a hard copy allows the user to select his study area.

LOCATION OF REMOTELY SENSED DATA BY EPHEMERIS NAVIGATION DATA AND GROUND CONTROL POINTS

Table 3 depicts the location accuracy achieved using SKYBET data in one of the time studies for Skylab S192 conical multispectral scanner (MSS). For certain applications, the location accuracy achieved from calculations using ephemeris data only may not be sufficient. For these applications, location accuracy may be increased by introducing ground coordinates or other known geometric properties of the data into the calculation.

The methods used to increase the location accuracy of remotely sensed data sets by utilizing ground coordinates and ephermis data are:

- Parametric modeling

- "Speculative" polynomial adjustment

- Combination method

Table 3. Location Accuracy of Skylab S192 Conical Scanner Data Using SKYBET

MISSION: SL-2

LOCATION OF DATA: Great Northern Valley Pass No. 3
(Lat: 40°20'N/Long: -122°40'W)

SCAN LINE NO.	TIME GMT H. M. S.	LAT. '	CENTER PIXEL DIFF.* "	LONG. '	DIFF.* "
400	19 22 45.465	0	38.3	0	07.2
500	46.519		36.4		06.6
600	47.574		33.6		09.7
700	48.629		32.6		11.3
000	49.684		32.9		11.8
900	50.738		34.1		12.1
1000	51.793		35.5		12.8
1100	52.848		38.6		12.8
1200	53.903		43.6		11.4
1300	54.958		49.8		9.1
1400	56.013		57.1		6.2
1500	57.068	1	05.7		2.6

*Difference = (S052-7 Product Value) - (True Value)

The parametric modeling is a dynamic study of the internal sensor motion and the vehicle motion with respect to the Earth. Ground control points and/or ephemeris data is then used to solve for the parameters associated with the dynamic math model (References 2 and 3). An example of the parametric modeling method is given in Table 4.

The "speculative" polynomial method uses a polynomial which is assumed to represent the mathematical function relating the remotely sensed data to the ground location. The coefficients (a,-f) of the polynomial (1) are solved for by utilizing the ground locations of points (X,Y) and their corresponding data sample positions (x,y) and by mathematically imposing a least-squares-fit condition. One example of this method is the first order bivariate polynomial often used to register ERTS-1 data to reference maps. The polynomial is as follows:

$$X = a + bx + cy \quad -1(a)$$

$$Y = d + ex + fy \quad -1(b)$$

Table 4. Location (Registration) Accuracy of Skylab S192 Conical Scanner Data by Parametric Method

SITE/PASS	GMT START D-H-M-S	STOP M-S	R.M.S. ERRORS (PIXEL*) CONTROL PTS. X	Y	CHECK PTS. X	Y
Mississippi Delta (52A)	264-13-46 00	46-20	±3.8	±3.2	±4.0	±4.3
Wisconsin (14A)	217-15-00-15	01-15	±3.3	±3.2	±4.0	±3.7
Washington, D.C. (14A)	217-15-03-45	04-10	±4.0	±3.8	±3.7	±3.8

*Pixel = 80 X 80 meters

Table 5 is an example of output from the polynomial method.

In combining the parametric and speculative polynomial methods, geometric error in the sensor data is removed both by parametric modeling and a speculative polynomial. For example, NASA eliminates sensor errors from ERTS-1 MSS data by parametric modeling to compensate for accelerations in the scan velocity, panoramic effect, and constant roll within a scene. The remaining geometric errors in the data are then eliminated by a first order bivariate polynomial (Equation 1). This procedure is being used in support of the National Program of Inspection of Dams (Reference 4).

Table 5. Location (Registration) Accuracy of ERTS Data by Polynomial Method

ID PT.	LINE	SAMPLE	LATITUDE ° ' "	LONGITUDE ° ' "	X (METERS)	Y (METERS)
35	84	534	29 33 11	-95 09 13	-67	-104
32	123	468	29 35 10	-95 11 01	-6	-9
118	102	451	29 34 21	-95 11 55	-58	-26
125	110	474	29 34 33	-95 11 01	-103	-41
30	156	402	29 36 56	-95 12 56	50	-20
28	169	376	29 37 38	-95 13 42	19	20
31	166	495	29 36 50	-95 09 36	-154	-93
40	47	458	29 32 00	-95 12 20	-53	126
34	60	409	29 32 49	-95 13 53	-5	3
33	58	389	29 32 50	-95 14 35	27	-42
36	108	560	29 34 02	-95 07 54	-205	143
24	217	348	29 39 46	-95 14 05	26	35
25	224	392	29 39 51	-95 12 26	108	3
48	69	311	29 33 47	-95 17 08	-37	-25
17	199	327	29 39 07	-95 15 03	-44	33
				RMS ERROR	±84	±65

PRODUCTS OF LOCATED (REGISTERED) DATA

The products of located (registered) data are:

- o Non-display products (magnetic data tapes, discs)

- o Display products (thematic maps from located classified data displayed on line printers, pen plotters, CRT's, or film recorders)

The display products may be overlaid by a map projection grid, and non-display products may contain this overlay information within the recording medium.

CONCLUSION

The single most important factor for selecting the method of location of remotely sensed data is the desired accuracy. In addition, factors such as the quality of the data-take also dictate the method of location. In the case of stable platforms (i.e., ERTS), the data-take is smooth and continuous (without abrupt or frequent changes in the parameters effecting location) such that simple mathematical transformations (first order bivariate polynomial) suffice to locate the data. However, in the case of unstable platforms (i.e., aircraft), location of data may require a more complicated mathematical model which simulates the sensor and vehicle motions during data acquisition.

Locating remotely sensed data gives the distribution as well as quantitative estimates of Earth's resources and its environmental conditions, thereby forming an important step in the remote sensing technology.

REFERENCES

1. Flight Support Division, 1973. SKYBET Parameter Formulation Document. National Aeronautics and Space Administration, Johnson Space Center, Houston, Texas, August 9, 1973.

2. Malhotra, R. C., 1974. Absolute Spatial Registration of Skylab S192 Conical Scanner Imagery by Means of Dynamic Geometrical Modeling. Proceedings of the ASP 40th Annual Meeting, St. Louis, Missouri, March 10-15, 1973.

3. Rader, M. L., 1974. A General Solution for the Registration of Optical Multispectral Scanner. Proceedings of the ASP 40th Annual Meeting, St. Louis, Missouri, March 10-15, 1973.

4. Earth Observations Division, 1973. Procedures Manual for Detection and Location of Surface Water Using ERTS-1 Multispectral Scanner Data. National Aeronautics and Space Administration, Johnson Space Center, Houston, Texas, December, 1973.

REMOTE SENSING OF SMALL TERRESTRIAL TEMPERATURE DIFFERENCES

N. J. CLINTON and C. E. CAMPBELL

*Lockheed Electronics Company, Inc., Aerospace Systems Div.,
Houston, Texas, U.S.A. 77058*

ABSTRACT

The detection of small differences in temperature of the earth's surface is needed for geological, oceanographic and other purposes. Detection can be achieved with an infrared scanner optimized for a small noise equivalent temperature difference rather than high resolution. For oil exploration, it is also necessary to fly at a time when the earth's surface temperature is not masked by other effects. Flights before dawn minimize the effect of solar heating. An overcast sky suppresses the effects of emissivity differences. Although not available for display, imagery taken under these conditions, which has been viewed by the authors, showed both surface manifestations of the subsurface structure and the reduction of emissivity differences in data taken during overcast conditions.

APPLICATION

The detection of small differences in temperature by remote sensors may be applied to geological, oceanographic, and forestry studies. Geological subsurface structures are often differentiated from the surrounding regions by small temperature differences at the surface. The geological information obtained from temperature differences at the surface detected by remote sensing is useful in locating petroliferous and mineral deposits.

Present infrared scanners and operational methods are far from optimum in detecting these small temperature differences. By designing an infrared scanner for high thermal resolution and flying it during favorable meteorological conditions, a considerably increased sensitivity to minute temperature differences can be achieved.

The subsurface heat flow and temperature distribution is modified by the geological subsurface structure or shape of the formations. In the late twenties Van Orstrand developed the use of terrestrial heat flow and temperature distribution to obtain additional information about the subsurface structure. Van Orstrand measured temperatures, corrected for equipment error, from considerable depth up to 50 meters deep in wells. His study showed that the earth's heat which reaches the surface, as modified by the subsurface structures, is manifested in anomalous patterns roughly proportional in size and shape to the structural conditions beneath. Thus, surface thermal patterns, manifested in radiant temperature patterns, can be analyzed for subsurface structural conditions such as anticlines, domes, and fault traps. An anticline or dome is an upwarp of strata above the corresponding surrounding strata. Oil and gas are found at the highest elevation of the permeable strata. Oil and gas rise through permeable layers until they meet an impermeable blockage and are trapped.

Temperature plotted as a function of depth results in contours of equal temperature called isotherms. The change of temperature across the isotherms is called the thermal gradient. Van Orstrand proved the concept that the thermal gradient parallels the folds of the strata. The thermal gradient is greater over an anticline or dome and results in a higher temperature at the surface. Over synclines, which are characterized as the valleys between the anticlines, the opposite situation occurs and the surface temperature averages slightly lower than the surrounding areas.

In the case of faults, water may run into and be held by the fractured zone so that it is moist when the surrounding areas have dried out. As it is more moist, the evaporation is greater and the temperature is slightly lower than the surrounding area. Thus the faults under these conditions have lower temperatures. More rarely, a fault may act as a conduit of heat up to the surface and the average surface temperature is warmer than the surrounding areas.

Admittedly, such geological structures are not always characterized by a corresponding temperature pattern. In particular, a large horizontal flow of ground water near the surface can wipe out the thermal patterns described above. Nevertheless, in most areas, the remote sensing of radiant temperature patterns would be valuable in analyzing the subsurface structure.

Surface temperature differences associated with subsurface structure are weak and usually overwhelmed by instrumental noise and solar insolation. However, by maximizing the instrumental signal-to-noise ratio and obtaining data under conditions which minimize the natural undesired signals, greatly increased sensitivity to the desired thermal effects can be obtained.

Obviously, after the aerial survey data is collected, it is necessary to explore further on the ground to substantiate any radiant temperature anomaly detected.

Such high thermal sensitivity as described here is helpful for detection of geothermal anomalies for use as geothermal power sources. In addition, thermal anomalies can be used in both oceanographic and forestry applications.

Thermal anomalies are used in locating areas of ocean upwelling; locating interfaces between areas of water from different sources, e.g., fresh water springs along island coasts; tracing thermal pollution and detecting small amounts of oil in rough water. Maximum thermal sensitivity to detect such phenomena is required.

Applied to forestry, thermal anomalies can be used to detect diseased vegetation. The temperature in the diseased forest is somewhat higher than that in the healthy foilage and usually begins to rise before there are visible signs of distress on the ground. (Small clearings under some conditions may introduce false signatures when the high thermal resolution scanner data is viewed but comparison with the high spatial resolution imagery will permit these false alarms to be detected.) Applications of thermal anomalies to detect stress in crops due to disease or soil conditions is also feasible.

SCANNER DESIGN

Most infrared scanners have been designed for high spatial resolution rather than high thermal resolution. For some applications high spatial resolution is desirable. For example, on those scanners which were originally designed for military purposes, the mirror has been designed to have as high a scan rate as possible and is therefore designed smaller than a mirror designed to operate at a slower speed and intercept sufficient radiation for a high sensitivity system.

To obtain high thermal sensitivity, i.e., a low noise equivalent temperature difference, it is necessary to sacrifice high resolution. This is acceptable since high thermal sensitivity is more important for the earth resource applications mentioned above than high spatial resolution.

The size of the detector which would optimize performance would be 12 to 20 mils across. At 10,000 and 20,000 feet a 15 mil detector would give a resolution of 150 to 300 feet, respectively. The optimum altitude would depend upon the particular application.

The number of photons per second hitting the detector is proportional to the area of the detector

$$\frac{dn}{dt} \sim A = W^2 \tag{1}$$

The area of the detector is the square of its width W.

Also, the dwell time dt of an optimized system may be shown to be proportional to the area.

$$dt \sim A \tag{2}$$

The resolution R (and reciprocal of the IFOV) for an optimized high signal-to-noise ratio system is inversely proportional to the detector width.

$$W \sim \frac{1}{R} \tag{3}$$

Thus the number of photons hitting the detector from any resolution element is

$$n \sim \frac{1}{R^4} \tag{4}$$

The signal-to-noise ratio for a photon noise limited system is proportional to the square root of the number of photons

$$\frac{S}{N} \sim \sqrt{n} \sim \frac{1}{R^2} \tag{5}$$

The sensitivity to radiation differences is inversely proportional to the square of the resolution. Thus, increasing the detector width from 3 to 15 mils increases the sensitivity to radiation differences by a factor of 25, although the resolution goes down by a factor of 5.

In addition to making the actual detector size larger, the effective detector area could be increased by a row of detectors staggered so that their centers will pass over the same ground point during a scan. The signals would be added after the appropriate time delays necessary to synchronize the signals of the several detectors. This technique has the advantage of increasing the sensitivity without decreasing the resolution, but at the disadvantage of increased complexity. A problem is that the direction of the line of detectors must vary with the V/H (velocity V divided by altitude H) although a design for a fixed altitude may be adequate.

To further increase the dwell time of the detector, and therefore increase the sensitivity, a slow aircraft would be desirable, possibly as slow as 120 knots.

The mirror would rotate slowly and the design of the mirror system would be simple. The slow mirror motion and low resolution requirement permits the use of large aperture optics which would further increase the signal strength.

The noise equivalent temperature difference of present scanners is often given by the manufacturer as 1/4° or 1/5° Centigrade. However, water temperature measurements are not generally accurate within a degree Centigrade, even at low altitude and close proximity to points where water surface measurements are taken for calibration.

Manufacturers' data for signal-to-noise ratios and noise equivalent temperature differences of scanners are often very optimistic. For example, one company measured the signal-to-noise ratio of a scanner with the scan motor off. This particular scanner had a four-sided mirror and there were obvious differences between the signals from each side of the mirror. Moreover, the aircraft, scan motor and cooler frequently cause considerable acoustical or electromagnetic interference.

Thus, there must be considerable care to eliminate all sources of interference. The relatively low data rate or band width of the system described here greatly reduces interference.

High fidelity of the scanner must also be preserved in the recorder and display. The strong signal output of the scanner may cause a problem with the dynamic range and noise of the display. To prevent this and provide enhancement, low frequency suppression and scale expansion may be necessary to detect the relevant detail. The recording may be either digital or analog. The final display would be on film.

The system can be made quantitative by using large blackbody sources at different temperatures on each side of the rotating mirror. A narrow field radiometer pointing at the nadir would give a section through the nadir locus which would give an independent check to the calibration by the internal sources. A simple method is to record the output of the scanner at the single pixel at nadir and graph the nadir value separately. This pixel would be blank on the film, showing exactly where the radiometric profile occurred.

Because of the low spatial resolution of the system, a simultaneous high-resolution system is required to show the location of the imagery. This "finder" system would be a high-resolution infrared system. Also, a low light level TV system is advantageous when the contrast of the IR imagery is low, as will be discussed later.

The format of the display is optional to the user. The display preferred by the authors is on 9 inch film divided into thirds. One third contains the high thermal resolution IR imagery, the second the high spatial resolution imagery, and third the graph of the radiometer, all synchronized in time. On each side of the IR imagery would be a strip showing the calibration sources. Time marks every 2 centimeters would be made on one side of the film. Spatial rectification is not necessary for interpretation.

A field of view as large as possible is desirable, preferably 160°. A repeated anomaly detected on each flight line, due to the wide field of view, is advantageous. Experience has shown that repetitive coverage of the anomaly is very useful for interpretation.

REQUIREMENTS FOR FLIGHT CONDITIONS

When used for geological purposes, high thermal sensitivity alone is not sufficient. The time of day, time of year, and meteorological conditions are also important, otherwise the relevant thermal anomalies may be masked by solar heating and other masking effects.

The best time for acquiring data is when the terrain is closest to thermal equilibrium with respect to these spurious effects. At such times the thermal contrast due to these spurious effects in the thermal infrared is low. Thus, a highly sensitive system can show the relevant thermal anomalies best with this type of data. Before dawn the effect of solar heating is at a minimum. Also the high-emissivity surfaces have radiated more than the low-emissivity surfaces, bringing their temperatures down so that the trend is to emit the same as low-emissivity surfaces.

Winter without snow cover is better than summer as there is less solar heating. The yearly effect as well as the diurnal effect is also suppressed. Moreover, in temperate climates the effects of foilage is minimized.

Rain produces a washout effect by suppressing temperature differences of the surface and giving the ground a more uniform emissivity. Some time after a rain the thermal anomalies for geological structures may be detected more easily.

A cloud layer suppresses the effect of surface emissivity differences in causing differences in the amount of radiation emitted.

For the case of a perfectly clear sky the spectral radiance is

$$N_\lambda = E(\lambda) \, B(\lambda,T) \tag{6}$$

where $E(\lambda)$ is the emissivity and $B(\lambda,T)$ is the Planck function which depends upon the wavelength λ and temperature T. The quantity which is actually detected is the voltage

$$V = \int R(\lambda) \, N_\lambda \, d\lambda \tag{7}$$

where $R(\lambda)$ is the sensor's responsivity.

For the hypothetical case of a cloud at the same temperature as the ground, the ground spectral radiance is

$$N_\lambda = E(\lambda) \, B(\lambda,T) + [1-E(\lambda)] \, B(\lambda,T) \tag{8}$$

$$N_\lambda = B(\lambda,T) \tag{9}$$

The second term in equation (8) is the radiance due to radiation originating at the cloud. The factor $[1-E(\lambda)]$ is the reflectance of the ground.

Obviously the radiance in the case of the clear sky (equation 6) is strongly emissivity dependent, whereas the radiance in the case of the flight under a cloud layer (equation 9) is independent of the emissivity and dependent only upon the temperature of the ground. This is the washout effect which greatly reduces the contrast of IR images, often making them useless for most purposes. It is similar to being inside a blackbody. However, the relevant geological thermal anomaly effects are not masked, although the sensor system must be highly sensitive to detect them.

For the real case, the cloud temperature is less than the temperature of the ground and the emissivity is only partly suppressed. Nevertheless, there is a strong improvement in removing the masking effects.

A disadvantage of flying under an overcast sky is that the altitude, and therefore the amount of ground coverage, is limited. Also, the higher the cloud the greater the ground coverage but the colder the cloud the greater the apparent differences of the surface emissivities.

Because the altitude may be limited, as described above, the scanner should have as wide a field of view as possible in order to maximize the ground coverage.

The reason that this method has not been exploited is the cost involved. An aircraft must be available on a standby basis to fly in the early morning hours when the meteorological conditions are favorable. One method of minimizing the cost would be to utilize an aircraft intended for other purposes but which is equipped for rapid installation of the scanner and a crew which is available on short notice.

There is little data to determine exactly how effective this method is, the meteorological effects, and how to optimize the method. It would be desirable to test this method over 3 years to obtain exact data on its effectiveness. Such a program would require shallow thermal measurements which would provide the necessary data on annual and diurnal fluctuations for bridging the gap between the dynamic temperatures of the earth and the radiometric temperatures recorded by the thermal infrared remote sensor.

BIBLIOGRAPHY

Call, P. D., S.A. Terry, and A. E. Pressman, 1969, Airborne infrared measurements applied to oceanographic, hydrologic, and water pollution problems: Trans. Marine Temperature Measurement Symp., Marine Tech. Soc., Miami Beach, p. 227-252.

Dibblee, T. W., Jr., 1962, Displacements on the San Andreas Rift Zone and related structures in Carrizo Plain and vicinity (California). in Guidebook Geology of Carrizo Plains and San Andreas Fault: San Joaquin Geol. Soc. and Am. Assoc. Petroleum Geologists-Soc. Econ. and Paleontologists and Mineralogists, Pacific Sec., p. 5-12.

Friedman, J. D., 1970, The airborne infrared scanner as a geophysical research tool: Optical Spectra, V. 4, no. 6, p. 35-44.

Gates, D. M., 1964, Characteristics of soil and vegetated surfaces to reflected and emitted radiation: Symposium Remote Sensing of Environment 3rd, Univ. Michigan, Ann Arbor, Infrared Physics Lab., Proc., p. 573-600.

Horai, Ki-iti, and Seiya Uyeda, 1969, Terrestrial heat flow in volcanic areas, in The Earth's Crust and Upper Mantle: Am. Geophys. Union Monograph 13, p. 95-109.

Lovering, T. S., 1965, Some problems in geothermal exploration: Mining Engr., V. 17, no. 9, p. 5-9.

Rowan, L. C., T. W. Offield, K. Watson, P. J. Cannon, and R. D. Watson, 1970, Thermal infrared investigations, Arbuckle Mountains, Okla.: Geol. Soc. Amer., Bull. V. 81, no. 12, p. 3549-3562.

Sabins, F. F., Jr., 1967, Infrared imagery and geologic aspects: Photogram. Engr., v. 33, no. 7, Jul. p. 743-750.

_____, 1969, Thermal infrared imagery and its application to structural mapping in southern California: Geol. Soc. Amer. Bull., V. 80, p. 397-404.

Van Orstrand, C. E., 1932, On the correlation of isogeothermal surfaces with rock strata: Physics, V. 2, no. 3, p. 139-153.

_____, 1934, Temperature gradients, in Problems in Petroleum Geology, Sidney Powers memorial vol.: Am. Assoc. Petroleum Geologists, p. 989-1021.

_____, 1934, Some possible applications of geothermics to geology: Am. Assoc. Petroleum Geologists, Bull., V. 18, no. 1, p. 13-38.

_____, 1935, Normal geothermal gradient in the United States: Am. Assoc. Petroleum Geologists, Bull., V. 19, no. 1, p. 78-115.

_____, 1937, On the estimation of temperature at moderate depths in the crust of the earth: Am. Geophys. Union Trans., 18th Ann. Mtg. pt. 1, p. 21-22.

Wallace, R. E., and R. M. Moxham, 1966, Use of infrared imagery in study of the San Andreas Fault system, California: U. S. Geol. Survey Prof. Paper 575-D, p. 147-156.

Wolfe, E. D., 1971, Thermal IR for geology: Photogram. Engr., v. 37, no. 1, p. 43-52.

Wood, C. R., 1972, Ground-water flow: Photogram. Engr., V. 38, no. 4, p. 347-352.

REMOTE SENSING APPLIED TO ENERGY-RELATED PROBLEMS

ALOIS SIEBER

*Deutsche Forschungs- und Versuchsanstalt
für Luft- und Raumfahrt e.V., Porz-Wahn, Linder Höhe, W.-Germany*

ABSTRACT

The German Remote Sensing Program is intended to investigate, for preservation of energy production and utilization, the prospects of new resources as well as the related environment effects. In experimental measurement programs, use is made of conventional and advanced technological methods to examine the economic prospects and exploration of the power producing resources coal, hydrocarbons, geothermics and hydraulics.

Simulataneously, investigations shall deal with the problems of economical energy utilization and the effects of the dissipated heat on hydrosystems, climate as well as with other ecological changes and influences.

In additon, remote sensing is employed to find the most favorable solutions to the problems of building and settling areas and the best possible locations for high power-stations.

For the accomplishment of all these remote sensing tasks, the German Remote Sensing Program provides for experimental measurement programs with conventional sensors and for the development and testing of new sensors and advanced technologies.

INTRODUCTION

Energy is the root of the efficiency of political economy and the criterion for the quality of life of the individual. This statement seems to be evident and nearly trivial to us today, as still in the middle of last year energy was - say from the consumer's point of view - nothing else but goods you could buy without limits in the shape of petrol, gas or electricity. But in the meantime it has become clear that our nergy supply is exposed to danger for the near and far future. The reasons are:

- The energy supply of the FRG is based at a percentage of over 95% on the natural products oil, gas, and coal. (Fig. 1) We know that the world reserve of these energy reservoires will be used-up by rising and even stagnating needs in a time which is within sight. A big problem is the dependence upon oil (55%) and gas (9%), the resources of which will definitely be exhausted first (Fig. 2).

- With respect to the almost exclusive import dependence of the German mineral oil supply (97%), especially on the supplying countries in the Near Est, the access to the most important energy source has become very critical.

- Consumption of energy, especially of energy derived from fossil reservoirs, leads to a critical impact on our environment and is restricting man's quality of life by heat waste and pollution.

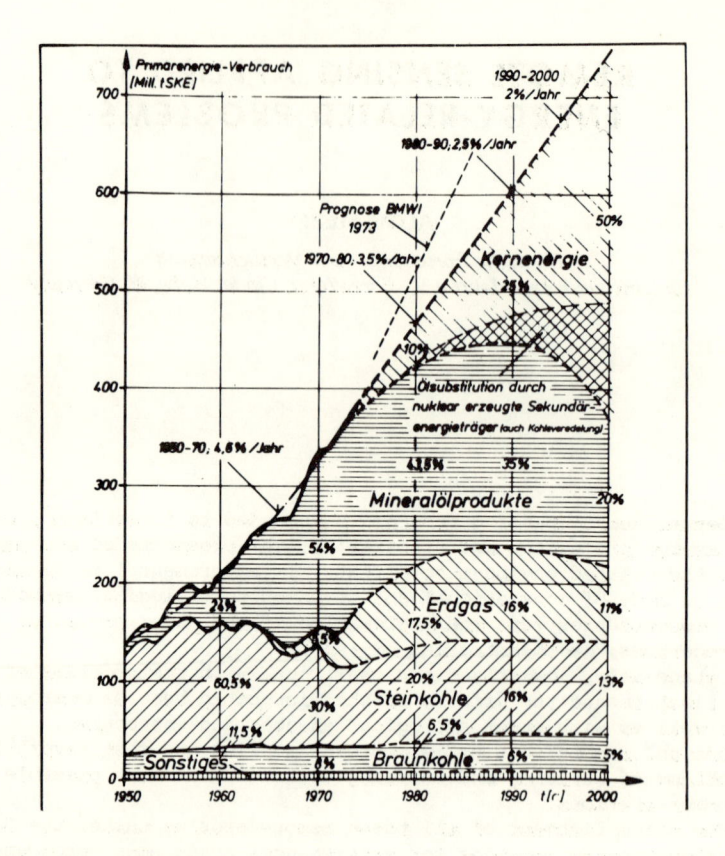

Fig. 1. - Use of Primary Energy in the FRG
(Bohn SfE Jülich)

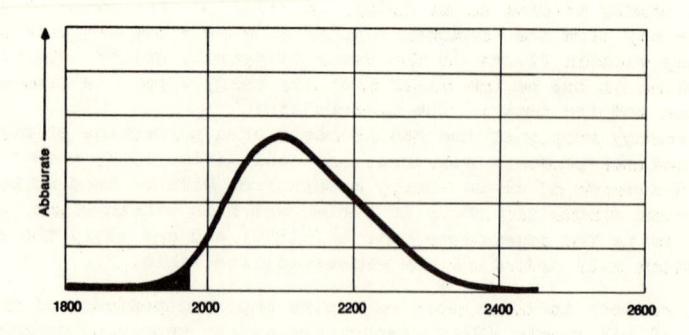

Fig. 2a. - Used and Predicted Exploitation of World Coal Resources
(BMFT: Rahmenprogramm Energieforschung 1974 - 1977,
- Referat für Presse und Öffentlichkeitsarbeit -
Bonn 1974)

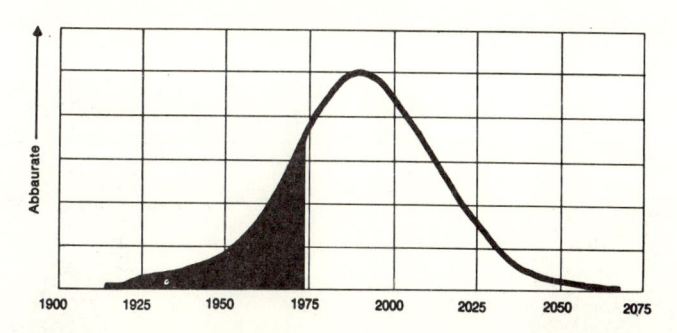

Fig. 2b. - Used and Predicted Exploitation of World Oil Resources
(BMFT: Rahmenprogramm Energieforschung 1974 - 1977
- Referat für Poresse und Öffentlichkeitsarbeit -
Bonn 1974)

The actual energy crisis has shown clearly that our economic system can easily be effected. Therefore the search for a substitution of oil products by equivalent energy reservoires, rendering possible an equally abundant application, has to be intensified. The past years have shown as well that it is not enough to secure an always sufficient and favourable offer of energy for the future. The hitherto existing abundance of energy or the energy being at disposal at least at a low price led to a relatively thoughtless use of energy, especially in daily life. According to our estimations, the relation of necessary to used energy is less than 45% in household, 55% in industry and in traffic hardly more than 2o%. Regarding an energy flow diagram of the FRG (Fig. 3) we see that 62,8% of the injected primary energy has to be realized as a dissipation. It is lost by transforming primary energy to usable energy, by the transport of energy and ba the use of the energy. As ineffective as this seems to be from an economic ooint of view, the more important attendant phenomenon is the stress on our environment by energy waste.

To realize an energy supply,

which will be adjusted to the requirements of all users in the FRG,
which will be secure in the near and in the far future,
which will be based on propitious costs and
which will include the requirements of the environmental influence

the Minister of Research and Technology published an energy research program with the following goals among others:
development of new technologies for the prospection and exploration of oil, gas, and coal;
energy loss reduction at the transformation, transportation, and accumulation of energy;
development of closed energy cycles to separate energy systems and environment;
search of new energy sources not known until today.

This proposed program will include a large number (nearly 45o) of individual tasks. The following list shows some of these tasks which have priority:

ENERGY-RELATED RESEARCH

Systems analyses of energy utilization
Economic use of energy on the domestic and commercial level
Utilization of solar energy and construction technology
Public awareness of energy problems and corresponding behaviour

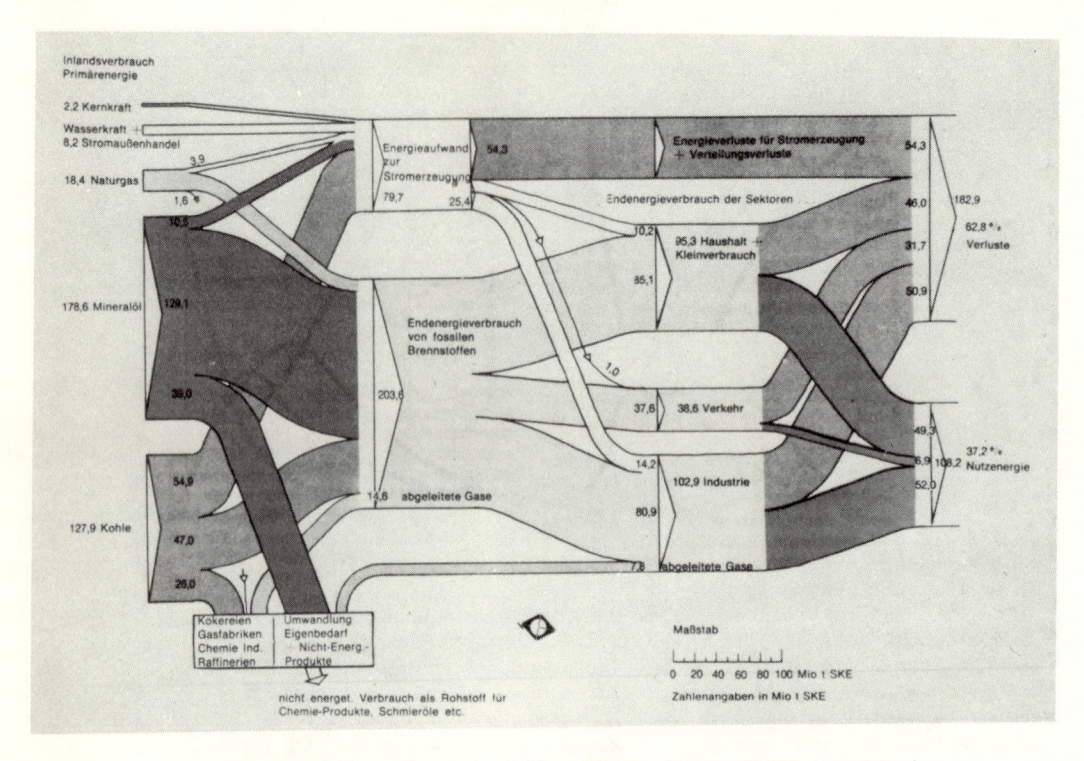

Fig. 3. – Energy Flow Diagram of the FRG in 1970 (62,8% are loss)
(BMFT: Rahmenprogramm Energieforschung 1974 - 1977
– Referat für Presse und Öffentlichkeitsarbeit –
Bonn 1974)

Increase of efficiency with respect to energy transformation and utilization
Industrial recycling
Utilization of waste heat of power-stations and processing facilities
Coupling of heat and power

Energy and ecosphere: environmental techniques (excl. traffic)
Energy and climate
Energy and sociosphere: admissible risks
Plans for use of water in general and for settlements
Local and industrial drainage
Ecologically beneficial recovery of resources
Pollution-control techniques in the traffic sector

Long-term forecasts, systems analyses, model developments
Storage and distribution of electricity, gases, liquid fuels
Production, storage and transport of heat
Refinement of coal: extraction, hydration, gasification
Substitution of fuel for traffic purposes (non-nuclear)
Hydrogene technologies

Exploration of oil, gas, and coal
Mining innovation
Nuclear propulsion systems
Nuclear-fuel supply
Development of fuel and breeder elements

Biotechnical energy transformation, photo-synthesis
Utilization of water and wind power
Fusion technology
Geothermal energy
Regional planning: site selection for power plants
Storage lake planning
Heat supply

Such a systematic registration of energy-related problems is very new in the FRG and has to be seen as a first step towards a comprehensive research. It must be the task of every scientist to look through critically the catalogue of problems and to examine his know-how whether he is able to participate in the finding of solutions.

Besides a number of applied sciences which seem to be more relevant here remote sensing could be on source to get information and data necessary for an energy distribution program.

Already today one knows a large number of investigations showing the usefulness of remote sensing, e.g. in geology, soil science, agriculture , and forestry. This success is based on the possibility to get sequential and synoptic data of reletively large areas. Therefore, besides single inventories, the monitoring of dynamic phenomena even over longer periods of time will be possible.

Now one has to examine these experiences critically to get ideas how useful remote sensing can be with respect to the complex field of energy supply.

It is the task of this paper to gather out of the far-reaching spectrum of groups of investigators in remote sensing in the FRG those results and their applicability, which can be classified belonging to the shown catalogue of individual problems. In the following some of these results will be presented. In detail will be seen the capacities of remote sensing applied to:

1. registration of the influence of warm cooling water on the ecosystem of a river

2. examination of the influence of cooling towers on the environment

3. examination of climatological problems caused by settlements

4. conception of a heat atlas

5. exploration of usable resources

INFLUENCE OF WARM COOLING WATER

At the planning of power plants there is one important problem - the impact on environment. Of special interest is to learn about the influence of warm colling water on the ecosystem of rivers.

In order to get some experiences in this field, the DFVLR (German Aerospace Research Establishment) performed an airborne measurement program together with the BfLR (Federal Institute For Land Use). This program should show how far remote sensing can contribute to find and to determine standard values, critical values and indicators for water loads.

As test area the middle part of the Saar valley was chosen (Fig. 4). This region is characterized by high density of settlements and of traffic. Especially the agglomeration of industry (mines, iron works, power plants) describes this area.

At the beginning of this measurment program there existed a comprehensive knowledge, gained by conventional field measurements, about thermal phenomena by fedding warm water into the river. Therefore, we have had an optimum basis

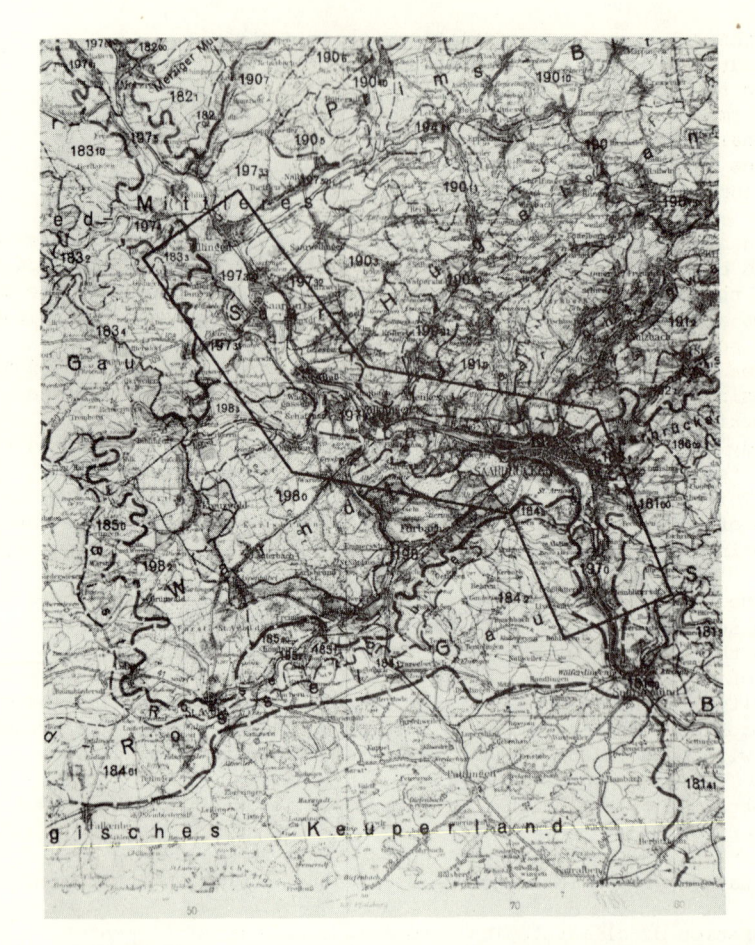

Fig. 4. - Saar Valley
(Landekundliche Luftbildauswertung im mitteleuropäischen
Raum - Selbstverlag der Bundesforschungsanstalt für
Landeskunde und Raumordnung - Bonn-Bad Godesberg 12,1974)

to examine the usefulness of remote sensing mehtods with respect to one aspect of energy problems. The sensor system consisted of an IR-Line-Scanner Reconofax VI and an IR thermometer PRT 5 respectively PRT 6 and was flown by a Dornier DO 27.

The spectral band width of the IRLS lies in the region of 8 to 14 µm (only qualitative measurments are possible), that of the PRT 5 between 8 and 14 µm too, whereas that of the PRT 6 lies between 9,5 and 11,5 µm (at lower flying altitudes the accuracy is about $\pm$ 0,4° C).

The thermal IR images obtained by an IRLS demonstrate the thermal behaviour of the water surface by feeding warm and cold water into the river.

One example for the distribution of warm water in the stream is shown in Fig. 5. On the right bank four warm inflows can be detected. This image demonstrates the fast heating effect within a small section of the river. The upstream located inflows can easily be detected. Downstreams the warm inflows do not show anymore a high contrast against their invironment because the temperature level of the river has risen.

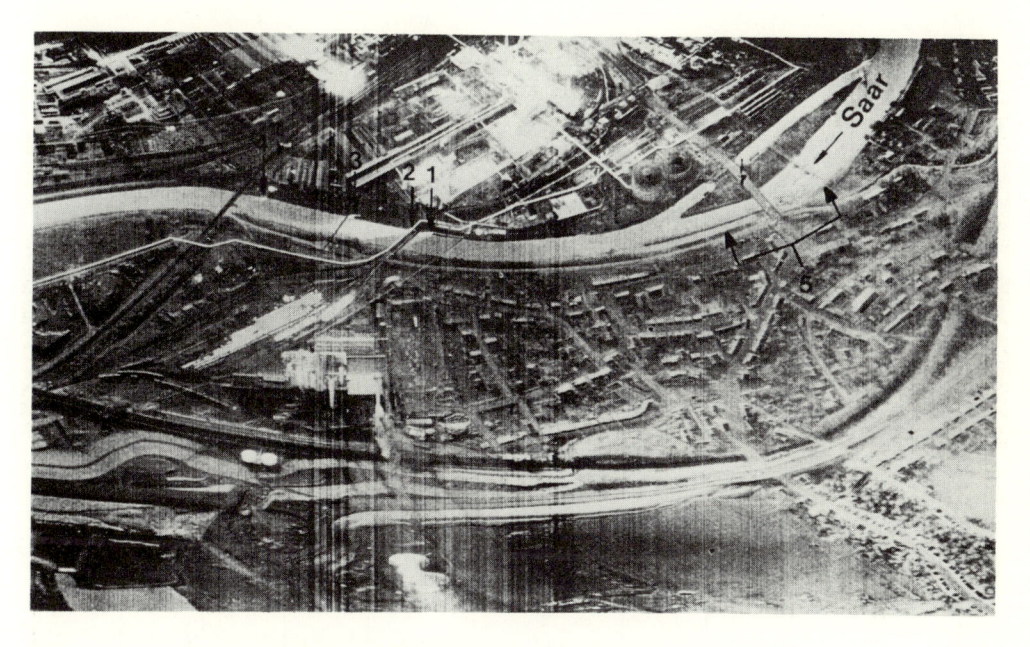

Fig. 5. - IR-Image of a Part of the river Saar,
1 - 4 Warm Inflows; 5 Foam; Image: DFVLR 2.6.1971,
5.o5 a.m. - 6.3o a.m.
(BfLR 12, 1974)

Fig. 6 shows another overflight of the same scene. Picture 5 is taken at sunrise, picture 6 in the early afternoon. Because of the high temperature difference between land and water in the afternoon, the sensitivity of the used IRLS is very rough. By this only the warmest inflows can be seen extremely clear. This may depend on the high quantity of warm water which is injected just at the time when the picture was taken.

There is another remarkable feature in picture 6: bright, which means relatively warm patterns below the dam. This can be related to foam which has a low heat conductivity and a very small heat capacity. Therefore, it appears warmer than the surrounding water, because of the sun's radiation heat during the day.

By the application of thermal images it was hoped to obtain informations on all warm water inflows. The result of this program has been very encouraging in spite of the necessity to derive the information from sequential images.

Unfortunately, the IRLS applied during this program has not the capability to resolve small and high tmperature differences simultaneously.

For comparison to above measurments the temperature readings of the PRT 6 radiation thermometer are listed in Fig. 7. In these diagrams the same inflows as in the IR-images are detectable. The most important information registrated by them is the heating-up in steps of the whole river. In a section of less than 35 km, the surface temperature rises by $10°$ C.

Thus it turned out that the methods of remote sensing provide qualitatively and quantitatively good informations on the thermal conditions of the river and about its hydrodynamics. If these results are applied on case studies aiming at the investigation of effects, which thermal pollution has on:

> water recovery, for example stronger stress with organic substances, increase of anaerobic reduction processes, higher concentration of iron and manganese;

Fig. 6. - IR-Image of the Same Part as in Fig. 5,
Image: DFVLR 22.3.1972, 2.o p.m. - 3.o5 p.m.
(BfLR: 12, 1974)

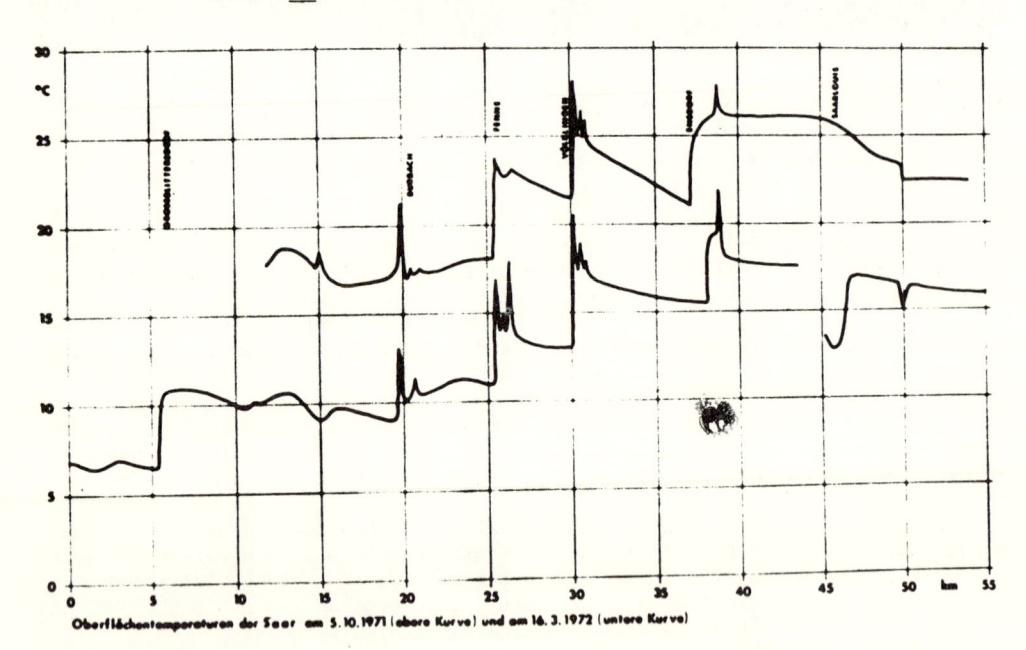

Fig. 7. - IR-Temperature of the Saar, Taken with the PRT VI
(BfLR 12, 1974)

concentration of ammonium and nitride ions, hydrogene sulfide as sulfate reduction;

purification (higher quantities of mud, of bacillus, etc.);

water distribution (less quantities of oxygene, more hydrogene sulfide,etc.)

supply with water for industrial use (more investitions for installing heat exchanger);

fishery (water temperature higher than 26° C seems to be critical for most of fish types),

it is possible to obtain an essential part of planning data for the conception of new power stations and industry over any chosen area. In a relatively short period of time also one important result is the possibility to test theoretical models of temperature distribution in a river under various frame conditions.

To proceed from the experience at the Saar the DFVLR and the BfLR flew over the river Rhine area from Karlsruhe to Worms to take thermal pictures. In this sector a great number of different types of power plants and industry are concentrated (Fig. 8).

In 1971, The Federal Workshops: Water (LAWA) published so-called thermal load plans of the river Rhine for the years '75 and '85 (Fig. 9/1o). These plans are based on theoretical models. They include in their calculations only nine warm water injectors. But the interpretation of the thermal images revealed 23 inflows (Fig. 11). Evidently, this difference shows the advantage of remote sensing techniques. Additional to these flights along the river Rhine more than 3o temperature profiles crossing the stream have been taken by using the PRT 6 IR-thermometer. With these results the author extrapolated the temperature distribution over the whole area between Karlsruhe and Worms (Fig. 12).

The comparison of the extrapolation with the theoretical results proves the high potential of thermal IR-images.

Only the systematic knowledge of the stream dynamics, its different degrees of thermal pollution, and its thermal behaviour allows together with theoretical models relevant statements.

How important it is to know about warm water inflows and to know the thermal behaviour of a river becomes evident when regarding a map of the FRG where all existing and planned power stations are listed (Fig. 13).

Biologists and botanists in the FRG warn insistently not to go on polluting the rivers. They published maximum values up to which the streams are still able to regenerate themselves. In some areas these values already have been exceeded. Therefore, one has to study new ways of getting rid of non-transformationable heat.

INFLUENCE OF COOLING TOWERS

One well-known possibility are the cooling towers. Except the possible occurrence of fog in their environement only few information on their effects

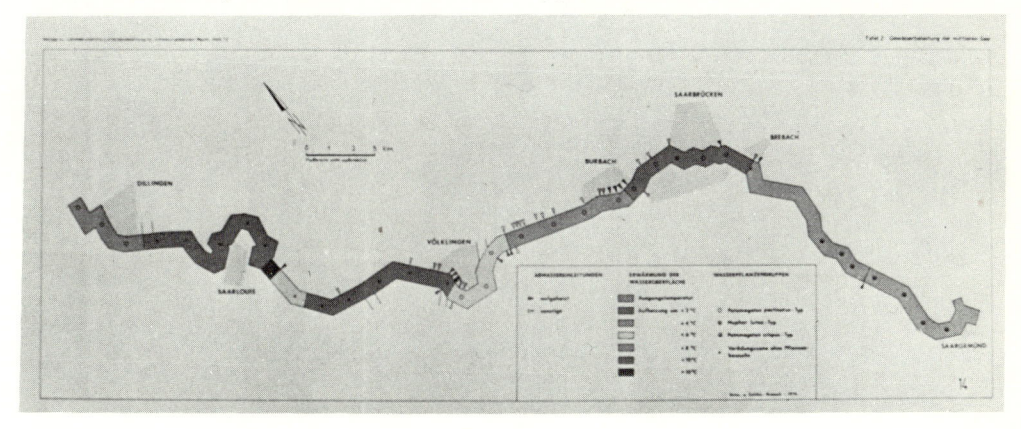

Fig. 8. - Thermal Map of the River Saar, the Different Gray Level Indicates Relative Temperature Differences Between the Sections of the River
(BfLR: 12, 1974)

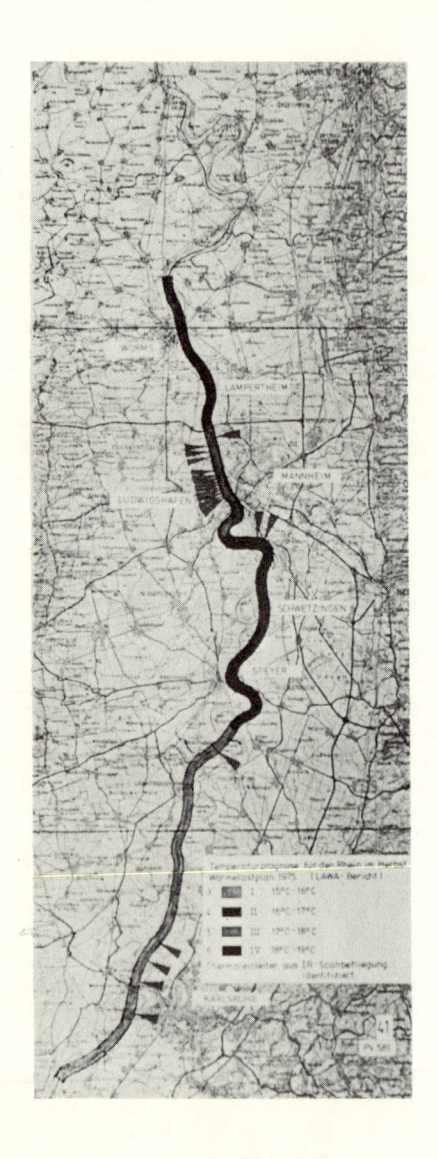

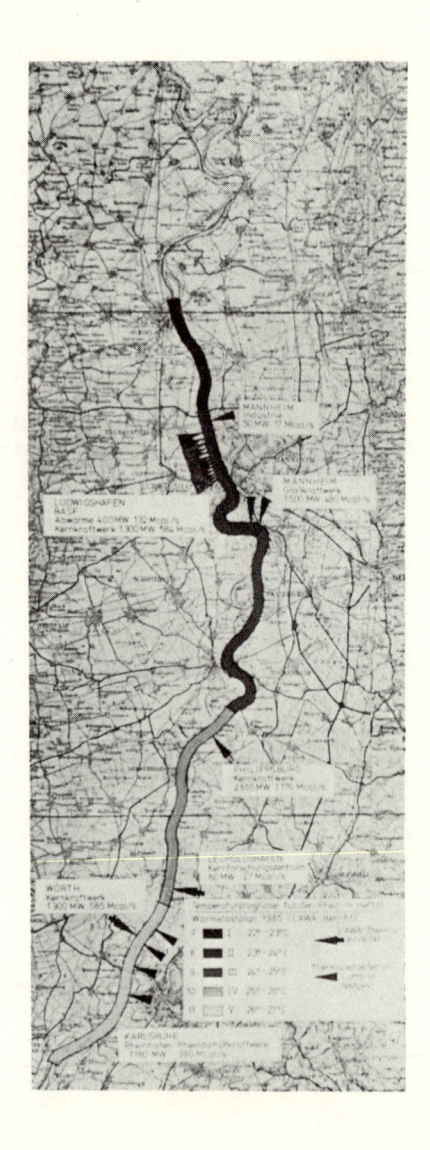

Fig. 9. - Thermal Map of the River Rhine (Between Karlsruhe And Worms) 1975
(Länderarbeitsgemeinschaft Wasser LAWA 1971)

Fig. 1o. - Thermal Map of the River Rhine 1985
(Länderarbeitsgemeinschaft Wasser LAWA 1971)

REMOTE SENSING APPLIED TO ENERGY-RELATED PROBLEMS   455

11

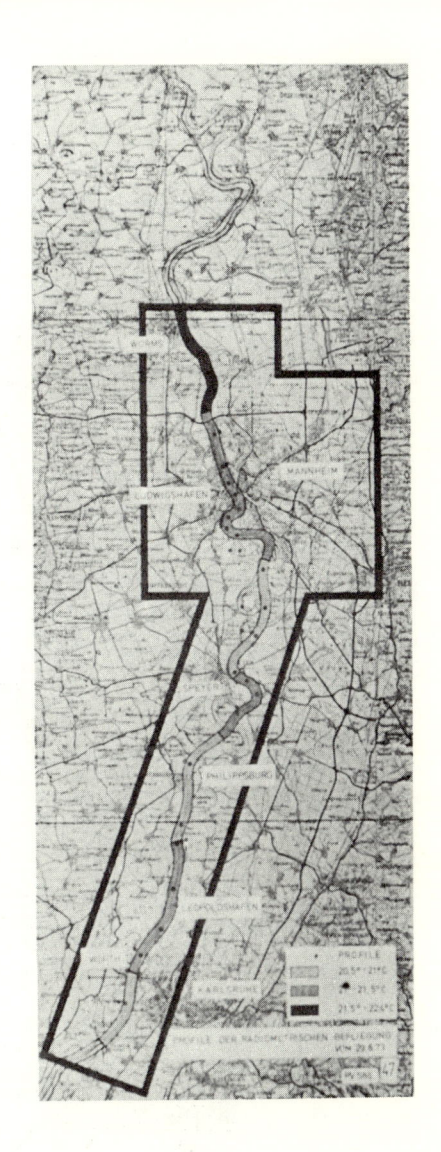

12

Fig.   11. - Industrial Areas, Thermal And Dirt Inflows, Recognized
            Out of IR-Images
            (Schneider, Kroesch BfLR 1974)

Fig.   12. - Thermal Map of the River Rhine 1973,
            the Points Indicate the Measurements of the Temperature
            Profiles Done with the PRT VI,
            (Measurements and Extrapolation of the Temperature: DFVLR)

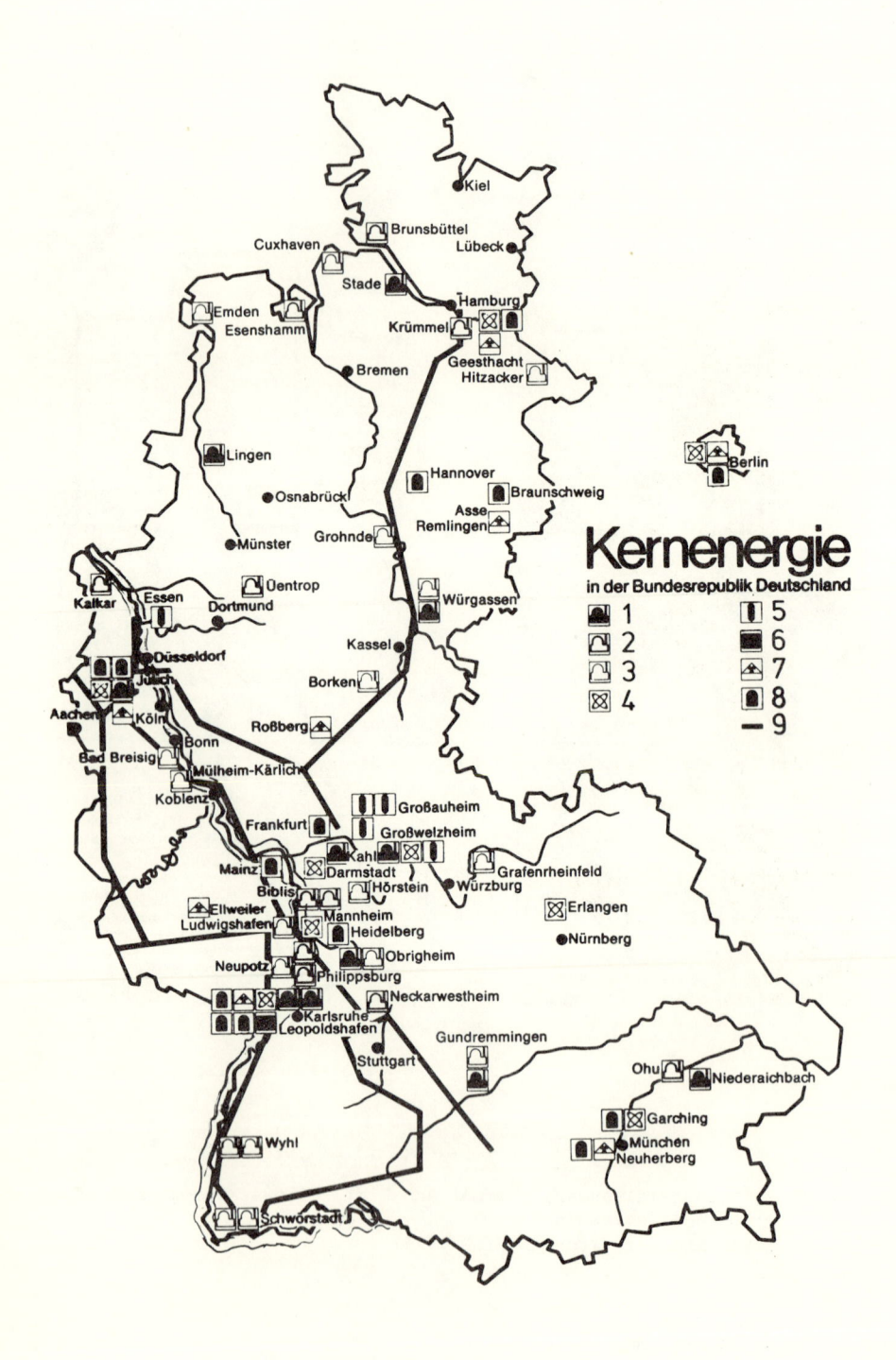

Fig. 13. - Map of All Existing and Planned Power Stations in the FRG

is known. Preliminary model measurements have been carried out by Professor Fortak in the DFVLR last year. At the Institute of Atmospheric Physics the theoretical behaviour of cooling tower plumes was calculated. These calculations have been verified by in situ measurements. For this purpose a small aeroplane equipped with thermometers and hygrometers was operated (Fig. 14-18).

For the next year DFVLR plans a measurement program at the same location but with remote sensing methods. We will take once more an IRLS and a microwave radiometer at 32 GHz. It is hoped to see if it is possible to get comparable informations as well.

CORRELATION OF CLIMATOLOGICAL PROBLEMS AND SETTLEMENTS

Without anticipating this program already today, we are able to answer questions about the possibilities of remote sensing for solving climatological problems. The influence of settlements and morphological features on air dynamics and thus on the local climate can be studied. Starting point for these questions have been problems of regional planning in the neighborhood of the

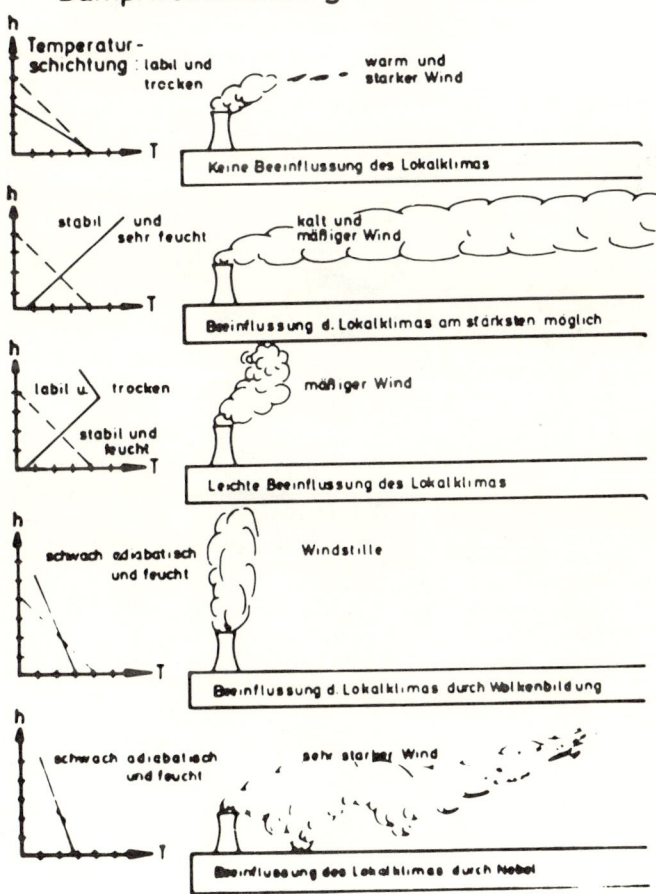

Fig. 14 - 18. - Theoretical Calculations of Relations Between Atmospheric Conditions And the Plumes of Cooling Towers (Fortak 1974)

agglomeration center Frankfurt/Offenbach (Fig. 19). In 1968, a plan for a new housing area has been established. This structure should be a new large city, including already existing villages.

The Rhine-Main region is one of the most dynamic con-urbation areas in the FRG. Topographically, the region is divided into three very different parts:

the Taunus with the Großer Feldberg (878 m) and the repeatedly intersected slopes descending to the southeast, towards the center of the region;

the Wetterau with the village areas which become very hilly in the south;

the southern plain of the river Main (alt. 1oo m) covered with wood.

The region under review contains a large number of cultivated areas.
The vegetation is multifarious corresponding to the orographical and climatic conditions.

The housing areas represent all kinds of buildings as well as trade and industry areas (refinery), conventional and nuclear power plants, airport.

The climate is determined by the basin-like shape of the region which is surrounded by the just explained topography. On 4o% of the days of a year there are atmospheric conditions with little exchange at airmasses. This leads in the critical seasons to prolonged inversions. On about 2oo evenings of a year the climate is improved by flows of cold air coming from the Taunus.

This flow of cold air was of priority importance for the regional planning, in particular with respect to the design of a new city. In order to take into

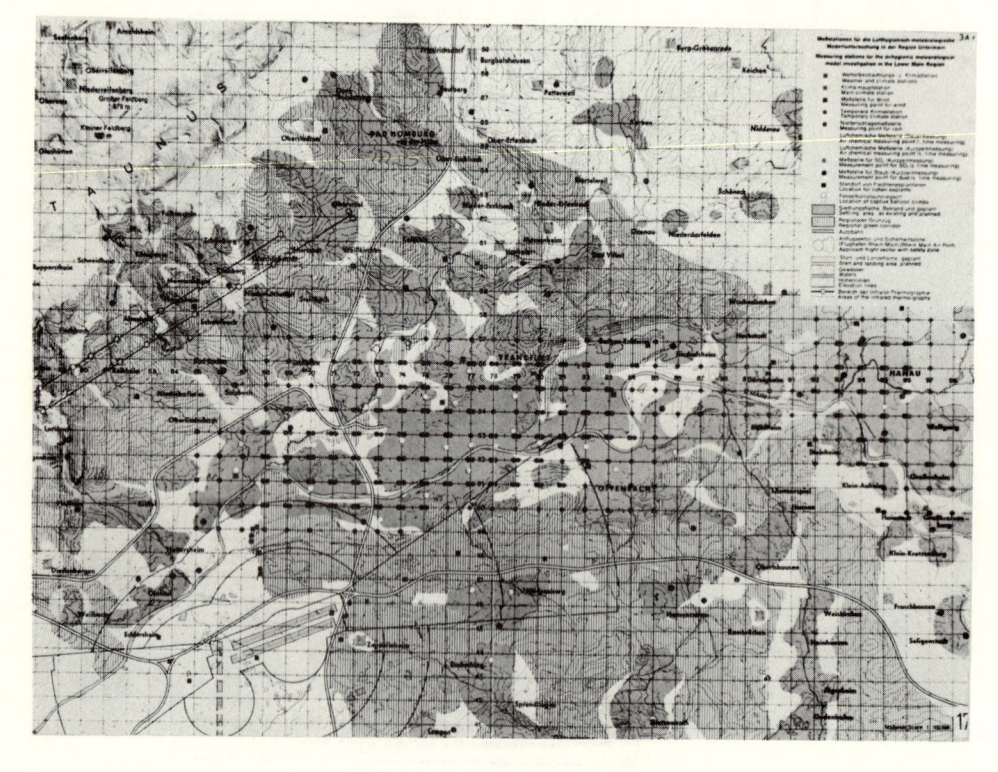

Fig. 19. - Rhine-Main Area Frankfurt/Offenbach
(Regionale Planungsgemeinschaft Untermain RPU,
3. Arbeitsbericht - Frankfurt/M. März 1972)

account the climatic phenomena and not to impede the outflow of the cold air, regional areas of green, which shall be left unsettled, have been provided for an alternative to this plan.

The registration of the air movements in a relatively large area - the region under review covers approximately 1,4oo square km - even with a very dense network of measuring stations presents substantial difficulties. Therefore it was obvious that this could not be achieved with conventional methods. For the performance of this task thermometric flights were carried out similar to those done in connection with the load on lake and rivers caused by heat waste. These measuring flights yielded the result that the formation and propagation of cold air can be detected by means of the infrared thermography. It should be noted, however, that the obtained imagery had to be interpreted with the utmost caution.

Such an image (Fig. 2o) will be demonstrated: it is a density slice representation of a thermal image which was taken in late October after 7.oo p.m. The area in question is a long, narrow valley with numerous wood-covered sidevalleys which implies a great inflow of cold air. This flow of cold air is hanpered by the narrowness of the valley, the low inclination of the bottom of the valley as well as by several bends, which entails low surface temperatures extending reletively far up the slopes, partly reaching the wooded sections. This image shows different shades of colour representing in a first approximation the surface temperatures; these shades go from red (warm) via yellow and green to blue and violet-black (cold).

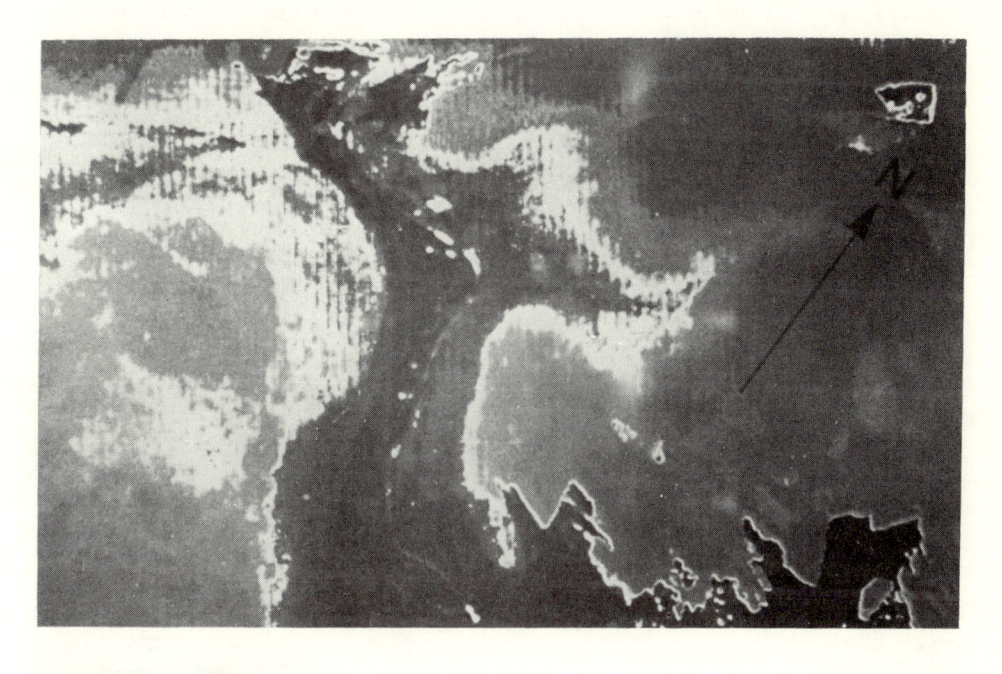

Fig. 2o. - Equi-Density Picture of the Schwarzbachtal
(Völger, Oktober 1971)
(RPU, 3. Arbeitsbericht - Frankfurt/M., März 1972)

This example stand for a large number of similar cases. Together with the data obtained from the meteorological measurements performed by the ground stations they led to the image 21. These results have been very important for the further work of the regional planning group. But it should not be the task of this report to analyse it in detail. The above results are summarized in the following map of the Rhine-Main region showing a midnight distribution of temperatures in summer at weather conditions characterized by little exchange Fig. 22).

After widespread experience had been gathered in the FRG with respect to the applicability of remote sensing methods, e.g. to the problem of the impact of men's need for energy on the ecosphere, a new project was defined last year, the so-called airborne measurement program.

This experimental program is intended to combine for the first time the following key activities with respect to the complex analysis of energy supply:

systematic data collection;

comparison of different sensors and/or selection of sensor combinations;

investigation of scale effects;

selection of suitable spectral ranges;

selection of favorable dates and hours for the surveys as well as

correlation of the measured data as shown on the images and the values measured on the ground;

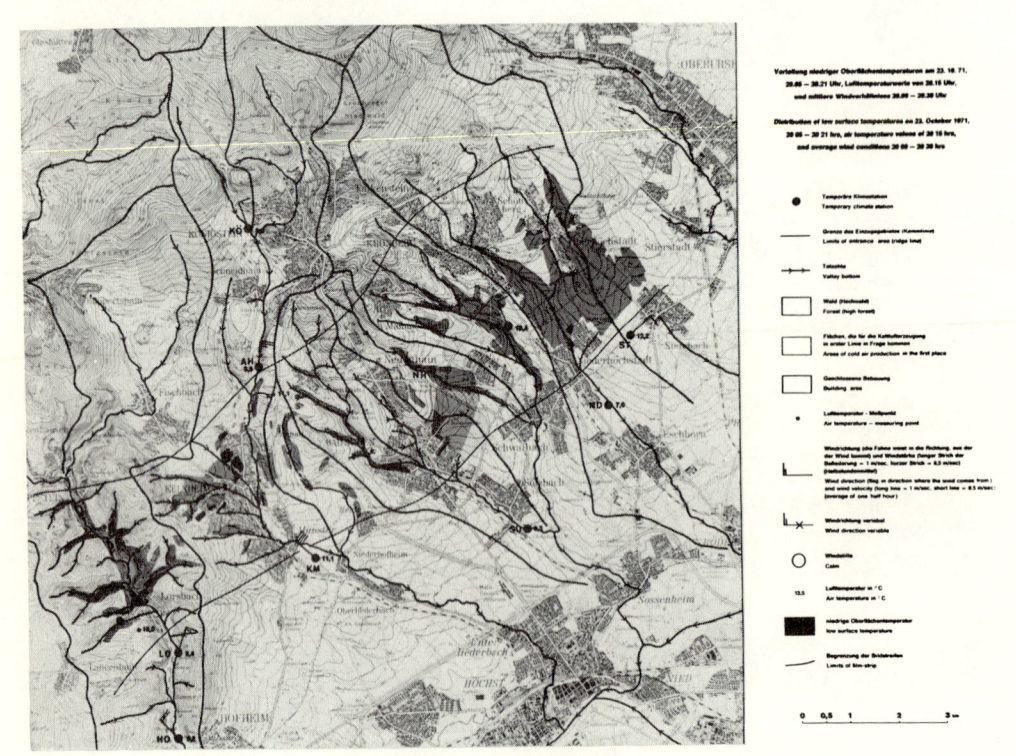

Fig. 21. - Distribution of Low Surface Temperature in October 1971
(RPU, 3. Arbeitsbericht - Frankfurt/M., März 1972)

Fig. 22. - IR-Thermogram of the Rhine-Main Area in October 1971
(RPU, 3. Arbeitsbericht - Frankfurt/M., März 1972)

determination of quantitative interpretation models for image evaluation by means of

analogue and digital image processing.

Therefore an aeroplane will be equipped with the following sensors:

six Hasselblad cameras 5oo EL/7o M,

one Hasselblad camera MK 7o,

one Zeiss cartographic camera RMK 8.5/23,

one 11-channel line scanner,

one 2-channel IR line scanner (Daedalus, Bendix),

one precision radiation thermometer PRT-5,

one ERTS-compatible radiometer (Exotech, Bendix),

one 14-track magnetic tape-recorder.

The Rhine-Main region around Frankfurt, already dealt with in this report, was again selected as test area. It is to hope that this experimental program will give answers to a great number of questions caused by the energy problems.

HEAT MAPPING

The possible application of IR-surveys include another very important sector in connection with the power supply, the thermal radiation of buildings, housing areas, cities. In the following, some examples will be given of this sector.

The first step of such IR-surveys may be that of a block of houses. The reason for this survey could be the tenant's complaint about too high heating costs. Thereupon the owner of the houses orders the taking of thermographs for the purpose of identifying possible insulation defects or leaks of the district heating.

The next step comprised IR-surveys of a whole quarter, here for example of the center of Rheinhausen (Fig. 23).

On this thermograph, the lighter innermost part of the town clearly contrasts with the other parts. A comparison with the Pan-film (Fig. 24) made of the same object shows the interrelationship between the building density and the heat radiated from houses and blocks.

Fig. 23. - IR-Image of the Center of Rheinhausen in October 1971 (Luftaufnahmen, Schriftreihe Siedlungsverband Ruhrkohlenbezirk SSR - Essen 44, 1972)

Fig. 24. - Pan-Film of the Same Scene as in Fig. 23. (SSR, Essen 44, 1972)

The logical continuation of these surveys are the so-called thermal cadastral surveys. They represent the thermal distribution of a large area, e.g. of the total FRG. If these thermal distribution surveys are taken in different seasons, as a function of climatic conditions, possibly even at different hours of a day, and joined in a so-called thermal atlas, the basis is established for a large field of projects, such as planning of district heatings, distribution of thermal power plants, plans for agricultural cultivation and for afforestation, housing projects, etc.

The following picture of the Ruhr region is an example of such a thermal cadastral suvey. A thermal cadastral survey of the FRG is just being started (Fig. 25).

The next picture (Fig. 26) shows how important the registration of thermal radiation can be. It is an ERTS image of the balley of the Rhine with the whole valley being covered with fog except for some black spots. A correlation of this image with a map yielded the result that these spots are identical with towns. The waste heat rising from these towns thus breaks through the fog.

EXPLORATION OF USABLE RESOURCES

Another task in the overall field of power supply is the examination of possible applications of remote sensing for the exploration of usable deposits of primary energy sources. In the past, deposits were localized by means of local inspections, soil analyses, surveys, and, at a more advanced stage, interpretations of aerial photographs with geological, tectonic, lithological and geobotanic informations serving as indicators.

Meanwhile, a large number of papers are available dealing with the identification of deposits by means of satellite images and/or data obtained from airborne measurements. In spite of these inspiring informations the judgement of the experts in Germany on the possibilities of remote sensing for the FRG is rather negative. The reasons are quite obvious: those systems which are

Fig.   25. - IR-Image of the Ruhr Area, 25.4.1973, 9.oo p.m.
             (SSR, Essen 44, 1972)

Fig. 26. - ERTS - 1 Image of the Rhine-Main Area, 21.9.1972

available so far describe but the state of the earth's surface; even the used microwave sensors penetrate only little beyond this surface. In order to localize deposits by the use of data obtained with these sensors one has to employ the indirect indicators mentioned above.

For a deposit-specific digital interpretation such a huge number of prior individual examinations have to be carried out that today a success does not seem probable.

In spite of above mentioned critical attitude, several groups of scientists in the FRG are trying to evaluate ERTS-images under more geological aspects. These efforts have shown that a new knowledge on lineament patterns could be derived.

For example ERTS data have revealed that the structural system of the "Rheinisches Schiefergebirge" is much more complicated than known until today (Fig. 27, 28). Many of the known ore bodies of the "Rheinisches Schiefergebirge" are closely related to tectonic structures; therefore it has to be examined if new ore bodies can be mapped by means of the new knowledge of structures.

In the region of the West German coal mining area (Münsterländer Bucht) also new lineament systems could be traced. With respect to their areal occurrence and their orientation they could partly be related to known fracture systems of the deeper carbonic layers.

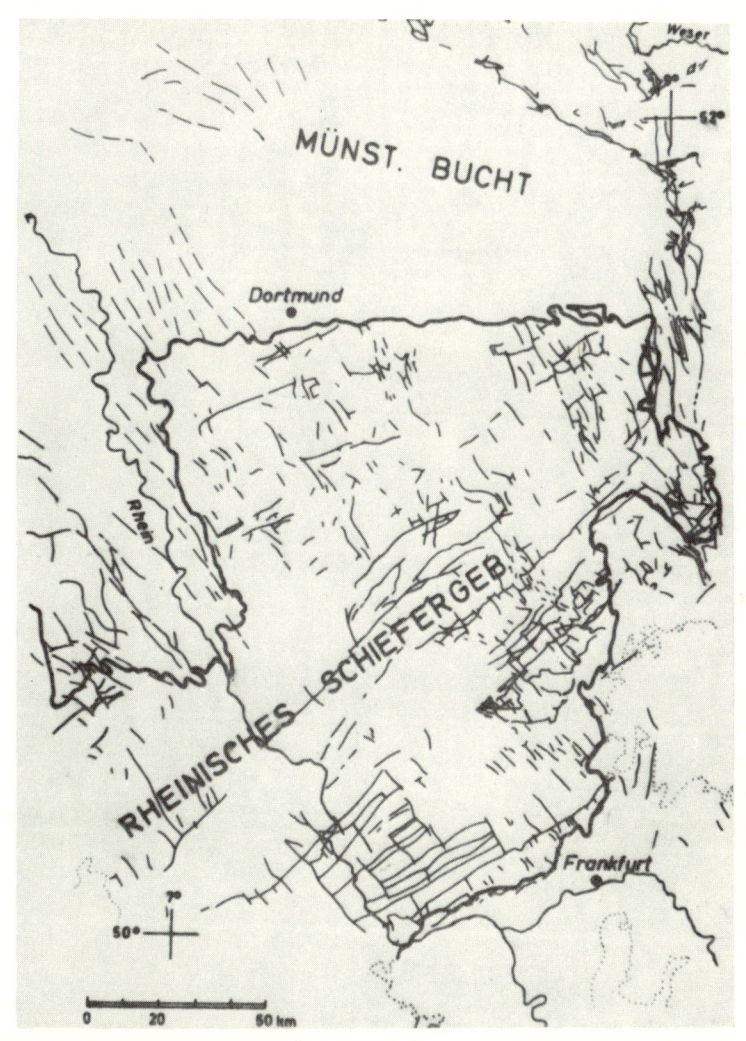

Fig. 27. - Known Tectonic Structure of the "Rheinisches Schiefergebirge"

By this, lineaments indicate structures through mesozoic and tertiary layers of 2ooo meters depth.

This example offers the high potential of synoptic satellite data for the application-oriented utilization if unknown fracture systems can be really identified.

Investigations on above relationship between lineament patterns and the predictions of fracture systems are carried out by the Geological Institute of Clausthal (Adler, Kronberg).

Also for hydrologic projects in the Ruhr valley (Fig.29),the knowledge of such structures is of great importance. The a priori knowledge of fractures is of high economic importance to the mining activities.

Besides the more or less conventional evaluation of ERTS images there are new aspects to detect the hydrocarbonic deposits. It is known that several hydrocarbonic deposits are degasing to a certain extent. By diffusing through the overlaying geological units traces of methan can be detected at the surface.

Fig. 28. - New Tectonic Structure of the Same Area from an ERTS - Image
(Interpretation: Kronberg, 1974)

Sometimes the methan concentration of the surface - near airlayer reaches the scale of percents (Klitsch).

For detection of trace gases the laser absorption is usable. The emission frequency has to be identical with the rotation frequency of the gas molecule to be detected. The absorption of the laserlight is a function of the gas concentration. For the detection of methan a helium-neon laser system is used. The emission of He-Ne-Laser corresponds to the absorption frequency of methan.

Thus a possibility is given to apply a direct measurement technique for the prospection of hydrocarbonic deposits. Research in this field is carried out by GfW. On the basis of this technique smimilar investigations with CO_2 lasers aim at the detection of spectral signatures.

If we compare the state of the art in remote sensing with the catalogue of problems concerning energy supply it becomes evident that remote sensing can contribute to a number of applications. Therefore, basic research in remote sensing has to be intensified in order to learn about its full potential.

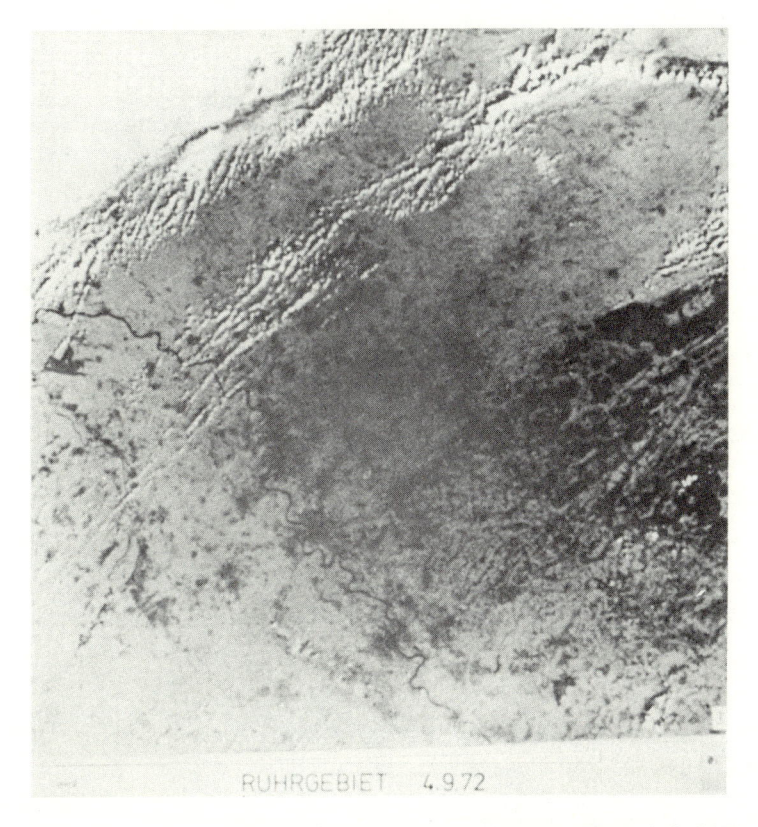

Fig.   29. - ERTS - 1 Image of the Ruhr Area, 4.9.1972

The consideration of sun energy as an energy supplier became more and more important in the last few years. Remote sensing, for example, could provide us with information on the global radiance balance as a basic information for the planning of sun power plants. A further consideration could be a world-wide mapping of wind field patterns for the construction of wind power stations.

The location of optimum sites for dams has been successfully carried out on the basis of remote sensing methods for some time.

OUTLOOK

The application-oriented examples mentioned in this report should demonstrate the multidisciplinary efforts of remote sensing activities in the FRG.

The multidisciplinary charcter of remote sensing in general led to a definition of programs which are supported by a large basis of scientists and potential users.

Furthermore, the coordination of all activities involved has to be performed by an institution which has the capability to plan and to implement application - oriented projects. In the field of German remote sensing activities the German Aerospace Research Establishment (DFVLR) assumes this task. This institution has the experience and the capability in the fields of sensor development and sensor handling, data acquisition (several aircraft programs) and data handling. It has already carried out several satellite projects, feasibility studies, and system analytical projects and therefore it is the optimum promoter for upcoming remote sensing programs.

The activities on remote sensing have to be regarded as one part integrated within the scope of the German energy supply program. By a close collaboration with German research institutes DFVLR is responsible for the realization of the frame conditions of a long-term energy supply program created by the Federal Ministry of Research and Technology. On the basis of feasibility studies the approach towards these problems will be:

to analyse various energy systems,

to perform several energy supply models for the FRG,

to evaluate model alternatives with respect to desired benefit and possible impact on human environment.

The realization of these efforts can be, will be, and has to be supported by modern remote sensing technology.

LITERATURE

BMFT: Rahmenprogramm Energieforschung '74 - '77

Hammond, A.L.; Metz, W.D.; Maugh II, T.H.; Schultze, H.: Energie für die Zukunft, Umschau-Verlag, Frankfurt/Main, 1974

Schneider, S.: Gewässerüberwachung durch Fernerkundung, Heft 12, 74

Kroesch, V.: dito

Lorenz, D.: dito

Schneider, S.; Kroesch, V.: unpublished work

Fortak: unpublished work

Hesler v., A.: Lufthygienisch-meteorologische Modellunteruchung in der Region Untermain,
Regionale Planungsgemeinschaft Untermain, Frankfurt/M., 1972

Völger, K.: dito

Lorenz, D.: dito

Länderarbeitsgemeinschaft Wasser (LAWA), 1971

Hirt, F.H.: Schriftreihe Siedlungsverband Ruhrkohle, Heft 44, 1972

Englisch, W.: Anwendungsmöglichekiten der Laser in der Raumfahrt, Battelle 1973

Steinmann, R.: unpublished information

Kronberg, P.: Umschau 74 (1974), Heft 15, S. 469 - 481

Projektplan Flugzeugmeßprogramm 1974

PART VI

WORKSHOP REPORTS

ACTIVE SENSOR APPLICATIONS-
WORKSHOP REPORT

C. L. WILSON
Bendix Aerospace Systems Division
Ann Arbor, Michigan, U.S.A.

HARRY V. SENN, *Co-Chairman*
Remote Sensing Laboratory
University of Miami
Coral Gables, Florida, U.S.A.

Participants: J. Byrne, P. Claybourne, H. Hiser, H. Senn and C. L. Wilson

RECOMMENDATIONS
A. RESEARCH AND DEVELOPMENT

1. Active microwave sensing techniques are at the same stage of development, as far as activity specifically directed toward remote sensing is concerned, as passive scanners were ten years ago. An R & D equipment similar to the development of the NASA 24 channel multispectral scanner should be developed in the active microwave portion of the spectrum to explore the information content available.

2. There is a need for interactive modeling and sensor development activity to simultaneously develop both technology and understanding of the phenomenology.

B. APPLICATIONS

To date, the field of remote sensing has coasted by extrapolating applications from a base of existing (military, etc.) hardware. What is needed is an assessment of information needs, not as hardware performance specs, but in the context of an "applications scenario" that defines requirements which can be reflected both into sensor needs, methods of extracting information and methods of disseminating information to end users.

ENVIRONMENTAL QUALITY MONITORING-
WORKSHOP REPORT

JAMES D. LAWRENCE, Jr., *Chairman*
NASA Langley Research Center
Hampton, Virginia, U.S.A.

SUBRATA SENGUPTA, *Co-Chairman*
University of Miami
Coral Gables, Florida, U.S.A.

Participants: J. DeNoyer, W.B. Foster, A. Ghovanlou, T.R. Heaton, K. Van Liers, R. Withrow

The group spent most of the time talking about future research directions. It was a free-wheeling session, without a definite agenda or fixed priority of topics.

AIR POLLUTION

In the area of air pollution we had a number of people interested in mathematical models. They quite properly influenced our discussions. Dr. Sengupta from the University of Miami pointed out that it was imperative that remote sensing efforts should be closely coupled with mathematical models. There are a large number of models which have been developed for various regions of the country. Unless these are calibrated, verified and tested any statement regarding their validity is premature. It was felt that in applying remote sensing to air pollution priority must be given to determination of vertical profiles of constituents. This is a very complex problem, but an important one.

The need for an extensive data base was recognized as very important. The modelling efforts under way cannot be verified without a nationwide data base.

The stratosphere and the effects on it of freon was discussed in detail. The development of remote sensors to identify and measure important factors like the chlorine and hydronyl radical was considered important.

WATER POLLUTION

The water pollution problem is far more complex than air pollution. The quality of water and its influence on ecosystems is difficult to define or measure. While safe levels of air pollutants have been defined the pollutants in water act more subtly. The effects on food chains and species distribution are often slow and long term though important.

For effective application of remote sensing techniques better understanding of theoretical properties of pollutants is important. Careful programs to identify the spectral signatures seem imperative. Oil slicks are fairly easy to detect but characterization of oil by type is required for better enforcement of standards.

Thermal pollution is the most easily sensed part of water pollution. Infrared radiometers are adequate in surface temperature measurements. For modelling efforts vertical profiles of temperature are important. Remote sensing techniques to determine underwater temperatures is still in an experimental state.

The need for more detailed models as well as an extensive data base was generally recognized.

LAND POLLUTION

It was difficult to define land pollution. The ERTS and aircraft multi-spectral scanners are quite suitable for monitoring land pollution associated with oil-shale mining etc. It was recognized that better techniques have to be developed for stone runoff water and the contamination of stone runoff water by land.

ITEMS NOT COVERED IN SYMPOSIUM

Some members of the working group expressed somewhat of a disappointment that more Skylab data was not presented at the conference. There was also a common feeling that data collection platforms and their relation to satellites should have been discussed in a paper. The three dimensional modelling effort of the University of Miami thermal pollution group merited a significant amount of attention. It was felt that a paper relating the modelling effort to remote sensing should have been there. Dr. Sengupta indicated that such a paper will be included in the final proceedings.

COMMENTS

The necessity for improved visual aids was pointed out. Dr. John DeNoyer expressed the point that we did not really stress strongly enough the complementary nature of in-situ sensors in relation to any remote sensing effort. Remote sensing cannot do it all and caution and intelligence have to be applied when using remotely sensed data.

LAND USE AND RESOURCE MONITORING WORKING GROUP REMOTE SENSING APPLIED TO ENERGY RELATED PROBLEMS SYMPOSIUM

A. T. JOYCE, *Chairman*

Chief, Land Remote Sensing Applications Group
Earth Resources Lab, NASA
Bay St. Louis, Mississippi

A. L. HIGER, *Co-Chairman*

Research Hydrologist
U.S. Geological Survey
Miami, Florida

Participants: M. Gallego, J.H. Senne, M.T. Hyson, W. Cheng, U.R. Barnett, A. Higer, C. Holladay, R.K. Stewart, O.R. Citron, J.D. Marino, R.H. Rogen, R. Hall, P. Poonai, W.J. Floyd, J. Gliozzi

This Committee has divided their topic into two categories: 1. Land-based raw material sources for energy, and 2. Comprehensive base-line information.

Addressing the Land-based raw material sources for energy, first; The statement of the problem is that it has two aspects: 1. the discovery and inventory of energy raw material sources (oil, shale, petroleum, coal, uranium, etc.) and 2. information needed for planning and design of exploitation and subsequent monitoring of reclamation sites and their restoration and environmental consequences. The group felt that we came, hoping to hear some startling revelations about the use of remotely sensed data leading to the immediate discovery of resources, but what we actually heard, that possibly the use of remotely sensed data is more applicable to the second aspect, information needed for planning and design of exploitation, and we use the phrase "planning and design" to cover both engineering and environmental considerations.

DISCOVERY AND INVENTORY OF ENERGY RAW MATERIAL SOURCES

Spacecraft acquired imagery is especially useful because it is of a synoptic and repetitive nature, thereby adding two dimensions that were not previously available to photo-geologists. The utility of the imagery is well documented in respect to editing and our revision of small-scale geological maps. Remotely sensed data should not be viewed as a panacea to exploration, but rather as a tool that can be used for basic studies in methodology and structural geology that logically lead to discovery. Although the full extent of the use of spacecraft acquired imagery and actual exploration is not known, the fact is, that the petroleum industry is the largest single purchaser of spacecraft acquired data. A computer-based method for enhancing ERTS imagery through band raticing to bring out effects of surface alteration has been developed which may prove to be a breakthrough in minerals prospecting. Remotely sensed data is not of great importance to planning and design of coal resource exploitation. Oil shale concentrations in western states are mainly in sparcely settled areas, but environmental base-line information can be partially furnished by remotely sensed data and it is, at the present time, scanty.

Of prime importance to western oil shale areas is availability in use of the water resource and the effects of the oil shale developments on cattle grazing activities and wildlife habitat. Although ground studies are ideal for environmental base-line data, the vastness of the western oil shale areas, coupled with the urgency of quantitative data, indicates that emphasis the need to be given to remotely sensed data. Computer implemented techniques using ERTS digital data have been useful for studying strip-mining activities and consequences of those. Multistate sampling concepts may also prove to be a useful means of using spacecraft acquired data to minimize aircraft and ground data collection for environmental baseline data. More work is needed for improvement of techniques for image enhancement of alteration zones, the identification of rock and soil types through thermal inertia techniques and the latter may also have some application to geothermal exploration.

Attention needs to be given to active microwave systems for advantages of power concentration and possibly of foilage concentration.

COMPREHENSIVE BASE-LINE INFORMATION

The statement of the problem is as follows: land use information alone is useful, but its utility can be greatly enhanced when it is integrated with other types of information, such as vegetation, soil characteristics, surface drainage, sub-surface drainage, slope aspect, landform, population, ownership, etc. There is a basic need for systems that are capable of integrating various types of information derived from both remotely sensed data and other data within a geographically referenced framework.

Some of the principal anticipated uses of information systems such as described in relation to energy problems are:

1. Facility Site Selection. This may be the site of nuclear plant, dam, resevoir, steam plant, routing of transmission lines, pipelines, etc. The group felt that oftentimes such site selection in the past has been made in consideration of political and economic consideration, but more emphasis should be given to furnishing the information that we have described, so that it could be also integrated into the decision of siting.

2. Input to Hydrological Models. Remote sensing information can be a useful tool in determining runoff calculations and snowpak extensiveness. Although the hydrological models, have always been used as a predictive tool - where the water is going to come from to fill reservoirs, the level of water in reservoirs, how the water is going to be used, etc. Remote sensing will take on a more important role in the use of these models, now, inasmuch as the oil shale development will also require us to face problems of transporting, storing and using water.

3. Determination of Land Capability, Ecological Units and Subsequent Land Use Alternatives. In the context of what was stated previously, relative to the planning, design and exploitation, you may have a question as to what does land capability have to do with energy problems. The past trends in agriculture have been to intensify the yield in given land unit areas and this intensification has come about because it is capital intensive and energy consuming intensive, especially with respect to heavy fertilizer imports as well as chemicals for herbicides and insecticides all of which, especially the fertilizer, are a large energy consuming technique. We are suggesting, that if the energy crisis becomes more serious, that there might be a reversal in this trend. There may be reversals to more extensive techniques. The question is: do we have the land for increasing food production in that manner, and if so, where is it?

4. Information for Environmental Impact Statements and Continued Monitoring. We can't do anything in the way of developing resources without an environmental impact statement.

5. Change Detection and the Use of Remote Sensing for Analysis of Up-Date Information. This relates to the first four points previously mentioned.

The Summary of the Statements are:

Most existing information systems are basically information storage and retrieval systems, therefore, principally involves the development of computerized information systems that are capable of integrating, correlating and combining data, supplying data for the operation of simulative and/or predictive models and the output of information from remotely sensed data, so the converse of this is a need for standardization in respect to classification in respect to systems, extraction systems, etc. However, we do suggest standardization relative to the criteria and the information input. Orbiting spacecraft acquired such data as ERTS and SKYLAB data is useable and especially applicable to coverage of extensive areas, both for the data base and base update and for change detection and analysis. Aircraft acquired data is most applicable to selected area coverage of high population size and intensly developed areas and/or special problem areas. Considerable advancement has been made during recent years in information extraction based on multiband imagery and image enhancement techniques, but the basic problem is that with these techniques, a second step, digitization, must be taken before such information can be handled by computerized information systems. Significant advancement has been made in computer implemented information extraction utilizing digital data collected by multi-spectral scanners. The advantage is that this technique, as opposed to the other is very compatible with input to computerized information system, the major limitation is that the necessary hardware and software is not yet widely used. More work is needed with computer implemented information extraction techniques to incorporate spacial parameters with spectral parameters and for simultaneous processing of two or more data sets.

Attention is needed for sun angles, scan angle and geometric corrections in aircraft acquired multi-spectral scanner data. Both image information and digital data information extraction techniques will be needed to get the most complete information from remotely sensed data. ERTS type spectral resolution is adequate for most information extraction purposes. There are indications that spectral data from the thermal region spectrum especially 8 microns to 14 would be useful for land-water categorizations. This would obviously require atmospheric correction to spacecraft acquired data at this stage. ERTS type spacial resolution is adequate for most information covering rural areas, but future increase of spacial resolution and spacecraft acquired data to the 10 to 30 meter category is deemed desireable after advancement in data handling techniques come about. Spacial resolution requirements in the less than 10 meter category are likely to be best satisfied with aircraft acquired data. The data collection platform as now is operating with ERTS institutes parameter measurements on the ground and the satellite is used as a relay to centralized receiving stations. This system should be used to compliment the multi-sensed data with ground-based point specific data. And, finally, some miscellaneous comments that do not fit either category, but both of them. There is a need for faster turn around in the delivery in spacecraft acquired data. There is a need for better than 18-day cycle coverage for many applications.

OCEANOGRAPHIC AND HYDROLOGIC MEASUREMENTS FROM SATELLITES- WORKSHOP REPORT

S. FRITZ, *Chairman*

National Environmental Satellite Center
National Oceanic and Atmospheric Administration
Suitland, Maryland, U.S.A.

N. WEINBERG, *Co-Chairman*

Professor, Department of Electrical Engineering
University of Miami
Coral Gables, Florida, U.S.A.

Participants: W.B. Briggs, J.D. Byrne, S. Ferdman, J.C. Gervin, K.L. Goldman, J.J. Horan, F.R. Krause, K. Krishen, J. Sansonetti, A. Sieber, K.W. Wanders, W.L. Webb, J.J. Woodruff

CALIBRATION OF SATELLITE MEASUREMENTS

Methods of calibration which are not routine even on the ground become more important and difficult when attempted from a satellite. Data calibration has a high priority and even when it is not as satisfactory as one desires, the data is still usable. However, the value of the data is reduced if the calibration has to be adjusted empirically after the launch which now is done mainly with soundings. Composition was not discussed much in this session since another group would discuss this, but ozone was mentioned to a certain extent. Aside from the question of ozone due to pollution, e.g., the supersonic transport, it would be important from the ecological point of view to measure the ozone distribution and its changes over time. To a certain extent, the ozone which absorbs ultraviolet determines the temperature distribution in the stratosphere, and the motions which are somewhat independent of that may be determined even more. Some interaction does give the basic properties of the stratosphere, and this should be measured from time to time. Over long periods the effect on the lower atmosphere of a change in the stratosphere is not known if it persisted a long time and had different distribution at one time or another. While it is expected to be investigated from a pollution point of view, we thought it was necessary to say something about this.

Another point in this discussion of satellite data is the investigation in the development of thunderstorms which occasionally give rise to tornados. These are relatively short-lived phenomena, of the order of a few hours duration, as seen on this cloud picture. Since this occurs in thunderstorms, we try using pictures to detect the beginning of thunderstorms early which sometimes develop tornadoes. Although the general picture is known, the large-scale synoptic picture in which tornadoes are likely to occur, we cannot pinpoint it. Hence, the forecast has to say "the liklihood of tornadoes over the midwest, within the next twelve hours is . . .". We would prefer to be able to state "it is going to occur in eastern Kansas within the next two hours." The satellite, focusing on this area where tornadoes are likely to occur, can look at more frequent intervals than 25 minutes, let's say every 5-10 minutes. We have that capability even now, and we do it. The development of clouds is found even before the radar detects it, and we start looking at the growth of the clouds as they start building and spreading on top with an attempt to give a better time and space forecast. Although this technique is now being used, the percentage of success is unknown. Looking at the future, Dr. Fujita, the University of Chicago, has been studying tornadoes by flying over big thunderstorms. He finds that there is a pulsating cloud at the top, with periods of just a few minutes, 5-10 minutes or less, with a change in a limited area. This change in the thunderstorm type cloud of a few kilometers which rises up and down is of the order of a few kilometers, about two kilometers. If this is important in tornado developing situations, it would be of interest to focus on these particular areas by using very high spacial resolutions and very high temperature resolutions to look for changes in the cloud top for short periods of a few minutes. If we could do this, obviously it would be of considerable significance since tornadoes are one of the disasters that strikes the midwest every year.

In regard to sea-surface temperatures some of the present techniques were brought out. We now produce maps every day, using the 10-12 micron window on the NOAA satellites. The clouds are a big problem, and when its overcast nothing can be done. Under partly cloudy conditions data can be obtained. Water vapor is a big problem. As you heard, if you get 4 centimeters of water, you may have 8 centimeters of reciprocal water overhead. This means a very large correction is necessary, about $8°$ or $10°$ to the measured radiation, since the radiation measured is colder than the sea-surface temperature even in the window. When such large corrections are needed, it seems obvious that the accuracy would not be very high although the objective is to obtain measurements within $1°$. By and large, the accuracy is not precisely known, but it is estimated to be about $1.5°$ RMS taking the world as a whole. This infers that a $3°$ or $4°$ error would not be infrequent but also wouldn't occur too often as the RMS would be $1.5°$. We would like to improve upon that if we could, and one of the methods mentioned by Dr. Christian in his talk involved using several channels in the microwave region. One would be temperature sensitive and the other would be sensitive to the wind and temperature on the surface. With these, the temperature and the sea surface properties other than temperature would be extracted from the microwave soundings.

OCEANOGRAPHIC RELATED TOPICS

One of the comments made by an attendee was that there had not been a concentrated attention on what the requirements are in oceanography. Although there were many papers on oceanography, it was rather diffuse, and he would like to have seen the requirements given for people in fisheries, people in ship navigation, etc. What are their requirements for oceanographic data? No doubt these exist in literature, but they were not brought out in this symposium. The possibility of doing something below the surface was mentioned from time to time, but it is not known if anything can be done in this area. Buoys were discussed, and although they are not indirect sensors, measurements can be made from them and information can be sent to a satellite which could then relay it. It would be possible to have many buoys, and that, indeed is the plan. We have a number of them, and there is a plan to install many more buoys which will communicate through the satellite. It has severe limitations since buoys don't last very long, maybe six months, and have to be replaced. It would be better if someone would find out how to measure the properties below the surface from the satellite. This would be a high-priority project if it could be accomplished.

COMMENTS ON SYMPOSIUM

I think a good point was made by someone that the title of the symposium was somewhat misleading. He had expected, from the title, and perhaps other people did too, that the symposium would deal with Remote Sensing as a means of discovering energy sources. There were a few papers on this, some optimistic, some not so, but these were a rather small part of the symposium. It could have scheduled a session on the problem of using remote sensors for the purpose of discovering energy sources. Rather, many authors got into the question of what is related to energy because of the title of the symposium, Remote Sensing Applied to Energy-Related Subjects. However, almost everything is energy-related. The ocean properties are important in determining what must be done to put platforms into the sea, weather forecasting, etc. A suggestion was made that it could have been organized to include a session on energy sources, location of these with remote sensing, and certain aspects of related problems. Pollutants were another energy-related subject, but that again could have been a sub-topic.

SUGGESTIONS ON SYMPOSIUM

The idea of the workshop seemed to be well received, but there was a feeling that there should be more workshops, perhaps at the end of each day. There should be a time set aside for them and it could be held in several different forms. One suggested that at the workshop there could be a separation, either on the basis of discipline or on the attendees' own activities to acquaint the group with his interests and his objectives. The extension of the workshop seemed to be well received, but most participants had diverse views on the details of its organization.

The consensus was that the idea of having 35-45 minutes for each paper was good since the topic could be covered more thoroughly than at meetings where you are allowed only 10-20 minutes. Another suggestion was that there should be a continuing dialogue between the people who attended the symposium. One possibility would be for the organizers of the symposium to keep in touch with the attendees to find out what they had gained from the symposium and what they needed to do further. It would also be useful to determine if there was a need for some of the members to meet to continue the work done here.

INDEXES

AUTHOR INDEX

Numbers indicate the first page of the chapter in which the authors appear.

Adams, W., 61
Akers, P., 335
Allison, J., 367
Amsbury, D. L., 395
Anderson, E. R., 335
Apel, J. R., 47, 129, 335
Arakawa, A., 335
Argyris, J. H., 335
Au, B. D., 61, 129

Baker, A. J., 335
Baltzer, R. A., 303
Barath, F. T., 129
Barclay, P. A., 103
Barnes, S. L., 103
Barrett, A. H., 129
Bates, H. E., 367
Battan, L. J., 103
Bennett, I., 247
Berdahl, P., 335
Bishop, C. J., 247
Blackmer, R. H., Jr., 103
Blais, R. N., 263
Bland, R., 335
Blythe, Richard, 303
Bowden, L. W., 263
Brandes, E. A., 103
Brooks, L. W., 129
Brown, G., 61
Brown, R. A., 103
Brown, W., 129, 367
Browning, K. A., 103
Bryant, G. W., 103
Byrne, James D., 157
Byrne, J. W., 335

Call, P. D., 437
Campbell, C. E., 437
Cannon, P. J., 437
Cardone, V. J., 61, 129
Chadda, T. B. S., 367
Chang, C. H., 157

Chang, T. C., 129
Cherry, W. R., 247
Chow, Ming-Dah, 303
Clanton, D. S., 395
Clinton, N. J., 437
Cochran, W. G., 291
Cole, H. S., 263
Collis, R. T., 113
Colvocoresses, A. P., 263
Colwell, R. N., 395
Conaway, J. W., Jr., 61
Connell, C., 403
Copeland, G. E., 263
Copeland, J., 129
Crosby, D. S., 303
Crowley, W. P., 335
Cummings, A. D., 103

Daley, J. C., 61
Derr, V. E., 263
Decker, M., 61
Defant, A., 335
de Sylva, D. P., 303
Dibblee, T. W., Jr., 437
Dicke, R., 61, 129
Dietrich, G., 303
Dooley, J. T., 103
Dooley, R. P., 129
Doviak, R. J., 103
Droppleman, J., 61
Drufuca, G., 103
Duda, R. O., 103
Dutton, E., 61
Dye, R. H., 403

Edinger, J. E., 335
Egami, R., 113
Egorov, S., 61
Eigen, D. J., 263
Eisenberg, R. P., 129
Ekman, V. X., 335
Elliott, W. P., 113

Elterman, L., 113
Englisch, W., 445
English, E. J., 103
Escher, W. J., 247
Evans, D., 61

Fabelinskii, I. L., 157
Fairbridge, R. W., 157
Falcone, V. J., Jr., 367
Faltermayer, E., 247
Finch, W. A., Jr., 233
Fischer, R. E., 129
Floyd, Walter J., 291
Friedman, J. D., 437
Fritz, S., 303, 479
Fu., K. S., 403
Fuller, W. H., 113
Fung, A. K., 61

Gates, D. M., 437
Geyer, J. C., 335
Glaser, Peter E., 367
Gloersen, P., 61, 129
Goubau, G., 367
Grant, C. R., 61
Gray, K., 103, 129
Greaves, James R., 303
Griggs, M., 263
Gross, E., 157
Guinard, N. W., 61
Gurvich, A., 61

Hach, J. P., 129
Hall, Royce, 291
Hammond, A. L., 445
Hammond, D. L., 61, 129
Hansen, Donald V., 303
Hardy, W. N., 129
Harleman, D. R. F., 335
Harrold, T. W., 103
Hay, H. R., 247
Hayne, G., 61
Hell, E. E., 303
Hesler v., A., 445
Hickey, H. N., 395
Hidy, G., 61
Higer, A. L., 475
Hilton, G. M., 263
Hindin, H. J., 61
Hindley, P. D., 303
Hirschberg, Joseph G., 157, 335
Hiser, H. W., 247, 335
Hirt, F. H., 445
Hoel, P. G., 291
Hofmann, D. J., 113
Hollinger, J. P., 61
Hollinger, J. T., 61
Hooke, R., 61
Horai, Ki-iti, 437
Hyatt, H. A., 61

Irbe, J. G., 303

Jansky, K., 61
Jeeves, T. A., 61
Jirka, G. H., 335
Jones, D. E., 129
Jones, W. L., Jr., 129
Joyce, A. T., 475

Keen, C. S., 263
Kemp, C. C., 291
Kendall, M. G., 291
Kennedy, J. M., 61
Kent, G. S., 113
Kerr, D. E., 61
Kerr, F. J., 61
Kessler, E., 103
Kindle, E. C., 263
Kinzer, G. D., 103
Krenkel, R. A., 303, 335
Krieger, F. J., 403
Krishen, K., 61
Kroesch, V., 445
Kronberg, P., 445
Kurath, Ellen, 303

Landgrebe, D. A., 403
Lawrence, G. F., 263
Lawrence, James D., Jr., 473
Lee, J. T., 103
Lee, S. S., 303, 335
Leendertse, J. J., 303
Lepley, L., 61
Lick, W., 335
Lilley, A. E., 129
Lindmayer, I., 367
Lockett, J. B., 303
Loomis, A. A., 129, 335
Lorenz, D., 303, 445
Love, A. W., 129
Lovering, T. S., 437
Lumb, F. E., 247
Lund, I. A., 247
Lynch, P. J., 61
Lyon, R. J. P., 233
Lyons, Walter A., 263

MacCallum, D. H., 263
McClellan, A., 263
McCoogan, J., 61
McCormick, M. Patrick, 113
MacDonald, F. C., 61
McFadden, J., 129
McGoogan, J. T., 129
McMillin, L. M., 303
Malhotra, R. C., 431
March, F. D., 303
Mathews, Charles W., 1
Maugh II, T. H., 445
Maul, George A., 303

AUTHOR INDEX

Massey, D. G., 303
Maynard, H. R., 103
Melfi, S. H., 113
Mennella, R., 61
Metz, W. D., 445
Meyer, Walter D., 303
Miller, L., 61
Miller, L. S., 129
Min, P. T., 403
Miner, R. M., 303
Moore, P. L., 103
Moore, R. K., 61, 129
Morse, F. H., 247
Moskowitz, L., 61
Moxham, R. M., 437
Munger, P. R., 233
Munk, W. H., 335
Murth, P. A., 263

Nagler, R. G., 129, 335
Nathanson, F. E., 129
Nicholass, C. A., 103
Nilson, B. W., 335
Nordberg, W., 61
Norris, D. J., 247
Northam, G. B., 113
Northouse, R. A., 263

O'Brien, G. F., 103
Offield, T. W., 437
Olsson, L. E., 113, 263
Oman, H., 247

Paddock, R. A., 335
Parasher, S. K., 61
Parker, F. L., 303, 335
Parker, H. Dennison, 171
Paris, J. F., 61
Peacock, K., 403
Pease, R. W., 263
Pease, S. R., 263
Pepin, T. J., 113
Pettyjohn, W. A., 403
Picton, W. L., 303
Pierson, W. J., Jr., 129
Platz, P., 157
Policastro, A. J., 335
Pooni, Premsuka, 291
Pressman, A. E., 437

Rader, M. L., 431
Ralph, E. L., 367
Reboh, R., 103
Reddy, S. J., 247
Reed, L. E., 403
Reid, W. M., 395
Richards, T. L., 303
Roache, J. P., 335
Rogers, R. H., 403
Rogers, R. R., 103

Rosen, J. M., 113
Ross, D., 61, 129
Rouse, J. W., 61
Rowan, L. C., 437
Russo, J. A., 103
Ryde, J. W., 103

Sabins, F. F., Jr., 437
Saunders, P. M., 303
Schmitt, Harrison H., 5
Schmugge, T. J., 129
Schneider, S., 445
Schooley, A. H., 61
Schultze, H., 445
Schwalb, A., 303
Sengupta, S., 303, 335, 473
Senn, H. V., 247, 471
Senne, Joseph H., 233
Shah, N., 403
Shapiro, A., 61
Shaw, R. W., 303
Sheets, M. M., 395
Short, Nicholas, M., 189, 395
Sidran, Miriam, 303
Sieber, Alois, 445
Sirmans, D., 103
Smith, D. L., 103
Smith, W. L., 303
Snedecor, G. W., 291
Starr, J. R., 103
Steel, R. G. D., 291
Steinmann, R., 445
Stogryn, A., 61
Stolzenbach, K. D., 335
Summers, W. R., 303
Swain, P. H., 403
Swift, C. T., 61, 129
Szego, G. C., 291

Taylor, Hugh P., 129
Taylor, V. C., 103
Tepper, Morris, 15
Terry, S. A., 437
Thomann, G. C., 61
Tien, C. L., 303
Tokar, J. V., 335
Tomiyasu, K., 129
Torrey, J. H., 291
Tsai, C. F., 335
Tuff, W. L., 113

Ulaby, F. T., 61
Urick, R. J., 157
Uyeda, Seiya, 437

Van Orstrand, C. E., 437
Van Siclen, D., 395
Vant-Hul, L. L., 247
Veziroglu, T. N., 303, 335
Vlcek, C. L., 103
Volger, K., 445

Wagner, R. J., 61
Wallace, R. E., 437
Ward, E., 263
Wark, D. Z., 303
Watson, K., 437
Watson, R. D., 437
Webster, W. J., 129
Weinberg, N. L., 303, 479
Weinreb, M. P., 303
Welander, P., 335
Wightman, J. M., 263
Wilheit, T., 61, 129
Wilk, K. E., 103
Williams, G. F., Jr., 61
Wilson, B. W., 303
Wilson, C. L., 403, 471
Wilson, J. W., 103

Wolfe, E. D., 437
Wolf, M., 367
Woodall, J. M., 367
Wood, C. R., 437
Wormser, Eric M., 303
Wouters, Alain, W., 157, 335
Wright, R. W. H., 113
Wu, S. T., 61
Wyatt, Phillip J., 303

Yaplee, B. S., 61
Young, J. O., 129
Young, L. A., 157
Yule, G. W., 291

Zarcaro, J. G., 129
Zelany, S. W., 335

SUBJECT INDEX

ACM, communications of, 68
Active sensor applications, 471
Airborne radiation thermometer, 314
Air Force Defense Meterological Program, 316
Altimeter, 82, 136
    operating modes, 144
Angstrom, 160
Anomalies
    closed, 220, 226
    hazy, 220
Apollo 9, 395
Applications Technology Satellite (ATS), 16
Atmospheric path-lengths, correction for, 326
Atmospheric window, 28

Backscattering
    cross section, 24, 25
    function model of the atmosphere, 118
Base-line information, 476
Bendix
    multiband scanner, 312, 313
    thermal mapper, 312
Black body, energy emission of, 158, 159
Boltzman's constant, 305
Brillouin scattering, 162

Climatological problems and settlements, correlation of, 457
Coal, 175
    distribution of, 176
Coefficient
    angle, comparison of, 77
    extinction, 123, 124
Conditions
    boundary, 338, 339
    initial, 338
Contrails, 283
    identification of, 263
Cooling towers, influence of, 453

Daedalus scanner systems, 314
Data
    display, 103
    located, products of, 435
    processing, 104
        techniques, 95
    remotely sensed, 356
        locating, 431, 433
    systems, multispectral, 403, 404
Derived wind field, 38
Directional spectrum, 54

Earth incidence angle, 133
Earth Resources Survey Program, 316
Earth Resources Technology Satellite (ERTS), 2, 264
    automated mapping from, 414
    data, 409
        machine processing of, 269
    multispectral scanner of, 293
Echo characteristics, 108
Electrically Scanning Microwave Radiometer (ESMR), 33
Energy
    conservation of, 6
    crisis in, 5-7
    development, integrated planning, 180
    future of, 5
    increase in, 5
    raw material sources, discovery and inventory of, 475
    related research, 447
    renewable, 8
    solar
        conversion, 371
        hours of, 255
    sources for, 475
    transition phase, 5, 7
Environmental modelling, application of, 345
Environmental models, data requirements of, 342
Environmental monitoring, 184, 473

SUBJECT INDEX

Flight conditions, requirements for, 441
Flood, analysis and warning, 105
Flood plain management studies, 245
Fossil fuels, 247
    exploration for, 189
Forecasting, radar applications to, 103

Geometric corrections, 410
Geostationary Operational Environmental Satellite (GOES), 247, 249, 250
    centralized data distribution system, 252
    coverage from, 251
Global Atmospheric Research Program (GARP), 15, 43
    first global experiment, 43
Global Observing System, 44
Ground footprints, 139

Heat mapping, 461

Image Dissector Camera Systems, 22
Imagers, 18
Imaging, infrared, measurement of water surface temperature by, 304
Information dissemination and application, 412
Infrared radiometer, 56
Intense storms, display of, 106
Isotherms, surface, 360, 361
Land use and resource monitoring, 475
Laser
    measure, 157
    probe, 160
    radar, 121
        data, 119
        scattering, 120, 121
    remote sensing probe, 161
    system, 114
Lidar, 113, 123
    compact system, 116
    theory, 115

Measurements
    atmospheric, evolution of, 115
    calibration data, 95
Meterological applications of radar, 103
Meterological phenomena, time and space scales of, 45
Meterological range, comparison of, 124
Meterological satellite, evolution of, 17, 46
    camera systems, 19
    scanners, 25
    sensors, 18
Meterologically measured windspeed, 149
Microwave
    attenuation, 337
    brightness temperatures, 80
    emissivities of polar seas, 83
    generation, 376, 378
    generators, assembly of, 378
    rectification, 376, 382
    sensing
        practical considerations, 129
        principles, 130
        techniques, 129, 152
    sensors, application of, 95
    sounding through a tropical cyclone, 42
    transmission, 376
    wind scatterometer, 53
Mie scattering, 268
Modelling, numerical, fundamentals of, 336
Models, application and variation of, 342
Multiple Docking Adapter (MDA), 134

National Oceanic and Atmospheric Administration Data, 317
NIMBUS
    G, 3
    2, 21, 23
    4, 24
NOAA-3 satellite data, digitized, 323
Nonblackness, correction for, 305
Nuclear fuels, exploration for, 189
Oceanographic measurement requirements, 148, 343
Oceanography experiments, physical, 151
Ocean surface
    phenomena, 61
    winds, observation of, 82
Oil shale, 171

Petroleum exploration from space, 228
Plant association borders, 299
Pollutants, air, satellite detection of, 263
Pollution
    air, 473
    episodes, synoptic scale, 275
    interregional transport, 272
    land, 473
    survey, 245
    thermal, 303
    water, 473
Power plant, site selection of, 420
Precipitation climatology, radar applications to, 103
Pulsed Radar Altimeter, 49

Radar
    calibration techniques, 95
    climatology, 108
    determined windspeed, 149
    imaging, 87
    precipitation detected by, 104
Radiation
    differences, sensitivity to, 439
    measurements, temperature profile inferred from, 41

SUBJECT INDEX

solar
    daily, 256
    measurement of, 258
Radiometer, 78
    microwave, 78
    scanning, 30
Radiometer/scatterometer, 136
    operating modes, 144
Radiosonde data, 323
Raman scattering, 117, 161
Rayleigh scattering, 120, 161, 268
RB-57, 401
Recreation area identification, 245
Remote sensing
    from aircraft, 243
    application of, 345
        to energy related problems, 445
        to numerical modelling, 335
        to thermal pollution, 303
    capabilities, 342
    experiment, 319, 320
    of oceans, 61
    for oil shale development planning, 171
    of small temperature differences, 437
    specific applications, 182
    study, flow chart for, 358
    surveillance of the Missouri River basin, 233
    systems, remote airborne, 312
    utility of, 177
    of vegetational resources under environmental
        stress, 291
    for western coal, ???
Resources, usuable, exploration of, 463
Rigid-lid results, 347, 350

Satellite Infrared Spectrometer (SIRS-A), 40
Satellite measurements, calibration of, 479
Satellite radiometer, surface temperature sensing
    from, 315
Satellite solar power station, 367
    design concept, 369
    economic considerations, 386
    environmental issues, 388
    flight control, 383
    orbit, location of, 368
    orbital assembly, 385
    principles,of, 368
Satellite system, evolution of, 16
Satellites, oceanographic and hydrologic measure-
    ments from, 479
Scan modes, 139
Scanner
    Bendix modular, 407
    design, 438
    multispectial, 405, 408
Scanning Multifrequency Microwave Radiometer
    (SMMR), 53
Scattered light, 166

Scattering, 160
    ratio, 122, 126, 127
Scatterometers, 64
SEASAT, 3, 60
    capability of, 59, 344
    configuration of, 48
    description, 49
    ground track for 24-hour period, 50
    objectives of, 47
    program, 146
    sensors, 58
    the total system, 56
    viewing the marine environment, 47
Skylab, 401
    in orbit, 135
    program, 138
Smoke plumes, 265
    dectection of, 263, 266
Solar power sites, determination of, 247
S-193, experiment performance specifications
    and characteristics, 142
Sounders, 33
    evolution of, 39
Space environment, effects of, 373
SSCC picture, 36
Storm warning, 105
Stratospheric measurements, 123
Sub-programs, interrelationships of, 347
Synchronous Meteorological Satellite/
    Geostationary Operational Environmental
    Satellite (SMS/GOES), 16
Systems design, improved, 95

Theme borders, 299
TIROS, 15, 18, 22, 190
    N, 6
Transmission, atmospheric, 310
Transportation
    Earth-to-orbit, 383
    ground to low-Earth orbit, 384
    LEO-to-synchronous orbit, 384
Tropospheric measurements, 117
Turbidity, 165

Vegetation
    identification, 243, 244
    spectral features of, 300
    types, 296
Vidicon tubes, 18

Warm cooling water, influence of, 449
Wave
    height, significant, 86
    refraction patterns, 54
Weather
    analysis, 103
    forecasting, 104
    modification, observations of, 278
    radar relationship, summary of, 104
Wind scatterometer and radiometer data, 55